XVI International Conference of Students and Young Scientists "Prospects of Fundamental Science Development" (PFSD-2019)

IOP Conference Series: Materials Science and Engineering
Volume 597

Tomsk, Russia
23 – 26 April 2019

ISBN: 978-1-5108-9467-9
ISSN: 1742-6588

Printed from e-media with permission by:

Curran Associates, Inc.
57 Morehouse Lane
Red Hook, NY 12571

Some format issues inherent in the e-media version may also appear in this print version.

Additional copies of this publication are available from:

Curran Associates, Inc.
57 Morehouse Lane
Red Hook, NY 12571 USA
Phone: 845-758-0400
Fax: 845-758-2633
Email: curran@proceedings.com
Web: www.proceedings.com

Table of contents

Volume 597

XVI International Conference of Students and Young Scientists "Prospects of Fundamental Sciences Development"

23–26 April 2019, Tomsk, Russia

Accepted papers received: 26 July 2019
Published online: 23 August 2019

Preface

Papers

IOP Conf. Series: Materials Science and Engineering **597** (2019) 011001 doi:10.1088/1757-899X/597/1/011001

Preface

XVI International Conference of Students and Young Scientists "Prospects of Fundamental Sciences Development" (PFSD-2019)

conf-prfn.org

23-26 April, 2019
Tomsk, Russia

The XVI International Conference of Students and Young Scientists "Prospects of Fundamental Sciences Development" (PFSD-2019), held on 23-26 April, 2019 in Tomsk, Russia, is an annual multidisciplinary conference gathering together students, young researchers, post-docs, and experienced researchers working in the different research areas in order to facilitate aggregation and sharing interests and results for a better collaboration.

The conference received more than 140 submissions. After a rigorous peer-review process, 80 papers were accepted. The accepted papers cover the following conference topics: (1) Chemistry: synthesis and properties of functional materials; physicochemical analysis methods for materials; nanomaterials; ecology; organic synthesis; catalysis and petrochemistry; (2) Construction methods and materials; (3) Physics: condensed matter physics; surface physics; accelerator physics; hydrogen energetic; geophysics; modeling of physical processes; (4) Fundamental Medicine: cell biology, biochemistry; microbiology, biophysics, biomaterials; (5) Mathematics & IT: mathematical modeling and data analysis; mathematical methods in physics, chemistry, medicine, computational mathematics, differential equations, automated data processing and control systems, electronics; optoelectronics and nanophotonics; plasma emission electronics.

We are very grateful to all reviewers, both local and international, for their time and effort in reviewing the papers. Their feedback to the authors has been excellent and useful to ensure the high quality of papers for the conferences.

The Organizing Committee of PFSD-2019 thanks Universities involved in the conference organizing and Sponsors for their financial support, as well as everyone who contributed to the organization of this event and to the Proceedings publication.

Chair of the PFSD-2019 Organizing Committee: Prof. Irina Kurzina

Editors of the PFSD-2019 Conference Proceedings: Dr. Mikhail Semenov and Anna Bogdan

IOP Conf. Series: Materials Science and Engineering **597** (2019) 011002 doi:10.1088/1757-899X/597/1/011002

Peer review statement

All papers published in this volume of *IOP Conference Series: Materials Science and Engineering* have been peer reviewed through processes administered by the proceedings Editors. Reviews were conducted by expert referees to the professional and scientific standards expected of a proceedings journal published by IOP Publishing.

IOP Conf. Series: Materials Science and Engineering **597** (2019) 012001 doi:10.1088/1757-899X/597/1/012001

Determining the conditions for oxidative desulphurization of oils

O Andrienko[1,2], N Kobotaeva[1,3], T Skorokhodova[1,3], E Marakina[1], V Sachkov[1] and R Medvedev[1]

[1]Laboratory of Chemical Technology, National Research Tomsk State University, 36 Lenin Avenue, Tomsk 634050, Russian Federation
[2]Quantum Electronics Laboratory, V.E. Zuev Institute of Atmospheric Optics of Siberian Branch of the Russian Academy of Science (IAO SB RAS), 1 Zuev Square, Tomsk 634021, Russian Federation
[3]Laboratory of catalytic processing of light hydrocarbons, Institute of Petroleum Chemistry of the Siberian Branch of the Russian Academy of Science (IPC SB RAS), 4, Akademichesky Avenue, Tomsk 634055, Russian Federation

E-mail: vioes@mail.ru

Abstract. The paper studies oxidative desulphurization of oils with an initial sulphur content of 1.98 wt% by hydrogen peroxide on a carbon-black (CB) catalyst coated by a combination of cobalt and nickel. The results demonstrate the effects of temperature and duration of the process, amount of the catalyst and the nature of the selected extractant on the removal efficiency of oxidized sulphur compounds. The extractant was represented by a solution of dimethylformamide (DMF) with water (10 vol%). The maximum removal degree of sulphur exceeded 70% and was achieved in 90 minutes with an amount of the catalyst of 1.0–1.3 wt.% and a temperature of 50 °C.

1. Introduction

The issues of heavy oil transformation into low-sulphur petroleum products are becoming increasingly high on the agenda. This is mainly due to the depletion of low-sulphur oil fields, which necessitates developing oil fields with moderate and high content of sulphur. Therefore, the share of high-sulphur petroleum products increases [1]. The presence of sulphur compounds in oil gives rise to corrosion and rapid destruction of equipment used in oil extraction and processing [2]. Increasing sulphur content in Russian oil is one of the major reasons for its declining competitiveness. The cost of oil depends on its processing depth. Oil extracted from different fields differs in chemical composition and may have considerably variant quality. Crude oil which incurs minimal processing costs has the highest value. High-quality benchmark crudes contain approximately 0.5% of sulphur. Oil quality can be improved by processing and removing the sulphur.

High-quality oil products are expensive and cover the processing costs. This justifies the development and implementation of advanced oil refining technologies, namely the extraction of sulphur products and the development of high-quality petroleum products.

IOP Conf. Series: Materials Science and Engineering **597** (2019) 012001 doi:10.1088/1757-899X/597/1/012001

One of the ways to reduce the sulphur content in oil and petroleum products is desulphurization which significantly improves the commercial and consumer appeal of oil and its products, reduces the environmental footprint, and increases the life of oil processing equipment.

Oil and its residues can be desulphurized—without decomposition of sulphur compounds—by oxidative desulphurization which modifies the functional groups of sulphur compounds to be removed. Due to its simplicity, the method has been actively used in oil refining for decades to separate oxidized sulphur compounds from hydrocarbons.

The most technologically favorable oxidizer is hydrogen peroxide (H_2O_2), a commercial product that is often used in industry. It is considered to be a green solvent, because its decay products include water [3]. Effective use of hydrogen peroxide in desulphurization of crude oil feedstock requires catalysts. Selective oxidation of sulphur compounds involves such metals as Ti, V, W, Mo, Co/Mn, Ag, Au, and Re and their oxides that serve as heterogeneous catalysts [4–7]. Alternatively, hydrocarbon-based systems are deemed to be attractive from the perspective of catalysis, due to specific physical and chemical properties, first of all extended surface and chemical inertness [8, 9]. The numerous research works—along with traditional carbon carriers, like active coals—also scrutinize other carbon-based materials, such as carbon nanotubes, graphitized nanofibers, carbon black (CB), fullerenes and various composite carbon-carbon materials [8–10].

Carbon materials are also used in petrochemistry to catalyze oxidation of sulphur compounds residing in heavy fractions of oil processing products [11–13].

The work was aimed at experimentally studying oil oxidative desulphurization using hydrogen peroxide in the presence of a CB-based catalyst coated by transition metals.

2. Experimental part

The object of the study was Usinsk oil field. The physical and chemical properties of its oil are shown in Table 1.

Table 1. Properties and composition of oil from Usinsk oil field.

Characteristic	Value
Density (kg/m^3)	966.7
Kinematic viscosity at 20 °C (mm^2/s)	3852.39
Molecular weight (AMU)	365
Elemental composition (wt%):	
• carbon	84.94
• hydrogen	11.98
• oxygen	0.47
• **sulphur**	**1.64**
• nitrogen	0.63
• N/A	1.68
Content (wt%)	
• Paraffin wax	1.24
• N-alkanes	1.29
• chlorides (mg/dm^3)	36.92
Component composition (wt%):	
• resins	18.3
• asphaltenes	8.1
• oils	73.9
Pour point (°C)	-22.5

The metal-carbon composite was made from P354 CB synthesized at the Institute of Hydrocarbon Processing Problems of the Siberian Branch of the Russian Academy of Sciences (Omsk, Russia). The CB specimen contained 96.4% of C, 2.5% of O, 0.5% of S, and 0.3% of N and H each [14]. The surface of the CB specimen was coated by nanoparticles of transition metals (nickel and cobalt) by chemical reduction of the metals from solutions of their salts using sodium hypophosphite (Ni–Co–CB). This method is widely used for chemical deposition of nickel on small metal products [15]. The advantages are final fineness of the coating and the uniformity of metal deposition on the substrate.

Oxidative desulphurization of oil was carried out in a 500-ml KGP-3-1-500-29/32-14/23-14/23 three-neck flask with a mechanical stirrer, a TL-2 N2 laboratory mercury in-glass thermometer and KhSh-1-400-29/32 ball reflux condenser. The flask was placed in a LOIP LB-140 water bath for thermal conditioning at temperatures of 35–95 °C. Thirty grams of crude oil feedstock were heated in the flask up to the necessary temperature after which a portion of the catalyst and 40 ml of oxidant were simultaneously added to the solution. The amount of the catalyst varied from 0.16 to 1.65 wt%. The oxidant was a 15-percent aqueous solution of hydrogen peroxide derived by dilution from a high-purity 30% solution (Yugtekhkom Company, Russia). The moment of the oxidant contacting with the feedstock was considered as the beginning of the reaction. The durations of sulphur compound oxidation were 30, 60, 90, 120 and 150 minutes.

After the reaction ceasing, the layer of hydrocarbons was separated from the aqueous phase in the separating funnel. To extract oxidized sulphur compounds, a solvent was added to the hydrocarbon layer; the solution was stirred for 10 minutes and kept under a layer of extractant for 60 minutes at a temperature of 35–40 °C. The mass ratio of the feedstock to the extractant was 1:2. Chemically pure acetonitrile (AN) or N,N-dimethylformamide (DMF) (EKOS-1 Company, Russia) with 10 vol.% of water were used as the solvent to extract oxidized sulphur compounds.

Total sulphur content in the feedstock and desulphurized product was determined by the method of burning in a lamp in the presence of oxygen (GOST 19121-73) [16]. The method entails burning an oil sample in a platinum sample carrier mounted on a stopper of a Schoniger combustion flask containing an oxygen-saturated absorption solution. After the combustion ceasing, the absorption solution was titrated by a barium nitrate solution in the presence of an indicator until reaching stable color. A blank experiment (without oil sample) was carried out under the same conditions. Using the results obtained, total sulphur content in the oil sample was calculated.

3. Results and discussion

The experiments on the oxidative desulphurization of oil allowed studying the influence of temperature and duration of the process on the sulphur removal degree (Table 2). The amount of the catalyst was 1 wt%; the molar ratio of sulphur in the feedstock to oxidant was 1:4. The extractant was DMF with 10 vol.% of water.

Table 2. Influence of oxidation time and temperature on sulphur residue in oil.

Oxidation time (min)	Temperature (°C)	Sulphur residue (%)
30	35	0.75
60	35	0.61
90	35	0.52
120	35	0.50
150	35	0.53
30	50	0.70
60	50	0.60
90	50	**0.50**
120	50	0.55
150	50	0.49
30	95	0.90
60	95	0.80
90	95	0.72
120	95	0.70
150	95	0.71

IOP Conf. Series: Materials Science and Engineering **597** (2019) 012001 doi:10.1088/1757-899X/597/1/012001

According to Table 2, the minimum amount of sulphur in oil remains after 90-minute oxidation at a temperature of 50 °C.

At a lower temperature, the feedstock (oil) is relatively viscous, while the reagents (oxidant and catalyst) are unevenly distributed in the raw material. Therefore, the degree of sulphur compounds during the first 30 minutes is low. However, with increased duration of the process and a reaction time of 90–150 minutes, the residual sulphur content in the feedstock drops to 0.5–0.53 wt%. Increased temperature to 95 °C reduces sulphur removal extent, probably due to the partial decomposition of hydrogen peroxide into oxygen and water and, consequently, fewer active ·OH-particles. Even at short heating time (30 mins) the residual sulphur content (0.9 wt%) exceeds that reached at 35 and 50 °C at different processing times.

The O–O bond is fragile, which makes hydrogen peroxide (H_2O_2) an unstable compound that easily decomposes. This may be also caused by the presence of transition metals, such as cobalt and nickel which are a part of the catalyst.

The work also studied the effect of the catalyst content on the oxidation of oil sulphur compounds (Table 3) at 50 °C during 90 minutes. Oil sulphur compounds were extracted by DMF or AN diluted by 10 vol.% of water.

Table 3. Influence of catalyst content in reaction mixture on residual sulphur content in oil.

Catalyst content (CB-Co-Ni) (wt%)	Residual sulphur (wt%)		Sulphur removal degree X (%)	
	DMF	**AN**	**DMF**	**AN**
0.16	1.3	1.45	20.7	11.6
0.33	1.15	1.28	29.9	22.0
0.48	0.8	0.9	51.2	45.1
0.64	0.65	0.7	60.4	57.3
1.0	0.5	**0.6**	69.5	**63.4**
1.3	**0.48**	0.66	**70.7**	59.8
1.65	0.52	0.63	68.3	61.6

The degree of sulphur removal from the feedstock was calculated by eq. (1):

$$X = \frac{(\omega_0 - \omega_k)}{\omega_0} \cdot 100\%, \tag{1}$$

where X is the degree of sulphur removal (%); ω_0 is the initial sulphur content (wt%); ω_k is the sulphur content in the feedstock after desulphurization (residual content) (wt%).

Evidently from Table 3, the residual sulphur content in oil steadily falls with increasing catalyst content, while the optimal value in the reaction system was 1–1.3 wt%.

At a lower catalyst content, the residual sulphur concentration under similar conditions is higher, which may be due to insufficient number of active sites on the catalyst surface. A catalyst content above 1–1.3 wt% demonstrates increased residual content of total sulphur in the processed feedstock, which may be caused by the shielding of the active sites on the catalyst surface by oxidation products.

The extraction efficiency of oxidized sulphur compounds by selected extractants is shown in Figure 1.

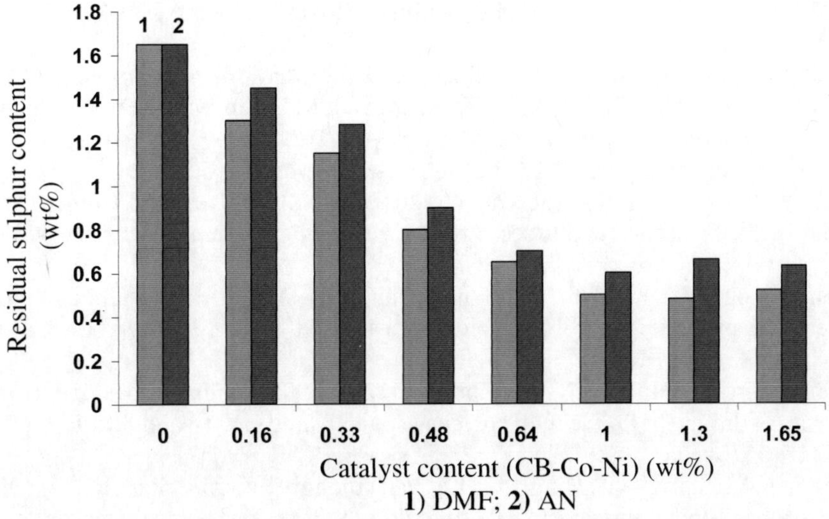

Figure 1. Influence of extractant type on the removal of oxidized sulphur compounds from oil.

Obviously, the most effective extractant is DMF with 10 vol.% of water. The residual sulphur content in oil achieved was 0.48 wt% at a process time of 90 minutes, temperature of 50 °C and catalyst concentration of 1.3 wt%.

Probably, the larger effectiveness of the DMF-water system v. AN-water system is conditioned by the formation of more sulphoxides than the sulphones during oxidative desulphurization, while the DMF-water system is more selective to sulphoxides then to sulphones. In the case of the AN-water system, sulphones are preferably extracted [17].

Thus, following the experimental data, oxidative desulphurization using carbon black coated by nickel and cobalt (CB-Ni-Co) is feasible under mild conditions: a temperature of 50 °C, a process time of 90 minutes, a catalyst content of 1–1.3 wt%, and a molar ratio of sulphur-containing feedstock to oxidant of 1:4. The DMF-water system (with 10 vol.% of water) is more effective for the extraction of oxidized sulphur compounds from oil to a high degree of more than 70%.

4. Conclusion

The research has discovered that the carbon black-based catalyst coated by nickel and cobalt shows high catalytic activity in oxidative desulphurization of oil. At a temperature of 50 °C, the duration of the process is 90 minutes, the amount of catalyst is 1-1.3 wt% and the extraction degree of oxidized sulfur compounds by an aqueous solution of dimethylformamide can exceed 70%.

Acknowledgment

This work was supported by the Ministry of Education and Science of the Russian Federation, the unique identifier of the work (project) is RFMEFI57817X0225.

References

[1] Schepalov A A 2012 Heavy oil, gas hydrates and other promising sources of hydrocarbon raw materials *Teaching and methodical manual.* - Nizhny Novgorod: UNN 93p

[2] Kundik A A 2017 Ensuring the reliability of pipeline systems. Method for diagnosing internal anticorrosion coating of pipelines // Oil. Professionally about oil. **1** 74–76

[3] Campos-Martin J M , Blanco-Brieva, J M and Fierro J L G 2006 Hydrogen peroxide synthesis: An outlook beyond the anthraquinone process *Angew. Chem. Int. Ed.* **45** 6962–84

[4] Kong L et al. 2006 Oxidative Desulphurisation of Organic Sulphur in Gasoline over Ag/TS-1 *Energy & Fuels* **20 (3)** 896–902

[5] Xiaojuan Si S C et al. 2007 Oxidative Desulphurisation of Model Oil over Au/Ti-MWW *Catalysis Letters* **122 (3–4)** 321–324
[6] Rakov E G, Baronin I V, Anoshkin I V 2008 *Catalysis in industry. Special edition.* pp 83–86
[7] Likholobov V A, Surovikin V F, Plaksin G V, Tsekhanovich M S, Surovikin Yu V, Baklanova O N 2008 *Catalysis in industry. Special edition.* pp 63–69
[8] Yang G W, Gao G Y, Wang C, Xu C L, Li H L 2008 *Carbon.* **46** 747–752
[9] Kobotaeva N S, Skorokhodova T S, Ryabova N V 2015 Catalytic Systems of Cumol Oxidation Based on Multiwalled Carbon Nanotubes *Russian Journal of Physical Chemistry* **89 (3)** 462–468
[10] Guoxian Yu, Shanxiang Lu, Hui Chen, Zhongnan Zhu. 2005 Diesel fuel desulphurisation with hydrogen peroxide promoted by formic acid and catalysed by activated carbon *Carbon.* **43** 2285–94
[11] Guoxian Yu, Shanxiang Lu, Hui Chen, and Zhongnan Zhu 2005 Oxidative desulphurisation of diesel fuels with hydrogen peroxide in the presence of activated carbon and formic acid *Energy & Fuels* **19** 447–452
[12] Jing Xiao, Luoming Wu, Ying Wu, Bing Liu, Lu Dai, Zhong Li, Qibin Xia, Hongxia Xi 2014 Effect of gasoline composition on oxidative desulphurisation using a phosphotungstic acid/activated carbon catalyst with hydrogen peroxide *Applied Energy.* **113** 78–85
[13] Razdyakonova G I, Likholobov V A, Moiseevskaya G V, Petin A A, Karavayev M Yu 2014 Innovative dispersed carbon. From idea to technology: monograph *Omsk: Omsk State Technical University Publishing House* **312** 244–245
[14] Rakov E G 2004 *Journal of the Russian Chemical Society. DI. Mendeleev.* **48 (5)** 12–29
[15] Petrova T P 2011 *Chemical coatings. // Soros Educational Journal.* **6 (11)** 57–62
[16] GOST RF 19121-73 Petroleum products. Determination of sulphur content by lamp method
[17] Zarifyanova M Z, Kozin V G 2007 Investigation of the selectivity of solvents in the separation of oil sulphoxides and sulphones from oxidates of oil fractions *Bulletin of Kazan Technological University.* **5** 111–116

IOP Conf. Series: Materials Science and Engineering **597** (2019) 012002 doi:10.1088/1757-899X/597/1/012002

Effect of the nature of silicon source on physicochemical properties of high-silica zeolites and the activity of Zn-pentasils prepared on their basis in the course of aromatization of lower alkanes

A V Vosmerikov[1,2], O S Smoliakova[2], A A Vosmerikov[1] and L N Vosmerikova[1]

[1]Institute of Petroleum Chemistry, Siberian Branch, Russian Academy of Sciences, 4, Akademichesky Ave., 634055, Tomsk, Russia
[2]National Research Tomsk State University, 36, Lenin Ave., 634050, Tomsk, Russia

E-mail: pika@ipc.tsc.ru

Abstract. The effect of the nature of the silicon source on the physicochemical properties of synthesized zeolites and the activity of Zn-containing catalysts prepared on their basis in the course of conversion of the propane-butane fraction into aromatic hydrocarbons is studied. It is shown that the silicon source plays an important role in the hydrothermal synthesis of zeolite. It has an effect on the duration of synthesis but its effect on the physicochemical characteristics of the resulting product is minor. It is established that the activity and selectivity of Zn-containing catalysts for the aromatization of lower alkanes are practically independent on the nature of the silicon source used in the synthesis of the high-silica zeolite, which is an acid carrier of these catalysts.

1. Introduction

The solution to the problem of the use of associated petroleum gas as a feedstock for the chemical and petrochemical industries is of great importance for both economy and environment. According to estimates of various sources, the amount of annually flared associated gas ranges from 15 to 35 billion m^3. This leads to extremely negative economic, environmental and social consequences. Restriction on the extensive use of natural and associated petroleum gases as a feedstock for chemical processes is due to the difficult activation of the highly stable molecules of lower alkanes present in gases. Given the huge reserves of natural gas in Russia, light alkanes can be considered as an important potential source of feedstock for the chemical industry. Of particular interest is the catalytic synthesis of highly demanded aromatic compounds from low molecular weight paraffins over pentasil-containing catalysts [1-4]. Its practical implementation will predetermine a substantial increase in their production. The basis for manufacturing high-silica zeolites (HSZ) is the method of hydrothermal crystallization of alkaline alumina silica gels. The nature and type of feed reagents, the composition of the reaction mixtures, and the conditions of synthesis have a targeted effect on the crystallization process, causing changes in the structure, morphology, phase, dispersion and chemical composition of crystals, sorption capacity, strength and concentration of acid sites, and catalytic activity. The appearance of modifications of the same phase but of various stability in the course of synthesis means, that the least stable modification crystallizes first. It is then replaced by a more stable phase and so on. This continues until the product is formed that is most stable under these conditions.

However, even with due consideration of this rule, it is not possible to fully disclose the laws of synthesis. This is due to the fact that the composition of synthesis products is strongly influenced by the rate of nucleation, which depends on the conditions of preparation of the reaction mixtures, their state and structure, the size of cations, and the presence of other additives [5]. The uncertainty in the choice of synthesis conditions makes it necessary to use a wide variety of compounds as sources of aluminum and silicon cations, and also to prepare mixtures on their basis, which differ in both chemical composition and physical properties.

The purpose of this work was to study the effect of the nature of the silicon source on physicochemical properties of zeolites and the activity of Zn-pentasils prepared on their basis in the course of conversion of lower alkanes into aromatic hydrocarbons.

2. Materials and methods

The main object of the study was the synthesized high-silica zeolite of the pentasil family with a molar ratio $SiO_2/Al_2O_3 = 40$. Liquid glass (19% SiO_2), silicic acid ($nSiO_2 \cdot mH_2O$), white carbon (86% SiO_2), and sol (40% SiO_2) were used as a source of silicon.

Zeolite K-1 based on liquid glass was prepared by adding an aqueous solution of aluminum nitrate Al (NO_3) 3 • $9H_2O$ and hexamethylenediamine (HMDA) as a structure-forming agent to the liquid glass (19% SiO_2, 7% Na_2O, 74% H_2O) under intensive mechanical stirring. The acidity of the reaction mixture was adjusted to pH = 10.8 by the addition of 1 N nitric acid solution.

Zeolite K-2 was prepared by mixing white carbon (WC) (86% SiO_2) with an alkali solution of NaOH under intensive mechanical stirring. Then an aqueous solution of aluminum nitrate Al (NO_3) 3 • $9H_2O$ and HMDA were added. The acidity of the initial reaction mixture 11.2-11.4 required no adjustment.

In the course of synthesis of K-3 zeolite, 40% sol of silicic acid was used as a source of silicon. It was added by aqueous solutions of alkali, aluminum nitrate and HMDA. The acidity of the initial reaction mixture 11.2-11.4 required no adjustment.

Zeolite K-4 based on silicic acid was manufactured by adding $SiO_2 \cdot nH_2O$ silicic acid powder to solution of NaOH alkali under intensive mechanical stirring to obtain a 29% SiO_2, 19% Na_2O, 62% H_2O liquid glass, to which a HMDA solution and an aqueous solution of aluminum nitrate $Al(NO_3)_3 \cdot 9H_2O$ were sequentially added. The acidity of the reaction mixture was adjusted to pH = 10.8 by the addition of 1 N nitric acid solution.

In all cases, a high-silica zeolite in the amount of 1-1.5 wt% of SiO_2 used in the synthesis was added as seed to the reaction mixture in order to accelerate the crystallization process. The crystallization of HSZ was carried out in steel autoclaves for 1–6 days at 175 °C. At the end of the crystallization, the solid phase was separated from the liquid phase by filtration and then it was washed from excess of alkali with distilled water and dried at 100 °C in air for 8 hours. To remove the structure-forming additive, the samples were calcined at 550 °C for 16 hours.

Conversion of zeolites to the active H-form was carried out by decationization with a 25% aqueous NH_4Cl solution for 2 hours at 90 °C, followed by washing with distilled water and calcining at 550 °C for 6 hours. Based on the resulted zeolites, Zn-containing zeolite systems were prepared via the impregnation with a solution of zinc nitrate ($Zn(NO_3)_2 \cdot 6H_2O$). The concentration of zinc (expressed as metal) in them was 3.0 wt%.

The quality indices of the synthesized HSZ were monitored using IR spectroscopy and X-ray phase analysis (XRF). The IR spectra of the samples under study prepared as tablets with KBr (1.2 mg of zeolite per 300 mg KBr) were recorded using a 5700 Nicolet IR Fourier spectrometer in the range of 2000-400 cm^{-1} in air. X-ray phase analysis of zeolites was performed using a Bruker DISCOVER D8 diffractometer. The acidic properties of Zn-containing zeolite catalysts were studied by temperature-programmed desorption of ammonia. The strength of acid sites of the catalyst was estimated from the temperature maxima on the thermal desorption curve, their concentration was determined from the amount of ammonia desorbed at the moment of fixation of the desorption peaks. It was expressed in μmol per 1 g of catalyst.

To investigate the process of converting the propane-butane fraction (PBF) into aromatic hydrocarbons and to study the properties of catalysts, a bench-scale flow-type reactor was used. The catalytic conversion of PBF (propane – 77.4; isobutane – 5.8; and n-butane – 16.8 vol%) was carried out at atmospheric pressure, the reaction temperature varying from 450 to 600 °C and the feed space velocity 250 h^{-1}. The reaction products were analyzed by GLC using a Chromatec-Crystal 5000.2 chromatograph. During the experiments, the degree of conversion of the initial lower alkanes and the yield in gaseous and liquid products were determined. The selectivity towards formation of the reaction products was also calculated.

3. Results and discussion

Data on the composition and degree of crystallinity of the synthesized zeolites are given in Table 1. According to IR spectroscopy data, it was found that the crystallization rate of the catalysts under study depends on the silicon source used during the synthesis of zeolite. For example, the synthesis of zeolites on the basis of liquid glass and silicic acid lasts for 5 days. The use of WC in the course of synthesis makes it possible to reduce the synthesis time to 2 days.

Table 1. Characteristics of zeolite catalysts.

Catalyst	Source of silicon	Synthesis time (day)	Degree of crystallinity (%)
K-1	Liquid glass	5	100
K-2	White carbon-100	2	100
K-3	Ludox HS-40 sol	3	93
K-4	Silicic acid	5	100

In Figure 1 are shown X-ray diffraction patterns of zeolites synthesized with different sources of silicon. XRD data confirm that the synthesized samples belong to the zeolite of ZSM-5 structural type. They are characterized by the presence of the most intense diffraction maxima in the range 2ϑ = 6-50°: a triplet at 23.0, 24.0 and 24.4°, peaks at 29.3 and 29.6°, and a doublet at 45.0- 45.5° were observed. The results obtained indicate a high phase purity of the synthesized zeolites.

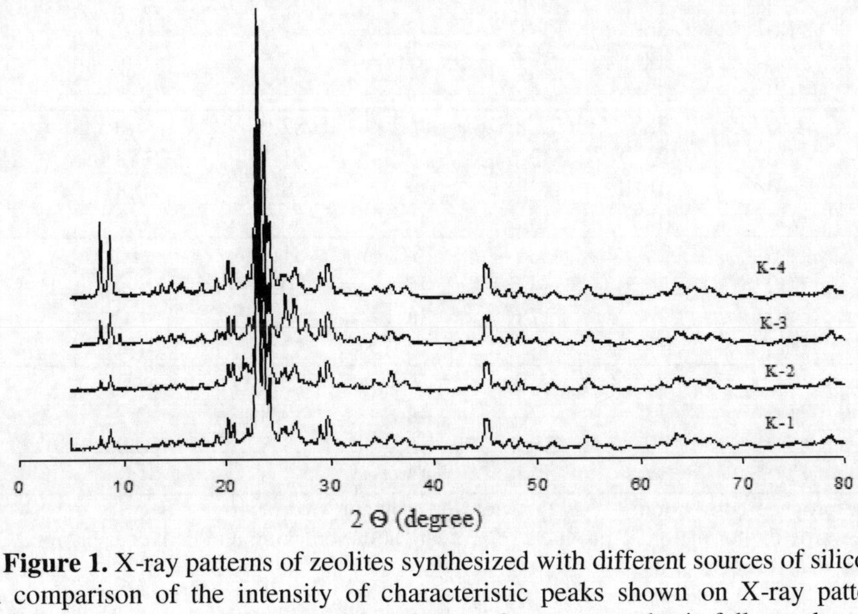

Figure 1. X-ray patterns of zeolites synthesized with different sources of silicon.

From a comparison of the intensity of characteristic peaks shown on X-ray patterns with the intensity of the peaks on the X-ray pattern of the reference sample, it follows that the degree of crystallinity of zeolites obtained with liquid glass, silicic acid and white carbon is 100%, while with sol – 94% (Table 1).

Acidic characteristics of Zn-containing zeolite catalysts are presented in Table 1. The presence of two peaks on the thermal desorption curves of the synthesized samples suggests the presence of two types of acid sites - weak and strong ones. The highest concentration of acid sites of both types is characteristic of Zn-pentasil, obtained on the basis of a zeolite synthesized from a sol. Their total concentration is 1327 μmol/g. The use of silicic acid as a silicon source in the course of zeolite synthesis results in a manufacture of Zn/K-4 catalyst with similar acid characteristics. The Zn/K-1 catalyst is characterized by the lowest concentration and strength of weak acid sites, while the Zn/K-2 sample – by the lowest concentration of strong acid sites.

Table 2. Acid characteristics of Zn-containing zeolites.

Catalyst	T_{max} (°C)		Concentration (μmol/g)		
	T_I	T_{II}	C_I	C_{II}	C_Σ
Zn/K-1	210	405	980	125	1105
Zn/K-2	215	405	1002	107	1109
Zn/K-3	215	410	1161	166	1327
Zn/K-4	220	400	1074	170	1244

Note. T_I, T_{II} are the temperatures of peak maxima for form I and II; C_I, C_{II} and C_Σ are the concentrations of acid sites in form I, II, and total concentration, respectively.

Thus, investigation of the acidic properties of Zn-containing catalysts obtained on the basis of zeolites synthesized with different sources of silicon showed that the catalysts slightly differ from each other by the distribution and ratio of acidic sites of different nature.

The results of studies of catalytic properties of Zn-containing zeolites prepared using various sources of silicon are presented in Table 3.

Table 3. Composition of the products of PBF conversion over Zn-containing catalysts.

Catalyst	T (°C)	X (%)	Y_{ar} (%)	S_{ar} (%)	S_{cr} (%)	S_{dehydr} (%)	S_{H2} (%)
Zn/K-1	450	39	2.9	7.4	83.7	3.2	5.5
	500	86	37.2	43.4	52.7	1.2	2.7
	550	100	37.7	37.7	58.5	0.8	2.9
	600	100	44.2	44.2	51.3	1.1	3.3
Zn/K-2	450	38	2.8	7.2	82.9	4.0	5.6
	500	95	33.6	35.5	61.1	0.8	2.6
	550	100	37.9	37.9	58.3	0.8	2.9
	600	100	45.5	45.5	50.1	0.9	3.5
Zn/K-3	450	25	1.4	5.5	81.8	5.6	6.5
	500	76	34.2	45.1	50.1	1.9	2.9
	550	99	37.6	38.1	58.0	1.0	2.9
	600	100	43.2	43.2	52.1	1.4	3.2
Zn/K-4	450	21	1.7	7.8	75.8	8.2	7.4
	500	73	32.7	44.8	49.8	2.2	2.9
	550	99	36.4	36.9	59.5	0.9	2.7
	600	100	43.0	43.0	52.6	1.2	3.1

Note. X is the propane conversion; Y_{ar} is the yield in aromatic hydrocarbons; S_{ar}, S_{kr} and S_{dehydr} are the selectivities towards the formation of products of aromatization, cracking and dehydrogenation; CH_2 is the selectivity towards hydrogen formation.

IOP Conf. Series: Materials Science and Engineering **597** (2019) 012002 doi:10.1088/1757-899X/597/1/012002

As can be seen from the presented data, Zn-containing zeolites are characterized by almost the same total catalytic (measured by the degree of conversion of PBF) and aromatization activity. Some differences are observed only at the initial temperatures of the process. The catalysts are practically similar in their catalytic properties at 550 °C and above. In the presence of all the catalytic systems under study, a noticeable conversion of PBF begins already at 450 °C and at 500 °C and above the desired product (aromatic hydrocarbons) are formed. With an increase in the process temperature for all samples, an increase in the conversion of PBF is observed, and at a reaction temperature of 550 °C and above, the feedstock converts almost completely.

Studies of the composition of the resulted liquid and gaseous products of the conversion of PBF over Zn-containing zeolites prepared using various sources of silicon showed that they do not differ from each other in their qualitative compositions, but differ in their quantitative contents. In Table 4 are given the data on the composition of the gaseous products of PBF conversion over Zn-pentasils. The contents of hydrogen, ethylene and propylene in gaseous products obtained over Zn-containing samples under study differ slightly, while the differences in the content of methane and ethane are significant. Hence, the conversion over Zn/K-1, Zn/K-3 and Zn/K-4 results in a lower content of methane and higher of ethane, while the content of methane and ethane obtained over the Zn/K-2 sample is almost the same. It is possible to note a rather low content of olefins in the reaction products, which indicates that they are intermediate products that actively enter into further interactions leading to the production of aromatic hydrocarbons.

Table 4. Composition of the gaseous products of PBF conversion over Zn-containing zeolite catalysts (T = 550 °C).

Catalyst	H_2	methane	ethane	ethylene	propane	propylene
Zn/K-1	4.6	39.4	54.2	1.2	0.5	0.1
Zn/K-2	4.7	45.4	48.3	1.2	0.3	0.1
Zn/K-3	4.6	33.8	57.9	1.2	2.1	0.4
Zn/K-4	4.2	34.6	57.5	1.1	2.3	0.3

In Figure 2 are presented the data on the composition of liquid products of the PBF conversion over Zn-containing zeolites synthesized with different sources of silicon. It is evident that the quantitative compositions of aromatic hydrocarbons produced over these catalysts vary notably. The highest amount of benzene, toluene and xylene (BTX fraction) was formed over Zn/K-4 catalyst, obtained on the basis of zeolite synthesized with silicic acid, while the BTX fraction of the lowest concentration was obtained over Zn/K-2 catalyst, prepared on the basis of zeolite with WC. The compositions of products formed over Zn-containing zeolites synthesized with liquid glass and sol differs slightly, but the amount of alkyl naphthalenes formed over the latter catalyst is lower. Differences in the compositions of gaseous and liquid reaction products formed over the catalysts under study containing the same amount of modifying additive, are due to the different flow rates at particular stages of the process, i.e. the nature of the zeolitic carrier affects the catalytic properties of the catalytic system.

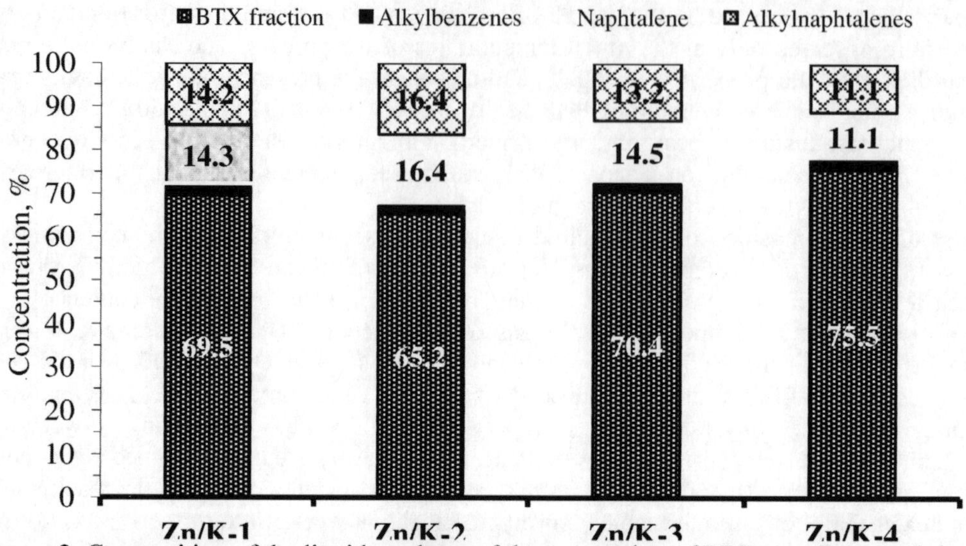

Figure 2. Composition of the liquid products of the conversion of PBF over Zn-containing zeolite catalysts (T = 550 °C).

4. Conclusion

As a result of the study, it was found out that Zn-containing catalysts prepared on the basis of zeolites synthesized from alkaline alumina silica gels using different silicon sources differ slightly in their physicochemical and catalytic properties in the course of PBF aromatization. This suggests the possibility of using different in their nature sources of silicon for the synthesis of zeolites.

In this case, the different durations of hydrothermal synthesis is the only significant factor that must be considered. The availability and price of the reagent should also be taken into account. Therefore, in the production of an effective catalyst for aromatization of gaseous hydrocarbons the quality indices of a zeolite carrier are critical, whose characteristics are not essentially affected by the silicon source.

Acknowledgments

The work was carried out as a part of the project No. V.46.2.1 of the Program for Basic Scientific Research of State Academies of Sciences.

References

[1] He Xiao, Junfeng Zhang, Xiaoxing Wang, Qingde Zhang, Hongjuan Xie, Yizhuo Hana and Yisheng Tan 2015 *Catal. Sci. Technol.* **5** 4081

[2] Xiaofang Su, Wang Zan, Xuefeng Bai, Gaoliang Wanga and Wei Wu 2017 *Catal. Sci. Technol.* **7** 1943

[3] Zaykovsky V.I., Vosmerikova L.N., Vosmerikov A.V. 2018 *Journal of Physical Chemistry* **92** **4** 589

[4] Caeiro G., Carvalho R.H., Wanga X., Lemosa M.A.N.D.A., Lemosa F., Guisnet M.,F. Ramˆoa Ribeiro 2006 *Journal of Molecular Catalysis A: Chemical* **255** 131

[5] R.M. Barrer 1985 *Hydrothermal chemistry of zeolites* (Moscow Mir) p 209

IOP Conf. Series: Materials Science and Engineering **597** (2019) 012003 doi:10.1088/1757-899X/597/1/012003

Spectroscopic properties of bis(BODIPY) complexes

Iu Aksenova[1], A Prokopenko[1], R Kuznetsova[1], L Antina[2] and M Berezin[2]

[1]Laboratory of Photophysics and Photochemistry of Molecules, Tomsk State University, Tomsk 634050, Russia
[2]Laboratory of Physical Chemistry of Solutions of Macrocyclic Compound, G.A. Krestov Institute of Solution Chemistry RAS, Ivanovo 153045, Russia

E-mail: alexpr898@gmail.com

Abstract. The spectral-luminescence characteristics of binuclear bis(BODIPY) complexes in cyclohexane and ethanol, which exhibit different photophysical and photochemical properties in these solvents, have been investigated. It has been shown that the difference in spectroscopic properties is due to intermolecular interaction of the excited molecules with solvation shell. The possibilities of practical use are discussed.

1. Introduction

Purposeful use of complex organic molecules and their derivatives for creating of various kinds of optical devices (temperature probes, optical sensors, nonlinear switches and limiters of high-power pulsed radiation) requires studying the relationship of spectroscopic characteristics with the structure of molecules, solvation shells and features of the media, in which are embedded these molecules. In recent years, interest in solutions and solid-state polymer media colored with organic dyes, which can be used as active media for tunable lasers, thin-film emitting layers for electroluminescent devices, and organic solar cells, has increased significantly [1, 2].

Boron fluoride dipyrromethene (BODIPY) coordination complexes can successfully solve the problems of practical application. Mononuclear BODIPY compounds have efficient absorption and emission in visible spectrum and depending on the quality and number of substituents in the ligand, their derivatives can change the efficiency of intercombination conversion. It makes possible successfully using such compounds in various fields of science and technology due to their unique spectral characteristics [3]. It is also known that binuclear bis(BODIPY) complexes exhibit impressive spectral properties, for example, the temperature dependence of fluorescence intensity, which can be used in the development of optical sensors to determine the temperature in the range of 300-80K. The creation of optical sensors for temperature in this range is important, for example, in instrument making when developing cryostats.

Targeted synthesis of appropriate structures for a particular application requires that the interrelation of the molecular structure to the properties of compounds of this class should be studied, as well as the influence on this interrelation of the medium in which this compound is placed. In this regard, this research is aimed to studying spectroscopic characteristics of a number of binuclear dipyrromethene complexes bis(BODIPY) in order to identify more efficient compounds to create modern materials for solving specific practical problem, namely the creation of effective optical devices.

IOP Conf. Series: Materials Science and Engineering **597** (2019) 012003 doi:10.1088/1757-899X/597/1/012003

2. Objects and methods of research

The objects of research were the series of binuclear coordination complexes of bis(BODIPY) containing a methylene (–CH_2–) spacer in different positions of proximal pyrroles. These compounds were compared with a complex 3,3-$CHCF_3$-bis(BODIPY) that has already been studied previously [4]. The structural formulas of investigated compounds are given in figure 1.

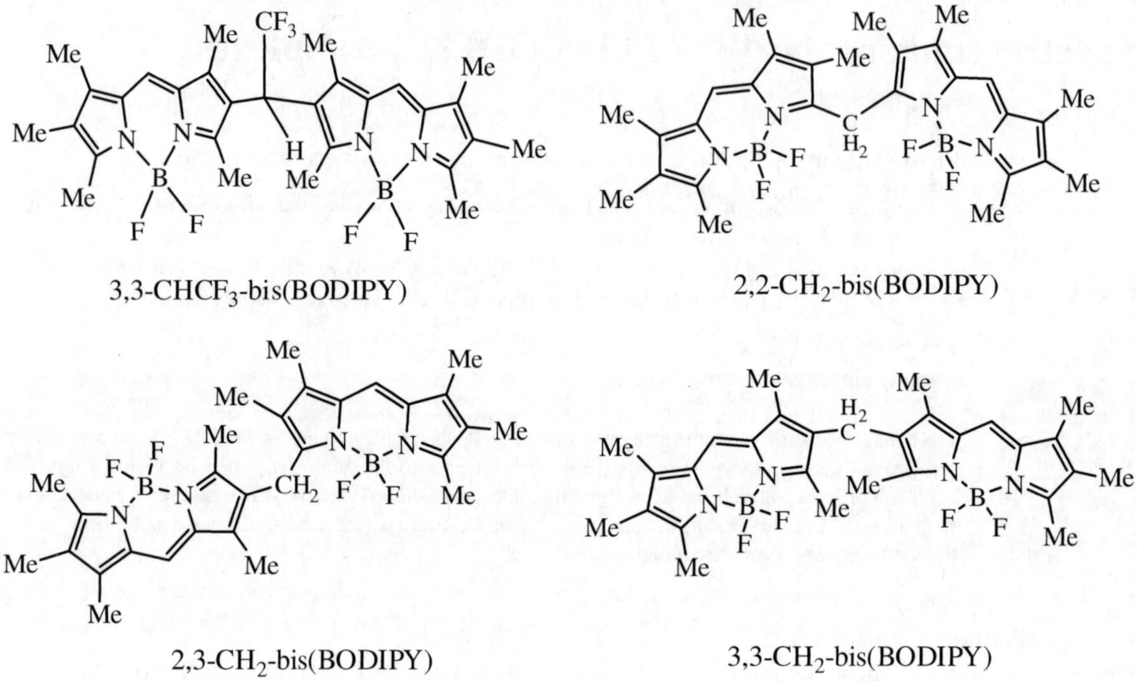

Figure 1. Structure formulas and denotation of investigated bis(BODIPY)s.

The compounds were synthesized according to the procedures described in [4, 5]. Their purity and identity were controlled by means of elemental analysis, mass spectrometry, and ^{1}H NMR spectroscopy. As solvents were used ethanol and cyclohexane of chemically pure grade.

The absorption spectra were measured by a SM2203 (SOLAR) spectrophotometer. The luminescence spectra at room temperatures were recorded on a Varian Cary Eclipse (Varian) spectrometer. The fluorescence quantum yield was measured at concentrations of 10^{-2}–10^{-3} mM (the absorption at the excitation wavelength was <0.1 cm^{-1}) by the relative method with respect to Rhodamine 6G with an error of 10%. All experiments were carried out in standard quartz cuvettes with an optical path length of 1 cm at a temperature of 298 K.

3. Results and discussion

In bis(BODIPY) molecules two BODIPY radicals are connected by –$CHCF_3$– and –CH_2– spacers and rotated relative to each other at a considerable angle, while retaining their most functional conformations. Such structural features manifest themselves in spectral-luminescent properties: there are three maxima in the electronic absorption spectra (figure 2). This is due to the actual existence of two chromophores in one compound.

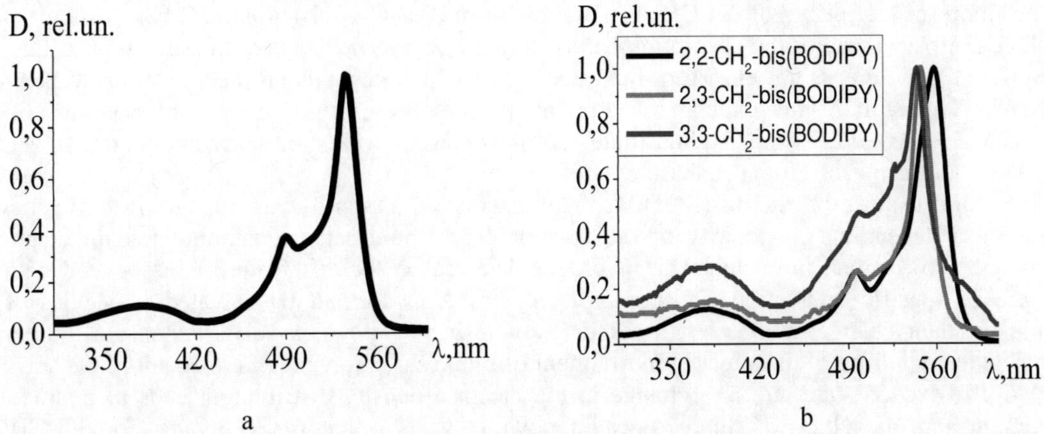

Figure 2. Normalized absorption spectra of 3,3-$CHCF_3$-bis(BODIPY) (a) and 2,2- , 2,3- , 3,3-CH_2-bis(BODIPY) (b) complexes in ethanol.

The 3,3-$CHCF_3$-bis(BODIPY) complex was studied in detail in [4] and has close spectral characteristics of absorption: 378, 494, 542 nm - in cyclohexane and 380, 490, 537 nm - in ethanol, respectively. In a series of newly studied compounds with (–CH_2–) spacer, complex 2,2-CH_2-bis(BODIPY) has the longest wavelength absorption and emission, followed by 2,3-CH_2-bis(BODIPY). The shortest wavelength absorption and fluorescence belongs to the 3,3-CH2-bis(BODIPY) complex (table 1).

Table 1. Spectral luminescent properties of bis(BODIPY) complexes.

Compound, solvent	λ_{abs} (nm)	λ_{fl} (nm) (λ_{ex} (nm))	γ_{fl}±10%
2,2-CH_2-bis(BODIPY), cyclohexane	382 500 558	538+580 (370,530)	0.99
2,2-CH_2-bis(BODIPY), ethanol	380 496 558	536+570 (370, 530)	0.1
2,3-CH_2-bis(BODIPY), cyclohexane	380 500 551	564 (370, 535)	0.99
2,3-CH_2-bis(BODIPY), ethanol	384 497 548	556 (370, 535)	0.06
3,3-CH_2-bis(BODIPY), cyclohexane	375 500 548	551 (370, 530)	0.99
3,3-CH_2-bis(BODIPY), ethanol	372 498 544	550 (370, 530)	0.1

The fluorescence spectra of 2,3-CH_2-bis(BODIPY) and 3,3-CH_2-bis(BODIPY) have a single band, which is a mirror image of the absorption bands. In the fluorescence spectra of the complex 2,2-CH_2-bis(BODIPY), two peaks are observed, the intensity of which depends on the excitation wavelength. Preliminary research results explain this fact by the display of two stable conformations of this compound in the equilibrium excited state. At this stage, work continues and aims to detailed explanation of the reasons for such features.

The main feature of the bis(BODIPY) complexes is a significant change in efficiency of fluorescence depending on polarity of the solvent. The fluorescence quantum yield in ethanol is significantly lower than in cyclohexane (table 1). The reason for this feature is associated with the presence of specific interactions in supramolecular systems formed by solvated molecules of the complexes, their change upon excitation and restructuring of complex solvate shells in the excited state. The restructuring of the solvate environment of binuclear complexes (movement in the solvation shell) in the excited state due to a change in the electron density distribution leads to nonradiative conversion and quenching of fluorescence in polar solvents (ethanol). Moreover, for 3,3-$CHCF_3$-bis(BODIPY) complex in frozen ethanol (77K), the fluorescence intensity increases by two orders of magnitude and approximate with intensity in cyclohexane.

4. Conclusions

The spectroscopic characteristics of binuclear boron fluoride complexes of dipyrromethenes were studied in various solvents depending on the wavelength of the exciting radiation. These complexes demonstrate a significant change in fluorescence efficiency, depending on the polarity of the solvent. This property can be used for creating sensors of the medium (solvent) polarity. In addition, weakly emitting ethanol solutions well limit the powerful radiation due to absorption of excited molecules, which is promising for creating laser limiters.

Acknowledgments

This study was supported by the Russian Foundation for Basic Research (projects nos. 18-33-00284), the State Task of Russian Ministry of Education and Science (project no. 4.6027.2017/8.9) and The Tomsk State University competitiveness improvement programme.

References

[1] Mueller T, Gresser R, Leo K, Riede M 2012 *Sol. Energy Mater. Sol. Cells.* **99** 176

[2] Zampetti A, Minotto A, Squeo B M, Gregoriou V G, Allard S, Scherf U, Chochos C L, Cacialli F 2017 *Sci. Rep.* **7** (1) 1611

[3] Kuznetsova R T, Aksenova Iu V, Bashkirtsev D E, Prokopenko A A, Pomogaev V A, Antina E V, Berezin M B, Bumagina N A 2018 *J. Photochem. Photobiol. A* **354** 147

[4] Kuznetsova R T, Aksenova Yu V, Solodova T A, Bashkirtsev D E, Kopylova T N, Tel'minov E N, Mayer G V, Berezin M B, Semeikin A S, Yutanova S L, Antina E V, Arabei S M, Pavich T A, Solovyov K N 2013 *Opt. Spectrosc.* **115** (5) 708

[5] Antina L A, Ksenofontov A A, Kalyagin A A, Antina E V, Berezin M B, Khodov I A 2019 *Spectrochim. Acta A.* DOI: 10.1016/j.saa.2019.03.117

IOP Conf. Series: Materials Science and Engineering **597** (2019) 012004 doi:10.1088/1757-899X/597/1/012004

Microwave synthesis and study of physicochemical properties of hydroxyapatite modified with silver and zinc ions

A S Gigilev, N M Korotchenko, A V Lariushina and V V Kozik
Department of Inorganic Chemistry, National Research Tomsk State University, 634050, Tomsk, Lenina, 36

E-mail: gigilev@mail.tsu.ru

Abstract. The authors obtained two-phase samples of silver- and zinc-modified hydroxyapatite (AgHA, ZnHA) by liquid-phase deposition under microwave exposure; the main phase for all samples was hydroxyapatite $Ca_5(PO_4)_3(OH)$ with hexagonal syngony. The parameters of the elementary lattice of HA were studied via X-ray phase analysis and confirmed the formation of solid substitutional solutions while modifying HA with silver and zinc ions. Solubility has been estimated at 20 °C for a 0.9% NaCl solution; specific surface area and porosity have been estimated for ZnHA powders obtained in presence of aminoacids. Surface morphology and dispersiveness have been studied for AgHA and ZnHA powders. Hammett method helped to establish the presence of acidic and basic centers on the surface of ZnHA and AgHA. Koch biotesting showed HA, AgHA, and ZnHA acting as inhibitors thus affecting *E. coli* tested samples. HA, AgHA, and ZnHA powders can be recommended for medical purposes.

1. Introduction

Creating biomaterials for recovering human bone tissue is currently an apparent problem. For this purpose, materials based on hydroxyapatite (HA) are widely used not only in surgery, dentistry, and injury treatment, but also in neurosurgery, cosmetology, and pharmaceutical industry. Hydroxyapatite with a composition $Ca_{10}(PO_4)_6(OH)_2$ is an inorganic component of bone and dental tissues; has certain chemical, biological, and mechanical properties [1, 2]. Mineral constituent of a bone includes HA crystals, which may contain various elements (e.g., potassium, zinc, silver, magnesium, fluorine, copper, etc.). Idealized implants, ceramics, and cements made of HA should have structure, composition, and morphology identical to human dental tissue [3-5].

Among a large number of cation- and anion-modified hydroxyapatite samples, HA containing silver (AgHA) and zinc ions (ZnHA) are of great interest to date. The ions are chosen because of their biological role and specific functions they perform in a human body [6, 7].

The purpose of the research project is to study the effect of Ag^+ and Zn^{2+} on the phase composition and on physicochemical and biological properties of hydroxyapatite powders obtained via microwave synthesis.

Synthesis of silver- and zinc-modified hydroxyapatite powders with initial ion content from 0.1 to 0.9 mol.% was carried out by liquid-phase deposition from aqueous solutions in a microwave field.

Physical, chemical, structural, surface, and biological properties for the obtained hydroxyapatite powders have been studied using a number of methods.

Phase composition for the isolated AgHA and ZnHA samples was determined by X-ray phase analysis using a Shimadzu XRD 6000 diffractometer, PCPDFWIN and JSPDS PDF 4+ databases, and POWDER CELL 2.4 full-profile analysis software.

Surface morphology for AgHA and ZnHA powders has been examined by SEM. Powder dispersiveness has been estimated using microphotography by a secant method in Adobe Photoshop CS5, as well as using specific surface area and pycnometric specific gravity of the powders.

Compleximetric titration has been employed to determine solubility of AgHA and ZnHA powders in 0.9% NaCl solution.

BET-method allowed the authors to estimate the specific surface area value for ZnHA. The acidic and basic properties of the latter were detected by Hammett method. These properties are studied in comparison with non-modified HA.

Zinc-modified HA powders underwent Koch biotesting.

The HA crystal structure, whose elementary lattice is shown in Figure 1, allows for substitution of certain elements with the other ones. Certain atoms can take different positions within HA structure both fully and partially, thus leading to diversity in the composition of natural phosphates with apatite structure. The ability of atoms or ions to enter the structure of another substance is pre-determined by the individual properties of these atoms or ions (i.e., size, charge, or electron structure) and by the peculiarities of a crystal lattice for substances forming solid solutions. The difference in atomic size for different constituents must be insignificant and should not exceed 8-15%.

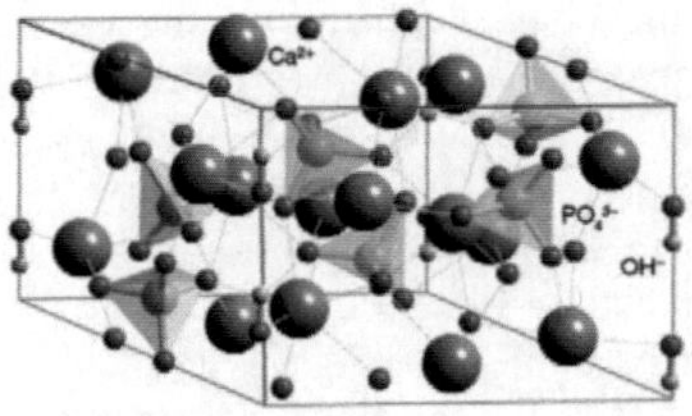

Figure 1. Elementary cell of hydroxyapatite crystal structure.

For a cation substitution, it is important to consider the charge and size of ions, as well as related electrochemical potential (φ). These values for Ca^{2+}, Ag^{+}, and Zn^{2+} cations are provided below.

Ion	Ca^{2+}	Ag^{+}	Zn^{2+}
Ionic radius r_i, nm	0.106	0.113	0.083
Potential $\varphi \cdot 10^{28}$, C^2/m	2.426	2.265	3.096

Silver and zinc ions can substitute calcium ions within a crystal lattice in an isomorphic manner on the basis of their ionic radius value. This results in the formation of mixed crystal of variable composition, otherwise known as substitutional solid solutions.

Cation substitutions within HA $Ca_{10}(PO_4)_6(OH)_2$ cause the changes in the parameters of crystal lattice an in the degree of crystallinity, which significantly affects HA solubility under physiological conditions and its chemical and biological properties [6]. However, this does not induce significant changes in the lattice symmetry.

Zinc is a microelement necessary to sustain many biochemical processes. Zinc takes part in osteogenesis and in collagen synthesis [6]. It is also involved in cell fission, in pancreatic gland insulin synthesis, and is an essential element for skin regeneration, for hair and nail growth, and for sebaceous gland secretion process. Additionally, it contributes to the vitamin E absorption and helps to maintain its normal concentration in blood. Zinc plays a key role in alcohol processing.

The antiseptic properties of silver have been known since ancient times whereas its biological function in a human body requires is not fully understood. It is the ability of silver ions to bring harm to pathogenic microorganisms that is currently used in medicine to cure numerous diseases including those attacking ENT-organs, bronchial tube, lungs, mouth cavity, digestive tract, and skin.

2. Experimental

Liquid phase synthesis of HA(1), AgHA(2), and ZnHA(3) powders was carried out using a stoichiometric ratio Ca/P = 1.67((Ca+M)/P=1.67) and the following reaction equations:

$$10Ca(NO_3)_2 + 6(NH_4)_2HPO_4 + 8NH_4OH = Ca_{10}(PO_4)_6(OH)_2 + 20NH_4NO_3 + 6H_2O; \quad (1)$$

$$(10–0.5x)Ca(NO_3)_2 + 6(NH_4)_2HPO_4 + xAgNO_3 + 8NH_4OH = Ca_{(10–0.5x)}Ag_x(PO_4)_6(OH)_2 + 6H_2O + 20NH_4NO_3; \quad (2)$$

$$(10–x)Ca(NO_3)_2 + 6(NH_4)_2HPO_4 + xZn(NO_3)_2 + 8NH_4OH = Ca_{(10–x)}Zn_x(PO_4)_6(OH)_2 + 6H_2O + 20NH_4NO_3; \quad (3)$$

where x = 0.1; 0.3; 0.5; 0.7; 0.9 (mol.%).

Aqueous solution of calcium nitrate was mixed with a solution of ammonium hydrophosphate, whose concentration made 0.5 M and 0.3 M, respectively. The synthesis of AgHA and ZnHA required introducing sample weights of silver or zinc into a solution of calcium nitrate. A pH value of 10-11 in a solution of reactants was reached with an aqueous solution of ammonia (25%, ρ = 0.9 g/ml). The reaction mixture underwent microwave exposure with 110 W during 40 minutes and then allowed to stand at room temperature during 48 hours. The precipitate was then filtered, rinsed with a diluted solution of ethanol, and dried until constant weight (~ 15 h) at 90 °C. After that, the sample was ignited during 4 hours at 800 °C [8].

3. Experimental results and discussion

To determine the phase composition for the obtained HA, AgHA, and ZnHA samples with different content of Zn^{2+} and Ag^{+} ions, the authors carried out X-ray phase analysis using PCPDFWIN database and POWDER CELL 2.4 full-profile analysis software.

X-ray patterns for AgHA and ZnHA samples show that all the powders are twophase with hexagonal syngony hydroxyapatite $Ca_5(PO_4)_3(OH)$ being the main phase. As the silver and zinc ions content increases in the initial solution (x values), there is a regular decrease in the amount of the main phase and an increase in the one for the second phase, that is, for β-tricalciumphosphate $Ca_3(PO_4)_2$ (Tables 1 and 2).

Table 1. The results of X-ray phase and X-ray diffraction analyses for HA and ZnHA powders.

Sample x (Zn^{2+}), mol.%	Phase	wt.%	Elementary cell parameters a, Å	c, Å	V, Å^3
HA	$Ca_5(PO_4)_3(OH)$	100.0	9.3690	6.8321	519.36
ZnHA 0.1	$Ca_5(PO_4)_3(OH)$	100.0	9.3750	6.8375	520.45
ZnHA 0.3	$Ca_5(PO_4)_3(OH)$	86.0	9.4601	6.8791	494.78
	$Ca_3(PO_4)_2$	14.0	–	–	–
ZnHA 0.5	$Ca_5(PO_4)_3(OH)$	85.0	9.4127	6.8714	527.24
	$Ca_3(PO_4)_2$	15.0	–	–	–
ZnHA 0.7	$Ca_5(PO_4)_3(OH)$	70.0	9.4132	6.8712	527.28
	$Ca_3(PO_4)_2$	26.0	–	–	–
ZnHA 0.9	$Ca_5(PO_4)_3(OH)$	41.0	9.2063	6.7408	494.78
	$Ca_3(PO_4)_2$	59.0	–	–	–

Table 2. The results of X-ray phase and X-ray diffraction analyses for AgHA powders.

Sample	Phase	wt.%	Elementary cell parameters

x (Ag^+), mol.%			a, Å	c, Å	V, $Å^3$
AgHA	$Ca_5(PO_4)_3(OH)$	98.0	9.3289	6.8248	593.95
0.1	$Ca_3(PO_4)_2$	2.0	–	–	–
AgHA	$Ca_5(PO_4)_3(OH)$	87.0	9.4144	6.8762	609.44
0.3	$Ca_3(PO_4)_2$	10.0	–	–	–
	$CaAgPO_4$	3.0	–	–	–
AgHA	$Ca_5(PO_4)_3(OH)$	86	9.4438	6.8942	614.86
0.5	$Ca_3(PO_4)_2$	14	–	–	–

Changing the parameters of HA elementary cell (Tables 1 and 2) for ZnHA and AgHA samples implicitly confirms that Zn^{2+} and Ag^{2+} ions are present in HA crystal lattice and thus ZnHA and AgHA can be considered as HA-based substitutional solid solutions.

The surface morphology for ZnHA (x = 0.1) and AgHA (x = 0.1) and the dispersivness were determined via micrographs (Figure 2) obtained with X-Max ShiftED 300 scanning electron microscope.

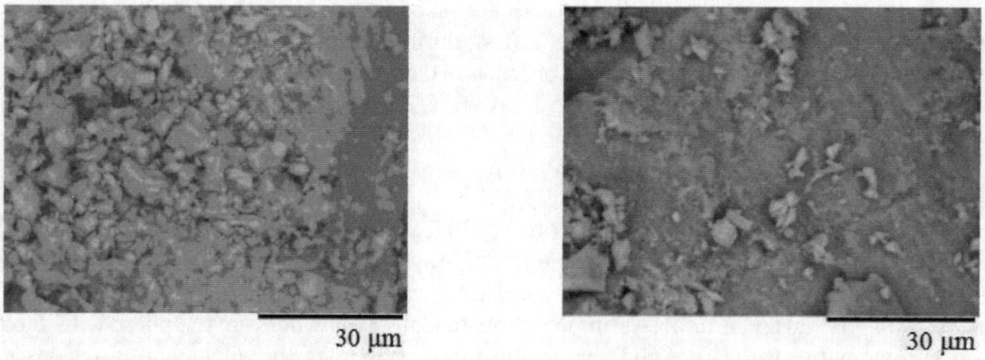

a) b)

Figure 2. Micrographs of the ignited powders' surface: a) ZnHA (x = 0.1); b) AgHA (x = 0.1).

The ZnHA and AgHA powders are agglomerates of various sizes: in ZnHA samples, 70% of the particles have a size of up to 5 µm; AgHA samples are dominated by larger agglomerates (7–20 µm). The SEM results show that the surface morphology of the powders is almost identical, while their dispersiveness differs.

Table 3. Solubility values for HA, ZnHA, and AgHA powders ignited at 800 °C (20 °C, 0.9% wt.%, NaCl solution; ammonia buffer, Eriochrome Black T indicator).

Content *x* (Zn^{2+}/Ag^+), mol.%	Solubility $C_M^{2+}\cdot 10^3$, mol / l		
	ZnHA	AgHA	HA
0	–	–	1.6 ± 0.08
0.1	0.50 ± 0.08	1.19 ± 0.06	–
0.3	0.69 ± 0.04	1.67 ± 003	–
0.5	0.45 ± 0.04	1.44 ± 0.09	–
0.7	0.38 ± 0.04	–	–
0.9	0.22 ± 0.04	–	–

To estimate the solubility of HA, ZnHA, and AgHA samples, the authors determined the total concentration of Ca^{2+} and Zn^{2+} (Ag^+) ions in the physiological solution (ω (NaCl) = 0.9%) where the

IOP Conf. Series: Materials Science and Engineering **597** (2019) 012004 doi:10.1088/1757-899X/597/1/012004

samples were kept for 3 days and stirred to reach complete saturation of the solution compared with a solid phase.

Compleximetric titration of calcium ions (together with zinc and silver ions) in the presence of Eriochrome Black T indicator helped to identify the solubility values (C_M^{2+}, mol / l; at 20 °C) for HA, ZnHA, and AgHA obtained by liquid-phase microwave synthesis and ignited at 800 °C.

The obtained solubility values are given in Table 3. It is noted that an increase in the content of Zn^{2+} and Ag^+ ions while synthesizing hydroxyapatite slightly reduces the solubility in a physiological solution compared with non-modified HA.

The porosity of calcium phosphate ceramics is a very important characteristic. The porosity of natural bone tissue is not a constant value, therefore, the methods for obtaining HA can be different depending on the potential application area of HA-based bioceramics [9, 10]. It is known that applying combustible organic additives in obtaining synthetic HA contributes to an increase in its porosity. So, in order to observe changes in HA porosity, the authors employed aminoacids (glycine and aspartic acid) and zinc salts containing the same aminoacids (Table 4, samples 3-6).

The specific surface area (SSA), volume, and pore size of HA-based powders were measured via adsorption with subsequent degassing under ~ 0.1 Pa at 200 °C for two hours on a TriStar 3020 gas-solid analyzer (BET method).

Applying aminoacids and their zinc salts does not contribute to an increase in porosity, therefore, it can be recommended only when synthesizing hydroxyapatite with low porosity. However, to obtain a highly porous HA, it is preferable to select other combustible additives.

Table 4. The results of determining the specific surface area and porosity of HA and ZnHA.

	Sample	S, m^2/g	Total pore volume, cm^3/g	Average pore size, nm
1	HA	15	0.037	10.0
2	ZnHA ($x = 0.5$)	21	0.046	8.8
3	ZnHA with HGly	18	0.039	8.6
4	ZnHA c H_2Asp	18	0.038	8.4
5	HA with $ZnGly_2$	20	0.041	8.2
6	HA with ZnAsp	20	0.041	8.2

A Koch biotesting of ZnHA samples ($x = 0.3$ and 0.7) was carried out with reference to their effect on *E. coli* bacteria (*Escherichia coli*, ATCC 25922) while cultivating the latter in standard *LB* nutrient medium. Sterilization conditions of the samples (25 min, 5 atm, 112 °C) do not decompose hydroxyapatite according to the literature data.

The results of biotesting (Table 5) show that HA and ZnHA powders inhibit the growth of *E. coli* under the cultivation conditions specified above, that is, they suspend the development of *Escherichia coli*, but do not completely decompose it. The ZnHA sample at $x = 0.7$ showed a more than twofold decrease in the number of bacteria compared with the ZnHA sample at $x = 0.3$. The inhibitory effect of HA sample is even more evident.

Table 5. The test sample count under different experimental conditions (significance value = 95%).

	Sample	Count (*E. coli*), cfu/ml
1	Reference sample	$(2.10 \pm 0.28)10^8$
2	HA	$(3.60 \pm 0.85)10^7$
3	ZnHA ($x = 0.3$)	$(9.70 \pm 0.65)10^7$
4	ZnHA ($x = 0.7$)	$(4.40 \pm 0.94)10^7$

Both acidic (Ca^{2+}, Ag^{+}, and Zn^{2+}), and basic (PO_4^{3-}, OH^{-}) Lewis centers might be present pn the surface of HA powders. Brønsted acid-base centers are possible as well. Therefore, the estimation of the relative distribution of acid-base centers on the surface of HA powders was carried out by the Hammett method. The results demonstrate that an increase in the content of the doping ion (Zn^{2+}, Ag^{+}) in HA changes the quantitative ratio of acid-base centers (Figure 3).

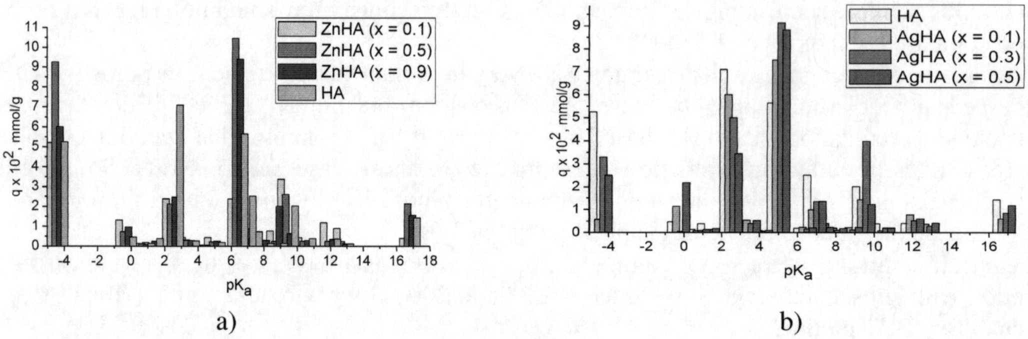

Figure 3. Histograms of the acid-base centers distribution on the surface of ZnHA powder (a) and AgHA powder (b).

4. Conclusions

1. Samples were obtained by liquid phase deposition within a microwave field at atmospheric pressure with different initial content of zinc and silver ions (from 0.1 to 0.9 mol.%). Hydroxyapatite $Ca_5(PO_4)_3(OH)$ with hexagonal syngony served as the main phase for all the samples.
2. According to XRD, the elemental composition of the powders obtained shows that the molar ratio of the elements (Ca + M) / P is 1.57-1.67 for ZnHA; 1.65-1.67 for AgHA, while the Ca / P elements ratio for bone tissue ranges from 1.5 to 2.0.
3. It has been established that ZnHA and AgHA powders are poorly soluble at 20 °C in NaCl. An increase in the content of Zn^{2+} и Ag^{+} ions leads to a slight decrease in the solubility of powders compared with non-modified HA.
4. The surface morphology, specific surface area, porosity, and dispersiveness of ZnHA powders do not undergo notable change when applying aminoacids and their zinc salts in the course of microwave synthesis.
5. The surface of ZnHA and AgHA powders is characterized by the presence of acidic (Ca^{2+}, Ag^{+}, and Zn^{2+} ions), basic (PO_4^{3-} and OH^{-} ions) Lewis centers, and Brønsted acid-base centers. Their ratio changes with increasing the content of biogenic Zn^{2+} и Ag^{+} ions in HA.
6. Biotesting of ZnHA powders showed its inhibitory effect on the growth of *E. coli* bacteria.
7. Phase composition, low solubility in physiological solution, and morphological characteristics of ZnHA and AgHA powders are similar to those of non-modified hydroxyapatite and allow the authors to recommend these powders for medical purposes.

Acknowledgements
This research is supported by the Russian Foundation for Basic Research (No. 17-03-00698\18).

References

[1] Barinov S M Calcium phosphate-based ceramic and composite materials for medicine 2010 *Russian Chemical Reviews* **79** **1** 13-29

[2] Kurzina I A, Churina E G, Shapovalova E G, Syusyukina V A, Kzhyshkowska J G 2018 Immunomodulatory properties of composite materials based on polylactide and hydroxyapatite *Bioceramics Development and Applications* **8 1** 109-1-109-7

[3] Shapovalova Ye, Syusyukina V, Churina E, Kzhyshkowska J, Kurzina I 2017 Qualitative assessment of polylactide and hydroxyapatite based composites. Features of the structure and biochemical properties *Materials Science and Engineering* pp 1-20

[4] Kurzina I A, Lytkina D N, Berezovskaya A A, Kalachikova P M, Heinrich L A, Churina E G, Korotchenko N M, Kozik V V 2018 Porous composites of hydroxyapatite and polylactide-co-glycolide acid: surface properties and in-vitro biocompatibility *Materials* **8** 1-14

[5] Lytkina D, Berezovskaya A, Kurzina I, Filimoshkin A, Kozik V 2016 Composites with a system of interconnected pores based on hydroxyapatiteand lactide-co-glycolide *BioNanoMat* **17 S1** 153-154

[6] Zhuk I, Rasskazova L, Korotchenko N, Kozik V, Kurzina I 2017 Synthesis and investigation of physico-chemical, antibacterial, biomymetic properties of silver and zinc containing hydroxyapatite *AIP Conf. Proc* **1899 1** 020017-1-9

[7] Lytkina D, Berezovskaya A, Korotchenko N, Kurzina I, Kozik V 2017 Preparation of composite materials based on hydroxyapatite and lactide and glycolide copolymer *AIP Conf. Proc.* **1899** 1 020015-1-6

[8] Rasskazova L A, Korotchenko N M, Kozik V V, Ivanov V K, Shilyaeva L P 2014 Study of the properties of silica-modified hydroxyapatite with various contents of silicate ions synthesized in microwave field *Theoretical Foundations of Chemical Engineering* **48 5** 682-6

[9] Syusyukina V A, Shapovalova Ye, Korotchenko N M, Kurzina I A 2017 Structural-phase state and surface properties of composite materials based on polylactide and hydroxyapatite *Russian Journal of Applied Chemistry* **90 1** 106-112

[10] Shapovalova Ye, Ly'tkina D N, Rasskazova L A, Zhuk I V, Gudima Alexandru, Filimoshkin A G, Korotchenko N M, Kurzina I A, Kzhy'shkovska Y G 2016 Preparation of Biocompatible Composites based on Poly-L-lactide/Hydroxyapatite and Investigation of their Anti-Inflammatory Activity *Key Engineering Materials* **683** 475-480

IOP Conf. Series: Materials Science and Engineering **597** (2019) 012005 doi:10.1088/1757-899X/597/1/012005

High entropy oxide systems based on rare earth elements

V I Sachkov[1], R A Nefedov[1] and I V Amelichkin[1]
[1]Innovation and Technology Center, National Research Tomsk State University, 36 Lenin Street, Tomsk 634050, Russian Federation

E-mail: amelichkinivan@gmail.com

Abstract. In this paper we synthesized high entropy oxide system $(Sc_{0.2}Ce_{0.2}Pr_{0.2}Gd_{0.2}Ho_{0.2})_2O_{3\pm\delta}$ by precipitation of hydroxides from REE nitrates aqueous solutions and annealing the mixture at 800 °C. It is possible to obtain the various forms of simple and complex composition based on the rare earth oxides, including oxides in the form of nanosized powders. The main idea of obtaining HEOS comes from the field of high-entropy alloys. The interconnection between configurational entropy and mixing enthalpy is a key factor for determining phase transformations. The phase composition, structure, and surface morphology of the obtained powders were investigated by scanning electron microscopy and X-ray phase analysis methods.

1. Introduction

High entropy oxide systems (HEOS) are systems that contain more than three components with a uniform distribution of ions in crystal lattice. Such systems are characterized by increased entropy of mixing [1,2].

The main idea of obtaining HEOS comes from the field of high-entropy alloys (HEA). HEAs have a number of important mechanical (strength, hardness, plasticity, wear resistance), physical (magnetic, conductivity) and chemical (corrosion resistance) properties compared to conventional alloys with one or two basic elements. In such metal systems, high mixing entropy thermodynamically stabilizes a single-phase solid solution by reducing the Gibbs free energy. By analogy with HEAs, it can be expected a similar entropy mixing effect in oxide systems. HEOS can be used as adhesive protective coatings, optical processors, waveguides, radiopaque substances, catalysts, luminescent compounds, multiferroics, acousto-optic, data storage and reading devices [3,4].

One of the important areas of modern research in the field of HEOS synthesis and materials based on them is the development and improvement of specific methods for their preparation. The choice of methods for the synthesis of such materials influences on the composition, structure, particle size and determines the technological effectiveness of the process for their preparation. Prospective compared with high-temperature methods of solid-phase synthesis are the methods of 'soft chemistry' or solution methods, for example, the sol-gel method [5]. In the latter case it is possible to obtain the various forms of simple and complex composition based on rare earth oxides, including oxides in the form of nanosized powders.

This work is focused to explore the possibility of obtaining of high entropy oxide powders based on rare earth elements (REE) by hydroxides precipitation from nitrates solution.

2. Experimental part

Synthesis of HEOS based on REE can be divided into five stages (figure 1). As starting materials were used: $Sc(NO_3)_3{\cdot}4H_2O$, $Ce(NO_3)_3{\cdot}6H_2O$, $Pr(NO_3)_3{\cdot}6H_2O$, $Gd(NO_3)_3{\cdot}5H_2O$, $Ho(NO_3)_3{\cdot}5H_2O$, NaOH.

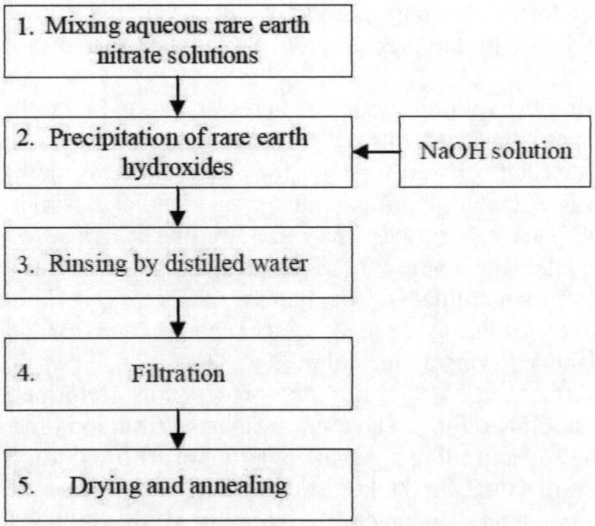

Figure 1. General flowchart of synthesis steps.

Consider the following example of synthesis HEOS $(Sc_{0.2}Ce_{0.2}Pr_{0.2}Gd_{0.2}Ho_{0.2})_2O_{3\pm\delta}$ with a theoretical yield of the product in an amount of 1.2 g.

The first stage is the mixing of aqueous solutions of REE nitrates. One of the conditions for obtaining HEOS is equal mole ratio of each component in the system. The aliquot volume (V_a) was calculated based on the concentrations of REE nitrates and the number of moles (0.002). The concentrations of initial solutions and the volumes of aliquot (V_a) are listed in table 1.

Table 1. The concentrations of the initial solutions and the volumes of the aliquot (V_a).

REE nitrates	Concentration (mmol/ml)	V_a (ml)
$Sc(NO_3)_3{\cdot}4H_2O$	0.330	6.060
$Ce(NO_3)_3{\cdot}6H_2O$	0.229	8.700
$Pr(NO_3)_3{\cdot}6H_2O$	0.230	8.680
$Gd(NO_3)_3{\cdot}5H_2O$	0.230	8.660
$Ho(NO_3)_3{\cdot}5H_2O$	0.226	8.820

The solutions were mixed in a 250 ml beaker on a magnetic stirrer for one hour at the room temperature.

The second stage is the precipitation of hydroxides from a mixture of nitric acid solutions of REE. Precipitation process may be described by the chemical reaction:

$$Me(NO_3)_3 + 3NaOH \rightarrow Me(OH)_3\downarrow + 3NaNO_3,$$

where Me = Sc, Ce, Pr, Gd, Ho.

The precipitation was carried out with a solution of NaOH, which was intensively mixed on a magnetic stirrer. The volume of NaOH solution taken with a 10% excess for the complete precipitation of REE hydroxides mixture of this system was calculated based on maximum pH of precipitation of hydroxides from the REE nitrates solution in the system. The excess solution of NaOH was calculated by multiplying the stoichiometric weight amount by the accepted amount of its excess. To reduce the reaction time, it is necessary to pour a mixture of nitric acid solutions of REE into the NaOH solution with vigorous stirring. This method is conventionally called the "reverse" method of synthesis. The solution resulting from mixing with the precipitate of REE hydroxides was incubate for two hours at room temperature.

The third stage is rinsing the solution with distilled water to pH = 8. The pH measurement was performed by a potentiometric method on a pH-150MI device. The solution became turbid with stirring. The resulting suspension was left to settle for precipitation to the bottom. After settling, the upper transparent layer was decanted, distilled water was added and stirred. After precipitation, the upper layer was decanted again and rinsed. The necessity of such rinses is due to the presence of nitrate ions in the precipitate. The volume of distilled rinsing water was more than 500 ml. The presence of nitrate ions was determined by qualitative reaction with diphenylamine. In this case, initially, colorless diphenylbenzidine is formed, which upon further oxidation turns into a quinoid derivative of diphenylbenzidine having a blue color.

The fourth stage - filtering solution. The filtration process was performed using a Buchner funnel and a Bunsen flask on a blue-ribbon filter. The total rinsing and filtration time were 8 hours.

The fifth stage is drying and annealing a gel-like mixture of REE oxides. Drying was carried out in cupboards at a temperature of 70 °C for 12 hours. Annealing was carried out in muffle furnaces at a temperature of 800 °C for one hour. The practical yield of oxides was 97–99% of the theoretical. The method of "direct" synthesis included similar stages as the method of "reverse" synthesis, but in the second stage, on the contrary, the NaOH solution was poured into a mixture of nitric acid REE solution.

To study phase composition X-ray diffraction analysis (XRD) 'DRON-3M diffractometer' and scanning electron microscopy (SEM) 'QUANTA 200 3D FEI' with X-ray energy dispersive microanalysis system were used.

3. Results

The structural properties of $(Sc_{0.2}Ce_{0.2}Pr_{0.2}Gd_{0.2}Ho_{0.2})_2O_{3\pm\delta}$ were investigated by XRD over 2θ range from 15° to 95° using a small step size (0.2°) and Cu-Kα radiation with a wavelength of 0.154 nm (figure 2).

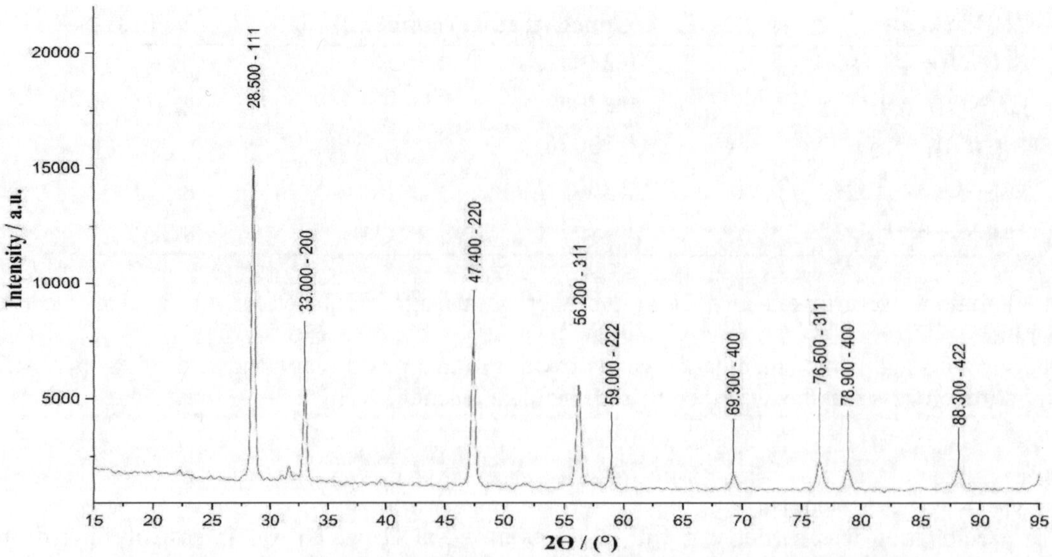

Figure 2. XRD pattern of HEOS $(Sc_{0.2}Ce_{0.2}Pr_{0.2}Gd_{0.2}Ho_{0.2})_2O_{3\pm\delta}$.

IOP Conf. Series: Materials Science and Engineering **597** (2019) 012005 doi:10.1088/1757-899X/597/1/012005

The XRD graph showed strong peaks at 28.5, 33.0, 47.4, 56.2, 59.0, 69.3, 76.5, 78.9, 88.3 degrees of 2Θ. All the synthesized oxide nanoparticles diffraction peaks corresponded to (111), (200), (220), (311), (222), (400), (311), (400) and (422) crystal planes. The average crystallite sizes of $(Sc_{0.2}Ce_{0.2}Pr_{0.2}Gd_{0.2}Ho_{0.2})_2O_{3\pm\delta}$ were determined based on the width of dominant peaks using the Scherrer equation, and they were in the range of 13.4 – 15.7 nm. The obtained XRD patterns obtained of the $(Sc_{0.2}Ce_{0.2}Pr_{0.2}Gd_{0.2}Ho_{0.2})_2O_{3\pm\delta}$ were compared with the standard data for CeO_2 (JCPDS file No. 34-0394). The characteristic peaks are very close to the face centric cubic CeO_2 indicating that all samples can be identified to ceria with the cubic fluorite structure. The lattice parameters of the prepared $(Sc_{0.2}Ce_{0.2}Pr_{0.2}Gd_{0.2}Ho_{0.2})_2O_{3\pm\delta}$ sample were determined to be 0.54181 nm. This lattice parameters are almost the same with the standard of CeO_2 (0.54111 nm) [6].

The chemical composition of the prepared $(Sc_{0.2}Ce_{0.2}Pr_{0.2}Gd_{0.2}Ho_{0.2})_2O_{3\pm\delta}$ oxide powders was determined using energy-dispersive X-ray spectroscopy (EDS); the results are shown in figure 3. The EDS spectra clearly indicated that all of REE ions were present in sample.

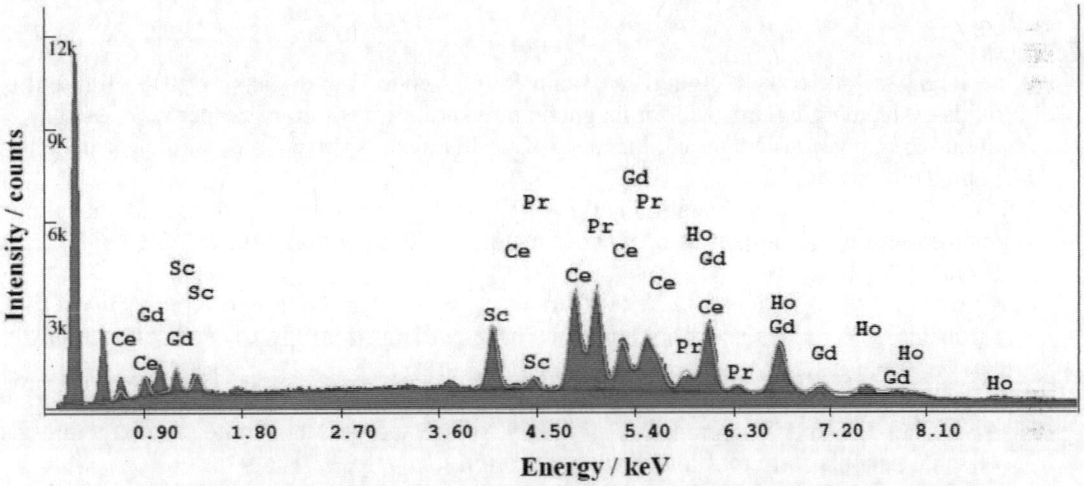

Figure 3. EDS spectra of HEOS $(Sc_{0.2}Ce_{0.2}Pr_{0.2}Gd_{0.2}Ho_{0.2})_2O_{3\pm\delta}$.

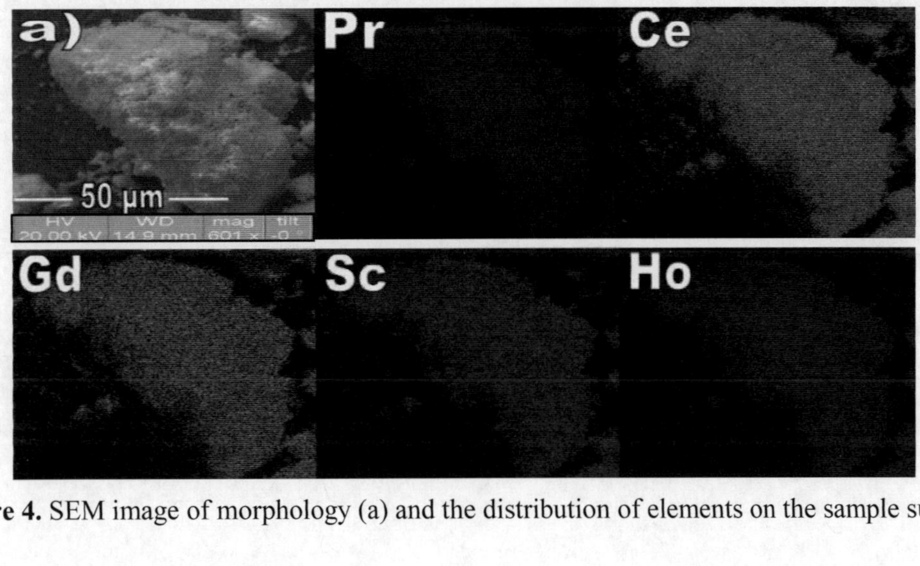

Figure 4. SEM image of morphology (a) and the distribution of elements on the sample surface.

According to the data of SEM, the sample under study is a loose powder with a developed surface morphology, which is represented by structural units in the form of spherical and oval particles, irregular polyhedrons and scales. All REE elements in the system are equally distributed over the sample surface (figure 4).

4. Conclusion

High entropy oxide system based on rare earth elements was successfully synthesized. The phase composition, structure, and surface morphology of the obtained powders were investigated by XRD, EDS and SEM methods. The lattice parameters of the prepared $(Sc_{0.2}Ce_{0.2}Pr_{0.2}Gd_{0.2}Ho_{0.2})_2O_{3\pm\delta}$ sample were determined to be 0.54181 nm. This lattice parameters are almost the same with the standard of CeO_2 (0.54111 nm).

Acknowledgments

The results were obtained as part of the fulfilment of the state task of the Ministry of Education and Science of the Russian Federation, project No. 10.3031.2017 / 4.6 and with the support of the Tomsk State University Competitiveness Enhancement Program.

References

[1] Witte R, Sarkar A, Kruk R, Eggert B, Brand R A, Wende H and Hahn H 2019 High entropy oxides: An emerging prospect for magnetic rare earth - transition metal perovskites

[2] Berardan D, Franger S, Dragoe D, Meena A K and Dragoe N 2016 Colossal dielectric constant in high entropy oxides

[3] Liu B, Xu L, Liu Y, Wang J, Wang J and Fang Q 2018 Effect of cold working and annealing on microstructure and properties of powder metallurgy high entropy alloy *Sci. China Technol. Sci.* **61** 197–203

[4] Karpets M V., Myslyvchenko O M, Makarenko O S, Gorban' V F and Krapivka M O 2015 Mechanical Properties and Formation of Phases in High-Entropy CrFeNiCuCoAl x Alloys *Powder Metall. Met. Ceram.* **54** 344–52

[5] Solovev Y V., Prilepskii A Y, Krivoshapkina E F, Fakhardo A F, Bryushkova E A, Kalikina P A, Koshel E I and Vinogradov V V. 2019 Sol-gel derived boehmite nanostructures is a versatile nanoplatform for biomedical applications *Sci. Rep.* **9** 1176

[6] Abbas F, Iqbal J, Tariq J, Badshah N and Ismail M 2016 Structural, morphological, Raman, optical, magnetic, and antibacterial characteristics of CeO_2 nanostructures *Int. J. Miner. Metall. Mater.* **23** 102

IOP Conf. Series: Materials Science and Engineering **597** (2019) 012006 doi:10.1088/1757-899X/597/1/012006

Solubility characteristics of the material based on porous hydroxyapatite ceramics coated with PLGA

D Lytkina[1], A Gutsalova[1], M Pozdniakov[1], L Heinrich[1,2] and V Kozik[1]
[1]National Research Tomsk State University, Russia
[2]University of Muenster, Germany

E-mail: darya-lytkina@yandex.ru

Abstract. Composites based on a porous hydroxyapatite carcass with a coating of PLGA were obtained. According to the SEM images, it was found that, the composites had a different coating character depending on the porosity of the HA framework. After 1 week, the dissolution rate of composites increases relative to pure PLGA, and the growth is proportional to the increasing of the amount of pore-forming agent in the material by 20-50%. The dissolution rate of HA in the composition of materials also increases since the second week, because the diffusion of the solvent in the HA increases as the result of PLGA hydrolysis.

1. Introduction

Currently, hydroxyapatite (HA) and a copolymer of lactide and glycolide (PLGA) are among the most popular components for obtaining bioresorbable materials for bone tissue repair [1].Typically, HA promotes of osteointegration [2], as is a component of native bone tissue, PLGA serves as a degraded carrier of HA, as well as a modifier of some of its properties. Conventionally, such materials can be divided into PLGA fibers with included HA particles [3,4], polymer films and molded blocks filled with HA [5,6], foams [7] and ceramic materials with polymer coating [8]. Solubility of such materials depends on the method of their preparation, their surface characteristics, crystallinity of the components and molecular weight of used PLGA. Previously, the approaches of preparation of a porous ceramic frame material covered with PLGA were made, and it was shown than such material is able to maintain mechanical strength during the process of dissolution for a long time. So, studying of the solubility characteristics of the materials based on HA-ceramics coated with PLGA and estimation of the time of their degradation in a model solution is the aim of this work.

2. Experimental

2.1. Preparation of *initial components and composite materials*

The original HA was obtained by a liquid-phase microwave method [2]. PLGA (Mw = 80 000) was obtained by coordination polymerization via ring opening polymerization of lactide and glycolide (lactide/glycolide ratio is 80/20) [9,10].

Materials were obtained in 2 stages. First one - obtaining a porous HA frame with usage of soluble NaCl as pore-forming agent. Varying of NaCl amounts, 3 types of HA frames with different porosity were obtained. Second stage – impregnation of the obtained scaffolds in the PLGA solution under ultrasound action. In details, materials preparation is described in [8]. The compositions of the materials are presented in Table 1.

IOP Conf. Series: Materials Science and Engineering **597** (2019) 012006 doi:10.1088/1757-899X/597/1/012006

Table 1. Composition of the materials

Samples	K1	K2	K3
Amount of pore-forming agent (NaCl), mass. %	10	25	50
PLGA content, mass. %	5	7	17

2.2. Study of the materials surfaces morphology

The morphology of materials surfaces were analyzed on a Gemini Zeiss 1540 EsB microscope at accelerating voltages of 5-10 kV using secondary and backscattered electron detectors. The survey was carried out with preliminary deposition of silver onto the surface of samples on Polaron SC 7640 sputter coater.

2.3. Determination of the materials solubility

To evaluate the solubility of materials experiment was conducted complex comprising HA solubility in the composition of materials (control of Ca^{2+} ion concentration), and the mass loss of the material over time. Solubility of pure HA and HA in the materials was determined by the total concentration of calcium ions Ca^{2+} by trilonometric titration in the presence of eriochrome black T with ammonia buffer (pH 9-10). For this purpose, samples were incubated in saline solution (ω (NaCl) = 0.9%) at 37 °C for 7 days to reach saturation with respect to solid phase. The mean values of the concentrations of Ca^{2+} ions(mol / l) and the confidence interval (with correlation index = 0.95) for them were calculated from three parallel measurements.

The materials degradation time were determined by measuring of the material mass loss after it was kept in sodium phosphate buffer for 28 days at 37 °C (weighing after drying).

3. Results and discussion

SEM images were obtained for composites (figure 1). All 3 composites are characterized by different coverage of HA surface with PLGA.

Figure 1. SEM images of composite materials a) K1, b) K2, c) K3.

PLGA coating of K1 samples were localized on the surface of HA particles in form of non-uniform single aggregates. For K2 samples uniform "island-like" coating with large number of phase boundaries between ceramics and polymer over its entire surface were formed. At the same time, K3 is completely covered with polymer, but the coating is non-uniform, as evidenced by large aggregates on the surface of ceramics.

According to the results obtained (figure 2), in the series from K1-K3, the solubility of HA decreases, probably due to the fact that the amount of PLGA increases, and the diffusion of calcium ions through the polymer layer becomes difficult.

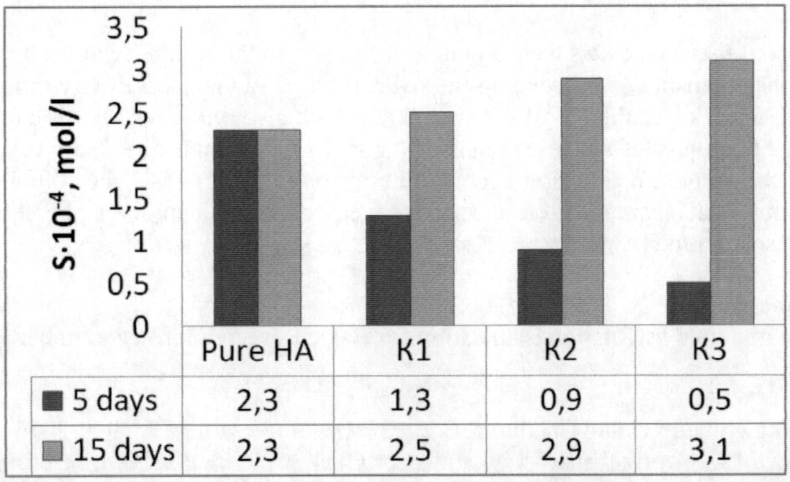

Figure 2. Solubility of pure HA and HA in the composition of composites.

However, with further incubation of samples in saline solution, polymer degradation occurs, which leads to a shift in the pH of the medium to a more acidic region and an increase in the solubility of HA. After 15 days of exposure, the opposite is true - in the K1-K3 series, the solubility of HA increases in accordance with an increase in the amount of PLGA by 2-6 times, which will positively affect of the material resorption (figure 3).

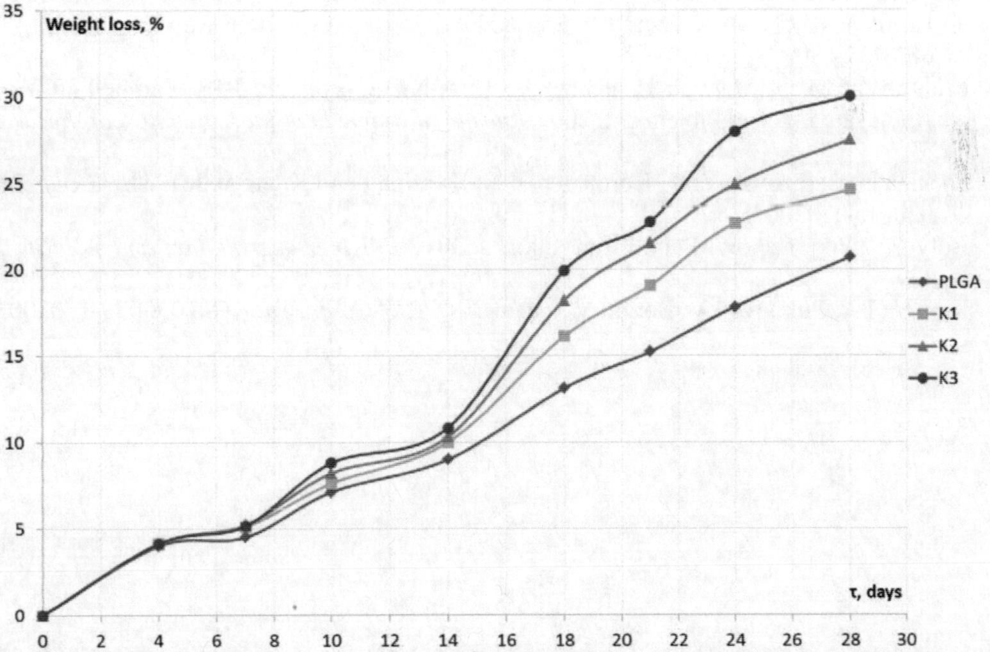

Figure 3. The dependence of the mass loss of materials from the time of incubation in PBS.

The dependences obtained show that the first five days the dissolution proceeds at the same rate, however, by the end of the second week an increase in the dissolution rate of the composites relative to pure PLGA is observed. The dissolution rate of composites increases from the second week, in proportion to the increase in porosity and the amount of PLGA in the material by 20-50%.

IOP Conf. Series: Materials Science and Engineering **597** (2019) 012006 doi:10.1088/1757-899X/597/1/012006

4. Conclusions

The components of the composites have a mutual influence on the solubility of each other, since HA is the main phosphate capable of influencing the hydrolysis of PLGA, and PLGA in turn, as a result of hydrolysis, and helps lower the pH of the medium, which also enhances the dissolution of HA. The dissolution rate of composites increases relative to pure PLGA from the second week, in proportion to the increase in the amount of blowing agent in the material by 20-50%. The solubility of HA in the composition of materials increases from the second week, since in the process of PLGA hydrolysis, the diffusion of the solvent to HA increases.

Acknowledgements

This work was supported by Russian Foundation for Basic Research (project number 17-03-00698\19).

References

[1] Zhou H, Lawrence J G and Bhaduri S B 2012 *Acta Biomaterialia* **8** 1999-2016

[2] Rasskazova L A, Korotchenko N M and Zeer G M 2013 *Russian Journal of Applied Chemistry* **86** 691-695

[3] Syusyukina V A, Shapovalova Ye, Korotchenko N M and Kurzina I A 2017 *Russian Journal of Applied Chemistry* **90** 112

[4] Kurzina I A, Churina E G, Shapovalova E G, Syusyukina V A and Kzhyshkowska J G 2018 *Bioceramics Development and Applications* **8** 109-1-109-7

[5] Rasskazova L A, Lytkina D N, Shapovalova E G, Botvin V V, Pozdniakov M A, Filimoshkin A G, Korotchenko N M and Kozik V V 2015 *Russian Journal of Applied Chemistry* **88** 669-675

[6] Shapovalova Ye, Ly'tkina D N, Rasskazova L A, Zhuk I V, Gudima A, Filimoshkin A G, Korotchenko N M, Kurzina I A and Kzhy'shkovska Y G 2016 *Key Engineering Materials* **683** 475-480

[7] Montjovent M O, Mathieu L, Schmoeke H, Mark S, Bourban P E, Zambelli P Y, Laurent-Applegate L A and Pioletti D P 2007 *Journal of Biomedical Materials Research Part A* **83** 41-51

[8] Lytkina D, Berezovskaya A, Korotchenko N, Kurzina I and Kozik V 2017 *AIP Conf. Proc.* **1899** 020015-1-020015-6

[9] Botvin V, Pozdniakov M and Filimoshkin A 2017 *Polymer degradation and stability* **146** 126-131

[10] Kurzina I A, Pukhova I V, Botvin V V et al 2015 *AIP Conf. Proc* **1688** 030033-1–030033-7

IOP Conf. Series: Materials Science and Engineering **597** (2019) 012007 doi:10.1088/1757-899X/597/1/012007

Synthesis of poly-ε-caprolactone/hydroxyapatite composite materials by *in situ* and mechanical mixing methods and investigation of their physico-chemical properties

S Karaseva[1], V Botvin[1] and A Filimoshkin[2]

[1]Laboratory of Polymers and Composite Materials, National Research Tomsk State University, 49 Arkadya Ivanova Street, Tomsk 634028, Russian Federation
[2]Faculty of Chemistry, Department of High Molecular Compounds and Petrochemistry, National Research Tomsk State University, 36 Lenina Avenue, Tomsk 634050, Russian Federation

E-mail: svetlana_karasyova_1997@mail.ru

Abstract. This work is dedicated both to investigation of the influence of different conditions on the molecular weight characteristics of poly- ε-caprolactone (PCL) and to the synthesis of composites based on PCL and hydroxyapatite (HA) by *in situ* and mechanical mixing methods. It was shown that *in situ* synthesis is accompanied by chemical interactions between PCL and HA that were confirmed by IR spectroscopy and thermal analysis. The morphology of *in situ* obtained composites has more homogeneous and regular distribution of HA particles in the polymeric matrix than composites synthesized by mechanical mixing of HA and PCL.

1. Introduction

Currently, one of the main tasks of modern materials science is the development of new functional materials for various fields, primarily for medicine. Among the huge variety of polymeric materials for medical purposes, the class of biodegradable polyesters, combining the advantages of synthetic and natural polymers, occupies a special place. One of the representatives of this class is poly-ε-caprolactone (PCL) [1]. Due to its physicochemical characteristics and biocompatibility, PCL, both independently and in the presence of inorganic filler, is successfully used in various branches of medicine, especially in bone tissue engineering [2]. Hydroxyapatite (HA), the most important inorganic component of bone tissue, most often acts as an inorganic filler [3]. It is known that the properties of composite materials principally depend on the molecular mass characteristics of the polymer matrix and the method of their preparation. In scientific works describing the production of composite materials for orthopedics, the emphasis is placed on polymers having a molecular weight (MW) of at least 50000 g/mol. Materials with such MW have optimal mechanical properties and time of degradation.

The method of obtaining also has a significant influence on the properties of composite materials. Methods for producing composites by mechanical displacement, when inorganic filler is added to the polymer solution with constant stirring, and then a composite material is formed from the resulting mixture, the most investigated [4, 5]. This method is universal; however, it has some disadvantages, among which are the low strength characteristics of the materials obtained and the irregular distribution of the filler in the polymer matrix. One of the most promising but poorly studied methods for producing composites is the *in situ* method, or synthesis in the reaction mixture. Using this approach, a large number of biodegradable polyester/inorganic filler systems were obtained [6,7]. Despite the fact that PCL is one of the representatives of the class of biodegradable polyesters, studies on the production of PCL/hydroxyapatite composites using the *in situ* method are currently presented only in a few publications [8, 9].

In this connection, this article is dedicated to the search for optimal conditions for the synthesis of high molecular weight PCLs, the preparation of composite materials based on PCL and HA by mechanical mixing and *in situ* and the investigation of some of their physicochemical properties.

2. Experimental part

2.1 Materials and methods

PCL was obtained by polymerizing of ε-caprolactone (99%) (Sigma-Aldrich, Germany) in the presence of an initiator – $Sn(Oct)_2$ (Sigma-Aldrich, Germany)) and co-initiators – alcohols of various functionalities: butanol C_4H_9OH (But), propylene glycol $C_3H_6(OH)_2$ (PG) and glycerol $C_3H_5(OH)_3$(GL) (all Vecton, Russia).

All commercial reagents are previously purified, mainly from traces of water, which contributes to the deactivation of the initiator and uncontrolled breakage of the chain. Commercial ε-caprolactone was distilled under reduced pressure over 4 Å molecular sieves. To dehydrate alcohols, they were subjected to azeotropic distillation with a Dean-Stark trap.

HA was obtained by liquid-phase microwave synthesis according to the method described in [10].

Molecular mass characteristics of PCL samples and composites obtained by the *in situ* method were analyzed by gel permeation chromatography (GPC) on an Agilent 1200 liquid chromatograph with a refractometric detector (eluent – chloroform).

The X-ray diffraction analysis of HA powders was performed on Shimadzu XRD 6000 diffractometer at $Cu_{K\alpha}$ radiation. The processing of the obtained data and phase identification were carried out using a database JSPDS PDF 4+ and full-profile analysis programs POWDER CELL 2.4.

The chemical and functional composition of the composites obtained by mechanical mixing and *in situ* was determined by IR spectroscopy in the range of wavenumbers 4000 – 500 cm^{-1} on IR-Fourier spectrometer Aglient Cary 630 FTIR.

Micrographs of PCL/HA composites were obtained by scanning electron microscopy (SEM) on a Hitachi TM-3000 instrument at an accelerating voltage of 15 kV under conditions of the charge-off mode (electron gun: $5 \cdot 10^{-2}$ Pa; camera for the sample: 30-50 Pa).

The thermal properties of composites based on PCL obtained by the methods of mechanical mixing and *in situ* were investigated by the methods of differential scanning colorimetry (DSC) and thermogravimetric analysis (TGA) on a combined NETZSCH STA 449F1 thermal analyzer in the range of 30-800 °C with a heating rate of 10 °C/min in argon atmosphere.

2.2 Synthesis of PCL as polymeric component of composites and studying the effect of various factors on its molecular weight characteristics

Polymerization of ε-caprolactone with a ring opening in the presence of $Sn(Oct)_2$ and alcohols of various functionalities was performed according to the scheme shown in Figure 1.

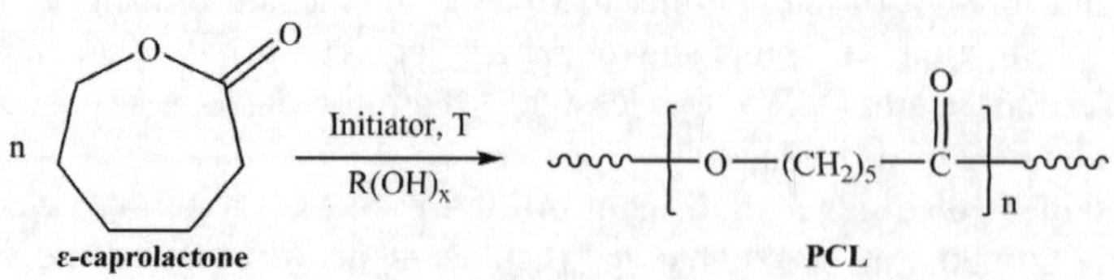

Figure 1. Ring opening polymerization of ε-caprolactone.

The synthesis of PCL was carried out in test tubes per 5 g of distilled ε-caprolactone. The amount of initiator was calculated from the ratio [monomer]: [initiator] 1:500 and all components were added into the tube with monomer. For the selection of optimal polymerization conditions for ε -caprolactone, co-initiators with different functionalities were used and the ratio [initiator]:[co-initiator] was varied. The polymerization reaction system was prepared in a glove box under a nitrogen atmosphere. After all the components were added, the tube with the monomer, the initiator and the co-initiator was placed in a Heidolph Hei-VAP Advantage rotary evaporator bath with constant stirring of the reaction mixture. The synthesis was carried out at 140-170 °C for 5 hours.

2.3 Preparation of composite materials based on PCL and HA

2.3.1 Mechanical mixing method. Composite materials of composition PCL/HA 90/10, 80/20 and 70/30 were obtained by mixing a solution of PCL in chloroform (c = 0.1 g/ml) and HA powder with constant stirring on a magnetic stirrer (IKA C - MAG HS 4). The resulting suspension was treated with ultrasound in an ultrasonic bath (JEIOTECH) for 20 minutes at a frequency of 40 kHz. The obtained samples were cast in the form of films on a Petri dish.

2.3.2 In situ method. When producing composites by the *in situ* method, reaction systems were prepared in a glove box. In four flasks were placed on 20 g of ε-caprolactone and $Sn(Oct)_2$, the amount of which was calculated from the ratio [monomer]:[initiator] 2000:1, and added 5 mass.% HA (composite 1 - C1), 10 mass.% HA (composite 2 - C2), 20 mass.% HA (composite 3 - C3) and 30 mass.% HA (composite 4 - C4). Suspensions of the monomer with different amounts of HA were pre-ground for 10 minutes to homogenize before adding to the flasks. Then the flasks were placed in a Heidolph Hei-VAP Advantage rotary evaporator bath at T = 170 °C with constant stirring. Synthesis was carried out for 5 hours. At the end of the synthesis, the selected samples of the composites were cast in the form of films on a Petri dish. To determine the MW of the composites obtained *in situ*, HA was removed by filtration on a Schott filter with a pore size of 1.6 μm.

3. Results and discussion

3.1 Studying the effect of various factors on the molecular weight characteristics of PCL
During the experiments, the effect of temperature, functionality of a co-initiator, and the ratio [initiator]: [co-initiator] on the molecular weight of PCL as a polymer component of composites synthesized by mechanical mixing was evaluated. It can be seen from the Figure 2 that for the ratio $[Sn(Oct)_2]$:[alcohol] = 1:1 the curves have the same profile for all three alcohols.

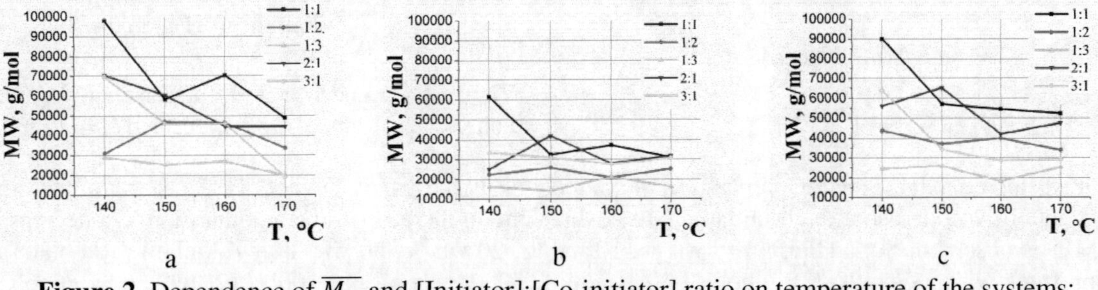

Figure 2. Dependence of $\overline{M_w}$ and [Initiator]:[Co-initiator] ratio on temperature of the systems: a) $[Sn(Oct)_2]$:[But]; b) $[Sn(Oct)_2]$:[PG]; c) $[Sn(Oct)_2]$:[GL].

It should be noted that at 140 ° C there is a maximum of MW value of about 100 000 g/mol for the system with But. A further increase of temperature leads to a decrease in MW. The appearance of the maxima on the curves at 160 ° C is associated with the boiling point of co-initiators (But – 117.7 °C, PG –187.4 °C, GL – 290 °C), which increases from But to GL and leads to a decrease in the amount of co-initiator, involved in the polymerization. An increase in the amount of co-initiator leads to a decrease in MW and to the broadening of the molecular weight distribution (MWD), which indicates that alcohol is not only a co-initiator of polymerization, but also acts as a regulator of MW, i.e.

involved in chain transfer reactions. For PG and GL the tendency to broadening of MWD can also be associated with the possibility of the formation of branched macromolecules due to the presence of two and three hydroxyl groups, respectively [11]. An increase in the amount of initiator (ratio [initiator]: [coordinator] = 2:1 and 3:1) leads to a decrease in MM at elevated temperatures due to adverse reactions of intra- and intermolecular transesterification occurring due to the active centers $Sn(Oct)_2$, which is an Lewis acid

Thus, to obtain PCL with MW close to 100 000 g/mol and with narrow MWD, polymerization should be carried out at 140 ° C in the presence of But and the ratio [initiator]:[co-initiator] = 1:1.

3.2 Composite materials based on PCL and HA

The phase composition of the synthesized HA was determined by XRD analysis. Its diffraction pattern is presented in Figure 3, and the general phase characteristics are shown in Table 1. The diffraction pattern contains both characteristic reflections of HA at the angles 2θ equal to 25.8, 32, 39.8, 46.7 and 49.5°, as well as low-intensity reflections of $Ca_3(PO_4)_2$ at 31.5, 34, 40.2, 46.7 and 49.5°.

The dependence of MW of *in situ* obtained composites is shown in Figure 4.

Table 1. Phase composition of HA and its crystal structure parameters.

Sample	Phase	Phase content (wt.%)	Lattice parameters (Å)	Average crystallite size (nm)
HA	$Ca_5(PO_4)_3(OH)$	88	a = 9.4065; c = 6.8732	69
	$Ca_3(PO_4)_2$	12	a = 10.4593; c = 37.5029	18

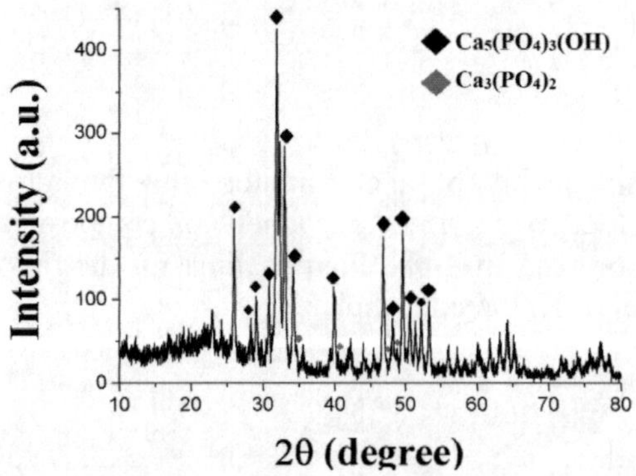

Figure 3. XRD patterns of HA.

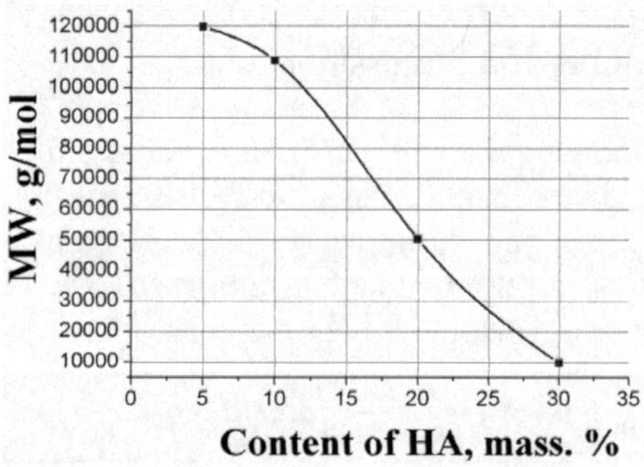

Figure 4. Dependence of $\overline{M_w}$ on content of HA for composites obtained by *in situ* method (T=170 °C).

It was noted that during the preparation of composites by the *in situ* method, HA acts not only as an inorganic filler, but also as a co-initiator. Moreover, with an increase in the amount of HA, a decrease in MW is observed. So, addition of 5 wt. %, 10 wt. %, 20 wt. %, 30 wt. % of HA allows to obtain a composites with MW 120000 g/mol, 109000 g/mol, 50000 g/mol and 10000 g / mol, respectively. Such regularity is associated with the structure of HA itself, containing hydroxyl groups, which, by analogy with alcohols, with increasing mass fraction in the reaction mixture, contribute to chain transfer reactions and, as a consequence, decrease in MW and broadening of MWD.

Along with the phase and molecular mass characteristics of the composites synthesized by *in situ* method, other physicochemical properties of the composites obtained by both methods were compared. The functional composition of the composites of both types was determined by IR spectroscopy. Figure 5 shows the IR spectra of both individual PCL, HA, and composite materials based on them: PCL / HA 90/10 and PCL / HA 80/20 (mechanical mixing); C2 and C3 (*in situ*).

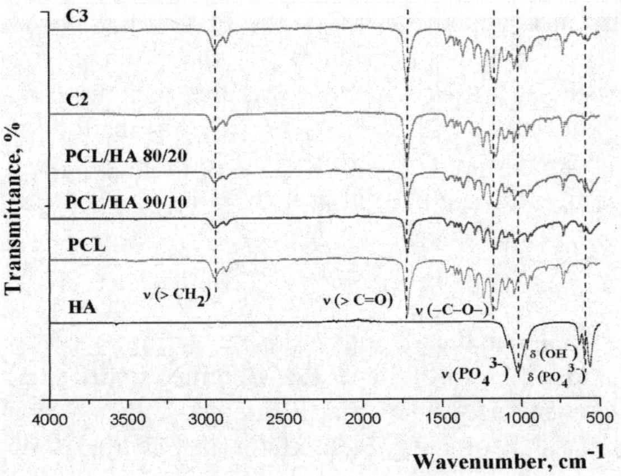

Figure 5. IR spectra of HA, PCL, PCL/HA 90/10, PCL/HA 80/20, C2 and C3.

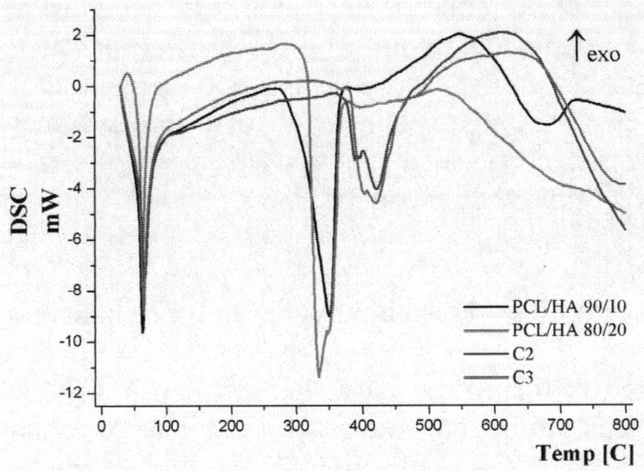

Figure 6. DSC curves of PCL/HA 90/10, PCL/HA 80/20, C2 and C3.

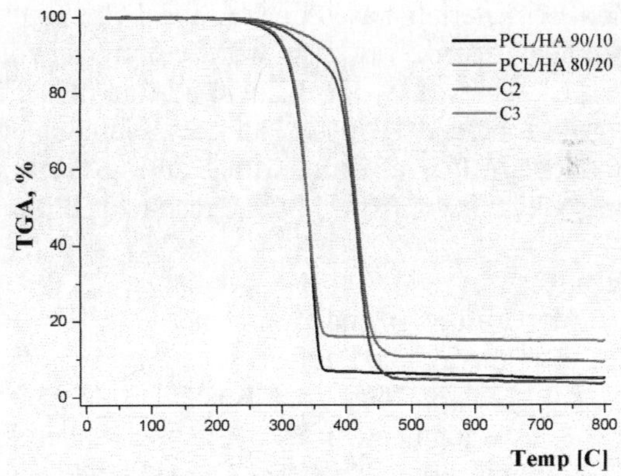

Figure 7. TG curves of PCL/HA 90/10, PCL/HA 80/20, C2 and C3.

The IR spectra of the composites obtained by both methods contain the absorption bands of HA and PCL. Moreover, in the case of composites synthesized by *in situ* method, there is a low-intensity band at 1624 cm^{-1}, related to the carboxylate anion, which indicates the chemical interaction between HA and PCL.

Another method confirming the chemical interaction between HA and PCL in composites obtained by *in situ* method is thermal analysis. So, DSC and TG curves of PCL/HA 90/10, PCL/HA 80/20, C2 and C3 composites are shown in Figures 6 and 7, respectively. From the presented dependences it can be seen that, despite the similarity in melting points of composites obtained by mechanical mixing and *in situ* method (~ 60 °C), the nature of their decomposition differs. This is characterized, on the one hand, by the appearance in the case of composites synthesized by *in situ* method of additional steps on the TG curves at a temperature of 280-360 ° C, on the other hand, by displacement of heat efficiency on the DSC curves at almost 100 °C towards of higher temperatures.

Comparison of the morphology of the composites obtained by the two methods was carried out on the basis of SEM micrographs, which are presented in Figure 8.

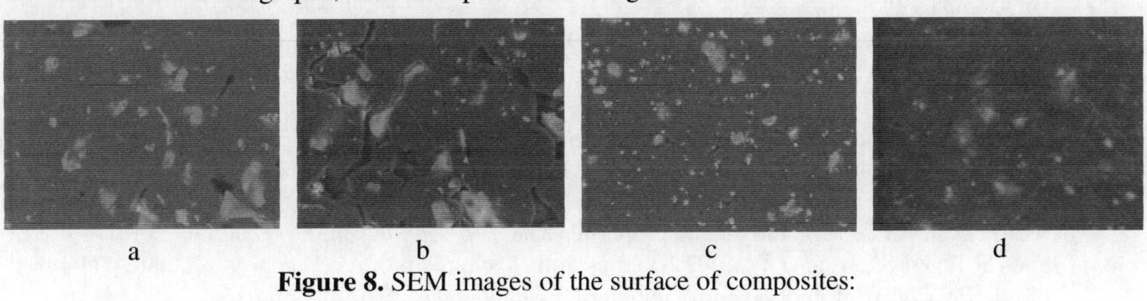

Figure 8. SEM images of the surface of composites:
a) PCL/HA 90/10, b) PCL/HA 80/20, c) C2 and d) C3(1000 times magnification)

IOP Conf. Series: Materials Science and Engineering **597** (2019) 012007 doi:10.1088/1757-899X/597/1/012007

Micrographs show that the composites obtained by mechanical mixing of the PCL solution and ground HA, the aggregation of HA particles and their irregular distribution in the polymer are observed (Figure 8, *a* and *b*). In composites obtained by *in situ* method, on the contrary, the distribution of HA particles in the polymer is homogeneous and regular, and the formation of agglomerates is not observed (Figure 8, *c* and *d*). At the same time, HA is well retained in the polymer matrix, which indicates a good adhesion of the filler particles to the matrix, probably, due to chemical interaction between HA and PCL.

4. Conclusions

It is shown that PCL with MW of about 100000 g/mol can be obtained during a ring opening polymerization at 140 °C in the presence of $Sn(Oct)_2$ and But with the ratio 1:1. Obtained under such conditions PCL can be used in the synthesis of composites by mechanical mixing method.

The *in situ* method should be considered as the most promising method for the synthesis of composite materials based on HA and PCL. It allows not only to obtain composites with a high MW of polymer matrix, but also with a homogeneous and regular distribution of HA particles, which cannot be achieved by mechanical mixing of the components. In the case of *in situ* polymerization of ε-caprolactone and HA, the latter acts simultaneously as a co-initiator and regulator of the polymer chains length. It is also noted that during the synthesis, the chemical interaction of the components occurs, which is confirmed by the results of IR spectroscopy and thermal analysis.

References

[1] Woodruff M A and Hutmacher D W 2010 The return of a forgotten polymer–Polycaprolactone in the 21st century *Prog. Polym. Sci.* **35** 1217

[2] Khoshroo K, Jafarzadeh Kashi T S, Moztarzadeh F, Tahriri M, Jazayeri H E and Tayebi L 2017 Development of 3D PCL microsphere/TiO2 nanotube composite scaffolds for bone tissue engineering *Mater Sci Eng C* **70** 586

[3] Li X, Zhang S, Zhang X, Xie S, Zhao G and Zhang L 2016 Biocompatibility and physicochemical characteristics of poly(ε-caprolactone)/poly(lactide-co-glycolide)/nano-hydroxyapatite composite scaffolds for bone tissue engineering *Mater. Des.* **114** 149

[4] Hao J, Yuan M and Deng X 2002 Biodegradable and Biocompatible Nanocomposites of Poly(ε-caprolactone) with Hydroxyapatite Nanocrystals: Thermal and Mechanical Properties *J. Appl. Polym. Sci.* **86** 676

[5] Causa F, Netti P A, Ambrosio L, Ciapetti G, Baldini N, Pagani S, Martini D and Giunti A 2005 Poly-ε-caprolactone/hydroxyapatite composites for bone regeneration: in vitro characterization and human osteoblast response *J. Biomed. Mater. Res. Part A* **76A (1)** 151

[6] Rasskazova L A, Lytkina D N, Shapovalova E G, Botvin V V, Pozdnyakov M A, Filimoshkin A G, Korotchenko N M and Kozik V V 2015 Bioactive composites produced in situ on the basis of calcium phosphates and lactic acid oligomers *Russ. J. Appl. Chem.* **88** 669

[7] Yi W, Li L, He H, Hao Z, Liu B, Chao Z and Shen Y 2018 Synthesis of poly(L-lactide)/β-cyclodextrin/citrate network modified hydroxyapatite and its biomedical properties *New J. Chem.* **42** 14729

[8] Helwig E, Sandner B, Gopp U, Vogt F, Wartewig S and Henning S 2001 Ring-opening polymerization of lactones in the presence of hydroxyapatite *Biomaterials* **22** 2695

[9] Fabbri P, Bondiolli F, Messori M, Bartoli C, Dinucci D and Chiellini F 2010 Porous scaffolds of polycaprolactone reinforced with in situ generated hydroxyapatite for bone tissue engineering *J. Mater. Sci. - Mater. Med.* **21** 343

[10] Rasskazova L A, Korotchenko N M and Zeer G M 2013 Microwave synthesis of hydroxyapatite and physicochemical study of its properties *Russ. J. Appl. Chem.* **86** 691

[11] Storey R F and Sherman J W 2002 Kinetics and Mechanism of the Stannous Octoate-Catalyzed Bulk Polymerization of ε-Caprolactone *Macromolecules* **35** 1504

IOP Conf. Series: Materials Science and Engineering **597** (2019) 012008 doi:10.1088/1757-899X/597/1/012008

Column simulation of Fe, Ti, V heap leaching from titanomagnetite ore

V I Sachkov, R A Nefedov and V V Orlov
Innovation and Technology Center, National Research Tomsk State University, 36 Lenin Street, Tomsk 634050, Russian Federation

E-mail: vvorlov92@mail.ru

Abstract. The results of the research of the process of metals extraction from titanomagnetite ore are presented. The experiments were conducted in PVC columns. The most efficient metal recovery is achieved at a hydrofluoric acid concentration of 4 mol/L and ammonium fluoride concentration of 0.42 mol/L. This represent approximately 10% for titanium, iron and vanadium. Selectivity on titanium during percolation of a solution with a hydrofluoric acid concentration of 1 mol/L and ammonium fluoride of 2.5 mol/L is observed.

1. Introduction

Deposits of titanomagnetite ores are considered as one of the promising industrial sources of iron ore, vanadium, titanium and other valuable elements [1]. For example, the Chineisk complex deposit of vanadium-containing titanomagnetites (Northern Transbaikal area, Russia) is one of the largest deposits in the world in terms of ore stocks (approximately 30 billion tons) [2].

Hydrometallurgical technologies for processing titanomagnetite ores are divided into acid and alkaline methods. Acidic processing methods are based on the dissolution of iron (II, III), vanadium and partially titanium (IV) and their transfer to the liquid phase with subsequent separation operations (precipitation, extraction). In the process of hydrochloric acid decomposition of ore, iron, vanadium, manganese pass into the solution, but titanium and silicon remain in the precipitate. The degree of ore decomposition essentially depends on the concentration and consumption of reagents, temperature conditions and the duration of the process. However, the known hydrometallurgical methods of processing of titanomagnetite ores have disadvantages such as multistage and high energy costs. Therefore, the development of cost-effective titanomagnetite ore processing technology is an important and urgent scientific and technical problem [1,3–5].

Heap leaching is well-established as a relatively low-cost and low-energy method of extracting metals (Au, Cu, U) from low grade ores, but it is practically not used for the processing of titanomagnetite ores[6].

Particular interest are hydrometallurgical technologies for ore processing, based on the effect of solutions containing ammonium and fluorine ions on the mineral material. The basis of the ore processing is that the oxides of transition and non-transition elements in contact with ammonium fluoride or solutions of ammonium fluoride and hydrofluoric acid form fluorometallates or oxofluorometallates favorable for further processing [7,8].

2. Experimental part

2.1 Ore

In the current work, we used a titanomagnetite ore from the Chineisk deposit (Chita region). The average particle diameter in the ore was approximately 20 mm (the ore was provided by public corporation Zabaikalstalinvest, Novaya Chara, Russia).

2.2 Leaching tests

The process of Ti, V and Fe leaching from the ore was simulated in PVC columns. The experiments were conducted in PVC columns with a height of 1 m and a diameter of 100 mm. Four parallel experiments were conducted with different leach aqueous solutions of ammonium fluoride and hydrofluoric acid. 15 kg of ore were added in each column. The volume of leach solution was 15 liters for each column and irrigation density was 250 ml/h. The percolation process was carried out in the circulation mode. The leach solutions were recycled four times and the fifth cycle was carried out with freshly prepared solutions. Samples of solutions were taken to analyze the content of Ti, Fe and V after each cycle. The column experimental conditions are presented in table 1.

Table 1. Column experimental conditions.

Column number	1	2	3	4
HF (mol/L)	4	1	1	1
NH_4F (mol/L)	0.5	2.5	0.5	1
Temperature (°C)	20			
Irrigation density (ml/h)	250			
Sample weight (kg)	15			
Leach solutions volume (L)	15			

2.3 Use of analytical techniques

The titanium, vanadium and iron concentration in the solution, and the elemental composition of the ore were determined using inductively coupled plasma mass spectrometry on a mass spectrometer ICP-MS ELAN-9000 DRC-e (Perkin Elmer USA).

3. Results and discussion

3.1 Characterization of the ore

According to the results of elemental analysis, the sample of titanomagnetite ore contains: Fe – 55.00 wt %, Ti – 6.84 wt %, V – 0.48 wt %. Table 2 reveals that the ore from the Chineisk deposit contains: aluminum (2.08 wt.%), manganese (0.21 wt.%), nickel (0.024 wt.%), tin (0.040 wt.%), chromium (0.011 wt.%), silicon (0,78 wt.%), copper (0,038 wt.%), zirconium (0.0024% wt.), strontium (0.0031 wt.%), scandium (0.0021 wt.%), gallium (0.0039 wt.%).

Table 2. The elemental composition of the ore (ICP).

Element	Mass Fraction (wt %)	Element	Mass Fraction (wt %)	Element	Mass Fraction (wt %)
Fe	55.0	Al	2.08	Mn	0.21
Ti	6.84	Ni	0.024	Ga	0.0039
V	0.48	Sn	0.040	Zr	0.0024
Si	0.78	Cu	0.038	Sc	0.0021
Cr	0.011	Sr	0.0031	Sb	0.0029

Thus, the titanomagnetite ore of the Chineisk deposit has a high titanium and vanadium content (V_2O_5 0.8-1.0 wt.%, TiO_2 11-12 wt.%) and can be considered as a huge resource for obtaining vanadium iron and titanium dioxide. However, the pyrometallurgical processing of such ore will be difficult due to the high content of TiO_2 (>4%). The relatively high content of some non-ferrous and rare metals makes it possible to extract them.

3.2 Leaching efficiency

This section presents the results of studying column simulation of Fe, Ti, V heap leaching from titanomagnetite ore.

The results of the experiments are presented in the table 3 and diagrams: extraction efficiency of Ti, V, Fe in column 1 and column 2 (figure 1) and extraction efficiency of Ti, V, Fe in column 3 and column 4 (figure 2).

Table 3. Comparative results obtained after leaching of titanomagnetite ore from the Chineisk deposit.

Column no.	The content in the solution after four cycles of percolation (g/L)			Extraction of metals into the solution after 4 cycles of percolation (%)			Extraction of metals into the solution after the 5th cycle of percolation (%)		
	Ti	V	Fe	Ti	V	Fe	Ti	V	Fe
1	4.46	0.35	41.95	6.52	7.29	7.63	10.90	11.29	9.99
2	0.65	0.01	0.45	0.95	0.23	0.08	1.26	0.17	0.12
3	0.71	0.08	10.70	1.04	1.69	1.94	1.05	2.15	1.99
4	1.35	0.14	12.69	1.98	2.92	2.31	1.35	2.33	1.78

As follows from the table 3 and diagram (figure 1 (a)), the highest degree of extraction of metals from the ore is achieved in column No. 1, where the concentration of HF is 4 mol/L and NH_4F is 0.5 mol/L. Thus, after 4 cycles of percolation, extraction to solution is 6.52 % Ti, 7.29 % V and 7.63 % Fe. After the fifth percolation cycle with a freshly prepared solution, these values are 10.90 %, 11.29 % and 9.99 % respectively. Therefore, ore is most efficiently processed using repeated percolation of solution No. 1

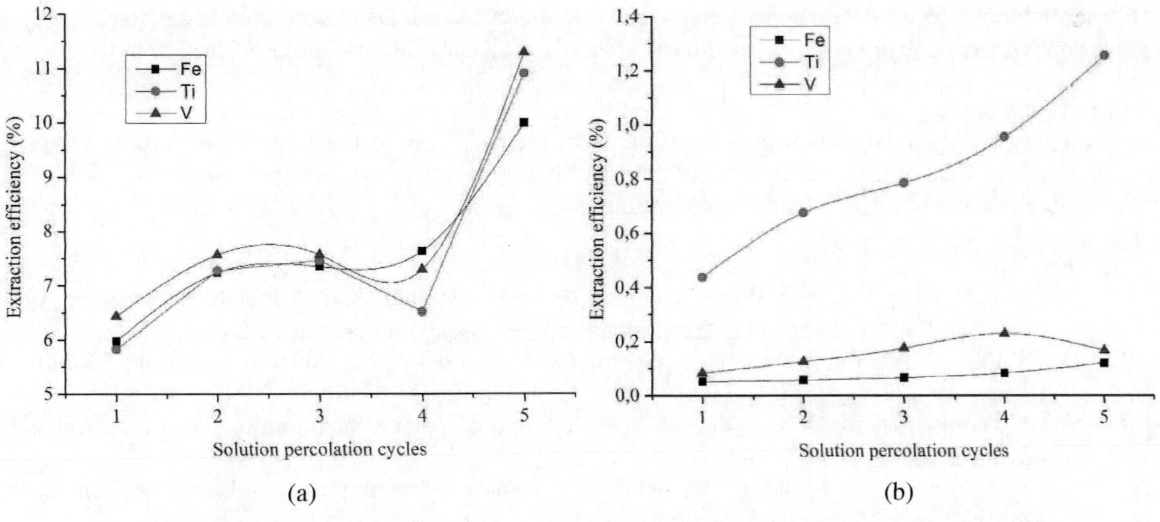

Figure 1. Extraction efficiency of Ti, V, Fe in column 1 (a) and column 2 (b).

In column No. 2, when using a solution with a concentration of HF - 1 mol/L and NH_4F - 2.5 mol/L, a low degree of metals extraction is observed, however, the content of titanium predominates in the enriched process solution (figure 1 (b)). The degree of extraction in the solution is: Ti - 1.26 %, V - 0.17 % and Fe - 0.12 %. Thus, with repeated percolation of solution No. 2, the method allows to achieve the selective extraction of titanium and the enrichment of titanomagnetite ore with iron and vanadium.

A low degree of extraction of metals is observed in columns No. 3 and No. 4; moreover, there is no selectivity on one of the elements. (figure 2 (a), (b)).

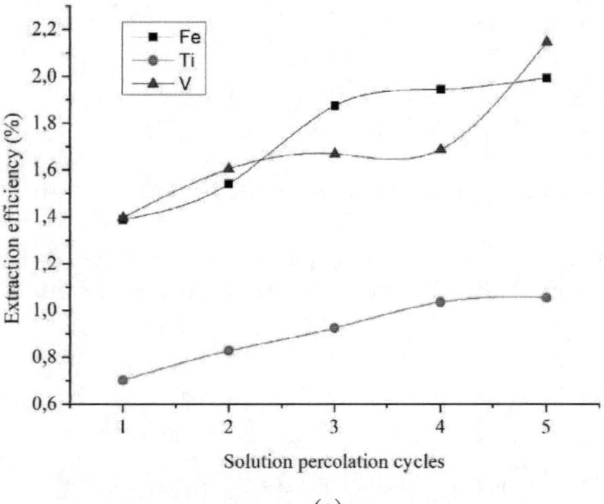

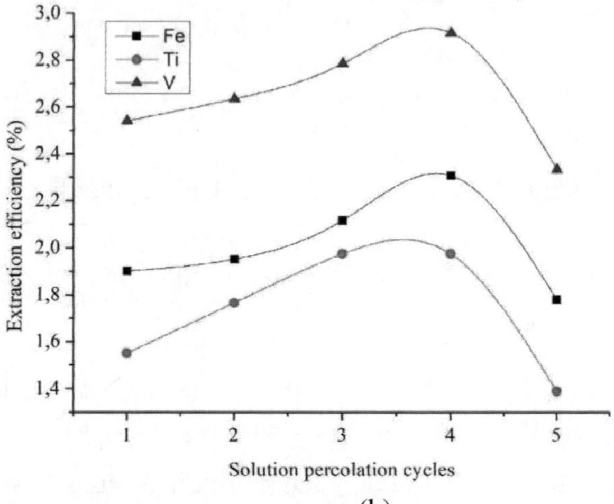

Figure 2. Extraction efficiency of Ti, V, Fe in column 3 (a) and column 4 (b).

4. Conclusion

This study confirms the potential of heap leaching of titanomagnetite ore for the extraction of Ti, V and Fe. The titanomagnetite ore of the Chineisk deposit contains Fe – 55.00 wt %, Ti – 6.84 wt %, V – 0.48 wt %. and can be considered as a huge resource for obtaining vanadium, iron and titanium dioxide. Experimental results show that the most effective extraction of ore components is achieved at a concentration of hydrofluoric acid - 4 mol/L and ammonium fluoride - 0.5 mol/L and is approximately 10% for titanium, iron and vanadium. Selective extraction of titanium is observed during percolation of a solution with a concentration of hydrofluoric acid of 1 mol/L and ammonium fluoride 2.5 mol/L.

Acknowledgments

The results were obtained as part of the fulfilment of the state task of the Ministry of Education and Science of the Russian Federation, project # 10.3031.2017/4.6 and with the support of the Tomsk State University Competitiveness Enhancement Program.

References

[1] Zhang Y, Wang L, Chen D, Wang W, Liu Y, Zhao H and Qi T 2018 A method for recovery of iron , titanium , and vanadium from vanadium-bearing titanomagnetite **25**
[2] Deryabin Y A 2007 Equilibrium Composition of Metal and Slag in Reductive Smelting of Binary Batch Based on Chineisk and Korshunovo Concentrates **37** 83–8
[3] Sun H, Wang J, Dong X and Xue Q 2012 A literature review of titanium slag metallurgical processes *Metal. Int.* **17** 49–56
[4] Jain S, technology P J-I journal of and 1977 undefined Hydrochloric-acid leaching of Gopalpur ilmenite *Counc. Sci. Ind.*
[5] Ogasawara T and Veloso De Araújo R V 2000 Hydrochloric acid leaching of a pre-reduced Brazilian ilmenite concentrate in an autoclave *Hydrometallurgy* **56** 203–16
[6] McBride D 2019 Heap leach modeling: Analysis and optimization with CFD technology
[7] Bakeeva N G, Gordienko P S and Parshina E V. 2010 Mutual influence of ammonium hexafluorotitanate and hexafluorosilicate in ammonium fluoride solutions *Russ. J. Gen. Chem.* **80** 223–6
[8] Laptash N M and Maslennikova I G 2013 Fluoride Processing of Titanium-Containing Minerals *Adv. Mater. Phys. Chem.* **02** 21–4

IOP Conf. Series: Materials Science and Engineering **597** (2019) 012009 doi:10.1088/1757-899X/597/1/012009

Influence of ZrO_2 modifier and Mo precursor on the state of active component in MoO_3/Al_2O_3 catalysts

Rumyantsev P A, Smirnov M Yu and Vodyankina O V

Tomsk State University, 36, Lenin ave., 634050 Tomsk, Russia

E-mail: rumyantsev9@gmail.com

Abstract. The effects of precursor and the method of support modification on the properties of supported metathesis catalysts for reaction of ethylene and trans-butene-2 into propylene are considered. The objects of the research are the catalyst comprising γ-Al_2O_3-supported molybdenum oxide as well as the modified alumina-based supports. The Mo precursors used is shown to significantly affect the state of the supported component. The distribution and morphology of the particles of active component inside the pores of the support differ for the ZrO_2-modified samples prepared by impregnation and wet mixing methods.

1. Introduction

The demand in olefins, especially ethylene and propylene, constantly increases and is connected with the rise of the consumption of the polymers on the basis thereof [1 - 2]. Currently, cracking and catalytic dehydrogenation are the main sources of olefins. Although these processes are widely used, the intensive studies on the development of alternative processes for production of light olefins are being carried out [1 - 3].

The metathesis reaction is considered a promising method for propylene production and comprises the redistribution of the alkylidene substitutes of the double bond [1, 4]. In case of propylene production, one molecule of trans-2-butene and one molecule of ethylene participate in the process yielding two propylene molecules. Besides propylene, pentenes and hexenes of various structure can be formed as a result of oligomerization and isomerization of propylene [5].

The catalysts for propylene production by ethylene and trans-butylene metathesis include the oxides WO_3, MoO_3, Re_2O_7 supported on γ-Al_2O_3, SiO_2, zeolites and some other supports [6 - 8]. Due to high cost of Re_2O_7 and its low stability towards catalytic poisons, the industry implements the WO_3-based catalysts. Such catalysts posses a relatively low activity, but are stable for long time. The MoO_3-based catalysts are promising systems, since they demonstrate the catalytic activity at lower temperatures than the WO_3-containing samples [9 – 10]. The activity of these samples depends on the degree of polymerization of the surface MoO_x particles, and the isolated dimeric and dispersed polymeric particles are active in the given reaction [11 – 14], while the crystalline MoO_3 is not catalytically active [14 – 20]. Moreover, the support properties affect the activity of the MoO_x particles in the metathesis process, e.g., the catalysts show higher catalytic activity, when supported on more acidic supports [21].

The acidity of the catalyst surface should be quite high facilitating the olefin sorption, however, too high acidity induces oligomerization of the adsorbed olefin molecules [5]. The modification of alumina by zirconia allows regulating the acidity of the support surface that should result in a decrease of the impact of this process. In the present work the effect of modification of alumina support by zirconia and the nature of the precursor of Mo oxide on the state of active component in MoO_3-based catalysts for metathesis of ethylene and butene into propylene is considered.

2. Experimental

2.1. Methods of catalyst preparation
To prepare the catalysts, the unmodified and ZrO_2-modified alumina supports were used. The unmodified support represented the industrial granulated sample Al_2O_3 (CJSC "Industrial catalysts", Ryazan, Russia). The modified supports were obtained by two methods. The support designated as ZrO_2/Al_2O_3 was obtained by wetness impregnation of alumina with the citric acid solution of zirconium nitrate followed by drying at 120° C and calcination at 500 °C. The support designated as ZrO_2-Al_2O_3 was obtained by the wet mixing using the reprecipitated aluminum hydroxide of type C (γ-Al_2O_3) (CJSC "Industrial catalysts", Ryazan, Russia) as a powder with a fraction of no more than 0.25 mm and citric acid solution of $ZrO(NO_3)_2$ (Aldrich). The mixed wet composite was dried at 120 °C and then calcined at 500 °C in nitrogen flow.

The active component was supported by the wetness impregnation method from water solution of molybdenum oxide precursor. $(NH_4)_6Mo_7O_{24}{\bullet}4H_2O$ (CJSC "Vecton") and $H_3PMo_{12}O_{40}{\bullet}17.26H_2O$ (CJSC "Vecton") were used as a Mo precursor. To improve the solubility of the ammonium heptamolybdate (AHM), citric acid was used with the n(Mo) : n(CA) = 1:1. The samples after the impregnation were dried at 110 °C during 48 h. The dried samples were calcined in air at 500 °C during 2 h in the flow of argon.

2.2. Methods of sample characterization
Textural characteristics of the samples were determined by the method of low-temperature N_2 adsorption on the TriStar 3020 (Micromeritics, US). For all samples, a preliminary degassing was carried out at 200 °C for 2 h. The specific surface of the samples was determined from the adsorption data according to the BET method, total pore volume and pore size distribution were determined according to the BJH method.

Phase composition of the samples was studied using X-ray diffraction (XRD) analysis on the XRD-6000 diffractometer ("Shimadzu", Japan) with a monochromatized Cu Kα radiation (λ=1.5418 Å) in the range of angles of 10-70 (2θ) and a scanning rate of 2 °/min. Analysis of phase composition was carried out using the PDF2 database.

To study the X-ray amorphous state of the supported modifier and/or active component using the temperature-programmed reduction (TPR) method, Raman spectroscopy and UV-vis DR spectroscopy were used. The TPR of the samples was carried out using the Micromeritics ChemiSorb 2750 (USA) with the gas mixture 10% H_2/Ar, 20 ml/min. The experiment was carried out in the temperature range of 25-900 °C with a heating rate of 10 °C/min. The UV-vis DR spectra were taken using the Evolution 600 Thermo Scientific (USA) spectrometer in the scanning range of 200-900 nm, with MgO used as a standard. Raman spectra were taken at ambient conditions on the inVia confocal Raman microscope (Renishaw, UK) using the 785 nm line.

3. Results and discussion
Table 1 shows the results of elemental analysis and textural characteristics of the prepared support samples and the catalysts on the basis thereof.

Chemical analysis shows that the amount of modifier in ZrO_2/Al_2O_3 and ZrO_2-Al_2O_3 supports is 4.5 and 6.1 % mass., respectively. Textural characteristics of ZrO_2/Al_2O_3 and ZrO_2-Al_2O_3 supports differ due to addition of ZrO_2 modifier using different methods. Despite close values of specific surface area, the maxima of bimodal pore size distribution are 7 and 14 nm for ZrO_2/Al_2O_3 support, while the one for ZrO_2-Al_2O_3 is 4 nm. Thus, the total pore volume for ZrO_2-Al_2O_3 support is practically two times lower as compared to ZrO_2/Al_2O_3 samples (Table 1).

Introduction of active component also results in reduction of specific surface and pore volume of the samples that is caused by the increase of their mass, and in case of $H_3PMo_{12}O_{40}$ is also connected with a slight sintering caused by a two-fold thermal treatment at 500 °C. The latter can also be connected with the effect of phosphorus oxide as a mineralizing agent. According to chemical analysis, the Mo content in the samples was 13.4–21.9 % mass.

Table 1. A Composition and textural characteristics of the samples

Samples	Content, % mass.		SSA, m^2/g	V_{pore}, cm^3/g
	Zr	Mo		
Al_2O_3	-	-	230	0.57
ZrO_2/Al_2O_3	4.5	-	224	0.52
ZrO_2-Al_2O_3	6.1	-	246	0.28
MoO_3/Al_2O_3	-	13.4	191	0.45
$MoPO_x/Al_2O_3$	-	13.9	173	0.47
$MoO_3/ZrO_2/Al_2O_3$	3.6	14.2	184	0.42
$MoPO_x/ZrO_2/Al_2O_3$	3.6	13.5	157	0.41
MoO_3/ZrO_2-Al_2O_3	5.9	21.9	167	0.20
$MoPO_x/ZrO_2$-Al_2O_3	5.2	17.3	163	0.20

Figure 1 shows the XRD patterns for the alumina supports and the catalysts on the basis thereof. Analysis of the XRD data shows that there is a crystalline phase of γ-Al_2O_3 in all support samples. The corresponding reflections in the XRD patterns are strongly widened that evidences the presence of low-crystalline phase. The crystalline phases of ZrO_2 in the modified supports were not detected that indicates their presence in a highly dispersed or X-ray amorphous state.

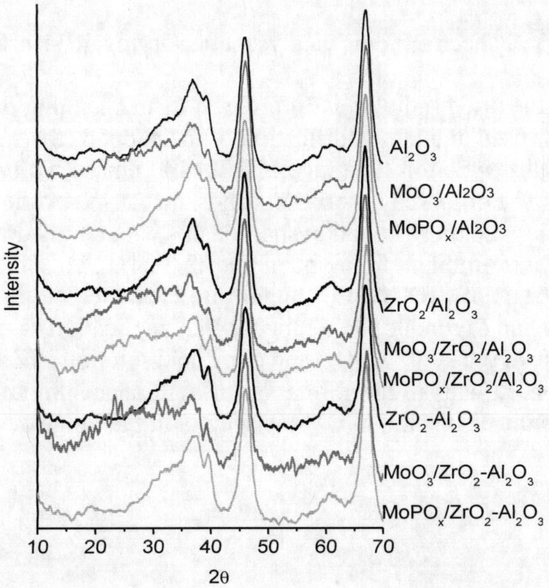

Figure 1. XRD patterns for the supports and catalysts.

The XRD patterns for the catalysts do not show the reflections of the crystalline phase of active component MoO_3. However, the XRD images of the samples obtained using the ammonium heptamolybdate with citric acid as well as for the $MoPO_x/ZrO_2/Al_2O_3$ sample in the range of 20–30° (2θ) there are additional wide reflections that can be connected with the presence of highly dispersed X-ray amorphous phases of active component.

Figure 2 shows the UV-vis spectra for the prepared catalysts samples after drying in air at 110 °C. Independently on the support used, the spectra for the catalysts prepared using the citric acid solution of ammonium heptamolybdate are characterized by the absorption band with a maximum at 255 nm corresponding to the $O^{2-} \rightarrow Mo^{6+}$ charge transfer in the isolated MoO_4 structures [22 – 23]. The spectra for the catalysts prepared using the heteropolyacid solution contain a shoulder at ~350 nm (along with the one at 255 nm) characteristic for the $O^{2-} \rightarrow Mo^{6+}$ charge transfer in the MoO_6 structures [22 – 24]. The intensity of the band at ~350 nm in the spectra of the catalysts prepared over the zirconia-modified supports is higher than for those prepared over the unmodified alumina. The presence of the band at 255 nm in the catalyst samples prepared using the heteropoly acid solution evidences a partial destruction of the heteropoly ion that can be connected with the interaction of the Mo precursor with the support surface. Modification of the Al_2O_3 surface with zirconium oxide preserves the structure of the anion in heteropoly acid in the dried samples that evidences the increase of the intensity of the band at 350 nm in the corresponding spectra.

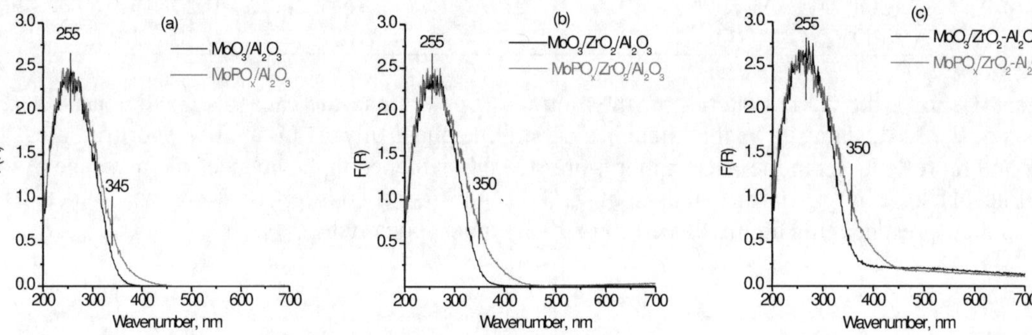

Figure 2. UV-ViS spectra for the catalysts after drying at 110 °C.

The subsequent calcination of the dried samples at 500 °C results in changing of the spectra (Figure 3). Independently on the precursor used and the modification of the support, the spectra of all calcined samples are characterized by the absorption in the range of 200–400 nm with a maximum at 253-257 nm and a shoulder at 295-302 nm. The observed absorption bands are caused by the $O^{2-} \rightarrow Mo^{6+}$ charge transfer in the isolated MoO_x species and those bound in short chains with tetrahedral (250–290 nm) and octahedral (300–320 nm) oxygen surrounding of molybdenum [22 – 23].

All catalysts prepared on the basis of ZrO_2-Al_2O_3 support are characterized by the increase of the absorption background (Figure 2c and Figure 3c). We suppose that application of wet mixing using the reprecipitated aluminum hydroxide of type C (γ-Al_2O_3) and citric acid solution of $ZrO(NO_3)_2$ at the stage of preparation of ZrO_2-Al_2O_3 support leads to the formation of high concentration of surface defects. Such a defective surface increases the dispersion in UV-vis spectra in the range of 400-700 nm.

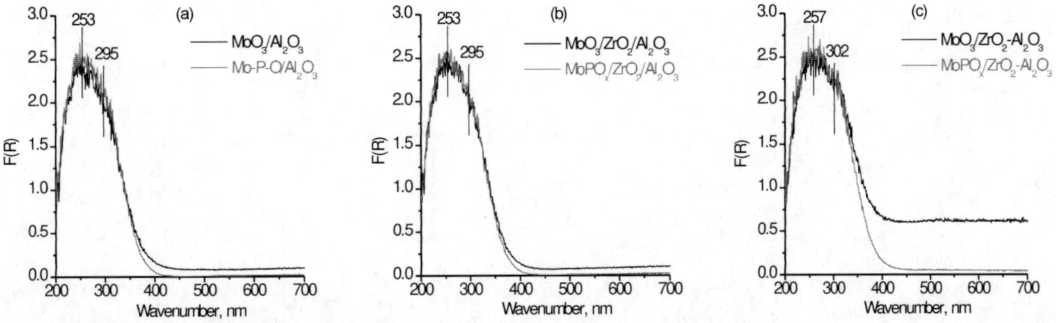

Figure 3. UV-vis spectra for the catalysts after calcination at 500 °C.

Figure 4 shows the Raman spectra of the synthesized catalysts obtained under the specific conditions at an excitation wavelength of 785 nm. The spectra of all samples do not contain the bands characteristic for crystalline molybdenum oxide (820, 996 cm^{-1}) that is consistent with the XRD data. The absence of the characteristic bands for phosphomolybdic acid at 883 and 983 cm^{-1} [25] in the spectra of the samples $MoPO_x/Al_2O_3$, $MoPO_x/ZrO_2/Al_2O_3$ and $MoPO_x/ZrO_2\text{-}Al_2O_3$ indicates a complete decomposition of the Mo precursor. The absorption band in the range of 900-1000 cm^{-1} with a maximum at ~970 cm^{-1} in the spectra of the samples $MoPO_x/Al_2O_3$, $MoO_3/ZrO_2\text{-}Al_2O_3$ and $MoPO_x/ZrO_2\text{-}Al_2O_3$ correspond to the surface molybdate structures formed by $(O=)_2MoO_2$ dioxy species [26 – 27]. A shifting of the maximum of this band to the range of larger wavenumbers in the spectrum for the MoO_3/Al_2O_3 sample can be connected with the presence of Mo in the $O=MoO_4$ monooxo species along with the $(O=)_2MoO_2$ dioxy species [26 – 27]. The absence of the bands in the spectra taken in air for the samples prepared on ZrO_2/Al_2O_3 support can be caused by the presence of Mo predominantly as the surface $O=MoO_4$ monooxo species that do not show bands in the Raman spectra in a hydrated state.

Figure 5 shows TPR profiles for the catalyst samples. Analysis of the state of active surface of the synthesized catalyst samples shows that the method of introduction of ZrO_2 modifier into the support as well as the Mo precursor used significantly influences the state of the active component.

For the MoO_3/Al_2O_3 sample, prepared from the AHM a maximum of reduction is observed at 462 °C. According to Ref. [28], a reduction of octahedral species Mo^{6+} to Mo^{4+} occurs in this range. For Mo^{6+} in a tetrahedral coordination of oxygen, a stronger interaction with the support surface is observed, thus, its reduction occurs at higher temperatures [29]. For the sample $MoPO_x/Al_2O_3$, hydrogen consumption with a maximum at 523 °C takes place. The observed shifting of the maximum of hydrogen consumption corresponds to the reduction of Mo^{6+} in a tetrahedral oxygen coordination into Mo^{4+}. This can be caused by different interaction of the used Mo precursors with the support surface at the stages of impregnation and subsequent calcination of the samples that results in the formation of different surface phases. One cannot also exclude the effect of phosphorus in the $MoPO_x/Al_2O_3$ sample on the reduction temperature of MoO_x species.

In case of the samples prepared over the modified ZrO_2/Al_2O_3 support, the reduction profile contains one peak in the range of 530-550°C that can be connected with the reduction of the tetrahedral Mo^{6+} to Mo^{4+} as well as with the effect of zirconium oxide. With that, no significant difference is observed in the reduction behavior of Mo(VI) oxide species on the precursor used. Given the Raman spectroscopy data, the observed shifting of the reduction maximum in comparison with the samples prepared using the unmodified Al_2O_3 is connected with effect of the modifier that changes the reaction ability of the $O=MoO_4$ monooxo species in the MoO_x phases.

For the samples prepared over $ZrO_2\text{-}Al_2O_3$ support, there are two reduction peaks in a high-temperature range. For the $MoO_3/ZrO_2\text{-}Al_2O_3$ and $MoPO_x/ZrO_2\text{-}Al_2O_3$ samples, the peaks with the maxima at 532 and 608 °C and at 518 and 715 °C, respectively, are observed. In case of MoO_3/ZrO_2-

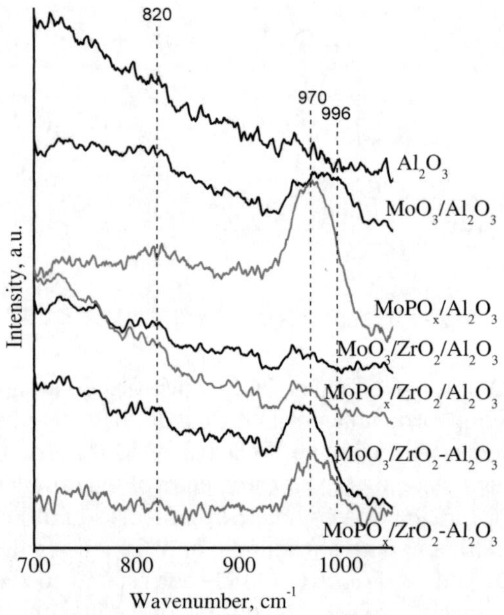

Figure 4. Raman spectra of the investigated catalysts calcined at 500 °C.

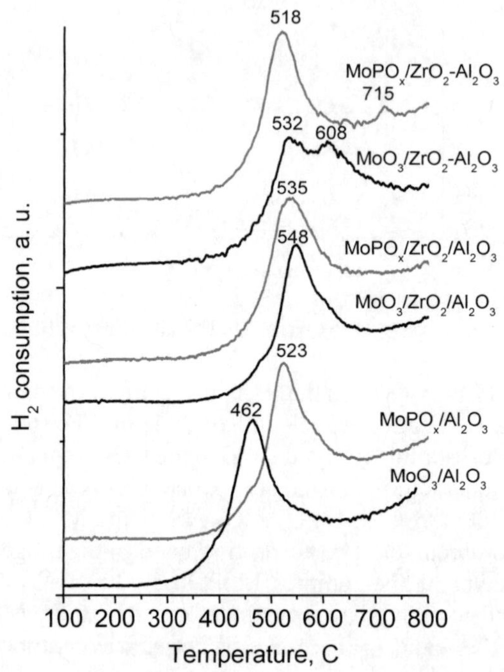

Figure 5. TPR H_2 profiles for the Mo-containing catalysts prepared with unmodified and Zr-modified Al_2O_3 supports calcined at 500 °C.

Al_2O_3 sample, the first peak at 532°C is similar to the one for the reduction of the $MoO_3/ZrO_2/Al_2O_3$ sample and can be attributed to the reduction of Mo^{6+} in octahedral coordination to Mo^{4+} over zirconia. The second peak at 608 °C can be assigned to reduction of Mo^{6+} in a tetrahedral coordination to Mo^{4+} over zirconia. For the $MoPO_x/ZrO_2$-Al_2O_3 sample, the first peak is similar to the one observed at the reduction of the $MoPO_x/Al_2O_3$ sample and can be attributed to the reduction of Mo^{6+} in a tetrahedral coordination to Mo^{4+} over the alumina surface. The peak at 715 °C can be connected with the reduction of highly dispersed MoO_3 and MoO_2 phases [30]. The observed difference in the TPR profiles of the samples prepared over the ZrO_2/Al_2O_3 and ZrO_2-Al_2O_3 samples, is probably connected with different distribution of the modifier in the support, which result in different distribution and state of the active component.

4. Conclusion

Thus, based on the XRD, UV-vis, Raman and TPR data it is possible to conclude that all catalyst samples are characterized by the presence of highly dispersed X-ray amorphous molybdenum phases. With that, the modification of the support by ZrO_2 results in the changing of the state of surface MoO_x phases in the samples. Introduction of ZrO_2 modifier into alumina support by impregnation leads to more homogeneous distribution of the modifier in a porous space of the support that provides the formation of surface MoO_x phases predominantly as $O{=}MoO_4$ monooxo species independently on the nature of the Mo precursor. The use of the wet mixing method to introduce the modifier results in the formation of separate particles of aluminum and zirconium oxides that is accompanied by different distribution of active component in the catalysts prepared with different precursors.

Acknowledgements
Financial support by the Ministry of Science and Higher Education of the Russian Federation is acknowledged. Unique project identifier RFMEFI60717X0169.

References

[1] Lavrenov A V, Saifulina L F, Buluchevskii E A and Bogdanets E N 2015*Catalysis in Industry* **7** 175–87
[2] Amghizar I, Vandewalle L A, Van Geem K M and Marin G B 2017 *Engineering* **3** 171–78
[3] Grant J T, Venegas J M, McDermott W P and Hermans I 2018 *Chem. Rev* **118** 2769–15
[4] Fürstner A 1998 *Alkene Metathesis in Organic Synthesis* (Heidelberg: Springer Berlin) pp. 1–36
[5] Zhang D, Li X, Liu S, Huang S, Zhu X, Chen F, Xie S and Xu L 2012 *J. Appl. Catal. A* **439-440** 171–78
[6] Oikawa T, Ookoshi T, Tanaka T, Yamamoto T and Onaka 2004 *Microporous and Mesoporous Mater.* **74** 93—103
[7] Li X J, Zhang W P, Liu S L, Xu L Y, Han X W and Bao X H 2007 *J. of Catal.***250** 55–66.
[8] Mol J C 2004 *J. of Mol. Catal. A: Chem.* **213** 39–45
[9] Ivin K J, Mol J C 1997 *Olefin Metathesis And Metathesis Polymerization* (London: Academic Press), pp. 397 –410
[10] Mol J C and Moulijn J A 1975 *Adv. Catal.* 131–71
[11] Xie S, Chen K, Bell A T and Iglesia E 2000 *J. Phys. Chem. B* **104** 10059–68
[12] Topka P, Balcar H, Rathousky´ J, Zilkova N, Verpoort F and Cejka J 2006 *Microporous Mesoporous Mater.* **96** 44–54
[13] Handzlik J 2007 *Surf. Sci.* **601** 2054–65
[14] Debecker D B, Schimmoeller B, Stoyanova M, Poleunis C, Bertrand P, Rodemerck U and Gaigneaux E M 2011 *J. Catal.* **277** 154–63
[15] Ono T, Anpo M and Kubokawa Y 1986 *J. Phys. Chem.* **90** 4780–84
[16] Debecker D P, Stoyanova M, Rodemerck U and Gaigneaux E M 2011 *J. Mol. Catal. A* **340** 65–76
[17] Debecker D P, Stoyanova M, Rodemerck U, Lonard A, Su B-L and Gaigneaux E M 2011 *Catal. Today* **169** 60–68
[18] Handzlik J, Ogonowski J, Stoch J and Mikołajczyk M 2004 *Appl. Catal. A* **273** 99–104
[19] Zhang B, Liu N, Lin Q and Jin D 1991 *J. Mol. Catal.* **65** 15–28
[20] Gordeev A V and Vodyankina O V *Petroleum Chemistry* **2014** 54 452 – 8
[21] Debecker D P, Hauwaert D, Stoyanova M, Barkschat A, Rodemerck U and Gaigneaux E M 2011 *Appl. Catal. A* **391** 78–85
[22] Hahn T, Bentrup U, Armbrster M, Kondratenko E V and Linke D 2014 *ChemCatChem* **6** 1664–72
[23] Fourier M, Louis J C, Che M, Chaquin P and Masures D 1989 *J. of Catal.* **119** 400-14
[24] Popa A, Sasca V, Verdes O, Barvinschi P and Holclajtner-Antunovic I 2014 *Mater. Res. Bull.* **50** 312–22
[25] Kale S S, Armbruster U, Eckelt R, Bentrupa U, Umbarkar S B, Dongare M K and Martin A 2016 *Appl. Catal. A* **527** 9–18
[26] Tsilomelekis G and Boghosian S 2013 *Catal. Sci. Technol.* **3** 1869-88
[27] Chakrabarti A and Wachs I E 2018 *ACS Catal.* **8** 949–59
[28] Meng D, Wang B, Yu W and Wang W 2017 *Catalysts* **7** 151
[29] Liu C, Wang W, Xu Y, Li Z, Wang B and Ma X 2018 *Appl. Surf. Sci.* **441**
[30] Wang B, Yu W, Wang W, Li Z, Xu Y and Ma X 2017 *Chin. J. Chem. Eng.***26**

IOP Conf. Series: Materials Science and Engineering **597** (2019) 012010 doi:10.1088/1757-899X/597/1/012010

Ag-Pd nanoparticles prepared by laser ablation for selective oxidation of propylene glycol to lactic acid

S Ten, E A Gavrilenko, V A Svetlichnyi and O V Vodyankina*

Tomsk State University, Tomsk, Russian Federation

*E-mail: vodyankina_o@mail.ru

Abstract. Silver/palladium composite nanoparticles (NPs) obtained by pulsed laser ablation in liquid (PLAL) technique catalyze oxidation of propylene glycol (PG) to lactic acid (LA) in alkaline media. The PLAL technique allows obtaining silver/palladium NPs with different silver/palladium molar ratio. Addition of palladium increases PG conversion (up to 52%) and LA selectivity (up to 88%). The reaction temperature and silver/palladium molar ratio strongly affects the catalytic activity of the silver/palladium composite NPs.

1. Introduction

Recently, the challenge of reduction of carbon sources by chemical synthesis from fossil hydrocarbons is widely discussed. Propylene glycol (PG) can be synthesized by catalytic hydrogenolysis of biomass-derived polyols, such as glucose, glycerol, sorbitol, etc. [1]. One of the most valuable products that can be obtained through PG oxidation is lactic acid (LA), which finds in a wide range of fields, especially in production of biodegradable plastics [2]. There are a lot of valuable products that can be obtained by PG catalytic oxidation: methylglyoxale [3], hydroxyacetone [4], lactic acid [5], etc.

Silver NPs are promising alternative to noble metals used as catalysts in reaction of catalytic oxidation of polyols [6]. Recently, Xue et al. [5] developed a green chemical route for selective LA synthesis through aerobic oxidation of PG over Ag-Pd NPs in alkaline medium. Tentative reaction mechanism consists of 2 steps: in the first step PG oxidizes to methylglyoxal (MGO) on the metal NPs, and in the second step MGO converts into sodium lactate via intramolecular Cannizarro reaction in an alkaline media. The catalysts were prepared by wet chemical reduction method. However, in this case to form the NPs, both template and toxic reducing agent hydrazine were used that led to additional steps of centrifugation, washing and drying of Ag-Pd NPs prior to the application.

Pulsed laser ablation in liquid (PLAL) attracts growing attention as a facile and "green" one-step method to synthesize nanomaterials to be used in catalysis [7]. High-energy PLAL allows obtaining "pure" and active NPs directly as colloid solutions in pure solvents without the addition of any surfactants or stabilizers. Additional advantages of this method are relative simplicity, a broad variation of materials for ablation and the opportunity to produce composite NPs [8]. PLAL technique allows obtaining bimetallic Ag-Pd NPs [9].

In the present work, we aimed to use PLAL to prepare monometallic (Ag and Pd) and bimetallic (Ag-Pd) NPs and estimate their catalytic activity in selective oxidation of PG with oxygen in alkali media.

2. Methods

IOP Conf. Series: Materials Science and Engineering **597** (2019) 012010 doi:10.1088/1757-899X/597/1/012010

2.1, Catalyst preparation.
The PLAL was carried out using Q-switch Nd:YAG laser source operating at a repetition rate of 20 Hz, with a laser duration, wavelength and energy of 7 nanosecond, 1064 nm and 150 mJ, respectively. Silver and palladium targets with purity of not less than 99.97 % were used. Focused radiation was introduced through the side wall of a 100 ml cylindrical glass reactor. Detailed experimental conditions of PLAL are described in Ref. [10]. The concentration of NPs in the dispersion was 50 mg/L determined by the loss of target mass. The operation was repeated several times to obtain the desired amount of dispersion.

To synthesize the silver/palladium NP composite in water, the as-prepared PLAL dispersions of silver and palladium in water were mixed in the given ratio, placed in a 100 ml cylindrical glass reactor with a diameter of 50 mm and then irradiated through the side wall by focused laser irradiation (lens with F = 50 mm) of the same Q-switch Nd:YAG laser. The focus of the lens was in the center of the reactor. Irradiation was carried out during 30 min with a continuous magnetic stirring.

2.2. Catalysts characterization.
SEM images were obtained by TEM CM12 (Phillips, Netherlands). Dispersions were deposited on copper grids with an amorphous carbon coating and dried. UV-vis DR spectra were obtained by Carry 100 (Varian, Australia) in the transmission mode in a quartz cuvette with a thickness of 2 mm in the spectral range of 200-800 nm.

2.3. Catalytic tests.
In the first mode (opened reactor) 20 ml of water dispersions of the metal NPs (50 mg/ml) was placed in a three-necked flask equipped with a condenser, then the calculated amount of propylene glycol (99.3 wt%) and sodium hydroxide were added to the reaction mixture. The reaction flask was placed in an oil bath with a thermocouple and heated up to the given reaction temperature. Oxidation reaction was started by pumping of molecular oxygen throughout the reaction mixture ($P(O_2)$ = 1 bar, oxygen flow rate = 80 ml/min). At the end of the reaction, metal NPs were separated by centrifugation and the reaction mixture was analyzed by GC and HPLC. In the second mode of testing (closed reactor), 20 ml of water dispersion of the metal NPs (50 mg/ml) was placed in a glass reactor, then the calculated amount of PG and sodium hydroxide were added to the reaction mixture. The reactor was purged with oxygen ($P(O_2)$ = 3 bar), placed in an oil bath with a thermocouple and heated up to the desired temperature.

3. Results and discussion

3.1. Catalyst characterization.
According to TEM data, the bimetallic NPs with a different Ag/Pd ratio are formed as a result of the focused laser beam treatment of the mixture of silver and palladium NPs (Fig. 1). To study the structure and morphology of the prepared NPs, TEM method was used. Both composite NPs have spherical shape with a diameter below 15 nm for Ag and 60 nm for Pd.

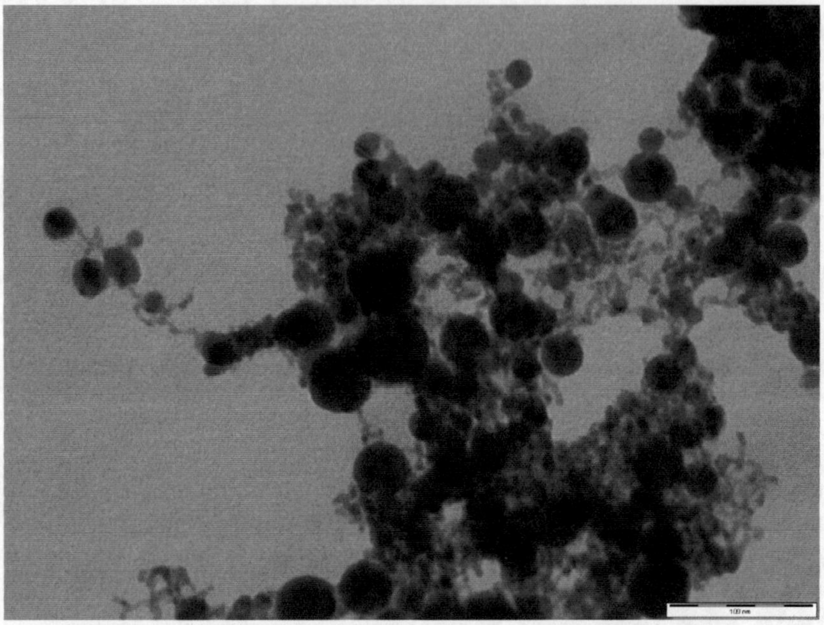

Figure 1. TEM image of Ag-Pd NPs with 50/50 ratio.

In the present work we studied the optical properties of the obtained metal NP dispersions (Fig. 2). Basic dispersions obtained by the PLAL technique (50 mg/ml) from monometallic silver and palladium targets in water possess the characteristic peaks of the surface plasmon resonance with a maximum at 410 nm for silver colloid solution (curve 1) and 205 nm for palladium solution, respectively (curve 2). In addition to a pronounced narrow peak at 205 nm, a wide band with the maximum at 250–400 nm appears in the palladium colloid probably due to the incorporation of hydrogen [11] into the palladium NPs as a result of high-energy synthesis and repeated interaction of palladium particles with the laser radiation [6]. The increased optical density in the long-wavelength range results from the scattering of the particles due to their agglomeration.

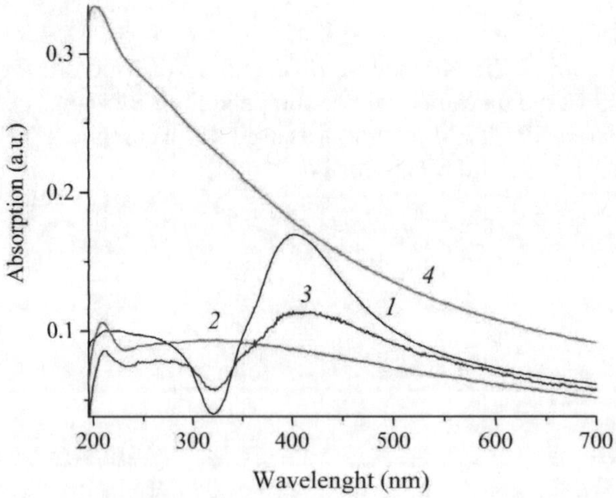

Figure 2. UV-Vis absorption spectra of NPs dispersions.
1 – Ag dispersion, 2 – Pd dispersion, 3 – a mixture of Ag and Pd dispersions before laser irradiation, 4 – a mixture of Ag and Pd dispersions after laser irradiation.

IOP Conf. Series: Materials Science and Engineering **597** (2019) 012010 doi:10.1088/1757-899X/597/1/012010

UV-vis DRS spectra of the initial mixture of silver and palladium NP dispersions (25 mg/ml of each component) are close to the superposition of the spectra of monometallic dispersions. Intensity of the characteristic peaks are lower due to decreasing of the concentration of Ag and Pd NPs in the resulting mixture (curve 3). Optical properties of the mixture change after the powerful laser irradiation (curve 4). The intensity of absorption (scattering) is increased in the entire investigated range. Characteristic peak of the silver surface plasmon resonance completely disappears, while the characteristic peak position of the palladium surface plasmon resonance does not change. According to Ref. [12], the presence of palladium in the bimetallic alloy suppresses plasmon resonance of silver NPs. Thus, the disappearing of the characteristic peak of silver plasmon resonance after laser irradiation confirms the formation of Ag-Pd alloy. Visually, the solution becomes more saturated with a dark brown tint.

3.2. Catalytic properties.
The result of PG catalytic conversion strongly depends on the catalyst nature. Low PG conversion in runs 1-4 (Table 1) may be due to low O_2 pressure. Ag/Pd molar ratio in the catalyst composition strongly affects the PG conversion pathway. The composite of Ag-Pd NPs synthesized by addition of 30% (mol) Pd dispersion to the Ag one results in a sharp increasing of the LA selectivity in comparison with the individual Ag NPs (Table 1, runs 1-2). Further enhancement of palladium content in the catalyst up to 100 % leads to increasing of both PG conversion and LA selectivity. Xue et al. [5] obtained similar dependence in the case of PG conversion and an opposite dependence in the case of LA selectivity. This fact may be due to different catalyst preparation methods used, which leads to the formation of another type of active surface. The overall carbon mass balance was poor in runs 1-2, probably, due to the formation of undetected lactaldehyde and carbon oxides. Hydroxyacetone is not found in the products that allow suggesting the PG oxidation only through the primary hydroxyl group.

Table 1. Catalytic properties of bimetallic silver/palladium composite NPs with different ratio in the PG to LA oxidation reaction.

Reactor type	O_2 flow rate/pressure, (ml/min / bar)	Reaction temperature (°C)	Ag/Pd ratio	PG conversion (%)	Selectivity (%)			
					LA	FA	AA	PA
Opened	80	75	100/0	11	10	1	2	1
Opened	80 ml/min	75	70/30	8	53	0	2	0
Opened	80 ml/min	75	50/50	13	84	0	0	1
Opened	80 ml/min	75	30/70	16	92	1	5	0
Opened	80 ml/min	75	0/100	70	90	0	2	0
Closed	3 bar	75	50/50	13	85	0	<1	0
Closed	3 bar	85	50/50*	21	85	1	3	0
Closed	3 bar	90	50/50*	27	85	1	11	1
Closed	3 bar	100	50/50*	58	81	1	11	0

Reaction time = 4 h, C(NaOH) = 0.6 M, C(PG) = 0.3 M, C (Ag/Pd) = 50 mg/ml
FA – formic acid, AA – acetic acid, PA – pyruvic acid, * - reaction time = 7 hours

In order to compare the influence of the oxygen pressure on the catalytic activity of the Ag-Pd NPs we tested the composite catalyst with a silver/palladium ratio of 50/50 under the same reaction conditions (T = 75 °C, C(PG) = 0.3 M, C(NaOH) = 0.6 M). Table 1 shows the results of the experiments (runs 6-9). PG conversion is higher at the increased O_2 pressure. We conclude that application of the closed reactor is more effective in the case of PG to LA oxidation on the silver-palladium NPs as a catalyst.

Increasing of the reaction temperature from 75°C to 90 °C leads to growth of PG conversion and LA selectivity (Table 1, runs 6-8). Further enhancement of the reaction temperature up to 100 °C results in LA selectivity decrease at much increased PG conversion (Table 1, run 9). The amount of AA rises as the temperature increases. However, the selectivity to another product of C-C cleavage, FA, does not change. This can result from the high susceptibility of FA to decomposition at elevated temperatures.

4. Conclusions

Powerful laser irradiation of the mixtures of silver and palladium water NP dispersions allowed forming the bimetallic NP composites. The obtained Ag/Pd composites demonstrated high activity in the PG oxidation to LA in alkaline media. Increasing of the reaction temperature up to 90 °C led to increase of both PG conversion and LA selectivity. However, further rising of the reaction temperature up to 100 °C resulted in the decrease of the selectivity towards LA up to 80 % due to the oxidative cleavage of C-C bound in the propylene glycol molecule. The highest catalytic activity and selectivity towards LA was demonstrated by monometallic Pd, the addition of silver in Ag/Pd ratios from 30/70 to 70/30 leads to a decreasing of the catalytic activity of the bimetallic catalyst. The enhancement of oxygen pressure up to 3 bar and reaction temperature up to 90 °C led to increasing in the LA selectivity.

Acknowledgments

This work was supported by Russian Science Foundation (project number 19-73-30026).

References

[1] F. Cavani, S. Albonetti, F. Basile and A. Gandini 2016 *Chemicals and Fuels from Bio-Based Building Blocks*, Wiley, New Jersey
[2] Martinez F A C, Balciunas E M, Salgado J M, González J M D, Converti A, de Souza Oliveira R P 2013 *Trends Food Sci. Technol.* **30** 70-83
[3] Ebert D Yu, Dorofeeva N V, Savel'eva A S, Kharlamova T S, Salaev M A, Svetlichnyi V A, Magaev O V and Vodyankina O V, *Catal. Today*, in press https://doi.org/10.1016/j.cattod.2018.07.029
[4] Torbina V V, Nedoseykina N S, Ivanchikova I D, Kholdeeva O A and Vodyankina O V, Catalysis Today, in press https://doi.org/ 10.1016/j.cattod.2018.11.063
[5] Xue W, Feng Yo, Yin H, Liu Sh, Wang A and Shen L 2016 *J. Nanosci. Nanotechnol.* **16** 9621-33
[6] Torbina V V, Vodyankin A A, Ten S, Mamontov G V, Salaev M A, Sobolev V I and Vodyankina O V 2018 *Catalysts* **8** 447
[7] Zhang D, Liu J, Li P, Tian Z, Liang C 2017 *ChemNanoMat.* **3** 512–533
[8] Zhang D, Gökce B, Barcikowski St 2017 *Chem. Rev.* **2017** 3990-4103.
[9] Chen Y-H, Tseng Y-H, Yeh C-S 2002 *J. Mater. Chem.*, **12** 1419-22
[10] Slavinskaya E M, Kardash T Yu, Gulyaev R V, Stonkus O A, Lapin I N, Svetlichnyi V A and Boronin A I 2016 *Catal. Sci. Technol.* **6 17** 6650-66
[11] Ershov B G 1996 *Russ. Chem. Bull.* **45** 299-302
[12] Yang Ch-Ch, Wan Ch-Ch and Wang Yu-Yu 2004 *J. Colloid. Interface Sci.* **279** 433-439 https://doi.org/10.1016/j.jcis.2004.06.098

IOP Conf. Series: Materials Science and Engineering **597** (2019) 012011 doi:10.1088/1757-899X/597/1/012011

Investigation of adsorption of water vapor on porous aluminium oxide material

S Reshetnikov[1] and I Kurzina[2]

[1]Boreskov Institute of Catalysis SB RAS, 5 Lavrentieva Avenue, Novosibirsk 630090 Russian Federation
[2]Tomsk State University, 36 Lenina Avenue, Tomsk 634050, Russian Federation

[2]E-mail: reshet@catalysis.ru

Abstract. This paper deals with theoretical investigation and experimental verification of adsorption dynamics (kinetics) of water vapor on aluminium oxide material. Influence of vapor flow rate and grain size of the adsorbent on adsorption dynamics was carried out on a laboratory equipment using McBain-Bakr quartz balance. The flow rate at which there is no influence of external diffusion limitations was determined. The material was characterized by the BET method and X-ray phase analysis (XRD). The theoretical effective diffusion coefficient was predicted from the known structure of the adsorbent. Mathematical modeling of the mass transfer rates are based on a quasi-homogeneous model was performed. A comparison between theory and experiments for the various grain size adsorbent samples allows to determinate of the effective diffusion coefficient D_{eff}.

1. Introduction

Adsorbents based on aluminium oxide are widely used in chemical process as catalysts in cracking and hydrocracking processes of petroleum products, alcohol dehydration ets. [1-3], as well as adsorbents for industrial gases purification, to extract water vapors during drying of natural and associated petroleum gases [4-7]. Gas transportation makes special demands on its degree of drying and cleaning so far as the presence of water at low temperature leads to formation of hydrocarbon hydrates, which, partial or complete can blockage a gas pipeline. Advantages of aluminium oxide (high activity, thermodynamic stability, relative ease of production and significant mechanical strength) provide its wide application along with such adsorbents as zeolites and silica gels. In this regard, the search for new more effective adsorbents is still actually. Nowadays, effective adsorbents, based on low-temperature forms of aluminium oxide (η-, γ- and χ-), obtained by calcination of alkaline hydration products of thermally-activated aluminium hydroxide with a bayerite phase are actively studied [8].

One of the main parameters of adsorbents is maximum adsorption capacity, which depends on equilibrium process. The adsorption equilibrium of water vapor has been studied widely on different materials, such as activated carbon [9], silica gel [10], zeolite and activated alumina [11-13]. As well as the equilibrium characteristics of the adsorbent, an important role belongs to adsorption kinetics, which considers the dynamics of adsorption and diffusion in individual adsorbent's granules and the rate of filling the capacity of these granules. Since the adsorbent granules, used in industrial conditions, have as a rule a shape of cylinders and spheres of various sizes, an investigation on the influence of the grain size and water diffusion on the adsorption kinetics is important.

The aim of this paper is theoretical and experimental investigation of adsorption kinetics of water vapor on aluminium oxide material and the theoretical determination of the effective diffusion coefficient D_{eff} on the base a quasi-homogeneous model and experiments for the various grain size adsorbent samples.

2. Experimental part

2.1 Adsorbent characterization and testing

A high effective aluminium porous material, which was a product of centrifugal thermal activation of hydrargillite, was chosen as an object of research [8, 15]. The experiments on the study of kinetics on the different aluminium oxide samples (0.25-0.5 mm and 0.5-1.0 mm and a 3.7 x 6 mm granule) were carried out on a laboratory installation using McBain-Bakr quartz balance [14]. The amount of adsorbent is 0.02÷0.03 g. To carry out the process of water vapor adsorption with 100 % moisture, argon, infiltrated through two Drexel bottles filled with distilled water, was supplied to the sample.

X-ray phase analysis (XRD) was performed on a Rigaku Miniflex 600 diffractometer with the following shooting parameters: the measurement range is 2θ=10–90 with Cu Kα radiation; voltage is 40 kV; current is 15 mA; shooting rate is 2 deg./min. Diffraction pattern deciphering was carried out using the ICDD database, version PDF2.

The textural characteristics were determined by nitrogen adsorption isotherms at 77 K. The mesopore volume was calculated by analyzing the integral curve of pore volume distribution depending on their radius (along the adsorption branch).

3. Results and discussion

3.1. Material characterization

The adsorbent sample was characterized by physicochemical research techniques [15, 16]. According to the XRD results, the porous material under study represented a mixture of low-temperature modifications of aluminium oxide i.e. $(\gamma+\chi)$-Al_2O_3.

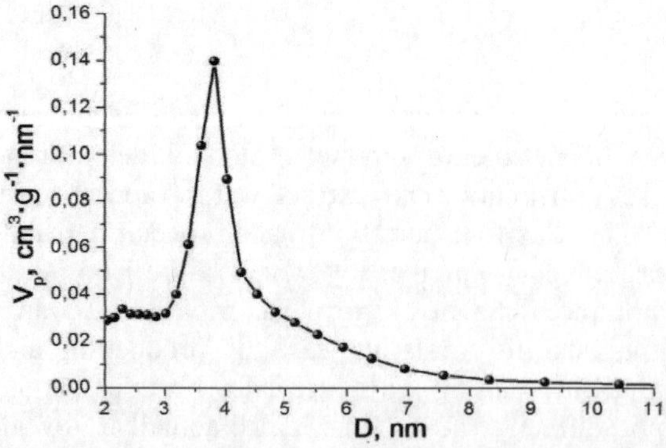

Figure 1. Pore size (*D*) distribution for the adsorbent (*V*p-pore volume).

Based on the isotherms of low-temperature adsorption and desorption of nitrogen, the specific surface area, porosity and pore size distribution were determined. Adsorption isotherms at relative pressures of 0.05-0.3 were described by the BET equation, which allowed calculating the specific surface size, which was ~305±30 m^2/g. Fig.1 shows the pore size distribution. The presence of fine mesopores (3÷7 nm) with a diameter of ~ 4 nm indicates that the adsorption process can be influenced by internal diffusion of water vapors in the adsorbent. Therefore, it is important to determine the influence of the grain size of adsorbent particles on adsorption kinetics.

3.2. Theoretical estimation of effective diffusion coefficient

The mass transfer in the adsorbent pores, depending on the pore size, can obey the laws of either molecular or Knudsen diffusion. The total diffusion coefficient of water in air is defined as [3]

$$D^* = \left(\frac{1}{D_{\mathrm{M}}} + \frac{1}{D_{\mathrm{K}}}\right)^{-1} \tag{1}$$

In addition, the diffusion path in the catalyst pores is tortuous rather than straight, and pores have irregular shape with a variable cross-section. As a result, the gas flow passing through a porous adsorbent depends on its porosity χ and tortuosity ξ of the pores. Therefore the effective coefficient of diffusion is expressed as $D_{eff} = \Pi \cdot D^*$. Here, the permeability coefficient $\Pi = \chi/\xi$. Considering that values of porosity are within 0.3-0.7, tortuosity ranges from 2 to 7, and their ratio is 0.04-0.35. The molecular diffusion coefficient D_m for binary gas diffusivities for water and air under the reaction conditions (T = 25°C, P = 1 bar) is calculated, according to the following equation [17]:

$$D_M = D_{H2O-Air} = \frac{10^{-7} \cdot T^{1.75} \cdot \sqrt{\dfrac{1}{M_{H2O}} + \dfrac{1}{M_{Air}}}}{P \cdot (V_{H2O}^{0.333} + V_{Air}^{0.333})^2} = 0.252 \cdot 10^{-4} \mathrm{m}^2 / c \tag{2}$$

The Knudsen diffusivity D_K ($m^2\ s^{-1}$) was calculated by:

$$D_{\mathrm{K}} = 9700 \cdot r_{\mathrm{p}} \cdot \sqrt{\frac{T}{M_{H2O}}} \tag{3}$$

The pore radius of catalyst was estimated from the Figure 1, $r_p \approx 2$ nm. The Knudsen diffusion coefficient and the total ones has the following values, $D_K = 0.79 \cdot 10^{-6}$ m²/s and $D^* = 0.77 \cdot 10^{-6}$ m²/s, $D_{eff} = \Pi \cdot D^*$, where $\Pi = 0.04\text{-}0.35$.

3.3. Effect of flow rate

Figure 2 shows the effect of the flow rate on water adsorption in the wide range of 3.6-36 l/h at the different time. The experiments were carried out to exclude the influence of the external diffusion.

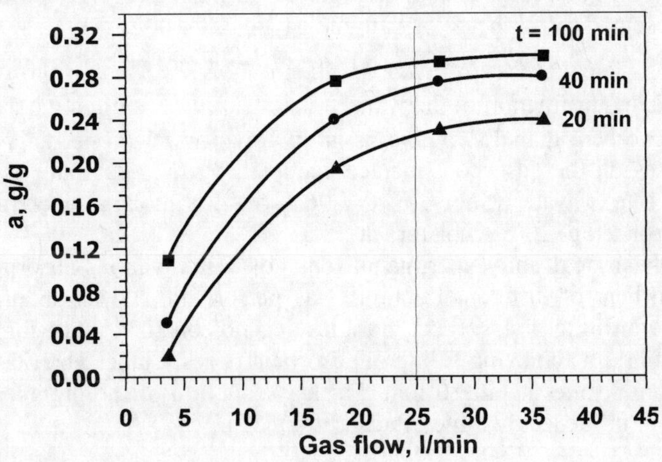

Figure 2. Influence of gas flow rate at different time on adsorption dynamics.

It was shown that at the water vapor flow rates less than 25 l/min the sample weight grows with an increase in the gas velocity for the each adsorption time. In these conditions, the external diffusion seems to be the rate-determining step of the water adsorption process. In the range of 25-35 l/min, the sample weight did not depend on the flow rate, and it can be concluded that the adsorption occurs in the kinetic regime. Therefore, all the experiments, presented on figure 3, were carried out at a flow rate of 30 l/h.

3.4. Effect of granules size on adsorption kinetics

At the same time with the capacities of the adsorbents, an important role play the adsorption kinetics. Adsorbent granules that are used in industrial conditions are shaped as cylinders and spheres of various sizes. To study the influence of the granule size on the dynamics (kinetics) of water vapor adsorption, the following fractions of the investigated aluminium oxide sample were taken: 0.25-0.5 mm and 0.5-1.0 mm and a 3.7 x 6 mm granule.

Mathematical modeling allows to obtain the quantitative estimation of parameters, such as adsorption rate constants and effective diffusion coefficient of the substance inside the grain. A quasi-homogeneous model is used most commonly. The mathematical model considering the adsorbent as a homogeneous absorbing medium includes the following equations:

- change in the concentration of the water (C_{H2O}) in the adsorbent grain:

$$\varepsilon_c \frac{\partial C_A}{\partial t} = D_{eff} \frac{1}{r^2} \frac{\partial}{\partial r}\left(r^2 \frac{\partial C_A}{\partial r} \right) - f_a R(C_{H2O}, \theta), \tag{4}$$

- change in the concentration of adsorption sites on the adsorbent surface occupied by the water molecules:

$$\frac{\partial \theta_{H2O}}{\partial t} = R(C_A, \theta). \tag{5}$$

Initial and boundary conditions:

$$t=0:\ C_{H2O}(0,r) = C_A^0,\ \theta_{H2O}=0. \tag{6}$$

$$r=0:\ \frac{\partial C_{H2O}(t,r)}{\partial r} = 0,\quad r=R:\ C_{H2O}(t,r) = C^0.$$

Here the adsorption-desorption process of water vapor on the surface of the aluminium oxide adsorbent can be represented by a scheme of Langmuir type

$$R = k_1 \cdot C_{H_2O} \cdot (1-\theta_{H2O}) - k_2 \cdot \theta_{H_2O} \tag{7}$$

In this case, C_{H2O} is molar fraction of the water; t is time, s; D_{eff} is effective diffusion coefficient of the adsorbate in the adsorbent granule, m^2/s; ε_c is grain porosity; r is grain radius coordinate, m; θ_{H2O} is concentration of sites on the adsorbent surface occupied by the adsorbate; k_1, k_2 are constants of adsorption/desorption rate of water vapors; C^0 (=3.06 % vol.) is molar concentration of water vapors in the gas flow corresponded to 100% moisture at 25 °C.

A comparison of theoretical and experimental data of water vapor adsorption on a sample of aluminium oxide adsorbent of different fractions was performed. It was obtained that the time for achieving an equilibrium filling of adsorbent granules depends on size. Thus, for the granule, the time necessary for the equilibrium achieving filling of adsorbent pores it takes about 250 minutes, and for a fraction of 0.25-0.5 mm it takes about 30 min. Further reduction of the granule size to a fraction of 0.125-0.25 mm does not influence the time of adsorption [15].

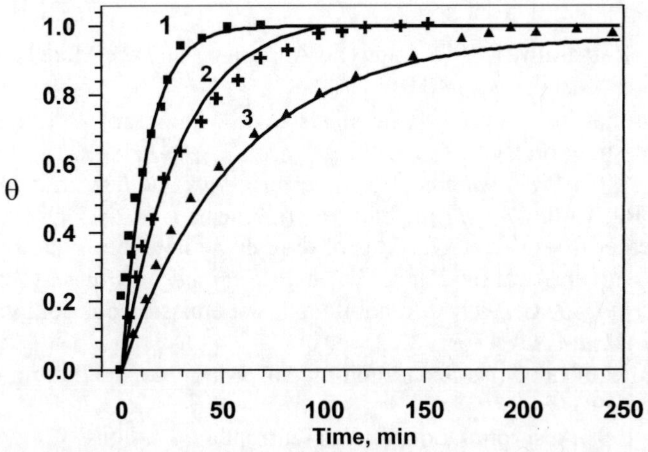

Figure 3. A comparison of theoretical and experimental data of water vapor adsorption on a sample of aluminium oxide adsorbent of different fractions: 0.25-0.5 mm (1), 0.5-1.0 mm (2) and a 3.7 x 6 mm granule (3). Symbols – Experiment; lines – calculated.

Mathematical model (4)-(7) describes quite well the experimental data (Figure 3) on the adsorption kinetics of water vapor on the aluminium oxide adsorbent. The theoretically effective diffusion coefficient was predicted from the known pore structure of the adsorbent. This lead to the value of $(0.04\text{-}0.35)\cdot10^{-6}$ m^2/s. The experimentally determined value is $1.2\cdot10^{-6}$ m^2/s. This value is correlated with effective diffusion coefficient ($3.6\cdot10^{-6}$ m^2/s) obtained during the study of the kinetics water adsorption on alumina [18]. It should be noted that according to the literature data, the experimentally determined diffusion coefficient obtained by various authors can be differs by more than 10 times [18].

4. Conclusion

On the base of a quasi-homogeneous mathematical model, a description of the dynamics of the adsorption of water vapor on a porous material was made. A high effective aluminium porous material, which was a product of centrifugal thermal activation of hydrargillite, was chosen as an object of research. The influence of the adsorbent granule size on the adsorption kinetics of water vapors on fractions of 0.25-0.5 mm and 0.5-1.0 mm and a 3.7 x 6 mm granule was studied. The experiments were carried out on the adsorption equipment using McBain-Bakr quartz balance. On the base the isotherms of low-temperature adsorption and desorption of nitrogen, the specific surface area, porosity and pore size distribution were determined. Adsorption isotherms were described by the BET equation, which allowed calculating the specific surface size, which was ~305±30 m^2/g. The presence of fine mesopores with a diameter of ~ 4 nm indicates that the adsorption process can be influenced by internal diffusion of water vapors in the adsorbent. It was shown that when the fraction size is greater than 0.25-0.5 mm, the water vapor adsorption rate is controlled by internal diffusion. Mathematical modeling allowed obtaining the quantitative estimation of parameters, such as adsorption rate constants and effective diffusion coefficient of the water inside the grain. Mathematical model describes quite well the experimental data on the adsorption kinetics of water vapor on the high active aluminium oxide adsorbent and can be used for adsorber modeling.

Acknowledgments
The study was financially supported by project RFMEFI57517X0139 of the Ministry of Education and Science.

IOP Conf. Series: Materials Science and Engineering **597** (2019) 012011 doi:10.1088/1757-899X/597/1/012011

References

[1] Knözinger H, Ratnasamy P 1978 Catalytic Aluminas: Surface Models and Characterization of Surface Sites *Catal. Rev.* **17** 31
[2] Misra C 1986 Industrial Alumina Chemicals. ACS Monograph 184. *American Chemical Society*. Washington DC
[3] Setterfield Ch 1984 *Practical course of heterogeneous catalysis*. M.: Mir, p 112
[4] Kurzina I A, Reshetnikov S_I, Karakchieva NI, Kurina L_N 2017 Direct synthesis of dimethyl ether from Direct synthesis of dimethyl ether from synthesis gas: experimental study and mathematical modeling *Chemical Engineering Journal* **329** 135.
[5] Belton D N, Taylor K C 1999 Automobile exhaust emission control by catalysts *Curr. Opin Solid State Mater. Sci.* **4** 97
[6] Sircar S, Applications of Gas Separation by Adsorption for the Future *Adsorption Science and Technology* **19** 347
[7] Fleming H L 1999 Adsorption on aluminas-current applications **120** 561
[8] Danilevich V V, Isupova L A, Danilova I G, Zotov R A and Ushakov V A 2016 Optimization of properties of aluminium oxide desiccants obtained based on the product of centrifugal thermal activation of hydrargillite *Zhurnal prikladnoy khimii* **89** 289
[9] Fletcher A.J., Yüzak Y., Thomas K.M. 2006 Adsorption and desorption kinetics for hydrophilic and hydrophobic vapors on activated carbon *Carbon*, **44** 989
Xia Zai Z, Chen Chuan J, Kiplagat Jeremiah K, Wang Ru Z and Hu Jin Q 2008 Adsorption Equilibrium of Water on Silica Gel *J. Chem. Eng. Data* **53** 2462
[10] Kim Jong-Hwa, Lee Chang-Ha, Kim Woo-Sik, Lee Jong-Seok, Kim Jin-Tae, Suh Jeong-Kwon, and Lee Jung-Min 2003 Adsorption Equilibria of Water Vapor on Alumina, Zeolite 13X, and a Zeolite X/Activated Carbon Composite *J. Chem. Eng. Data* **48** 137
[11] Adsorption on Activated Alumina. I1 - Kinetic Behaviour Desai R., Hussain M. and Ruthven D. M. 1992 *The Canadian Journal of Chemical Engineering* **70** 707
[12] Marcussen Lis 1974 The influence of temperature oneffective diffusivity and adsorption kinetics for humid air-poros alumina *Chemical Engineering Science* **29** 2061
[13] Reshetnikov S.I., Livanova A.V., Meshcheryakov E.P., Kurzina I.A., Isupova L.A. 2017 Kinetic regularities of adsorption on aluminium oxide desiccants doped with alkali metal cations *Zhurnal prikladnoy khimii* **90** 1451
[14] Reshetnikov S, Budaev Z, Livanova A., Meshcheryakov E., Kurzina I. 2019 Effect of particle size on adsorption kinetics of water vapor on porous alumenum oxide material *J. Phys Conf Ser.* **1145** 012033
[15] Livanova A, Meshcheryakov E, Reshetnikov S, Kurzina I 2017 Study of Water Vapour Adsorption Kinetics on Aluminium Oxide Materials *AIP Conference Proceedings*. **1899** 020018
[16] Reed R, Prausnitz J, Sherwood T 1982 *Properties of gases and liquids*. L.: Chemistry. p 591
[17] Marcussen L 1970 The kinetics of water adsorption on porous alumina *Chem. Eng. Sc.* **25** 1487

Development of a scratch test method to determine adhesion and strength properties of coated steel sheets

V K Belov[1], E V Gubarev[1, 3], A V Papshev[2], N G Gofman[2] and D O Begletsov[1]
[1]Microtopography Research Centre, Nosov Magnitogorsk State Technical University, Lenin street, 38, Magnitogorsk, 455000, Russian Federation
[2]Kirov street, 93, Magnitogorsk, 455000, PJSC MMK, Russian Federation

E-mail: john_g@mail.ru

Abstract. A simple method to assess adhesion properties of organic coatings by Ω parameter was developed and tested, where Ω is a ratio of indentation speed at a linearly increasing normal load to a resulting force on the indenter at the end of plastic deformation, when the indenter reaches a substrate. A relationship between Ω parameter and a coating hardness H determined by Oliver and Pharr method is shown.

1. Introduction

Presently, many goods are produced with various coatings. These coatings serve as anti-corrosive, decorative, wear resistant, tribological (for low or high friction purpose), galvanic, biocompatible, optical, magnetic and conductive coatings, as well as chameleon coatings for military machinery, waterproof coatings, intumescent fire-resistant coatings, transparent coatings for optics and glassware, highly antibacterial coatings, etc. [1].

The main requirements to coatings include provision of expected performance, low cost of a coating material and that of the process involving its application onto a substrate.

The most important performance parameters of any coating are adhesion properties, i.e. resistance to damaging and peeling off a substrate. There are plenty of methods used to determine these properties [2]. Most common are: a bend test [3-5], a cross-cut test [6], a pencil test [7] and a pull-off test [8].

A lot of attention is now paid to evaluation of adhesion properties of various coatings using a scratch test [2, 9, 10, 11]. The test involves scratching of a coating (i.e. making a 'track') with a small-radius tip indenter (see figure 1) in two measurement modes: 1) at a constant vertical speed of the indenter, and 2) at a constantly increasing load on an indenter. The first mode is generally used more frequently. Horizontal speed of the indenter is not high and is usually about 3-4 cm/s for both modes.

After scratching, a sample is analyzed with a microscope. Use of a 3D-surface analyzer as a microscope makes it also possible to determine the surface characteristics along the scratch, such as its depth, width, curvature radius etc. This allows to evaluate a coating failure pattern alongside the scratch (local cracking, wedging, recovery, spallation, etc.) [12, 13] and to calculate a load which results in this failure.

A scratch test is most advantageous for being able to reproduce very realistic coating failures (i.e. scratches, pin holes, galling, etc.). In some studies, scratch test results are correlated with physical parameters of a coating (surface energy, mechanical properties of a surface layer, friction properties) [1, 11, 14-16]. Another benefit of a scratch test is that the results can be expressed numerically to exclude any subjective estimation.

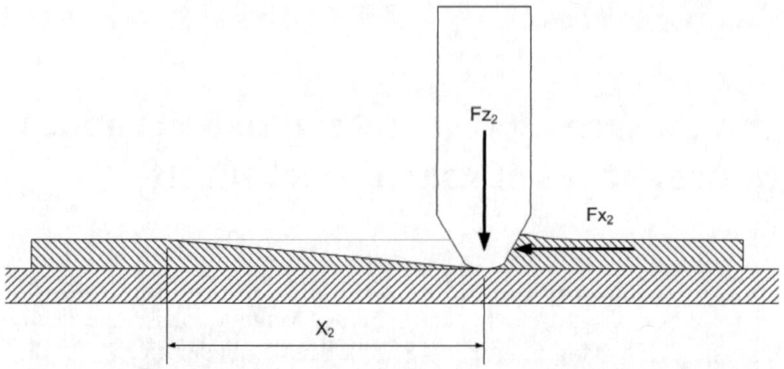

Figure 1. Coating thickness and adhesion measurement principle.

A scratch test, however, has got a disadvantage which lies in the fact that its results strongly depend on the initial frictional properties of a surface to be analyzed [17]. It requires a proper pre-cleaning of a surface from absorbed molecules. Unfortunately, the cleaning procedure is specific for every type of coating and is therefore to be thoroughly described while representing the test results.

The purpose of this work is to develop a simple scratch test procedure for numerical evaluation of adhesive properties of coil coatings. This method shall allow for an unbiased selection of suppliers of various raw materials for production of coated steel at PJSC MMK (Magnitogorsk, Russia), as well as improved application techniques to achieve better coating adhesion along with lower consumption of raw materials.

2. A simple method for evaluation of adhesive and strength properties of organic coatings

The research was done using the following equipment of Microtopography Research Center, MSTU: 1) an optical surface analyzer Contour GT K1 by Bruker, USA; 2) a stylus surface analyzer MarSurf XR20 with XT20 by Mahr, Germany; 3) a measurement station UMT-1 by Bruker, USA, for determination of physical and mechanical properties of materials.

Figure 2 gives an example of the coating thickness and adhesion measurement report. Upper image shows the results of scratch test track measurements made with Contour GT K1. Lower image shows the measurements results obtained by UMT-1. Both images are in the same horizontal scale.

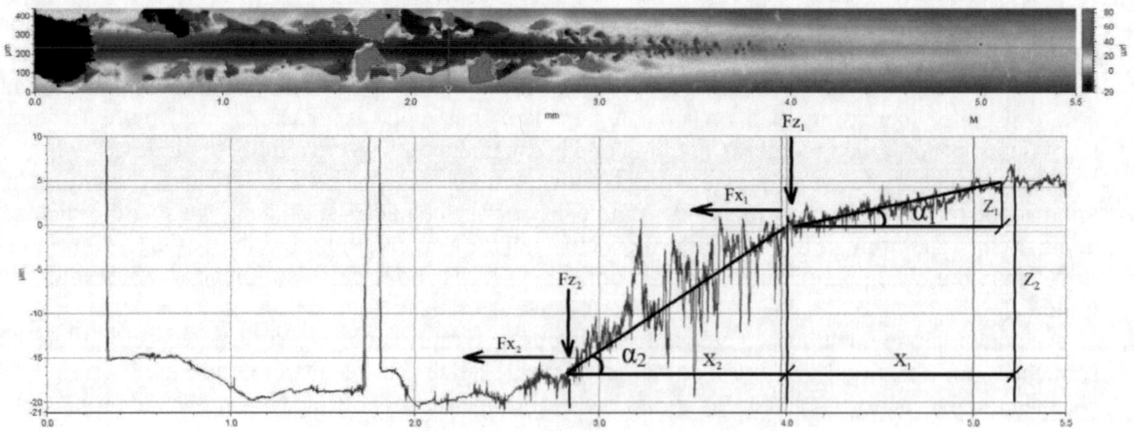

Figure 2. Results of a scratch test track measurement.

X_1 – track length of elastic deformation;
X_2 – track length of plastic deformation;

Z_1 – track depth at the end of elastic deformation;
Z_2 – track depth at the end of plastic deformation;
F_{z1} – normal force on indenter at the end of elastic deformation;
F_{z2} – normal force on indenter at the end of plastic deformation when indenter reaches a substrate;
F_{x1} – tangential force on indenter at the end of elastic deformation;
F_{x2} – tangential force on indenter at the end of plastic deformation when indenter reaches a substrate.

Adhesion properties of coatings have been measured under the following conditions:

- 50x50 mm samples were glued onto a steel substrate and fixed in a measurement station;
- vertical loading mode was applied, i.e. linearly increasing load F_z from 0.2 N up to 15 N with testing time t=720 s;
- indenter speed is V = 0.007 mm/s on 5 mm track length;
- type of indenter: Rockwell C scale (spheroconical diamond tip with 120° angle and 0.2 mm radius);
- the tests are done at a room temperature;
- measurement is made at least 5 times;
- cracking start (F_{z1}, Z_1) and reaching of zinc substrate by indenter (F_{z2}, Z_2) are determined by optical images and load diagrams (see figure 2).

Table 1 gives characteristics of samples submitted for analysis done by Coatings Laboratory, PJSC MMK.

Table 1. Samples of coatings for analysis.

Sample number	Main coating	Primer	Organic coating
1	Zn	Basf	no coating
2	Zn	Basf	Valspar RAL 9003 (top)
3	Zn	Basf	Prime top (back)
4	Zn	Basf	DYO (back)
5	Zn	DYO	DYO RAL 6005 (top)
6	Zn	DYO	DYO (back)
7	without Zn coating		no coating
8	without Zn coating	Basf	Valspar RAL 9003
9	without Zn coating	Basf	DYO (back)

Main purpose of the analysis was to compare adhesion properties of the following samples pairs: 1–7; 2–5; 2–8; 3–4; 3–6; 8–9

Producers of coated steel were concerned about coating damage between coil laps during handling and transportation, whereas processors/customers were concerned about possible coating damage during manufacture and service life of finished parts.

Plastic properties of coating materials were determined by a tangent of α_2 (see figure 2), i.e. by the following parameter:

$$K = \frac{Z_2 - Z_1}{X_2} \tag{1}$$

This parameter indicates the indentation speed at a linearly increasing normal load on an indenter. Coated samples with smaller K values are less damageable.

The load when plastic deformation is considered to be complete and the indenter reaches a substrate may be defined as follows:

IOP Conf. Series: Materials Science and Engineering **597** (2019) 012012 doi:10.1088/1757-899X/597/1/012012

$$F = \sqrt{F_{x2}^{2} + F_{y2}^{2}} \tag{2}$$

Coated samples with greater F values are regarded as less damageable.

Therefore, the adhesive properties of a coating can be estimated by its resistance to damage (failure) as follows:

$$\boldsymbol{\Omega} = \frac{\boldsymbol{F}}{\boldsymbol{K}} \tag{3}$$

Coated samples with greater Ω values are regarded as less damageable.

Results of adhesion measurements for tested samples are presented in table 2 (see also figure 2).

Table 2. Adhesion characteristics of analyzed samples.

	Sample number	**1**	**2**	**3**	**4**	**5**	**6**	**7**	**8**	**9**
Sample characteristics	Main coating	Zn	Zn	Zn	Zn	Zn	Zn	no Zn	no Zn	no Zn
	Primer	Basf	Basf	Basf	Basf	DYO	DYO	Basf	Basf	Basf
	Paint coating	–	Valspar RAL 9003 (top)	Prime top (back)	DYO (back)	DYO RAL 6005 (top)	DYO (back)	–	Valspar RAL 9003	DYO (back)
Measured characteristics	X_1, mm	0.45	1.22	1.24	1.50	1.12	1.16	0.62	1.13	1.64
	Z_1, μm	2.36	1.61	3.70	5.30	6.50	4.46	2.50	2.98	5.50
	Fz_1, N	1.6	3.6	4.0	4.6	3.6	3.6	2.0	3.5	5.13
	Fx_1, N	0.24	0.96	0.7	1.0	1.0	0.8	0.4	1.0	1.2
	X_2, mm	1.34	2.29	2.44	2.35	2.71	2.14	1.53	2.60	2.49
	Z_2, μm	7.6	19.6	17.1	14.4	23.4	13.9	7.4	22.2	16.5
	Fz_2, N	4.2	6.8	7.2	7.2	8.3	6.6	4.4	8	7.7
	Fx_2, N	1.1	2.8	2.5	2.2	2.8	2.5	1.2	3.3	2.8
	Oliver and Pharr hardness, GPa	–	1.13	1.13	2.31	0.75	1.74	–	1.37	3.77
Coating characteristics	Thickness, μm	7.60	19.60	2.44	2.35	23.40	13.90	7.40	22.20	16.50
	K	3.91	7.86	5.49	3.87	6.23	4.41	3.22	7.39	4.42
	F, N	4.3	7.35	7.61	7.53	8.76	7.06	4.56	8.65	8.19
	Ω, N	**1.10**	**0.94**	**1.38**	**1.95**	**1.40**	**1.60**	**1.42**	**1.17**	**1.85**

Ω parameter allows for numerical representation of adhesion properties without any subjective estimation.

Based on obtained values of Ω parameter, the coil coating technique at PJSC MMK was updated.

IOP Conf. Series: Materials Science and Engineering **597** (2019) 012012 doi:10.1088/1757-899X/597/1/012012

3. Correlation of *Ω* parameter with Oliver and Pharr hardness

The obtained data made it possible to assess any relationship between *Ω* parameter and Oliver-Pharr coating hardness [14, 16].

Coating hardness has been measured under the following conditions:

- 50x50 mm samples were glued onto a steel substrate and fixed in a measurement station UMT-1;
- vertical loading mode is applied, i.e. linearly increasing load F_z from 0.1 N up to 1 N with testing time t=90 s;
- exposure for 90 s until 1 N is reached;
- measurement is made at least 10 times;
- type of indenter: Vickers diamond tip;
- distance from a sample edge to indentation point is at least 20 mm;
- the tests are done at a room temperature;
- analysis of load-depth diagrams (*F*-*h*) was carried out according to Oliver and Pharr method in Viewer software by Bruker, USA.

Measurements were made in full compliance with standard [18]. Examples of Oliver and Pharr hardness measurements are presented in figure 3(a), where the left section corresponds to loading and right section corresponds to unloading. The resulting hardness value was averaged on 10 measurements taken at different places of the sample. Dispersion of these measurements is shown in figure 3(b).

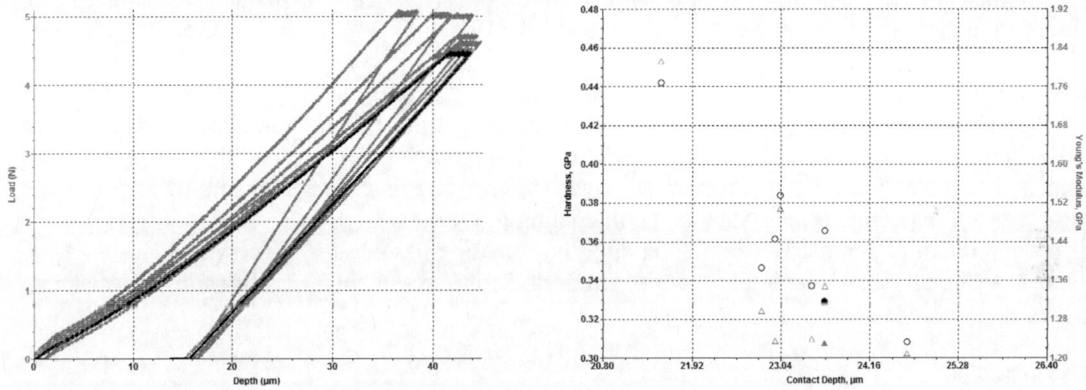

(a) (b)

(a) Indentation load-depth diagram of loading and unloading according to [18],
(b) Martens hardness (GPa) and Young's modulus (GPa) versus contact depth.

Figure 3. An example of Oliver and Pharr hardness measurement report.

Figure 4 shows a *Ω* parameter versus Oliver and Pharr hardness *H*.

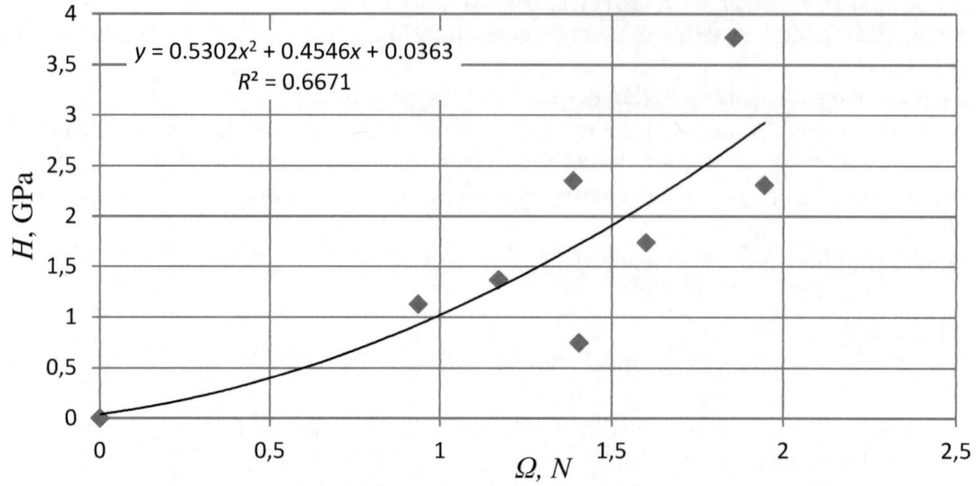

Figure 4. Ω parameter vs. Oliver and Pharr coating hardness H.

According to the diagram, the coating failure with vertical (i.e. normal) indentation (Oliver and Pharr method) and that with simultaneous normal and tangential indentation (a technique suggested to assess the Ω parameter) have a nonlinear relationship. This nonlinearity can be explained by different failure modes used in hardness measurements and scratch tests [19].

4. Conclusions

A simple method to assess adhesion properties of organic coatings by Ω parameter was developed and tested, where Ω is a ratio of indention speed at a linearly increasing normal load to a resulting force on the indenter at the end of plastic deformation, when the indenter reaches a substrate. A higher value of Ω parameter means a higher resistance of a coil coating to failures.

A correlation of Ω parameter with Oliver and Pharr coating hardness H is presented.

Based on the above, Ω parameter can be suggested for assessment of coil coatings' resistance to damages (i.e. scratches, etc.).

References

[1] Lee L H 1991 *Fundamentals of Adhesion* (New York: Plenum Press) pp. 454
[2] Mittal K L 2001 *Adhesion Measurement of Films and Coatings* vol 2 (Utrecht-Boston-Koln-Tokyo: VSP) pp 351
[3] ISO 17132:2007 Paints and varnishes - T-bend test
[4] ISO 1519:2002 Paints and varnishes - Bend test (cylindrical mandrel)
[5] ISO 6860:2006 Paints and varnishes - Bend test (conical mandrel)
[6] ISO 2409:2013 Paints and varnishes - Cross-cut test
[7] ISO 15184:1998 Paints and varnishes - Determination of film hardness by pencil test
[8] ISO 4624:2002 Paints and varnishes - Pull-off test for adhesion
[9] ISO 1518-1:2011 Paints and varnishes - Determination of scratch resistance - Part 1: Constant-loading method
[10] ISO 1518-2:2011 Paints and varnishes - Determination of scratch resistance - Part 2: Variable-loading method
[11] Bull S J, Rickerby D S 1990 New developments in the modelling of the hardness and scratch adhesion of thin films *Surface and Coatings Technology* **42** 149–164
[12] Suresh Kuiry Advanced Scratch Testing for Evaluation of Coatings *www.bruker-axs.com* Bruker Nano Surfaces Division Tribology and Mechanical Testing, 1717 Dell Ave, Campbell, CA 95008, U.S.A

[13] Bull S J 1997 Failure mode maps in the thin film scratch adhesion test *Tribology Internation* **30.7** 491-498

[14] Oliver W C, Pharr G M 1992 An improved technique for determining hardness and elastic modulus using load and displacement sensing indentation experiments *Journal of Materials Research* **7 iss. 6** 1564–1583

[15] Friedrich K, Schlarb A K (eds.) Tribology of Polymeric Nanocomposites: Friction and Wear of Bulk Materials and Coatings 2nd Edition. — Butterworth-Heinemann, Elsevier, 2013. 832 p. — ISBN–13: 978-0-444-59455-6 — (Tribology and Interface Engineering).

[16] Ollivier B, Matthews A 1994 Adhesion of diamond-like carbon films on polymers: an assessment of the validity of the scratch test technique applied to flexible substrates, *J.Adhesion Sci.Technol.* **8, Nr. 6** 651-662

[17] Blees M H, Winkelman G B, Balkenende A R, and den Toonder J M J 2000 The Effect of Friction on Scratch Adhesion Testing: Application to a Sol–gel Coating on Polypropylene *Thin Solid Films* **359** 1–13

[18] ISO 141577-2:2002 Metallic materials - Instrumented indentation test for hardness and materials parameters. Part 2. Verification and calibration of testing machines

[19] Kurkcu P, Andena L, Pavan A 2012 An experimental investigation of the scratch behaviour of polymers: 1. Influence of rate-dependent bulk mechanical properties *Wear* **290–291** 86-93

IOP Conf. Series: Materials Science and Engineering **597** (2019) 012013 doi:10.1088/1757-899X/597/1/012013

Systematic Approach to the Elaboration of a Structured Model of a Sulfur Production Unit under Uncertainty

B B Orazbayev[1], T S Kenzhebaeva[1], V L Goncharov[2] and K N Orazbayeva[3]
[1]L.N. Gumilyov Eurasian National University, Astana, Kazakhstan
[2]National Research Tomsk Polytechnic University, Tomsk, Russia
[3]Kazakh University of Economy, Finance and International Trade, Astana, Kazakhstan

E-mail: togzhan_kenzhebaeva@mail.ru

Abstract. The paper proposes a systematic approach to the elaboration of a structured mathematical model and modeling of chemical and technological systems under uncertainty exemplified by a sulfur production unit. The novelty of the suggested method lies in the choice of a suitable model type for each element of the system based on the systematic analysis, development of the selected model type and their merger into a unified system of models. The method was successfully implemented to construct a system of models for major units of the catalytic reforming plant at the Atyrau Oil Refinery (Republic of Kazakhstan). The study compares existing results, modeling results based on the suggested method and experimental results obtained using an LG-setup of the Atyrau Oil Refinery. The deliverables demonstrate high efficacy and excellence of the approach to the modeling of interconnected equipment (chemical and technological systems).

1. Introduction.

Urgent tasks for any manufacturing enterprise are effective management of the plant, improving quality and efficacy of technological and production processes, increasing output and characteristics of the product. One of the promising methods for solving these tasks is enhanced proficiency in managing production facilities based on scientifically justified methods of systematic analysis, elaborating and making decisions resting upon mathematical models built using the systematic approach [1–4].

The subject of the study is successful solution of modeling problems and multicriteria optimization problems occurring in the process of management of production facilities under uncertain and fuzzy initial information, which requires elaborating a methodology for constructing mathematical models of chemical and technological systems in such conditions.

The optimal technological regime requires establishing relations between input and output parameters of separate units, i.e. developing mathematical models of objects. These are impossible without special means and mathematical apparatus, therefore, a lot of research works towards their solution have been undertaken [6–9].

The goal of the present work is to develop structured system of models for the chemical and technological system under uncertainty of its probabilistic and fuzzy nature on the basis of various initial information.

IOP Conf. Series: Materials Science and Engineering **597** (2019) 012013 doi:10.1088/1757-899X/597/1/012013

The construction algorithm of the mathematical models of the chemical and technological system under uncertainty based on various information is depicted in Figure 1.

The construction of mathematical models for different production technologies often suffers from insufficient information, i.e. demands organizing expert assessments [1–3]. Therefore, the expert assessments—in this case—are a method for organizing the work with experts and processing their opinions expressed in quantities and qualities to gather initial information. This will be used to develop the mathematical models of the subject under study and prepare information for decision makers [4].

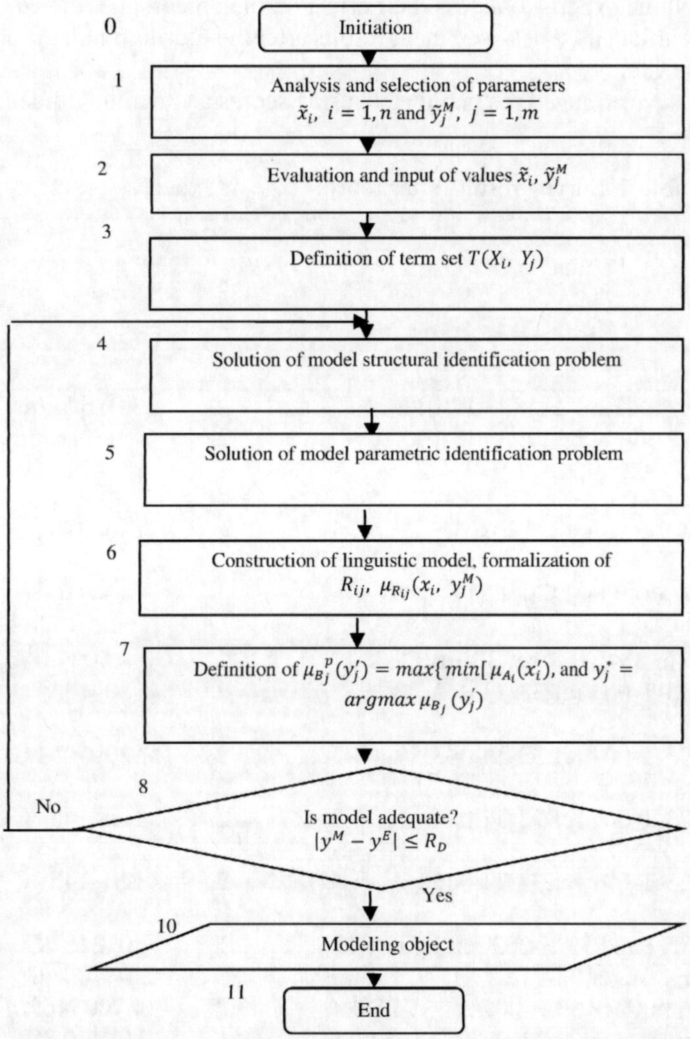

Figure 1. Flow chart of mathematical model construction for a chemical and technological system under uncertainty based on various information:
$\tilde{x}_i,\ \tilde{y}_j$ fuzzy input and output parameters; $T(X_i,\ Y_j)$ are term sets of fuzzy parameters; X, Y are universal subsets; R_{ij} are fuzzy mappings that relate fuzzy input $\tilde{x}_i$ and output $\tilde{y}_j$ parameters; $\mu_{R_{ij}}(\tilde{x}_i,\ \tilde{y}_j)$ is fuzzy matrix; $\mu_{B_j}^p(\tilde{y}_j)$ is the membership function of parameters $\tilde{y}_j$ in fuzzy set $\tilde{B}_j$ characterizing the output values; p is quantum number; y_j^c are specific numerical output values; R_D is admissible discrepancy level.

Let us consider the main results of expert assessments that were to collect necessary information for the development of mathematical models of sulfur production units at the Atyrau Oil Refinery.

The expert survey included researchers, decision makers (product engineers), senior operators and operators maintaining the unit. The experts were technical engineer of the unit, four senior operators that maintain the unit and work in shifts, two instrumentation and control engineers, sulfur production unit supervisor and two researchers studying sulfur production management problems. In total, ten experts were involved into the survey. The expert assessment was performed using the Delphi method [4, 5].

We composed a questionnaire with a list of input and operational parameters of the main sulfur production units and asked the experts to assess their effect on the process, i.e. the amount and quality of the product (sulfur, in particular). The assessment results after the fourth round, i.e. adjustment of the list, are presented in Table 1.

Evidently, the experts have arranged the parameters using scores and mainly used numbers: more 1s and 2s, and less 3s and 4s.

Table 1. Survey results after fourth stage of expert assessment

No.	Parameter	Rank	Interval (min-max)	At the moment
1.	Pressure in boilers 33-E-002 ref.33-PI-011	2	3.0–6 kg/m	4.5 kg/m
2.	Pressure in boilers 33-E-002 ref.33-PI-012	2	2.0–3.5 kg/m	2.5 kg/m
3.	Output from 33-E-002 ref.33-TI-039	3	130–210 °C	150–200 °C
4.	Output from 33-E-002 ref.33-TI-040	4	130–210 °C	150–200 °C
5.	Output from 33-E-002 ref.33-TI-041	4	130–210 °C	150–200 °C
6.	Air consumption in 33-T-001 ref.33-FI-022A	1	0–150 Nm^3/h	70 Nm^3/h
7.	Temperature in 33-T-001 ref.33-TI-042	2	120–145 °C	125–130 °C
8.	Temperature in 33-T-001 ref.33-TI-043	2	120–145 °C	125–130 °C
9.	Temperature in 33-T-001 ref.33-TI-044	2	120–145 °C	125–130 °C
10.	Temperature in 33-T-001 ref.33-TI-045	2	120–145 °C	125–130 °C
11.	Input into 33-D-004 and 33-F-002 ref.33-TI-041	1	0–210 °C	130 °C
12.	Furnace gas 33-F-002 ref.33-TI-047	3	0–100 °C	60 °C
13.	Air combustion in furnace 33-F-002 ref.33-FI-030	2	0–700 Nm^3/h	300 Nm^3/h
14.	Temperature at reactor input 33-R-001 ref.33-TI-021	1	0–410 °C	290 °C
15.	Catalyst temperature 33-R-001 ref.33-TI-23	1	200–360 °C	300 °C
16.	Catalyst temperature 33-R-001 ref.33-TI-024	1	200–360 °C	300 °C
17.	Catalyst temperature 33-R-001 ref.33-TI-025	1	200–360 °C	300 °C
18.	Catalyst at reactor output 33-R-001 ref.33-TI-026	1	300–735 °C	345 °C
19.	Combustion air consumption 33-F-001 ref.33- FI-07	1	400–1220 Nm^3/h	500 Nm^3/h
20.	Combustion air consumption 33-F-001 ref.33- FI-08	1	400–500 Nm^3/h	200 Nm^3/h
21.	Furnace output temperature 33-E-001 ref.33-TI-013	1	150–300 °C	250 °C
22.	Furnace output temperature 33-E-001 ref.33-TI-050	1	300–760 °C	645 °C

IOP Conf. Series: Materials Science and Engineering **597** (2019) 012013 doi:10.1088/1757-899X/597/1/012013

The experts were proposed a five-point scale, and 5s were used almost never. They are excluded from the table as having almost no effect on the process.

Therefore, the expert assessment has defined the most important parameters that affect sulfur production process, i.e. input and output parameters necessary for elaborating the models. Thus, after the fourth stage of the expert assessment, the input and operating parameters affecting the sulfur production process were ranked. At the second stage of the assessment, the results of the first stage were used to estimate the interrelation between chosen input and output parameters of sulfur production unit.

2. Results and Discussion

Below, we summarize the results of the studies above and use them for developing the system of mathematical models of the main constituents of the sulfur production unit at the Atyrau Oil Refinery.

The unit is a complex chemical and technological system consisting of interconnected elements that simultaneously affect a lot of different parameters. The main elements of the unit are reactor (F-001, R-001, R-002, R-003), condenser (E-001, E-002, E-004), furnace (F-002), separators (D-001, D-004) and pumps (B-001, B-002).

The results of investigated specific processes and elements of the unit, experimental data, expert assessment and analysis of approaches to modeling of similar units [8, 9, 10] were used to estimate possible types of models for each element of the unit. The result of the analysis (model estimation) is given in Table 2. To estimate the model types, a five-point scale was used.

Table 2 shows the estimations for each type of basic model of sulfur production unit resting upon the analysis results. The information provided in the table above can be used to choose the type of sulfur production unit model according to a given criteria.

The study of the operating elements of the sulfur production unit at Atyrau Oil Refinery show that due to the complexity of the equipment and processes in them and impossibility of obtaining adequate data, the construction of deterministic models for the reactors is virtually impossible or not economically feasible. According to the assessment, the statistic and fuzzy models of these sulfur production unit elements received relatively high scores. However, the most effective reactor models—in terms of total score—are combined models.

Therefore, to construct the models for complex industrial facilities, namely the sulfur production units and other oil refining installations, this paper proposes a new approach based on the study of each unit operation and consequent construction of a model on the basis of the data collected. Then, these models are combined to describe the whole process by a single system of models.

3. Comparison of simulation results and reliability evaluation

The simulation of thermal production reactor of sulfur production unit at the Atyrau Oil Refinery based on the proposed models was compared with experimental production data. The main results of the comparison are shown in Table 2.

Table 2. Results of proposed models v. experimental data of sulfur production unit at Atyrau Oil Refinery

Defined parameters	Modeling results	Experimental data
Sulfur yield [t/h]	26.0	25.3
Mass fraction of sulfur [%]	99.98	$(99.96)^{lab}$
Mass fraction of ash [%]	0.018	$(0.02)^{lab}$
Mass fraction of organic substances [%]	0.01	$(0.01)^{lab}$
Mass fraction of water [%]	0.15	$(0.18)^{lab}$

Note: input and operation parameter of the process taken are similar; $()^{lab}$ means laboratory data.

IOP Conf. Series: Materials Science and Engineering **597** (2019) 012013 doi:10.1088/1757-899X/597/1/012013

The data in the table demonstrates that the modeling results fairly well comply with real (experimental) data. The models can be used to determine qualitative indicators of the product in fuzzy conditions that cannot be determined by usual modeling methods.

Table 3. Analysis and assessment of models of sulfur production unit elements at Atyrau Oil Refinery

Main elements	Criterion	Model types			
		Deterministic	Statistical	Fuzzy	Combined
Reactors (F-001, R-001, R-002, R-003)	Availability of necessary information	2.5	4.5	4.5	5.0
	Development cost	1.0	4.0	3.5	3.0
	Accuracy	4.5	3.0	2.5	4.0
	Applicability as designed	3.5	4.0	3.5	5.0
	Suitable for bundling	4.0	3.5	3.5	3.5
	Adequacy	3.0	3.5	3.5	4.0
	Total	**18.5**	**22.5**	**21.5**	**24.5**
Condensers (boilers) (E-001, E-002, E-004)	Availability of necessary information	3.5	4.0	4.5	4.0
	Development cost	1.5	4.0	4.0	3.0
	Accuracy	4.5	3.5	3.5	3.5
	Applicability as designed	4.0	4.0	4.0	4.5
	Suitable for bundling	4.5	4.0	4.0	4.0
	Adequacy	3.5	3.5	3.5	3.5
	Total	**21.5**	**23.0**	**23.5**	**22.5**
Furnace (F-002)	Availability of necessary information	4.0	5.0	4.5	4.0
	Development cost	3.0	5.0	4.0	4.0
	Accuracy	4.5	4.5	3.5	4.0
	Applicability as designed	4.0	4.0	4.0	4.0
	Suitable for bundling	4.0	4.5	4.0	5.0
	Adequacy	4.0	4.5	4.5	5.0
	Total	**23.5**	**27.5**	**24.5**	**25.0**
Separators (D-001, D-004)	Availability of necessary information	4.5	5.0	4.0	4.5
	Development cost	3.0	5.0	4.0	2.5
	Accuracy	4.5	4.5	2.0	4.0
	Applicability as designed	4.0	4.5	4.5	4.5
	Suitable for bundling	3.5	4.0	3.5	4.0
	Adequacy	4.0	4.0	3.5	4.0
	Total	**23.5**	**27.0**	**21.5**	**23.5**
Pumps (B-001, B-002)	Availability of necessary information	4.5	4.0	4.0	4.5
	Development cost	5.0	4.5	4.0	4.0
	Accuracy	5.0	4.0	4.0	4.5
	Applicability as designed	4.5	4.5	4.0	4.5
	Suitable for bundling	4.5	4.0	4.0	4.0
	Adequacy	4.5	4.0	4.0	4.5
	Total	**28.0**	**25.0**	**24.0**	**26.0**

Note: the scale-based evaluation (ranking) from 1 to 5, where 1 is the lowest estimate, 5 is the highest one. The estimates can be fuzzy.

IOP Conf. Series: Materials Science and Engineering **597** (2019) 012013 doi:10.1088/1757-899X/597/1/012013

4. Conclusions

The novelty of the work results is conditioned by the proposed method of mathematical modeling of a chemical and technological system under uncertainty, being applicable to construct an adequate model of a complex facility consisting of a variety of interconnected units. To evaluate the practical applicability of the proposed approach, the paper has analyzed and evaluated possible models of basic sulfur production unit elements at the Atyrau Oil Refinery. A system of mathematical models for sulfur production unit reactors has been developed.

Thus, the work justifies the systematic approach to the development of a set of mathematical models for interconnected technological units of a chemical and technological system. This allows solving basic problems in construction of mathematical models and simulating operating regimes of the processing facility units under conditions of fuzzy initial information. The proposed comprehensive method of model development differs from other approaches in implementation of various information available, including fuzzy information. Different models of the studied objects were constructed that were then combined into a single system of models. This approach was implemented in the construction of the system of models for the main sulfur production units at the Atyrau Oil Refinery. The results helped identifying the structure and parameters of the mathematical models of thermal reactor F-001 and Claus reactor R-001 of the unit. Moreover, the simulation results of the thermal reactor operation were compared and comply well with the experimental and industrial plant data. Thus, the models can be used to evaluate the quality of products in fuzzy environment, which cannot be done by traditional modeling.

References

[1] Mohamed A S, Sapuan S M, Megat Ahmad M M H, Hamouda A M S and Hang Tuah Bin Baharudin 2012 Modeling the technology transfer process in the petroleum industry: Evidence from Libya. *Mathematical and Computer Modelling* **55** 451-470

[2] Orazbaev B B 2016 *Methods of modeling and decision making for production management in a fuzzy environment* (ENU im. L.N.Gumileva, Astana) p 398 [in Russian]

[3] Hatcher L 1994 *A Step-By-Step Approach to Using the SAS System For Factor Analysis and Structural Equation Modeling*, SAS Publishing

[4] Dupljakin V K 2007 Modern problems of the Russian oil refining and individual tasks of its development. *Russian Journal of General Chemistry* **4** 11-22 [in Russian]

[5] Seif Mohaddecy S R, Zahedi S, Sadighi S and Bonyad H 2006 Reactor modeling and simulation of catalytic reforming process. *Petroleum & Coal* **48 (3)** 28-35

[6] Ospanov Y A, Orazbayev B B, Orazbayeva K N, Mukataev N S and Demyanenko A I 2016 Mathe-matical modeling and decision-making on controlling modes of technological objects in the fuzzy environment *Proceedings of the Words Congress on intelligent Control and Automation, 2016* (WCICA).IEEE Catalog Number: CFP16496-ART ISBN: 978-1-4673-8414-8, pp 103-108

[7] Dvinin V A 2007 Possible scenarios for the modernization of refineries to produce high-quality fuels *Neftepererabotka i neftehimiia* **3** 12-22 [in Russian]

[8] Aliev R A, Cerkovnyj A J E and Mamedova G A 1991 Production management with fuzzy source information (Moscow: JEnergoatomizdat) p 250 [in Russian]

[9] Serikov T P, Orazbaev B B, Urazbaev K K and Kozyrev D V 2008 *Atyrau Oil Refinery new installation: Sulfur production unit* (Er-tostik: Atyrau: A-Poligrafija: Aktobe) p 135 [in Russian]

[10] Orazbaev B B, Orazbayeva K N and Utenova B E 2014 Development of Mathematical Models and Modeling of Chemical Engineering Systems under Uncertainty *Theoretical Foundations of Chemical Engineering* **48 2** 138-147

Synthesis and properties of photocatalytic active composite materials based on TiO_2 and modified by Ag and SiO_2

A A Buzaev, A O Rogacheva, O S Khalipova, A S Brichkov and V V Kozik
Faculty of Chemistry, National Research Tomsk State University, 36 Lenin Street, Tomsk 634050, Russian Federation

E-mail: buzaev92@icloud.com

Abstract. Nowadays air and water pollutions are becoming a serious problem for urban environment. One of the efficient methods of purifying water and air is photocatalysis on TiO_2 films. In this work, we obtained composite materials based on TiO_2 and modified by Ag and SiO_2 on the surface of the fiberglass material ($Ag/SiO_2/TiO_2$/FGM) by a sol-gel method. Complex studies were carried out to identify the phase compositions, morphology and textural characteristics of obtained samples by X-ray diffractometey, scanning electron microscopy, micro-X-ray spectral analyses and analysis of N_2 adsorption-desorption isoterms. Surface area and average pore size of this sample is 8.2 m^2/g and 11.4 nm respectively. Prepared composite Ag(0.5%)/SiO_2(3%)/TiO_2/FGM material revealed photocatalytic activity in model reaction of methyl orange azo dye photodegradation under UV-irradiation.

1. Introduction

Titanium dioxide (TiO_2) is one of the widely used semiconducting metal oxides. Titanium dioxide exhibits unique physicochemical properties such as close thermodynamic parameters of polymorphic structures, high strength Ti-O bonds, relatively low cost, non-toxicity, chemical stability, stable optical properties. These advantages make materials based on TiO_2 very attractive for researchers. Polyfunctional materials based on titanium dioxide are promising candidates for application as self-cleaning surfaces, solar and fuel cells, protective and optical coatings [1].

The photocatalytic activity of TiO_2 is one of its most interesting features. Photocatalysis is a recognized highly efficient method of purifying water and air, which has obvious advantages over traditional methods, including chlorination. Scientists all over the world are actively searching the effective photocatalytic materials, which can become a key to important environmental problems, including the elimination of organic pollutants contained in sewages. Nowadays filters based on TiO_2 powders are used for photocatalytic purification of water and air from toxic organic impurities [2, 3]. Thin films based on TiO_2 also can be used as photocatalyst [4]. The photocatalytic activity of TiO_2 is largely determined by properties such as specific surface area, porosity, degree of crystallinity and the ratio of anatase and rutile crystalline phase [5]. One of the route to change these characteristics and improve photocatalytic activity is realized by addition of the other metal such as Ag [6–8].

The sol-gel synthesis using film-forming solutions (FFS) is one of the recognized methods to obtain functional composite oxide films with the required properties [9]. In this technique, the properties and composition of the initial sol significantly influence the obtaining of uniform oxide coatings. In the previous work [1] was shown that fiberglass materials of silicate origin are promising materials as a substrate for deposition TiO_2 film. However, to create a homogeneous coating of TiO_2 is difficult, because fiberglass is hydrophobic. It was found, that the introduction of tetraethoxysilane into the sol composition made it possible to obtain uniform oxide coatings on substrates of various types [10, 11]. Moreover, it has been reported, that the presence of silica in SiO_2-TiO_2 materials contributes to the formation of the anatase phase [12].

In the present study, composite Ag/SiO2/TiO2 coatings on the surface of the fiberglass were prepared and thoroughly characterized. The prepared material revealed a catalytic activity in a model reaction of methyl orange oxidation under UV irradiation.

2. Experimental part

2.1. *Materials and experimental procedure.*

A fiberglass material (FGM) manufactured by JSC "Stekloplastik" (Russia) was used [1]. The composite Ag/SiO_2/TiO_2/FGM materials on the fiberglass were prepared using sol-gel method. At the first stage, to prepare the film-forming solution, three components were mixed: butanol-1 (99.9%, Ekos-1, Russia) as a solvent, bidistilled water as an initiator of alkoxide hydrolysis reaction, and HNO_3 (UralPromDelivery, Russia). After, the tetrabutoxytitanium (TBT, extrapurity, Acros, USA), (TEOS, extrapurity, Ekos-1 , Russia) and $AgNO_3$ (99.9%, LenReactive, Russia) were added to the C_4H_9OH–H_2O–HNO_3 mixture. The content of $AgNO_3$ and TEOS was based on the need to obtain composite TiO_2/FGM, Ag(0.5%)/TiO_2/FGM, Ag(1%)/TiO_2/FGM, Ag(1.5%)/TiO_2/FGM, Ag(2%)/TiO_2/FGM, SiO_2(1%)/TiO_2/FGM, SiO_2(2%)/TiO_2/FGM, SiO_2(3%)/TiO_2/FGM, Ag(0.5%)/SiO_2(3%)/TiO_2/FGM materials, respectively (in the work molar percentages are given). Freshly prepared solutions were kept at room temperature for three days. At the second stage, the deposition of film-forming solutions on the surface of fiberglass substrate was performed by impregnation technique. Thermal treatment was performed in two stages: drying at 60 °C for 1 h and calcination at 600 °C for 1 h at a heating rate of the muffle furnace 5 °C/min. The calcination temperatures of samples were chosen using previous studies [10, 11] and literature [13] data.

2.2. *Research methods.*

The morphology of the prepared samples was investigated by scanning electron microscopy (SEM) on a TM-3000 (Hitachi, Japan) operated at an accelerating voltage of 15 kV. The X-ray microanalyzer (Shift ED 3000) was used for energy-dispersive X-ray (EDX) spectroscopic analysis.

The phase composition of the samples was determined by X-ray diffraction (XRD) on a Rigaku MiniFlex 600 diffractometer (Japan) with CuKα source in the range of reflection angles (2θ) from 10 to 80°. Interpretation of the obtained XRD patterns of samples was performed using the international data bank PDF-2.

BET surface area, pore volume and pore size distribution was determined by N_2 adsorption-desorption method using a TriStarII automatic gas-adsorption analyzer and Micromeritics3Flex(USA) instrument. Prior to adsorption-desorption measurements, the samples were degassed for 2 h at 200 °C in a vacuum (10^{-2}Torr).

Prepared materials after calcination were used as photocatalyst for the model reaction of methyl orange azo dye photodegradation. Stock solution methyl orange was prepared by dissolving required amount of methyl orange in distilled water (concentration $1.5 \cdot 10^{-4}$ mol/L). For all the experiments, photocatalysts with the same size were added to 50 ml aqueous methyl orange. Then the reactor was irradiated by UV light with a radiation power of 20 mW/cm^2. All the experiments were performed in presence of air at atmospheric pressure and constant magnetic stirring condition. The concentration of the dye was monitored directly in course of the photocatalytic reaction by use photoelectric concentration colorimeter, at a wavelength of $\lambda = 463$ nm. Measurements were performed every 30 min. Resulting dependencies of dye concentration on the duration of UV-irradiation were used further as a measure of photocatalytic activity of samples.

3. Results and Discussion

The structure parametrs for all prepared materials are presented in Table 1. Pure fiberglass is a non-porous system with very low surface area (0.3 m^2/g) as reported in the previous work [1]. Deposition of TiO_2, SiO_2/TiO_2, Ag/TiO_2 and $Ag/SiO_2/TiO_2$ on fiberglass surface results in increase in pore volume.The presence of a hysteresis loop in nitrogen adsorption-desorption isotherms indicates the presence of mesopores in these samples. According to the results of the BET study all the samples have low surface area since fiberglass is the main component.

The increase of Ag content leads to decrease the surface area from 7.9 to 6.6 m^2/g, total pore volume and average size of composite $Ag/TiO_2/FGM$ materials (Table 1). The maximum of structural parameters are observed for Ag(0.5 %)/TiO_2/FGM sample. The surface area, total pore volume and average size of $SiO_2/TiO_2/FGM$ materials increases with the growth of SiO_2 content. The maximum value of structural parameters are observed for SiO_2(3%)/TiO_2/FGM sample (Table 1). According to the results of the BET study (Table 1) the composite material based on TiO_2 and modified SiO_2(3%) and Ag(0.5%) on the fiberglass was chosen for future investigation. The specific surface area of the $Ag/SiO_2/TiO_2/FGM$ sample is 8.2 m^2/g, that is 27 times higher than that of pure fiberglass.

Table 1. Structural properties of composite materials.

Composition	Average poresize, nm	Total pore volume, cm^3/g	S_{BET}, m^2/g
TiO_2/FGM	11.21	0.27	8.3
Ag(0.5%)/TiO_2/FGM	10.16	0.22	7.9
Ag(1%)/TiO_2/FGM	9.67	0.18	7.1
Ag(1.5%)/TiO_2/FGM	7.43	0.14	6.6
Ag(2%)/TiO_2/FGM	8.36	0.17	7.4
SiO_2(1%)/TiO_2/FGM	12.23	0.31	8.8
SiO_2(2%)/TiO_2/FGM	13.88	0.36	9.1
SiO_2(3%)/TiO_2/FGM	14.82	0.42	9.5
Ag(0.5%)/SiO_2(3%)/TiO_2/FGM	11.44	0.29	8.2

Figure 1 presents the XRD patterns of prepared samples. XRD patterns of all composites are characterized by intense and narrow reflexes indicating a high crystallinity of the samples. The XRD pattern of TiO_2 cleary shows well resolved and sharp peaks at 25.29° and 27.54°, 37.93°, 41.35°, 54.53° which confirm the presence of two different crystalline phases: anatase (tetragonal) and rutile (tetragonal) (PDF-2 Map No. 00-21-1276). The XRD patterns of $SiO_2/TiO_2/FGM$, $Ag/TiO_2/FGM$ and $Ag/SiO_2/TiO_2/FGM$ samples have a similar appearance (Figure 2). X-ray analysis of these samples showed the absence of signals related to anatase and the presence of the main peaks at 27.54°, 37.93°, 41.35°, 54.53° corresponding to rutile structural phase. It should be noted that for TiO_2 samples, the intensity of main signal related to rutile is 732 a.u. at an angle of $2\theta = 41.35°$. The intensity of this peak for Ag(0.5%)/TiO_2/FGM (Figure 3), SiO_2(3%)/TiO_2/FGM (Figure 4) and Ag(0.5%)/SiO_2(3%)/TiO_2/FGM composites are 956, 844 and 1197 a.u. respectively. This confirms that addition of SiO_2 and Ag contributes to formation of titanium dioxide in the rutile modification.

The XRD patterns of $SiO_2/TiO_2/FGM$, $Ag/TiO_2/FGM$ and $Ag/SiO_2/TiO_2/FGM$ samples showed no peaks that can be attributed to the silver or silicon dioxide. We suggest that SiO_2 and Ag content is below the instrument detection limit (less 5 wt. %).

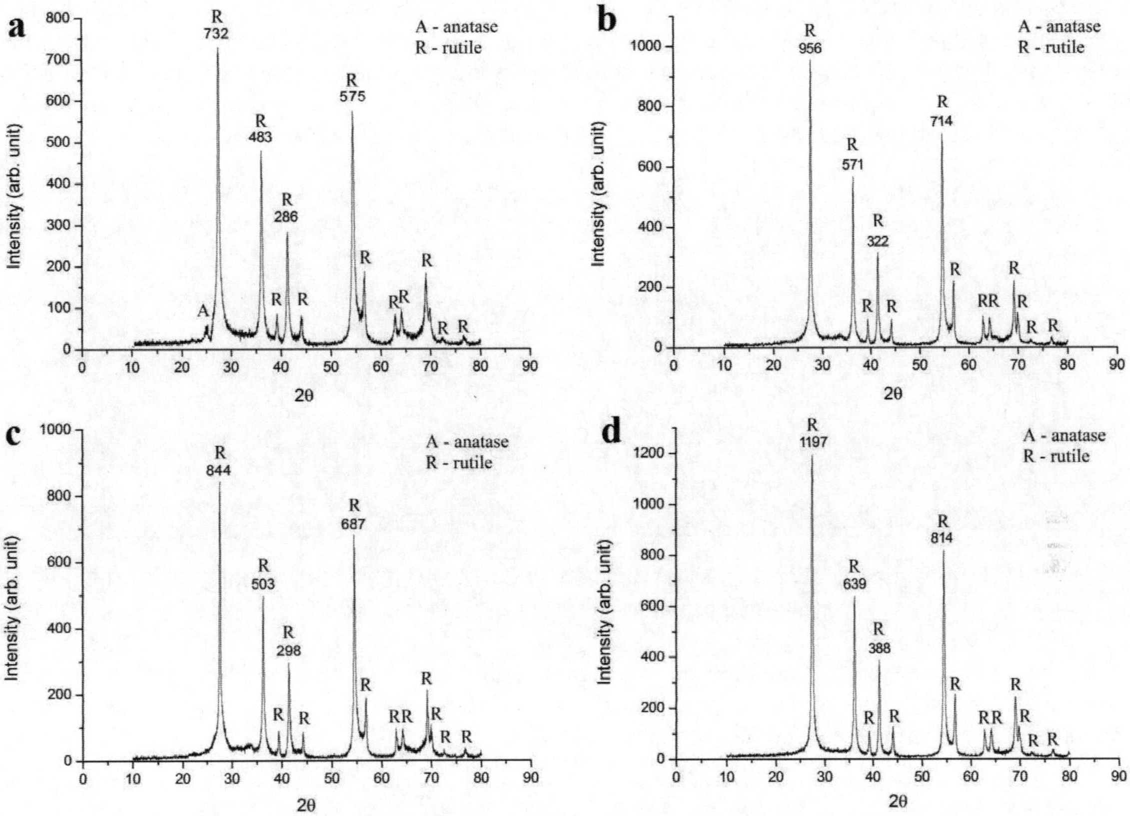

Figure 1. XRD patterns of samples:
a – TiO_2/FGM ; b – Ag(0.5%)/TiO_2/FGM ;
c – SiO_2(3%)/TiO_2/FGM ; d – Ag(0.5%)/SiO_2(3%)/TiO_2/FGM.

The results of a qualitative (Figure 2) X-ray spectral microanalysis confirm the presence of the the silver and silicon in the composition of final TiO_2/SiO_2(3%)/Ag(0.5%)/FGM sample. According to previous studies [10, 11], literature [13] and obtained results of EDX analysis, we assume the formation of Ag and SiO_2. The results of the quantitative EDX analysis show that the content of silicon compound in the obtained sample is less than 3 wt. % and silver content is less than 1 wt.%. This result agrees with the XRD data.

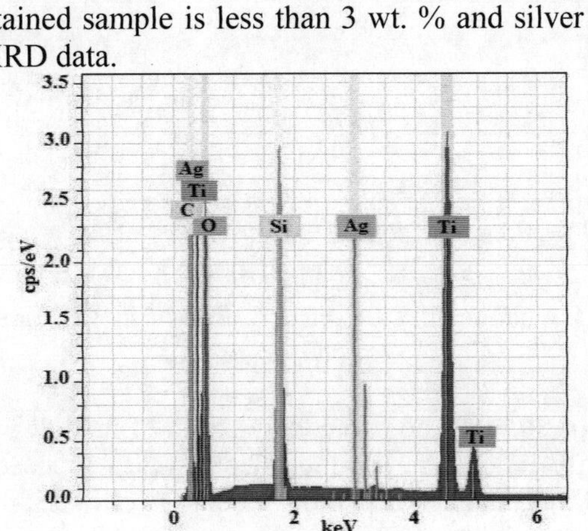

Figure 2. EDX images of Ag(0.5%)/SiO_2(3%)/TiO_2/FGM sample.

SEM images of TiO_2/FGM , SiO_2/TiO_2/FGM , Ag/TiO_2/FGM and Ag/SiO_2/TiO_2/FGM samples have a similar appearance. The morphology of the obtained composite Ag/SiO_2/TiO_2/FGM material is presented in Figure 3. It can be seen that the composite is attached to the surface of glass fibers. Each fiber is covered by Ag/SiO_2/TiO_2 films. As shown in Figure 3, Ag/SiO_2/TiO_2 film was not uniformly. The are some regions with cracks.

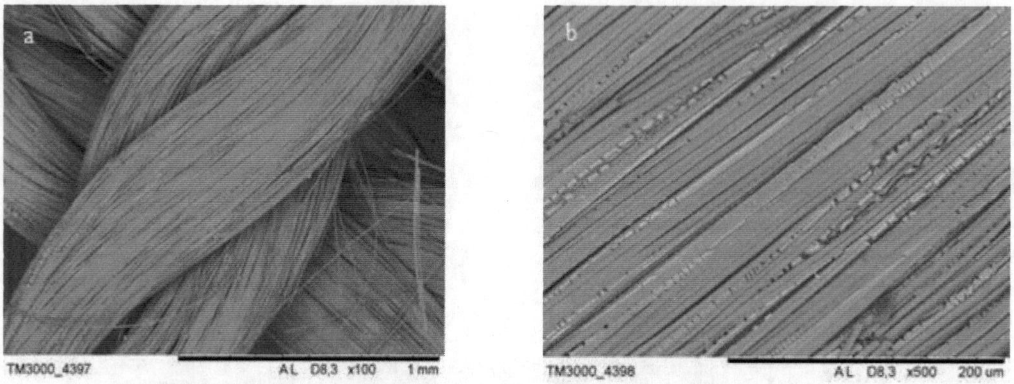

Figure 3. SEM images of Ag(0.5%)/SiO_2(3%)/TiO_2/FGM sample:
a – magnification x100; b – increase x500

Methyl orange was used as a representative organic pollutant to evaluate the photocatalytic activity of the prepared composites. Figure 6 represents the photodegradation of methyl orange as a function of irradiation time over different samples.

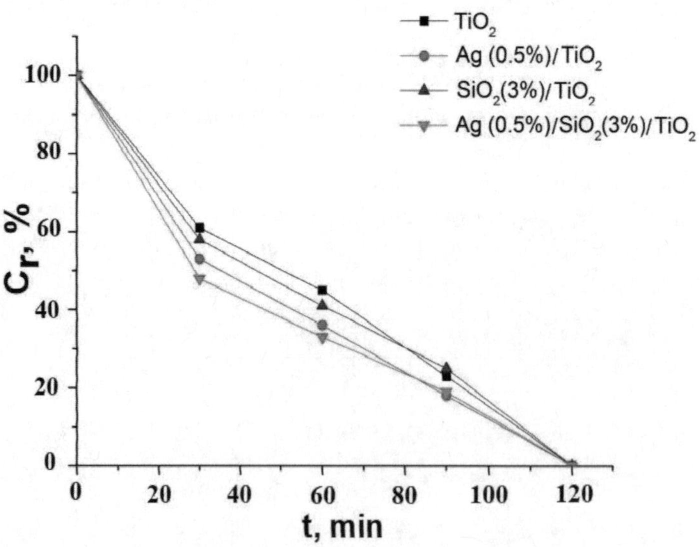

Figure 4. Changes in the concentration of methyl orange in time on photocatalysts.

After the UV irradiation for 30 min, the photodegradation of methyl orange on TiO_2/FGM sample was 40%. Photodegradation of the dye on the modified samples proceeds more efficiently. For the Ag(0.5%)/SiO_2(3%)/TiO_2/FGM sample the photodegradation efficiency reached 50% for the same period of illumination. Solutions were completely discolored after the irradiation for 2 hours.

4. Conclusion

The immersion of fiberglass material in the film-forming solution prepared by tetrabutoxytitanium, tetraethoxysilane, silver nitrate, and butanol, and subsequent heat treatment resulted in formation of the composite $Ag/SiO_2/TiO_2/FGM$ materials. The formation of TiO_2 phases and presence of the Ag and SiO_2 phases were confirmed by X-ray powder diffraction data and qualitative X-ray spectral microanalysis. XRD confirmed that the addition of SiO_2 and Ag contributes to formation of titanium dioxide in the rutile modification. It was shown that each glass fiber was covered by composite $Ag(0.5\%)/SiO_2(3\%)/TiO_2/FGM$ film, which led to an increase in the total surface area of material from 0.3 to 8.2 m^2/g, and the appearance of pores with an average size of 11.4 nm. Prepared composite $Ag(0.5\%)/SiO_2(3\%)/TiO_2/FGM$ materials revealed photocatalytic activity in model reaction of methyl orange azo dye photodegradation under UV-irradiation and showed the best degradation effect among $Ag/TiO_2/FGM$, $SiO_2/TiO_2/FGM$ and unmodified TiO_2/FGM samples. Obtained results confirm promising use of composite $Ag/SiO_2/TiO_2/FGM$ materials as photocatalyst.

Acknowledgments

The results were obtained within the framework of the state task of the Ministry of Education and Science of the Russian Federation, project No.10.2281.2017/4.6.

References

[1] Shamsutdinova A N, Brichkov A S, Paukshtis E A, Larina T V, Cherepanova S V, Glazneva T S and Kozik V V 2017 *Catalysis Communications* **89** 64–68
[2] Sadovnikov A A, Baranchikov A E, Zubavichus Y V., Ivanova O S, MurzinV Y, Kozik V V and Ivanov V K 2015 *Journal of Photochemistry and Photobiology A: Chemistry* **303-304** 36–43
[3] Selishchev D and Kozlov D 2014 *Molecules***19** 21424–41
[4] Gerasimenko Yu V, Logacheva V A and Hoviv A M 2010 *Condensed media and interphase boundaries* **12** 113–118
[5] Kittisakmontree P, Yoshida H, Fujita Sh, Arai M and Panpranot J 2015 *Catalysis Communications* **58** 70–75
[6] Deshmanea V G, Owena S L,Abrokwahc R Y and Kuila D 2015 *Journal of Molecular Catalysis A: Chemical* **408** 202–213
[7] Wang X, Xue J, Wang X and Liu X 2017 *PLOS ONE* **12** 1–16
[8] Ahmad W, Noor T and Zeeshan M 2017 *Catalysis Communications* **89** 19–24
[9] Vinogradov A V, Vinogradov V V, Gerasimova T V and Agafonov A V 2012 *Journal of Alloys and Compounds* **543** 172–175
[10] Kozik V V, Brichkov A S, Brichkova V Y, Tolbanov O P, Ivanov V K, Ogorodnikov V D and Tretyakov Y D 2012 *Dokl Chem* **445** 155–158
[11] Khalipova O S, Kuznetsova S A and Kozik V V 2014 *J Inorg Chem* **59** 913–917.
[12] Houmard M, Riassetto D, Roussel F, Bourgeois A, Berthome G, Joud JC and Langlet M 2007 *Appl Surf Sci* **254** 1405–1414
[13] Vahryshev A Yu, Gorbunova V V, Boytsova T B and Starogarov V M 2016 *Russian Journal of General Chemistry* **86** 603–608

IOP Conf. Series: Materials Science and Engineering **597** (2019) 012015 doi:10.1088/1757-899X/597/1/012015

Processing of heavy residual feedstock on Mo/Al_2O_3-catalytic systems obtained using polyoxomolybdate compounds

A S Akimov[1,2], N N Sviridenko[1], M A Morozov[1], T V Petrenko[1], S P Zhuravkov[3], S O Kazantsev[4] and S V Panin[4]

[1]Institute of Petroleum Chemistry, Siberian Branch, Russian Academy of Sciences, Russia, Tomsk, Akademichesky Av., 4
[2]National Research Tomsk State University, Russia, Tomsk, Lenin Av., 36
[3]National Research Tomsk Polytechnic University, Russia, Tomsk, Lenin Av., 30
[4]Institute of Strength Physics and Materials Science, Siberian Branch, Russian Academy of Sciences, Russia, Tomsk, Akademichesky Av., 2/4

Email: zerobox70@mail.ru

Abstract. The urgency of creating new efficient catalysts for the processes of deepening oil refining rises on the background of stricter requirements for the quality of motor fuels, as well as the deterioration of the quality of crude oil for processing, and an increase in the number of distillates of secondary processes involved in the production of commodity petroleum products. In this work, alumina-catalytic systems were synthesized using polyoxomolybdate compounds. The morphology, structure and phase composition of the synthesized catalytic systems were studied using the following analysis methods: scanning electron microscopy, microelement analysis, X-ray phase analysis, X-ray diffraction, electron spectroscopy. It has been established that the Mo / Al_2O_3 system is active in the process of thermal catalytic conversion of heavy residual raw materials.

1. Introduction

The world oil refining is currently characterized by a reduction in light and medium oil reserves, an increase in the share of extraction and processing of heavy oil and sour crude oil (total sulfur content is about 0.6-3.5% by mass), environmental protection requirements for oil products quality are becoming tougher [1-5]. These factors necessitate the restructuring and modernization of the oil refining industry, the construction of new facilities with an increase in refining depth and a reduction in fuel oil consumption, as well as the intensification of the production of high-quality motor fuels. Increasing demand for motor fuels of improved quality is applied by the development and implementation of technologies aimed at the deep processing of oil residues and the maximum removal of sulfur-containing compounds. Leading global companies rely on complex technology for processing heavy hydrocarbons using both thermal (thermal cracking, visbreaking, delayed coking) and catalytic (catalytic cracking, hydrocracking) processes. Using this approach allows to proceed the process with high flexibility and obtain products of the required quality [5]. At the same time, there are also disadvantages: the need for a fairly high investment, the presence of a hydrogen production, harsh technological conditions (high temperatures and pressures, high ratio of hydrogen to feedstock). Thus, the development of processes

IOP Conf. Series: Materials Science and Engineering **597** (2019) 012015 doi:10.1088/1757-899X/597/1/012015

for the processing of heavy hydrocarbon feedstock into motor fuels on existing equipment or with minimal reconstruction costs is an urgent task.

Polyoxometalates of molybdenum (polyoxomolybdates) belong to a very promising class of materials. They are very interesting from theoretical and practical points of view. The important properties of polyoxometalates include: the possibility of synthesizing various giant clusters (due to the presence of the Mo–O–Mo bond and the possibility of changing the coordination numbers from 4 to 7), high reactivity (due to the presence of metal centers of various electronic forms), good solubility (the presence of a large number of H_2O ligands) [6–9]. The above properties are found in molybdenum blue, which is a mixture of various oxo compounds of molybdenum, the oxidation state (Mo) of which is between +5 and +6. Along with the properties characteristic of polyoxometalate compounds, molybdenum blue has a number of distinctive features: monodispersity, extremely small particle size, structural analogy of fragments with catalytically active metal oxides. Despite the fact that molybdenum blue has been known for almost 200 years, the possibility of creating deposited materials (for example, catalysts/catalytic systems) on that basis is practically not considered.

Carriers of catalysts and catalytic systems are the most important components of the efficiency of their work. The role of the carrier is to create the dispersion of the active phase, mechanical strength, thermal stability; moreover, the carrier itself can catalyze some reactions of petroleum fractions. The absolute leader among carriers of hydroprocess catalysts is γ-Al_2O_3; other carriers, such as carbon materials, ZrO_2, SiO_2, TiO_2, did not find industrial application in heavy oil refinery. Therefore, the improvement of the properties of γ-Al_2O_3 is in the center of attention of scientists and industrialists.

The purpose of this work is to synthesize and study the activity of catalytic alumina systems obtained using polyoxomolybdate compounds in the processing of heavy hydrocarbon feedstock.

2. Experimental part

Synthesis of polyoxomolybdate compounds (molybdenum blue) was carried out according to the procedure described in detail in [10] by reducing mechanocomposite (obtained by mechanical activation of natural molybdenite in a planetary ball mill) with organic alcohol. γ-Al_2O_3 obtained by heat treatment of commercial pseudoboehmite (KNT group) was used as a carrier of the catalytic system. The application of molybdenum blue was carried out using traditional methods for obtaining supported catalytic systems by impregnating the support with an alcohol solution. The physicochemical properties of the samples were studied using the following instrumental methods: scanning electron microscopy with the possibility of microanalysis of the surface, X-ray analysis (X-ray diffraction and X-ray phase analysis), optical spectroscopy. Testing of the obtained catalytic system during the processing of heavy residual feedstock was carried out in a reactor-autoclave in a static mode under the following conditions: loading of feedstock - 5 g, temperature 420 °C, duration 1.5 hours. As feedstock for testing the synthesized Mo/Al_2O_3 system was used goudron (Novokuibyshevsk refinery) with the following characteristics: density - 0.9870 g/cm^3, sulfur content - 3.04 % wt., the initial boiling point - 343 °C, the ratio H/C = 1.56, fractional composition: 343 - 350 °C - 8.7%; 350 °C and higher - 91.3%. The activity of the catalytic system in the process of thermal catalytic conversion of heavy residual feedstock was evaluated by the content of light fractions (start of boiling - 350 °C) in the liquid products of the process.

3. Discussion of the results

Figure 1 shows electron micrographs of pseudoboehmite (KNT group) and γ-Al_2O_3 obtained on its basis. The morphology of both samples is a particle of spheroid (or close to it) shape.

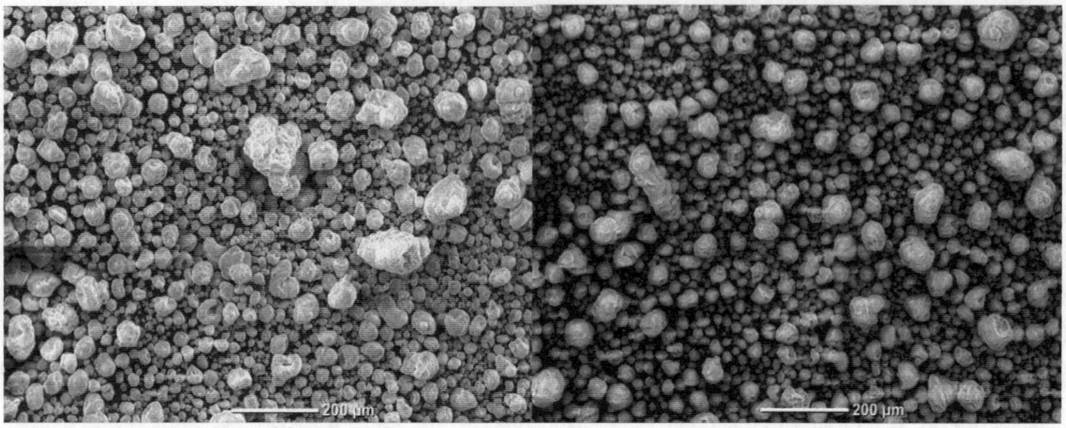

(a) (b)

Figure 1. Electronic micrographs of samples
a - initial pseudoboehmite; b - aluminum oxide (γ-Al_2O_3).

A micrograph of the catalytic system synthesized by the traditional method (impregnation method) is shown in Figure 2. The shape of the carrier particles (γ-Al_2O_3) did not change significantly. Thus, the morphology of particles in the series "pseudoboehmite → γ-Al_2O_3 → catalytic system" practically does not change.

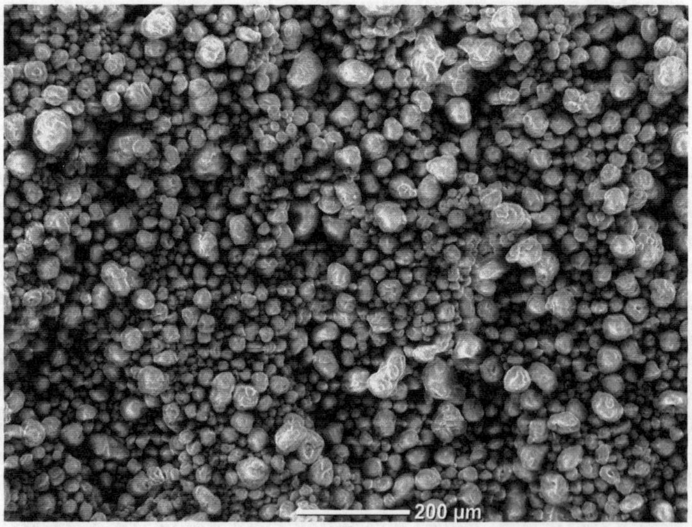

Figure 2. Typical electron micrograph of the catalytic system.

The particle size distribution (Figure 3) of the carrier and catalyst system is different. This is reflected in the difference in average particle sizes — 59 µm for γ-Al_2O_3 and 42 µm for the catalytic system. Particle size distribution is also different. The catalytic system is characterized by a narrower distribution of particles (0...50 µm) than the carrier (0 ... 200 µm). It should be pointed out that the maximum of the catalytic system is in the range of 0...20 µm, and the maximum of the carrier is 0...50 µm.

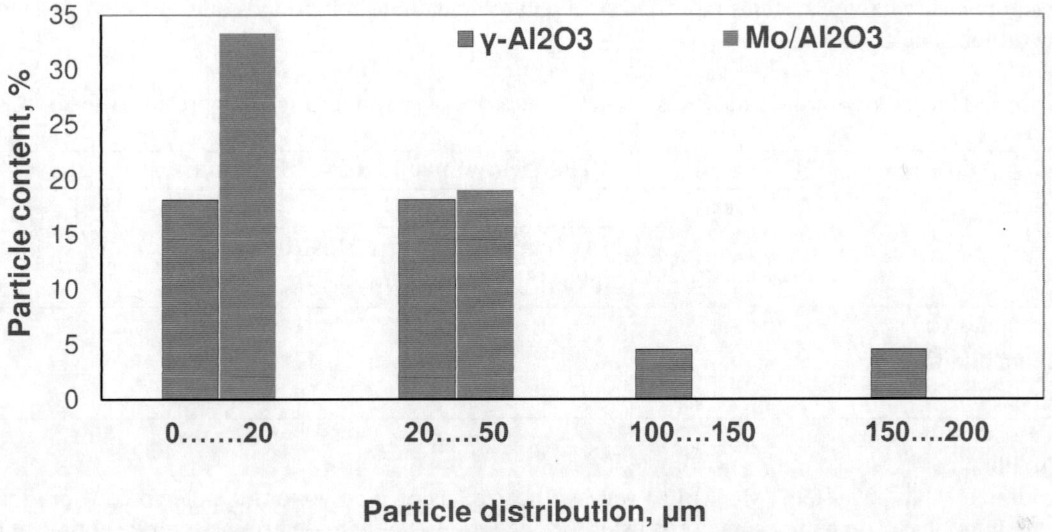

Figure 3. The particle size distribution of the samples.

Figure 4 shows, as an example, x-ray diffractograms of pseudo-boehmite (KNT group), aluminum oxide (γ-Al_2O_3) on its basis, and the catalytic systems. The pseudoboehmite diffractogram contains a set of well-defined characteristic reflexes. In the diffraction patterns of γ-Al_2O_3 and the catalytic system based on it, wide reflections are observed in the region of angles 2Θ 35-40 ° and 58-64 °. In addition to the wide reflections in both samples, the reflections characteristic of the relatively well-crystallized γ-Al_2O_3 phase of the cubic modification are fixed. The reflections specific to molybdenum containing phases (MoO_2, MoO_3, MoS_2) have not been found. Probably, it has to do with their small dispersion and low content.

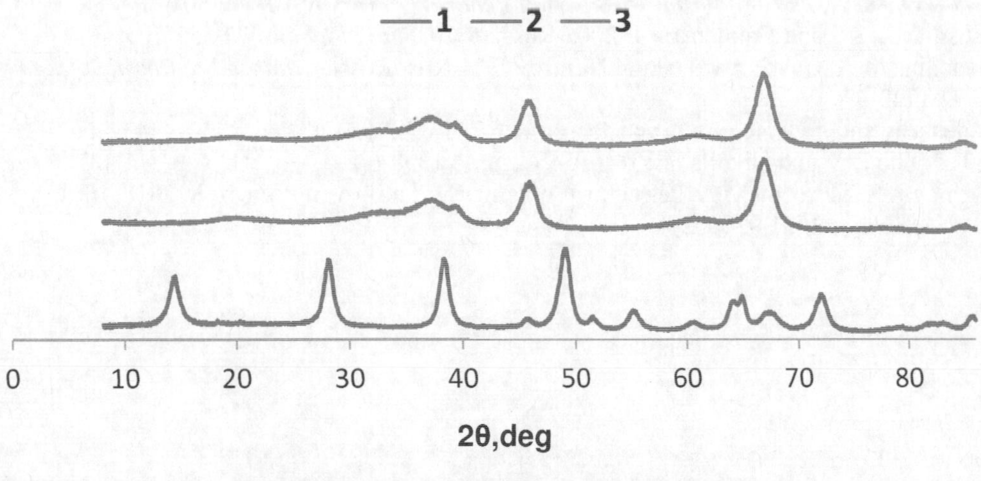

Figure 4. X-ray diffraction patterns of samples.
1 - initial pseudoboehmite; 2 - aluminium oxide; 3 - catalytic system.

Testing (table 1) of the obtained catalytic systems showed their activity in the process of processing heavy hydrocarbon feedstock - the conversion of the residual fraction (temperature of boiling over 350 °C) was about 68 % and the yield of light fractions was about 66%wt. Thus, it is shown that it is

IOP Conf. Series: Materials Science and Engineering **597** (2019) 012015 doi:10.1088/1757-899X/597/1/012015

promising to use catalytic systems modified with polyoxometallate compounds during the processing of heavy oil feedstock.

Table 1. The activity of the catalytic system in the process of converting heavy residual feedstock.

Additive	The yield of products,% of the mass.			
	Gas	Liquid		Solid
		Light fractions (up to 350 °C)	Residue (over 350 °C)	
Initial goudron	-	9	91	-
Without additive	7	46	42	5
Mo/Al_2O_3	7	66	19	9

4. Conclusion

The alumina catalytic system (Mo/Al_2O_3) was synthesized using a polyoxomolybdate precursor – molybdenum blue. The main characteristics of the catalytic system are determined: phase composition, morphology, particle size and activity in the processing of heavy hydrocarbons.

Acknowledgment

The work was performed as part of the basic project of the Russian Academy of Sciences No. 0370-2018-0004.

References

[1] Hart A, Greaves M and Wood J 2015 *Chem. Eng. J.* **282** 213-223
[2] Upreti S R, Lohi A, Kapadia R A and El-Haj R 2007 *Energy Fuels* **21** 1562-74
[3] Go S K, Lim S H, Kim Y K, Kwon E H and Nho N S 2018 *Catalysis Today* **305** 92-101
[4] Ortiz-Moreno H, Ramirez J, Sanchez-Minero F, Cuevas R and Ancheyta 2014 *Fuel* **130** 263-272
[5] Sosnin G A, Larichev Y V, Eletskii P M, Zaikina O O, Kukushkin R G, Gerasimov E Y, and Yakovlev V A 2017 *Journal of Siberian Federal University* **10** 489-501
[6] Talismanov S S and Eremenko I L 2003 Russian *Chemical Reviews* **72** 555-569
[7] Myachina M A, Gavrilova N N and Nazarov V V 2018 Russian *Journal of Physical Chemistry A* **92** 1743-47
[8] Neuenschwander U, Negron A and Jensen K F 2013 *J. Phys. Chem. A* **117** 4343-51
[9] Ju F, Velde D V and Nikolla E 2014 *ACS Catal.* **4** 1358-64
[10] Akimov A S, Morozov M A, Petrenko T V Panin S V and Vosmerikov A V 2018 *AIP Conference Proceedings* **2051** 020002-4

Petroleum Resin Influence on the Relaxation Process of Petroleum Paraffin Solutions Treated in Ultrasound

G I Volkova[1, 2] and A V Morozova[2]

[1]Tomsk State University, 36, Lenin Avenue, Tomsk, 634050, Russia
[2]Institute of Petroleum Chemistry Russian Academy of Sciences, Siberian Branch, 4, Akademichesky Avenue, Tomsk, 634055, Russia

E-mail: galivvol@yandex.ru

Abstract. This paper summarizes the experimental work done on the effect of petroleum resins on the process of relaxation the structural and rheological properties of petroleum paraffin solutions treated in ultrasonic field.
A combined treatment of petroleum paraffin solutions including 22 kHz ultrasonic exposure and addition of petroleum resins (0.3% 0.6% 0.9% by weight) was carried out. The addition of petroleum resins into the treated solution depresses the crystallization of hydrocarbons and favours the reduction of structural-mechanical characteristics of the system. The transition of the excited state of systems after the removal of the ultrasonic load to an equilibrium state occurs in time, and the nature of the change in the rheological characteristics and microstructure of paraffin crystals depends on the composition of the dispersion medium. It was revealed that trace amounts of resins introduced into paraffin oil systems treated with ultrasound will reduce their viscosity and improve transportation of hydrocarbons for 2 days at minimal cost.

1. Introduction

The increase in energy consumption worldwide and the decline of global light-oil reserves necessitates more attention to paraffinic oil reservoirs. Crude oils with a high content of high molecular n-alkanes and low content of resin-asphaltene components have high pour point. Lowering the temperature of the crude oil leads to a significant increase in viscosity due to formation of a dispersed phase. Formation of the dispersed phase is mainly associated with the presence of a multicomponent mixture of paraffin containing carbon atoms in the range of C_{18}-C_{65} in oils. Such formations have a solid gel-like structure formed by three-dimensional (3D) network of interconnected paraffin crystals and occluded dispersion medium. Novel methods and techniques are of great importance to remove barriers and increase production from such reservoirs. Developing new methods in the field of heavy oil is an issue of great importance, as the large volume of these reserves can help meet the world's ever-increasing energy needs.

Currently, the possibility of using ultrasound technologies for the oil industry needs is being widely studied. Theoretical and experimental studies were successfully carried out in the field of ultrasonic cavitation. These studies allowed the development of new technological processes occurring under the influence of ultrasound [1-5]. Ultrasonic treatment (UT) is one of the most environmentally friendly methods of influencing oil disperse systems. Also worth noting are the advantages such as ease of operation, high manufacturability and low cost.

A review of scientific studies has shown that the structural and mechanical properties of different components of crude oils after treatment in an ultrasonic field do not change unambiguously. In this regard, to predict the properties of crude oil systems after ultrasound exposure, it is necessary to evaluate the effect of oil components on their behaviour.

It is known that the process of hydrocarbons crystallization in petroleum in the presence of resins is slowed down. Resins are natural depressors of viscosity-temperature parameters and inhibitors of sedimentation. Oil resins are high-molecular condensed heterocyclic systems, in the hydrocarbon part of which aromatic and naphthenic cycles, paraffin chains are contained. The resins form fairly strong associative complexes with an ordered structure while interacting with paraffin hydrocarbons, thereby preventing the aggregation of crystal nuclei [6].

Identifying of the behaviour of different component composition of highly paraffin systems in an ultrasonic field can be simplified by using model systems.

The purpose of this work is to study the effect of petroleum resins on the process of relaxation of petroleum paraffin solutions after ultrasonic treatment.

2. Experimental methods and materials

Resins were isolated from a highly resinous crude oil via liquid-adsorption chromatography. At first, asphaltenes were extracted from oil in a 40:1 excess of n-hexane and then the deasphalted oil (DAO) was separated into saturates and resins by chromatographic separation. DAO was charged to silica gel and extracted with solvents of increasing polarity. Silica gel activation proceeded at 250 °C for 2 h. The silica gel-DAO slurry placed in vacuum oven at 50°C for 24 h. Chromatography columns (1 x 50 cm with 150 ml solvent reservoir) were initially filled with a mixture of 95:5 hexane:toluene (v/v). Specially cleaned activated silica gel was added until the depth reached ~ 10 cm. Finally, silica gel with adsorbed DAO was transferred to the column until full. A solvent mixture contained 95% (v/v) hexane and 5% (v/v) toluene eluted saturates, mono-, di-, and triaromatics from the silica gel. Once the saturates and aromatics were extracted, a more polar solvent (50:50 toluene:ethanol) was applied to elute the resins. The resin-solvent mixture was filtered to remove any silica gel fines and rotary evaporated until dry. The resins were transferred to vials and placed in vacuum oven at 60 °C for 48 h or until completely dry.

Acoustic treatment of the samples was carried out for 10 min in a thermostatted cell using an UZDN ultrasonic disintegrator at an operating frequency of 22 ± 1.65 kHz, radiation intensity 18 W/cm^2.

The effect of resins and acoustic treatment on structural and mechanical properties was investigated using 6 wt.% petroleum paraffin solution in decane (PP-d).

The rheological parameters of the solutions were determined using a HAAKE Viscotester iQ rotational viscometer.

The process of changing the structure of paraffin particles was investigated by optical microscopy on an AXIO LAB.A1 Carl Zeiss microscope in transmitted light.

3. Results and Discussion

Petroleum resins isolated from petroleum were investigated in order to reveal the structural features of resins capable of decreasing viscosity values (figure 1). Structural-group parameters of petroleum resins were calculated based on the data of molecular weight, elemental analysis, and proton magnetic resonance spectroscopy. Structural and group characteristics of the petroleum resins under study are given in the article [7].

IOP Conf. Series: Materials Science and Engineering **597** (2019) 012016 doi:10.1088/1757-899X/597/1/012016

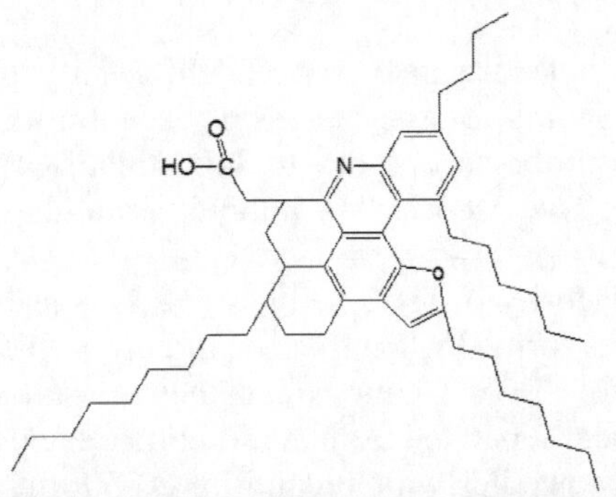

Figure 1. Structural formulas of petroleum resins.

Studies have shown that after ultrasonic treatment of the PP-d solution for 10 min, the viscosity increases 1.6-1.8 times, which is due to the intensive structuring of the system after removing the ultrasonic load. The viscosity of the treated sample reaches a maximum within two hours, and then relaxes to its initial value in 5 days (figure 2a).

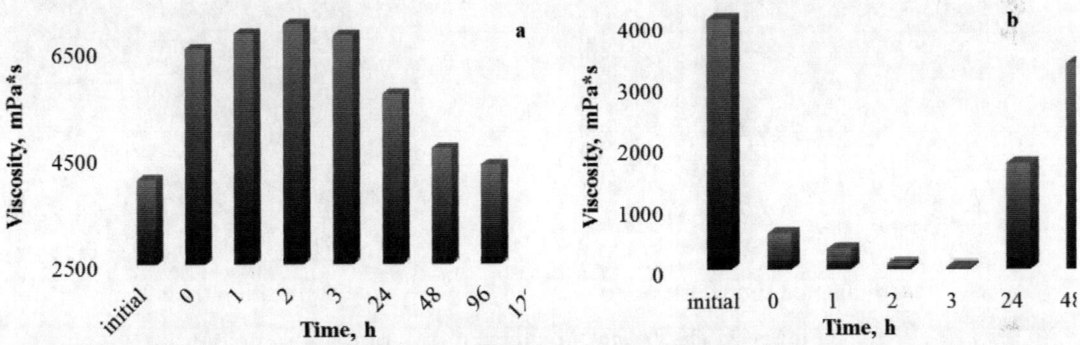

Figure 2. Effective viscosity of relaxing PP-d solutions: a - UT; b - UT + 0.3 wt.% resins.

Table 1 The effect of the petroleum resins concentration on the effective viscosity of the relaxing PP-d solution after ultrasonic treatment.

Concentration, wt. %	Relaxation time, h	Viscosity, mPa.s*
0	0	4085
0.3	0	590
	1	341
	2	96
	3	34
0.6	0	10
	1	8
	3	8
0.9	0	5
	1	4
	3	4

*shear rate 1 s^{-1}

A study of the effect of resins concentration that was injected into the solution of PP-d after ultrasonic treatment (table), on the viscosity of the relaxing system was made. Complex treatment in the presence of 0.3, 0.6 and 0.9 wt.% resins leads to a decrease in viscosity of 7, 408 and 817 times respectively.

IOP Conf. Series: Materials Science and Engineering **597** (2019) 012016 doi:10.1088/1757-899X/597/1/012016

Relaxation processes after complex treatment proceed differently in comparison with only ultrasonic treatment. The viscosity of the relaxing solution at a resin concentration of 0.3 wt.% continues to decline for 3 hours, and then increases on the second day to 3400 mPa·s, which, however, is lower than the initial value by 18 % (figure 2b). The viscosity of solutions containing 0.6 and 0.9 wt.% resins does not change for 3 hours.

The presented data on the viscosity of dispersed systems is possibly related to changes in the size and shape of the crystal structures formed when the temperature of the solution decreases.

Precipitation was isolated from PP-d solutions and their crystal structure was investigated using optical microscopy. In the sediment separated from the solution treated in the acoustic field (P_1), paraffin crystallizes in two modifications: needles with a length of 20-50 μm and spherulite formations with a diameter of 22-50 μm (figure 3). After relaxation for 24 hours, rhombic lamellar crystallites with a longitudinal diagonal of 40-80 μm and with a transverse diagonal of 30-65 μm are formed, and parallel-oriented needle-like crystals and spherulites of larger size are present in comparison with similar structures in the freshly treated solution.

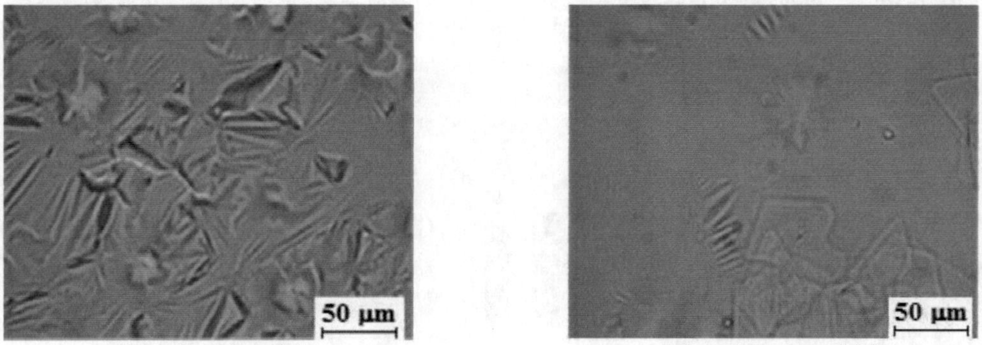

relaxation time 0 h relaxation time 24 h

Figure 3. The crystal structures micrographs of sample P_1.

After complex treatment of the PP-d solution, a precipitate is formed (P_2), the microstructure of which is represented by single crystals (figure 4). In the process of relaxation for 24 hours, the precipitate is recrystallized with the formation of needle-like crystalline formations with a length of 20-60 μm.

relaxation time 0 h relaxation time 24 h

Figure 4. The crystal structures micrographs of sample P_2.

4. Conclusion

Thus, the transition of the excited state of systems after the removal of the ultrasonic load to an equilibrium state occurs in time, and the nature of the change in the rheological characteristics and microstructure of paraffin crystals depends on the composition of the dispersion medium. The

IOP Conf. Series: Materials Science and Engineering **597** (2019) 012016 doi:10.1088/1757-899X/597/1/012016

introduction of resins in a paraffin-containing dispersed system, treated with ultrasound, leads to a decrease in viscosity after removing the acoustic load, and the depressant effect has prolonged nature. A significant effect of resins on the relaxation process is due to the inhibition of crystallization processes in paraffin-containing systems, which prevents the formation of a continuous crystalline network. Trace amounts of resins introduced into paraffin oil systems treated with ultrasound will reduce their viscosity. In this case, hydrocarbons are transported for 2 days at minimal cost.

Acknowledgments
This work was supported by the Ministry of Science and Higher Education of the Russian Federation (Project No. AAAA-A17-117030310198-4).

References
[1] Wang Z, Zeng J, Song H and Li F 2017 *Ultrason. Sonochem.* 36 162
[2] Wang Z and Xu Y 2015 *Energy* 89 259
[3] Anufriev R V and Volkova G I 2014 *Prospects of fundamental sciences development (Tomsk)* (Tomsk: TPU) p 281
[4] Mousavi S M, Ramazani A, Najafi I and Davachi S M *Petroleum Science* 2012 9(1) 82
[5] Salehzadeh M, Akherati A, Ameli F and Dabir B *Can. J. Chem. Eng.* 2016 94(11) 2202
[6] Beshagina E V, Yudina N V, Krutey A A and Loskutova Y V *Procedia Chemistry* 2014 10 229
[7] Morozova A V and Volkova G I 2018 *Problems of Geology and Subsoil Development* (Tomsk) vol 2 (Tomsk: TPU) p 337

IOP Conf. Series: Materials Science and Engineering **597** (2019) 012017 doi:10.1088/1757-899X/597/1/012017

The Study of Donor-Acceptor Equilibria in Oil-Displacing Compositions Based on Acid Coordinating Solvents

L K Altunina[1,2], V A Kuvshinov[1], L A Stasieva[1] and I V Kuvshinov[1]
[1]Institute of Petroleum Chemistry, Siberian Branch of the Russian Academy of Sciences (IPC SB RAS), 4, Akademichesky Avenue, 634055, Tomsk, Russia
[2]National Research Tomsk State University, 634050 Tomsk, Lenin Avenue, 36

E-mail: alk@ipc.tsc.ru

Abstract. The acidity of donor-acceptor complexes of boric acid with glycerol and sorbitol has been studied in aqueous solutions using pH-metry and electrical conductivity methods. A theoretical model of donor-acceptor acid-base equilibria in boric acid – polyol – water systems has been created. The constants of polyol boric acids formation and ionization were determined. The constants of glycerol and sorbitol boric acids formation were 0.05 and 0.06 $l^2 \cdot mol^{-2}$, and those of glycerol and sorbitol boric acids ionization were $7.1 \cdot 10^{-6}$ $mol \cdot l^{-1}$ and $3.2 \cdot 10^{-3}$ $mol \cdot l^{-1}$, respectively. The systems studied were used to create new oil-displacing compositions for enhanced oil recovery.

1. Introduction

Currently, the actual scientific and technical problem is the development of physicochemical methods to increase the oil recovery of heavy high-viscosity oil deposits [1, 2]. These deposits are predominantly located in carbonate reservoirs. Among physical-chemical methods of influencing carbonate oil-saturated reservoirs with the aim of increasing oil recovery, the most widely used is injection of solutions of acid oil-displacing compositions. To improve constantly the compositions, make them more complex and extend a range of their useful functions is a general trend. Along with individual acids, complex acids, which are formed as a result of donor-acceptor interaction of the components of the composition are used. The composition and properties of complex acids are regulated by the concentrations and ratios of the donor and acceptor. Complex acids have surface-active properties and can increase the oil recovery of a low-permeability matrix of the reservoir at the same time due to dissolution and capillary counter-current impregnation of fractured-porous rock.

In this work, the acidic properties of complex acids based on boric acid and coordination solvents – polyatomic alcohols (polyols), such as glycerol and sorbitol, are investigated. Inorganic polyacid-polyol systems form complex acids, which are much stronger than the initial acid, due to donor-acceptor interaction [3, 4]. Donor-acceptor interaction enhances the acidic properties of oil-displacing compositions and increases the duration of their action in the reservoir by increasing the buffer capacity and expanding the range of the buffer action in the acidic pH range. Donor-acceptor interaction proceeds in the medium of an aqueous solution of a polyol (polyhydric alcohol) – glycerol and sorbitol. Such a solution is a coordinating solvent the polyol in it is a Lewis base, an electron pair donor. Dissolved in a coordinating solvent, Lewis acid – boric acid – as well as aqua ions of some metals: calcium, magnesium, iron, and aluminum – are acceptors of the donor electron pair. A donor-acceptor chemical bond has the properties of a polarized covalent bond and is called a coordination or dative bond. The interaction of the donor and the acceptor leads to the formation of a donor-acceptor molecular complex called the coordination compound or adduct. The complex is a much stronger acid than the original Lewis acid. This fact aroused interest in Russia and abroad in the use of complex acids in physical-chemical technologies for enhancing oil recovery.

2. Donor-acceptor equilibria in solutions

Molecules of the complex acid are able to interact with metal aqua ions due to their hydroxyl alcohol groups. Figure 1 shows the reaction scheme, which reflects the stereochemical feature of the glycerol boric acid molecule – its ability to form soluble outer-sphere cyclic complexes with metal ions due to terminal hydroxyl groups. When the concentration of metal aqua ions in solution increases, along with cyclic structures, the formation of polymer-like associates is possible, in which metal aqua ions play the role of bridges connecting the complex acid molecules to linear and branched spatial associative structures. As a rule, such a structure rotation leads to a significant increase in viscosity. The method of regulating the viscosity and density by adding salts of these metals can be useful for regulating the physicochemical and rheological properties of the compositions. In addition, this interaction contributes to the compatibility of complex acids based on polyols with formation waters, especially with highly mineralized, with a high content of calcium and magnesium salts.

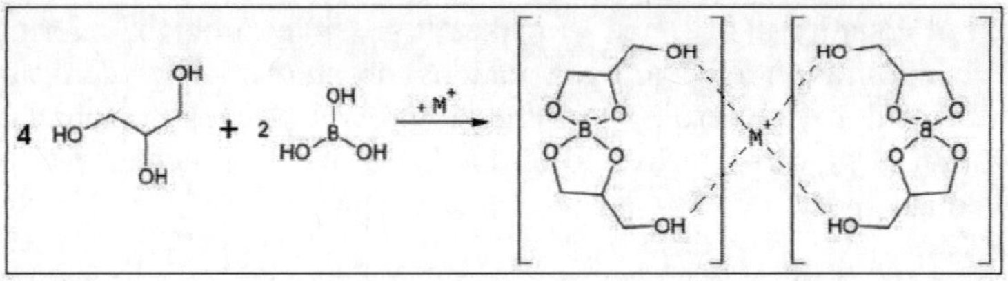

Figure 1. Diagram of the interaction of glycerol boric acid with a metal cation in solution with the formation of a soluble outer-sphere cyclic complex.

The quantitative characteristic of donor-acceptor interactions in solutions of polybasic acids and polyols are the complexation and ionization (dissociation) constants. Knowing of the constants allows one to calculate the acidity of oil-displacing compositions, their buffer capacity, to predict and model their behavior under reservoir conditions, in particular, their interaction with reservoir water and reservoir rock. The theoretical aspects of the calculation of the constants for the formation and ionization of donor-acceptor complexes of polybasic acids with polyols are summarized below using experimental data obtained by methods of pH-metry and electrical conductivity. The presentation is conducted on the example of glycerol boric acid. Calculations are given for all investigated acids and aqua metals.

Boric acid, when dissolved in an aqueous glycerol solution, almost instantaneously forms glycerol boric acid, dissociating into monobasic acid ions. The solution is established dynamic equilibrium of two consecutive stoichiometric reactions:

- formation of glycerol boric acid:
$H_3BO_3 + 2\ C_3H_8O_3 = C_6H_{12}O_6BH + 3\ H_2O$;
- ionization, acid dissociation of glycerol boric acid:
$C_6H_{12}O_6BH = C_6H_{12}O_6B^- + H^+$.

In symbolic form, these reactions can be represented as:
$HB + 2\ G = HBG_2$,
$HBG_2 = BG_2^- + H^+$,
where HB is boric acid, G is glycerol, HBG_2 is glycerol boric acid.

IOP Conf. Series: Materials Science and Engineering **597** (2019) 012017 doi:10.1088/1757-899X/597/1/012017

The formula for the formation constant K_c of glycerol boric acid is:

$$K_c = \frac{c_{hbg2}}{c_{hb} \cdot c_g^2}, \tag{1}$$

where c_{hbg2} is the concentration of glycerol boric acid, c_{hb} is the concentration of boric acid, c_g is the concentration of glycerol.

Since the concentrations of the anion of glycerol boric acid and the hydrogen cation are equal, the formula for the ionization constant Kd of glycerol boric acid is:

$$K_d = \frac{c_{bg2}^2}{c_{hbg2}} \quad \text{or} \quad K_d = \frac{c_h^2}{c_{hbg2}}, \tag{2}$$

where c_{bg2} is the concentration of the glycerol boric acid anion, c_h is the concentration of the hydrogen cation.

The material balance equation, which takes into account the formation and ionization of glycerol boric acid, has the form:

$$c_{hbo} = c_{hb} + c_{hbg2} + c_{bg2}, \tag{3}$$

where c_{hbo} is the stoichiometric concentration of glycerol boric acid.

The concentration of anions of glycerolboric acid cbg2, which is an organic acid, is much less than the concentration of molecular forms. Therefore, the material balance equation can be written in the form:

$$c_{hbo} = c_{hb} + c_{hbg2}, \tag{4}$$

The joint solution of equations (1, 2, and 4) makes it possible to obtain a formula expressing the dependence of the concentration of hydrogen cations or anions of glycerol boric acid on the concentrations of boric acid and glycerol. The formula makes it possible to calculate the constants of formation and ionization of glycerol boric acid on the basis of experimental data of methods of pH-metry and electrical conductivity:

$$h(c._{hbo}, c_{go}) = \sqrt{\frac{K_d \cdot K_c \cdot c_{hbo} \cdot c_{go}^2}{1 + K_c \cdot c_{hbo} \cdot c_{go}^2}} \quad \text{or} \quad bg2(c._{hbo}, c_{go}) = \sqrt{\frac{K_d \cdot K_c \cdot c_{hbo} \cdot c_{go}^2}{1 + K_c \cdot c_{hbo} \cdot c_{go}^2}}. \tag{5}$$

The pH value is the negative decimal logarithm of the concentration of hydrogen ions:

$$pH(c._{hbo}, c_{go}) = -\log\left(h(c._{hbo}, c_{go})\right).$$

Logarithmic equation (5) for the concentration of hydrogen ions, we get:

$$pH(c._{hbo}, c_{go}) = -\frac{1}{2}\log\left(\sqrt{\frac{K_d \cdot K_c \cdot c_{hbo} \cdot c_{go}^2}{1 + K_c \cdot c_{hbo} \cdot c_{go}^2}}\right). \tag{6}$$

In studies using the method of electrical conductivity, various concentration dependences of the specific electrical conductivity (SEC or κ – greek kappa) solutions were measured. The theoretical formula linking the SEC with the constants of formation and ionization of glycerol boric acid can be obtained on the basis of the theory of electrolytes. According to the theory, the SEC is related to the equivalent electrical conductivity λ_{hbg2} of glycerol boric acid by the formula:

$$\kappa = \lambda_{hbg2} \cdot c_{hbg2} = \lambda_{hbg2} \cdot \frac{c_{bg2}}{\nu}, \tag{7}$$

where $\nu = 2$ is the stoichiometric coefficient of the reaction of ionization of glycerol boric acid.

The equivalent electrical conductivity of glycerol boric acid is equal to the sum of the equivalent electrical conductivities of the hydrogen cation λ_h and the glycerol boric acid anion λ_{bg2}:

$$\lambda_{hbg2} = \lambda_h + \lambda_{bg2}. \tag{8}$$

The equivalent electrical conductivity of the ion λi according to the Walden – Pisarzhevsky rule is related to the viscosity of the solvent η and the radius of the hydrated ion r_i by the formula:

$$\lambda_i = \frac{e_0 \cdot N_F}{6 \cdot \pi \cdot \eta \cdot r_i}, \tag{9}$$

where e_o is the electron charge, N_F is the Faraday number.

The combination of formulas (5, 7-9) allows to obtain the equation linking the experimental values of SEC and viscosity with the constants of the formation and ionization of glycerol boric acid:

$$\kappa(c_{\cdot hbo}, c_{go}) = \sqrt{\frac{K_d \cdot K_c \cdot c_{hbo} \cdot c_{go}^2}{1 + K_c \cdot c_{hbo} \cdot c_{go}^2} \cdot \frac{e_0 \cdot N_F}{6 \cdot \pi \cdot \eta \cdot v} \cdot \left(\frac{1}{r_{bg2}} + \frac{1}{r_h}\right)}. \quad (10)$$

The radius of the hydrated hydrogen cation r_h = 0.25 nm is calculated by formula (9) using the reference value of the limit equivalent electrical conductivity of the hydrated hydrogen ion λ^0_i=330S·cm²·mol⁻¹ and viscosity η = 1 MPa·s at 22 °C [5]. The radius of the anion of glycerol boric acid is calculated by the formula

$$r_{bg2} = \sqrt[3]{\frac{mw_g}{\rho_g \cdot N_A}}, \quad (11)$$

where mw_g = 92.1 g/mol is the molecular weight of glycerin, ρ_g = 1.26 g/cm³ is the density of glycerol, N_A is the Avogadro number.

3. Materials and methods

For laboratory studies of acid-base equilibria in systems with donor-acceptor interactions: polybasic inorganic boric acid and polyhydric alcohol (glycerol and sorbitol) were prepared solutions of boric acid with concentrations of 1, 5 and 10% wt. in water-alcohol solvents with concentrations of polyhydric alcohols from 10 to 70% wt. Chemically pure reagents and distilled water were used. The dependence of pH, specific conductivity, viscosity and density of solutions on the concentration of components was determined. The pH values of the solutions were determined by a potentiometric method using a microprocessor laboratory pH meter manufactured by HANNA Instruments, the specific electrical conductivity using an ANION 4120 conductometer, and the density by a pycnometric method. The viscosity of solutions of polyatomic alcohols was determined using the vibration "Reokinetic" viscometer with a tuning fork sensor. The measurements were carried out at a temperature of 20–23 °C. The research results are presented in Figures 2-6 and in Table 1.

4. Results and discussion

Figures 2 - 5 show the experimental graphs of pH and SEC solutions of the polyol concentration at a constant concentration of boric acid and calculated by equations (6) and (10). The graphs in Figures 3 and 5 demonstrate the strong influence of the viscosity of the solvent on the electrical conductivity. In the framework of the theory of nonequilibrium phenomena in electrolyte solutions, this is due to the dual effect of the polyol. On the one hand, an increase in the concentration of the polyol leads to an increase in the concentration of the electric current carriers – anions of polyol boric acid and hydrogen cations in accordance with the reactions of formation and ionization of the acid. On the other hand, an increase in the concentration of the polyol leads to an increase in viscosity, which causes inhibition of the movement of current carriers and, thus, to a decrease in SEC. Initially, the formation of current carriers dominates, then, after the point of extremum, viscous drag becomes the main factor. Extremum SEC is in the range of concentration 25-35% wt. glycerol and sorbitol.

From figures 2, 4 it can be seen that the decrease in pH with increasing concentration of the polyol is monotonic. This is due to the continuous shift of ionization equilibrium reaction, conjugate with a shift reaction to form a complex of boric acid with a polyol.

The extreme nature of the concentration dependences for SEC and monotonic for pH indicates changes in the structure of the coordinating solvent, starting from 30% wt. polyol.

The coordinated processing of experimental concentration dependences of pH and specific conductivity of glycerol boric and sorbitol boric acids solutions by equations (6) and (10) allowed us to obtain the values of the formation constants and ionization of glycerol boric and sorbitol boric acids in the polyol coordinating solvent (Table 1). As can be seen from the table, sorbitol boric acid is much stronger than glycerol boric acid. Therefore, of interest is the combined use of glycerol and sorbitol in oil-displacing compositions.

IOP Conf. Series: Materials Science and Engineering **597** (2019) 012017 doi:10.1088/1757-899X/597/1/012017

Table 1. Concentration constants of formation and ionization of glycerol boric and sorbitol boric acids in a polyol coordinating solvent.

Acid type	K_c, $L^2 \cdot mol^{-2}$	K_d, $mol \cdot L^{-1}$
Glycerol boric	0.05	$7.1 \cdot 10^{-6}$
Sorbitol boric	0.06	$3.2 \cdot 10^{-3}$

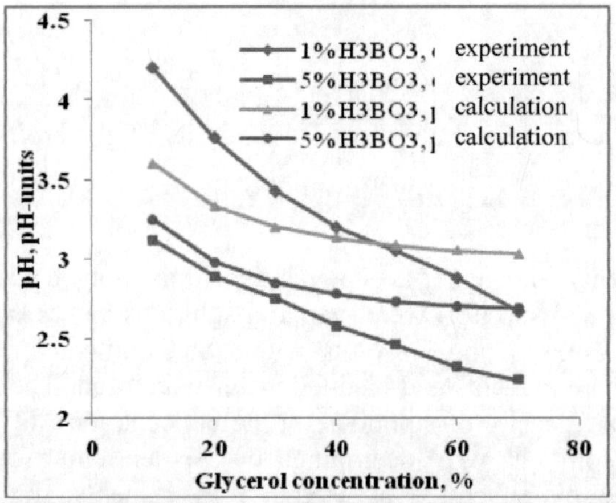

Figure 2. Solutions pH of glycerol boric acid, depending on the concentration of glycerol at a constant concentration of boric acid.

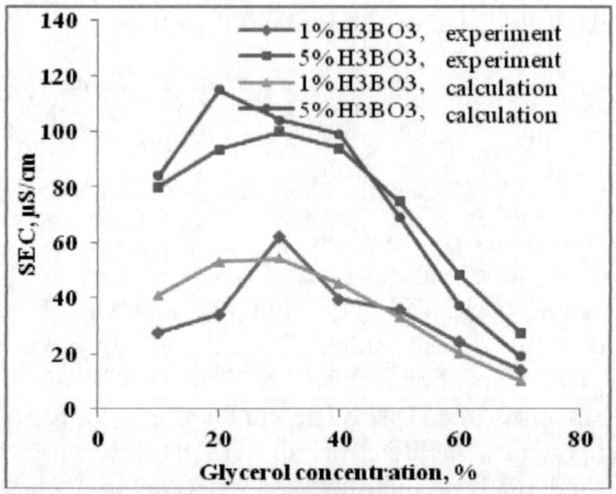

Figure 3. Specific conductivity of solutions of glycerol boric acid, depending on the concentration of glycerol at a constant concentration of boric acid.

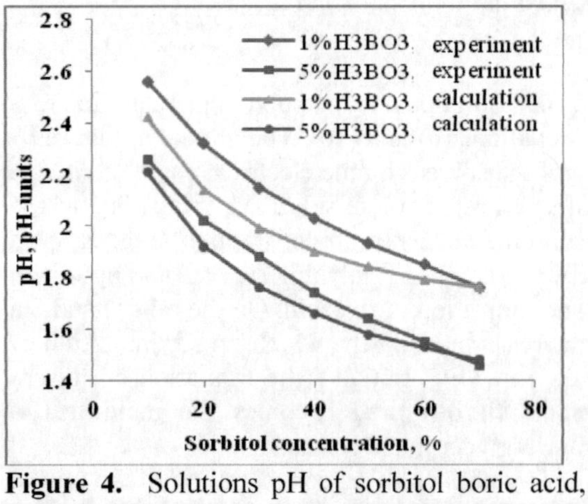

Figure 4. Solutions pH of sorbitol boric acid, depending on the concentration of sorbitol at a constant concentration of boric acid.

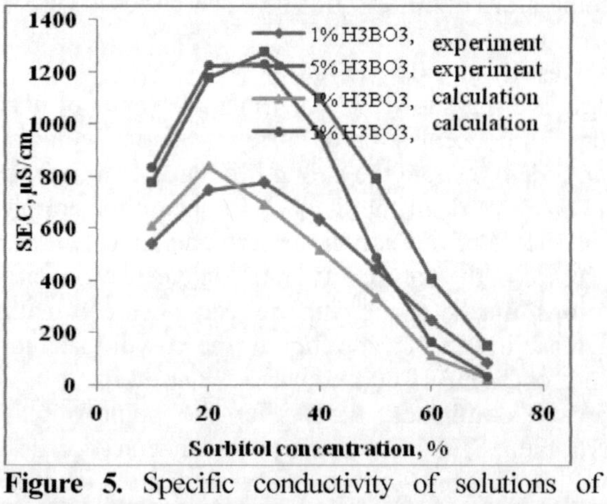

Figure 5. Specific conductivity of solutions of sorbitol boric acid, depending on the concentration of sorbitol at a constant concentration of boric acid.

The obtained values of the constants were used to calculate the buffer capacity of the compositions. Buffer capacity is the amount of acid or alkali, by adding which the pH of the buffer solution is changed by one. The higher the buffer capacity, the more the solution resists changing its pH with the addition of acids, alkalis or dilution. For a buffer system consisting of polyol boric acid and its salt, the analytical expression for the buffer capacity β is:

$$\beta(pH, c_a) = 2.303 \cdot \left(\frac{K_a \cdot c_a \cdot 10^{-pH}}{(K_a + 10^{-pH})^2} + 10^{-pH} + \frac{10^{-pK_w}}{10^{-pH}} \right), \quad (12)$$

where pH is a pH value; c_a is the acid concentration; K_a is the acid dissociation constant, pK_w is the ionic product of water.

Figure 6 shows the graphs of the buffer capacity of systems of glycerol and sorbitol acids separately and mixed with each other, depending on pH. The graphs are calculated by equation (12).

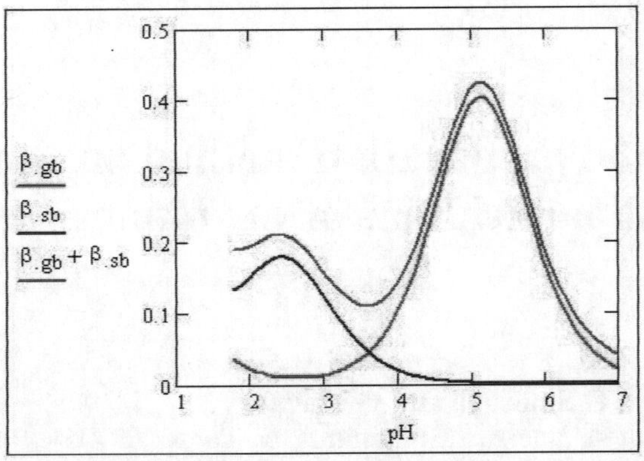

Figure 6. Buffer capacity of solutions for glycerol boric (βgb), sorbitol boric (βsb) acids and solution of their mixture in a ratio of 7: 3 (βgb + βsb).

As can be seen from the figure, range buffering acid mixture is a sum of individual acid ranges systems. This circumstance makes it possible to adjust the range of the buffer action by using in the compositions a mixture of polyol boric acids in different ratios.

5. Conclusions

A theoretical model of donor-acceptor acid-base equilibria in coordinating polyol solvents was created, which made it possible to determine the constants of the formation and ionization of polyol boric acids using pH-measurement and electrical conductivity methods. The obtained values of the constants were used to calculate the buffer capacity of the compositions. The possibility of regulating the range of the buffer action is shown by using a mixture of polyol boric acids in compositions in different ratios. The congruence of experimental data with the results of calculations using a theoretical model indicates the perspectives of developing oil-displacing acidic compositions based on surfactants, coordinating solvents and complex compounds in order to intensify the development and increase oil recovery of heavy high-viscosity oil.

Acknowledgments

The work was financially supported by the Ministry of Science and Higher Education of the RF, Subsidy Agreement No. 14.604.21.0176 from 09/26/2017, Unique identifier – RFMEFI60417X0176

References

[1] Altunina L K and Kuvshinov V A. 2008 *Oil & Gas Science and Technology* (Rev. IFP) **63** **1** 37
[2] Altunina L K, Kuvshinov V A, Kuvshinov I V, Stasieva L A, Chertenkov M V, Shkrabyuk L S and Andreev DV 2017 *Oil industry* **7** 26
[3] Schwartz E M 1968 Complex boron compounds with polyoxy compounds (Riga: Zinatne) p 120
[4] Troitsky G V and Azhitsky G Yu 1984 Isoelectric focusing of proteins in borate-glycerin pH gradient (Kiev: Naukova Dumka) p 220
[5] Handbook of Electrochemistry 1981 Ed. A.M. Sukhotina (L: Chemistry) p 488

IOP Conf. Series: Materials Science and Engineering **597** (2019) 012018 doi:10.1088/1757-899X/597/1/012018

Dependence of the preparation method on the phase composition and particle size of the binary NiO–ZrO_2 system oxides

S Galanov, O Sidorova and O Magaev
Tomsk State University, 36 Lenin Ave., Tomsk 634050, Russian Federation

E-mail: sidorova@xf.tsu.ru

Abstract. Was studied how preparation method influences phase composition, oxide particle size and catalytic activity of the binary NiO–ZrO_2 systems. The processes taking place under the thermal influence while NiO–ZrO_2 catalysts are formed from precursors, obtained using a variety of methods, were determinated using the methods of simultaneous TGA-DTG/DSC analysis and XFA. Was studied the influence of the precursor preparation method upon the catalysts' phase composition, sizes of the nickel oxide and zirconium dioxide particles. The research revealed that preparation of precursor using coprecipitation method makes it possible to obtain a binary system, where nickel oxide has minimal size, determined by CSR, and monoclinic phase prevails in ZrO_2, after heating it to 800 °C. The research unearthed that the catalyst exhibiting maximal catalytic activity by deep oxidation of methane is nickel oxide-zirconium dioxide, containing equal amounts of monoclinic and tetragonal phases of ZrO_2.

1. Introduction

Zirconium dioxide has a number of valuable characteristics, which make is suitable for use in manufacture of ceramic materials [1], microelectronics [2], fuel elements [3], also as a catalyst by partial oxidation of methane to syngas [4, 5] high-temperature shift catalyst for methane [6, 7]. Most of the studies is devoted to the investigation of the influence of the amount of the active component (metal) [8] or the second oxide introduced on the phase composition, metal dispersion, stability and catalytic activity of the systems obtained in the methane oxidizing conversion reaction [9, 10], yet they do not focus on the influence the catalyst precursor preparation method has on physical, chemical and catalytic properties of the zirconium dioxide-based systems. The research works [6, 7] show that changing synthesis methods for ZrO_2 can change characteristics of the catalyst and the carrier it is based on, but to the date, the influence of preparation methods on ultimate characteristics of zirconium dioxide-based binary oxide systems hasn't been studied yet.

The goal of this research work is to unearth the influence of the oxidic nickel-zirconium system precursor preparation stages and methods of nickel compound doping on the phase compounds, porous structure and catalytic activity of the synthesized binary oxides.

2. Experimental Part

The Acros reagents were used to obtain binary oxide nickel-zirconium systems with a mole ratio of NiO : ZrO_2 = 1 : 4.3 : $ZrO(NO_3)_2 \cdot 2H_2O$, Na_2CO_3, $Ni(NO_3)_2 \cdot 6H_2O$.

The catalyst precursors were prepared in the following way:

1) Sample No. 1 (hereafter referred to as $ZrO_2 \cdot nH_2O$).

Twofold excess of Na_2CO_3 (c = 0.5 mol/L) was poured into boiling $ZrO(NO_3)_2$ (c = 0.25 mol/L) solution; the suspension obtained was boiled for 0.5 hours until gas release stopped. The suspension was centrifuged, rinsed with double-distilled water till pH=7. The precipitation obtained was dried at T = 60 °C for 2 hours and at T = 110 °C for 4 hours. By fractionation of $ZrO_2 \cdot nH_2O$ two fractions were obtained: 0.20–0.25 mm and 0.020-0.025 mm.

2) Sample No. 2 (hereafter referred to as Ni-Zr(o)) was prepared by solvent distilling.

The 0.20–0.25-mm fraction of $ZrO_2 \cdot nH_2O$ sample was added to the measured amount of solution $Ni(NO_3)_2$ (c = 0,1 mol/L). Water was distilled off until the sample was dried in a rotary evaporator at T = 45 °C and at negative pressure of $48 \cdot 10^{-3}$ Bar. The precursor was dried the same way with 1.

3) Sample No. 3 (hereafter referred to as Ni-Zr(o)-950) was prepared by solvent distilling.

The sample was prepared the same way as Ni-Zr(o) sample following the method (2). The precursor was dried the same way with 1.

4) Sample No.4 (hereafter referred to as Ni-Zr (d-p)) was prepared by deposition-precipitation method.

The 0.020-0.025-mm fraction of $ZrO_2 \cdot nH_2O$ was added to a measured amount of $Ni(NO_3)_2$ (c = 0.0058 mol/L) solution and boiled up and added twofold excess of Na_2CO_3 (c = 0.5 mol/L) solution while stirring vigorously; the rest of the procedure was the same with Sample 1.

5) Sample No. 5 (hereafter referred to as Ni-Zr(co)), was prepared by coprecipitation.

Twofold excess of Na_2CO_3 (c = 0.5 mol/L) solution was poured into boiling solution of $ZrO(NO_3)_2$ (c = 0.25 mol/L) and $Ni(NO_3)_2$ (c = 0.058 mol/L) salts; the rest of the procedure was the same with Sample 1.

Baking of precursors for samples No. 1, 2, 4, 5 was carried out with increase in temperature from 20 to 800 °C with heat rise rate of 10 °/min in an air stream, with exposition time of 4 hours at 800 °C. Baking of Sample No.3 was carried out with increase in temperature from 20 to 950 °C with heat rise rate of 10 °/min in an airstream with exposition time of 4 hours at 950 °C.

Quantitative determination of nickel oxide in the baked samples was carried out using Oxford ED-2000 (Great Britain, Oxford Instruments) energy-dispersing X-ray-fluorescent spectrometer. Phase composition of the samples was determined using Rigaku Miniflex 600 X-ray diffractometer CuK_{α}-emission (l=1.5418 Å). Radiography specifications: scan velocity 2 degrees/min, scanning angular range 2q: from 10° to 100°. Average size (*D*) of the coherent scattering region (CSR) was determined by X-ray imaging based on Fourier analysis of diffraction peaks' profile. The average CSR size was evaluated according to Scherrer-Selyakov equation [8]. Thermal analysis of the samples was carried out using Netzsch STA-449 (Germany) simultaneous TGA-DTG/DSC analysis device, with heat rise rate of 10 °C/min. The samples' BET surface area (S_{sp}) was determined using TriStar II (3020) (Micromeritics, USA) device. Catalytic activity in the methane deep oxidation reaction was studied using a flow-through catalytic device in the temperature range of 400–700 °C, with contact time of 0.5 sec, the reaction mixture contained the following: CH_4 – 0.5 %, O_2 – 12.7 %, N_2 – 86.2 % vol.

3. Results and Discussion

Simultaneous TGA-DTG/DSC analysis of hydrated zirconium dioxide is marked by a wide endothermic peak (Figure 1. *a*) in the temperature range of 50–400 °C, which corresponds with removal of the hydration water, the weight reduction calculation makes it possible to calculate the formula of the obtained and dried zirconium oxide as $ZrO_2 \cdot 2H_2O$, which agrees with the data provided in the research work [11].

Peak exotherm at (T_{max}) 491.1 °C, can be assigned to crystallization process of the noncrystalline ZrO_2 into the tetragonal variety (t-ZrO_2), which corresponds with the data provided in the research work [11]; exo-effect at T_{max} = 701.4 °C indicates partial transition of metastable tetragonal phase into the monoclinic one (m-ZrO_2). Slight reduction of the sample weight in the temperature range

500–1000 °C may mark removal of hyperstoichiometric oxygen and residual water of crystallization [11].

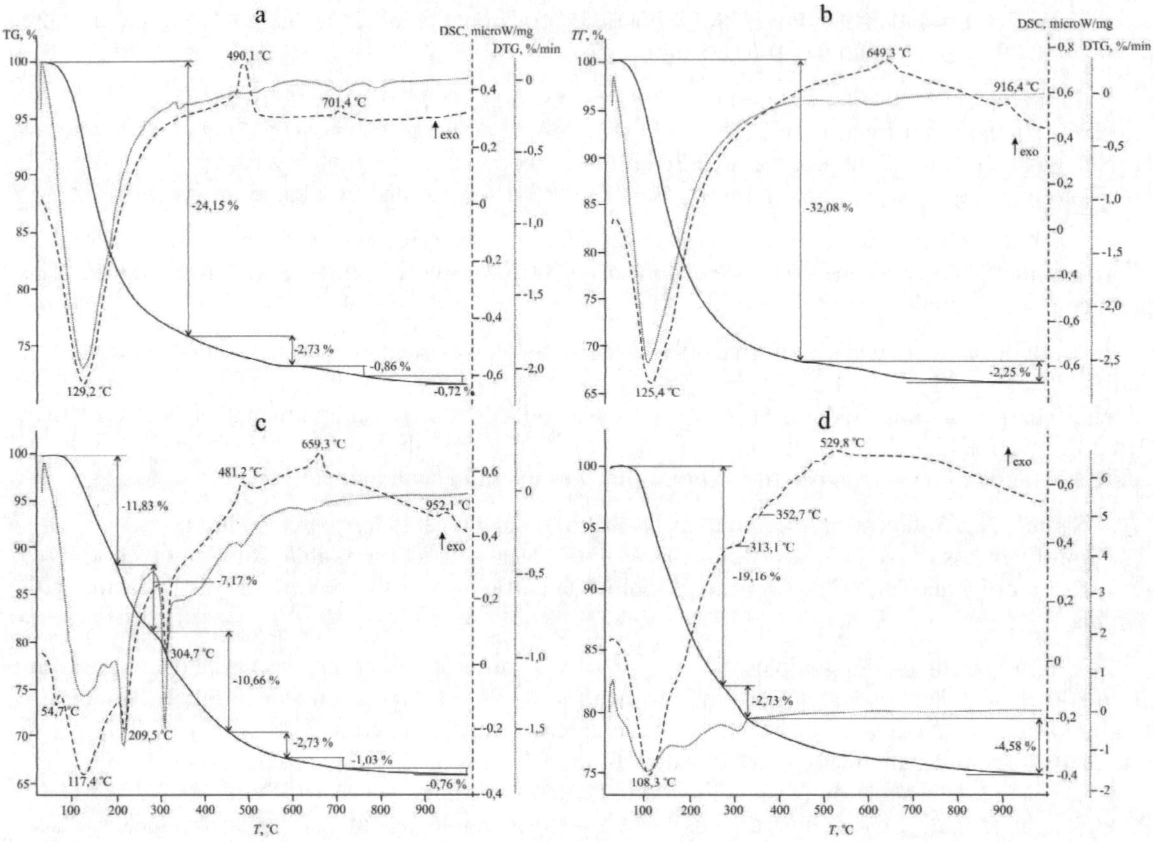

Figure 1. Synchronous TGA-DTG/DSC analysis of precursors of catalysts: (a) – $ZrO_2 \cdot nH_2O$, (b)– Ni-Zr(o), (c) – Ni-Zr(d-p), (d) – Ni-Zr(co).

According to the data obtained by RFA (Figure 2.*a*), two phases are present in the ZrO_2 sample heated up to 800 °C: the tetragonal one, t-ZrO_2 (2Θ = 30.14; 35.13; 50.56; 60.02; 62.75; 74.35) and the monoclinic one, m-ZrO_2 (2Θ = 17.39; 23.90; 28.12; 31.31; 40.62), containing 32% of m-ZrO_2. As we obtain zirconium dioxide in the annealing temperature range of 700–1000 °C, regardless of the initial precursor production method, by a) solution deposition [9, 10]; b) pyrogenic decomposition of zirconyl chloride, two phases of zirconium dioxide are produced: the monoclinic and the tetragonal ones.

According to TGA-DTG/DSC analysis statistics, Ni-Zr(o) sample (Figure 1.*b*), is marked by endothermic peak at T_{max} = 54.7 °C, which corresponds with melting of nickel nitrate hydrate on the surface of $ZrO_2 \cdot 2H_2O$ granules, the endothermic peak at T_{max} = 117 °C with weight reduction corresponds with removal of hydrated water from $ZrO_2 \cdot 2H_2O$. The overlapping weight reduction endothermic peaks of T_{max} = 209.5 °C and T_{max} = 304.7 °C correspond with removal of water from nickel nitrate hydrate (according to TGA, the design equation is $Ni(NO_3)_2 \cdot 4.5H_2O$) and decomposition of nickel nitrate with production of nickel oxide. The exothermic peak at T_{max} = 481.2 °C (Figure 1.*b*) marks crystallization of the X-ray amorphous ZrO_2 into t-ZrO_2, and at T_{max} = 659.3 °C it marks partial transition of ZrO_2 tetragonal phase into the monoclinic one [11]. According to XFA (Figure 2.*b*), Ni-Zr(o) binary oxide system heated up to 800 °C contains two phases of zirconium dioxide (tetragonal and monoclinic ones), as well as the cubic phase NiO (2Θ = 37.16; 43.18; 62.75). At that, the content of ZrO_2 monoclinic phase is lower than that for pure ZrO_2 (Table 1).

Table 1. Influence of a way of preparation of a precursor on phase structure, size of particles of oxides, catalytic activity of NiO-ZrO_2 systems.

Parameter	$ZrO_2 \cdot nH_2O$	Ni-Zr (o)	Ni-Zr (d-p)	Ni-Zr (co)	Ni-Zr (o)-950 °C
S_{sp}, m^2/g	1.9	6.3	4.6	4.1	1.3
NiO:ZrO_2, mol	-	1 : 4.32	1 : 4.33	1 : 4.30	1 : 4.32
quantity m-ZrO_2, %	32.1	13.3	49.4	99.5	99.3
D_{NiO}, nm	-	29	25	20	38
$D_{m\text{-}ZrO2}$, nm	20	20	29	31	41
$D_{t\text{-}ZrO2}$, nm	28	29	24	-	-
T_{80}, °C*	600	520	460	490	550

* T_{80},°C – temperature is 80% of conversion of methane

According to Figure 1.*b*, Ni-Zr(o) sample is marked by exothermic peak at T_{max} = 952 °C, which attests further phase transition of the remaining t-ZrO_2 to m-ZrO_2. The Ni-Zr(o)-950 sample heated up to 950 °C consists of m-ZrO_2 and NiO (Figure 2.*e*); however high-heat treatment leads to the following phenomena: increase in the CSR average size of nickel oxide and zirconium dioxide crystallites (Table 1). According to TGA/DSC analysis (Figure 1. *c*), Ni-Zr(d-p) system is marked by wide endothermic peak at T_{max} = 125.4 °C and sample weight reduction in the temperature range 50–490 °C which is representative of the process of water removal from the hydrated zirconium oxide and decomposition of nickel hydroxocarbonates.

The exothermic peak at T_{max} = 679.3 °C can be related to simultaneous processes of amorphous zirconium dioxide crystallization and phase transitions; further transition of t-ZrO_2 phase into the monoclinic one takes place at the temperature of 916.4 °C (Figure 1.*c*). According to XFA (Figure 2.*c*), this is a multiple-phase system and contains t-ZrO_2, m-ZrO_2 and NiO phase, but specific gravity of the monoclinic phase compared to the samples of Ni-Zr(o) and ZrO_2 obtained from $ZrO_2 \cdot 2H_2O$, increased (Table 1), which probably is due to the initial ten-fold reduction of the hydrated zirconium oxide particles and possibly, with the mineralizing effect of nickel oxide. Average sizes of the nickel oxide crystallites for Ni-Zr(o) and Ni-Zr(d-p) samples are similar (Table 1). NiO particles of the same size (27–29 nm) were obtained by deposition of nickel onto heat-treated ZrO_2 [7].

TGA-DTG/DSC analysis of Ni-Zr(co) sample also revealed analogous to the previous samples wide endothermic peak accompanied with sample weight reduction in the temperature range of 50–460 °C (Figure 1.*d*) with faint endothermic peaks at T_{max} = 313.1 and 352.7 °C, which are concerned with decomposition of nickel hydroxocarbonates in the bulk of catalyst granule (the same effect was observed in the research work [6]) and faint exothermic peaks at temperatures of 490 and 529.8 °C, concerned with crystallization and phase transitions of ZrO_2. According to Figure 2.*d*, in the Ni-Zr(co) sample heat-treated at the temperature of 800 °C, monoclinic phase of zirconium dioxide and cubic NiO prevail, the sizes of nickel oxide crystallites are smaller than those in the samples obtained from the precursors prepared using methods 2–4 (Figure 2.*d*, Table 1).

By adding nickel oxide and heat-treating it at T = 800 °C BET surface area of the samples grows larger compared to that of pure ZrO_2; Ni-Zr(co) sample has the smallest BET surface area among binary systems. Increase of the heat-treatment temperature up to 950 °C also promotes further shrinkage of the Ni-Zr(o)-950 catalyst BET surface area (Table 1). The size of m-ZrO_2 crystallites increases as the proportion of monoclinic phase in zirconium dioxide grows. Size of t-ZrO_2 crystallites is approximately the same in all systems: 26–29 nm (Table 1). Catalytic properties of a system depend not only on the range of particle dimensions of the NiO active component, as [7] shows, but also on the proportion of tetragonal and monoclinic phases in ZrO_2.

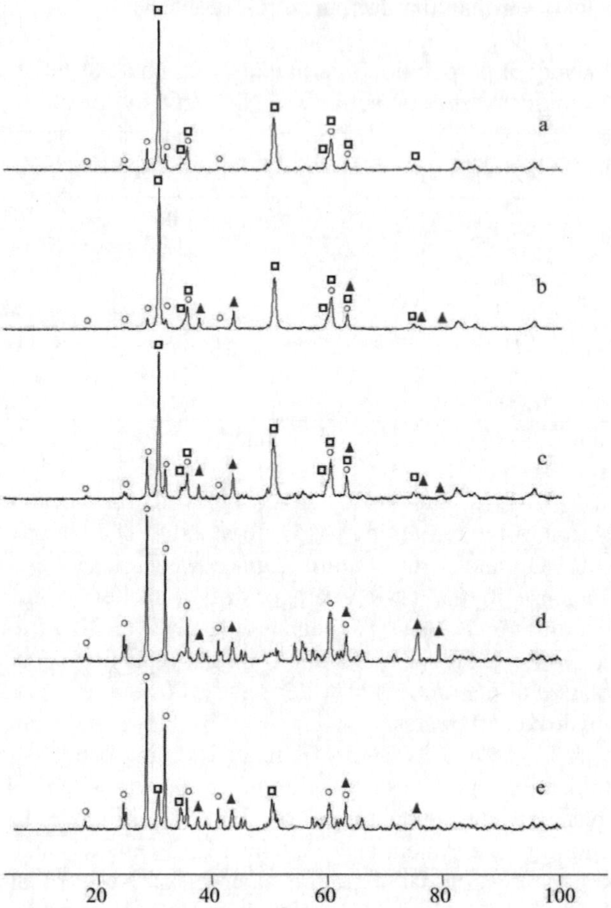

Figure 2. XRA ranges of binary NiO-ZrO_2 of systems: (a) – $ZrO_2 \cdot nH_2O$, (b) – Ni-Zr(o), (c) – Ni-Zr(d-p), (d) – Ni-Zr(co), (e) – Ni-Zr(o)-950. The reflexes relating to: □ – ZrO_2 is tetragonal, ○ – ZrO_2 monoclinic, ▲ – NiO cubic

According to [7], in the methane deep oxidation reactions, the temperatures at which methane conversion reaches a specific level, which is 80% in our case, is the catalytic activity criterion for the catalyst. Therefore, a catalyst's activity is the higher, the lower 80% methane conversion temperature is. According to the Table 1, Ni-Zr (d-p) catalyst with NiO particle size of 25 nm showed the highest catalytic activity compared to other systems, while equal amounts of t-ZrO_2 and m-ZrO_2 were present in zirconium dioxide (Table 1). We believe this leads to increase in oxygen exchange ability of zirconium dioxide, which is one of the key factors in its oxidation-reduction reactions proceeding involving gas-phase oxygen.

4. Conclusion

Thus, the studies have shown that the variation of the preparation method of the precursor and the temperature of heat treatment can produce NiO–ZrO_2 oxide binary systems with a variety of tetragonal and monoclinic phase proportions in zirconium dioxide. The nickel oxide particle size for binary systems can also be adjusted in the process of NiO–ZrO_2–system precursor formation. Activity of NiO–ZrO_2 catalysts in the oxidation-reduction reaction of methane deep oxidation involving gas-phase oxygen depends both on the active component's - nickel oxide's - range of particle dimensions and the proportion of monoclinic and tetragonal phases in the zirconium dioxide supporter, which provides for the binary oxide system's high oxygen exchange ability in general. The catalyst containing equal amounts of monoclinic and tetragonal phases of zirconium dioxide with NiO particle size of 25 nm exhibits maximal catalytic activity in oxidation-reduction reaction of hydrocarbon oxidation.

Acknowledgements
This work was supported by Russian Science Foundation (project number 19-73-30026).

References

[1] Dudnik E V 2009 Modern methods for hydrothermal synthesis of ZrO_2-based nanocrystalline powders *Powder Metallurgy and Metal Ceramics* **48** 3-4 238

[2] Koch T, Ziemann P 1996 Zr-silicide formation during the epitaxial growth of Y-stabilized zirconia films on Si(100) and its avoidance by ion beam assiated deposition at a reduced temperature *Applied Surface Science* **99** 1 51

[3] Xianshuang X, She L, Qingshan Z, Xiqiang H and Wenhui S 2007 Fabrication of dense YSZ electrolyte membranes by a modified dry-pressing using nanocrystalline powders *Journal of Mater. Chem.* **17** 16 1627

[4] Galanov S I, Sidorova O I 2014 Effect of a precursor on the phase composition and particle size of the active component of Ni-ZrO_2 catalytic systems for the oxidation of methane into syngas *Russian Journal of Physical Chemistry A.* **88** 10 1629

[5] Galanov S I, Sidorova O I, Batyreva, V A. 2015 Nickel-containing systems in reaction of partial oxidation of hydrocarbons *Key Engineering Materials.* **670** 33

[6] Song Yue-Qin, He De-Hua, Xu Bo-Qing 2008 Effects of preparation methods of ZrO_2 support on catalytic performances of Ni/ZrO_2 catalysts in methane partial oxidation to syngas 2008 *Applied Catalysis A: General.* **337** 19

[7] Otroshchenko T P, Turakulova A O, Voblikova V A, Sabitova L V, Kutsev S V, Lunin V V 2013 NiO and ZrO_2-based catalysts in the reaction of complete methane oxidation *Russian Journal of Physical Chemistry A.* **87** 11 1804

[8] Galanov S I., Sidorova O I 2014 Combined conversion of methane into syngas on Ni/MgO-Cr_2O_3 catalysts *Russian Journal of Applied Chemistry* **87** 10 1435

[9] Berrocal G P, Da Silva A L M, Assaf J M, Albornoz A, Rangel M D C 2010 Novel supports for nickel-based catalysts for the partial oxidation of methane *Catalysis Today* **149** 240

[10] Asencios Y J O, Bellido J D A, Assaf E M 2011 Synthesis of NiO-MgO-ZrO_2 catalysts and their performance in reforming of model biogas *Applied Catalysis A: General* **397** 138

[11] Ivanov V K, Kopitsa G P, Grigoriev S V, Polezhaeva O S, Garamus V M 2010 Investigation of the evolution of the hydrated zirconia mesostructure at different stages of heat treatment *Physics of the Solid State* **52** 5 957

IOP Conf. Series: Materials Science and Engineering **597** (2019) 012019 doi:10.1088/1757-899X/597/1/012019

Non-oxidative methane conversion over Mo/ZSM-5 catalysts with mesoporous structure

A V Vosmerikov[1,2], A A Stepanov[1], K N Zharnov[2] and L L Korobitsyna[1]

[1]Institute of Petroleum Chemistry, Siberian Branch, Russian Academy of Sciences, Tomsk, Akademichesky Ave. 4, 634055, Russia
[2]National Research Tomsk State University, Tomsk, Lenina Ave. 36, 634050, Russia

E-mail: pika@ipc.tsc.ru

Abstract. Physicochemical and catalytic properties of 4.0% Mo/ZSM-5 catalysts manufactured on the basis of zeolites with a hierarchical pore system are investigated. The effect of the conditions of synthesis of zeolites with an additional mesoporous structure on their textural and acidic characteristics is studied. The structure and morphology of crystals of synthesized zeolites and catalysts produced from them are determined by the methods of X-ray diffraction analysis and high-resolution scanning transmission electron microscopy. It is shown that 4.0% Mo/ZSM-5 catalysts produced from zeolites with a mesoporous structure exhibit higher activity and stability in the course of methane dehydroaromatization.

1. Introduction

Rational use of natural and associated petroleum gases, the main component of which is methane, is of great importance for the production of valuable petrochemical products, as well as for solving environmental problems. It is known that Mo-containing ZSM-5 zeolites exhibit the highest activity and selectivity in the course of methane dehydroaromatization [1-3]. One of the disadvantages of these catalysts is the rapid loss of activity as a result of the carbonization of their active sites. The presence of a mesoporous structure in the zeolitic matrix improves the egress of aromatic compounds from the volume of the zeolite to the catalyst surface, which leads to an increase in its activity and stable operation in the course of methane dehydroaromatization [4].

The purpose of this work was to obtain Mo/ZSM-5 catalysts based on zeolites with a mesoporous structure and to study their physicochemical and catalytic properties in the reaction of methane dehydroaromatization.

2. Materials and methods

Zeolites with a mesoporous structure were manufactured by the method of hydrothermal synthesis from alkaline alumina silica gels with a molar ratio $15Na_2O:5,4R:Al_2O_3:40SiO_2:2200H_2O$, where R is hexamethylenediamine. At the stage of synthesis, P354 grade carbon black powder (produced by the Center for New Chemical Technologies of the Federal Research Center Boreskov Institute for Catalysis, Omsk), whose content varied from 0.5 up to 5.0 wt% was added to the reaction mixture. Carbon black has the following characteristics: the specific surface area is 110 m^2/g, the pore volume is 0.034 cm^3/g; elemental composition: C – 96.4%, O – 2.5%, S – 0.5%, N and H per 0.3%. The carbon particles, being compositionally homogeneous, have a 30-60 nm globular shape.

IOP Conf. Series: Materials Science and Engineering **597** (2019) 012019 doi:10.1088/1757-899X/597/1/012019

The resulted suspension was stirred for 0.5 h, placed into an autoclave, and synthezised at 170-175 °C for 4 days. The solid phase obtained was filtered using a Buchner funnel, washed with distilled water to pH = 6-7, and dried at 100 °C in a drying cabinet for 4 hours. To remove the template and carbon black, the synthesized zeolite was calcined in air at 550 °C for 6 hours. To convert the zeolite from the sodium to the ammonium form, decationization with a 25% aqueous solution of NH_4Cl at a temperature of 90 °C for 2 hours was carried out. The samples prepared were referenced as follows: K1 was the initial zeolite obtained without carbon; K2, K3, K4, K5, and K6 were zeolites obtained by adding carbon black in the amount of 0.5; 1.0; 2.0; 3.5 and 5.0 wt%, respectively.

The quality of the synthesized zeolites was monitored using IR spectroscopy and X-ray diffraction. The IR spectra of the samples were recorded using a Nicolet 5700 IR Fourier spectrometer within the range 2000-400 cm^{-1}. The degree of crystallinity of zeolites was determined by IR spectroscopy according to the method described in [5]. X-ray diffraction analysis was performed using a Bruker DISCOVERD8 diffractometer in the angle range 2ϑ = 10-70 degrees. The parameters of the porous structure and the determination of the specific surface of the samples were evaluated using a TriStar 3020 automated gas-adsorption analyzer (Micromeritics, USA). The specific surface was calculated from the isotherm of low-temperature sorption of nitrogen vapors (BET method). Electron-microscopic studies of the morphology of zeolite crystals were performed using an LEO-1420 electron scanning microscope. The acidic properties of the samples were investigated by the method of temperature-programmed desorption of ammonia, which allowed to determine the strength and concentration of acid sites. The 4.0% Mo/ZSM-5 catalysts were prepared by dry mechanical mixing of zeolites in the NH_4 form, synthesized with various amounts of technical carbon, and nano-sized Mo powder obtained by the method of electrical explosion of a conductor in argon. The average size of Mo particles was 70-110 nm. The resulted mixtures were calcined in a muffle furnace at 550 °C for 4 hours.

The process of non-oxidative methane conversion (purity level of 99.99% by volume) was carried out in a flow fix bed reactor at a temperature of 750 °C, atmospheric pressure and a volumetric flow rate of methane 1000 h^{-1}. The volume of the catalyst loaded into the quartz reactor was 1.0 cm^3 and the size of its granules was 0.5-1.0 mm. The reaction products were analyzed by GLC using a Chromatec-Crystal 5000.2 chromatograph every 40 min of catalyst operation. To assess the catalytic activity of the samples, the degree of methane conversion and the yield in the resulted reaction products were determined.

3. Results and discussion

Figure 1 shows the IR spectra and X-ray diffraction patterns of zeolites synthesized with the addition of carbon black in the amount of 1.0 and 3.5% and without it. It is evident from the presented data that the use of carbon black during the synthesis of zeolite has no effect on its structural characteristics. All the zeolites obtained belong to the ZSM-5 structural type, and their crystallinity is 100%.

Figure 2 shows electron microscopic images of zeolite synthesized without the use of carbon black (Figure 2a) and with carbon black additives (Figure 2b and 2c). The particles of the initial zeolite K1 are quite homogeneous in their composition. They have the appearance of polycrystalline spheroids with sizes from 6 to 8 microns. A looser compound (junction) of crystals in polycrystalline spheroids is observed on the micrograph of zeolite synthesized with the addition of 1.0% carbon, after calcination at 550 °C, while their sizes vary over a wider range from 4 to 8 μm (Figure 2b). As the amount of carbon increases to 3.5% during the synthesis of zeolite and the subsequent calcination, the zeolite crystals become even more heterogeneous (Figure 2c). The size of polycrystalline spheroids varies from 4 to 9 microns. In addition, the formation of larger particles sized up to 12 microns, whose morphology differs from the structure of particles of the initial K1 zeolite.

IOP Conf. Series: Materials Science and Engineering **597** (2019) 012019 doi:10.1088/1757-899X/597/1/012019

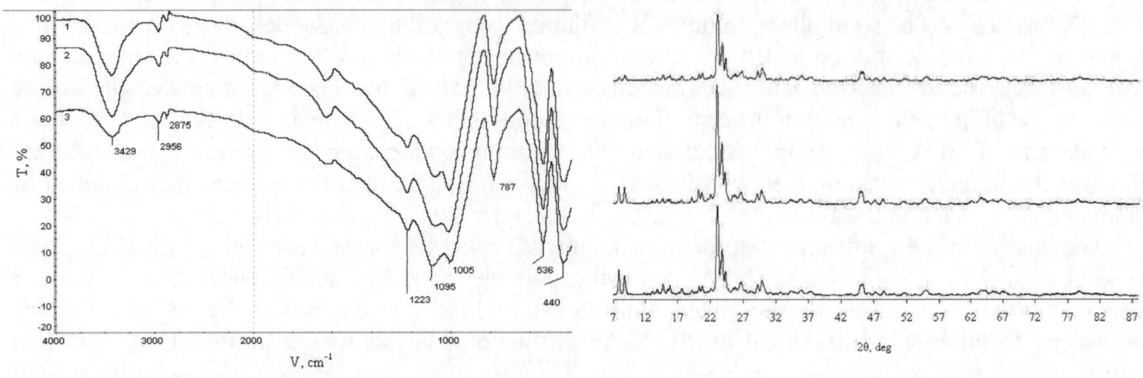

a b

Figure 1. IR-spectra (a) and XRD patterns (b) of zeolites synthesized without the use of carbon black and with the use of various amounts of carbon: 1 – 0%, 2 – 1.0%, 3 – 3.5%.

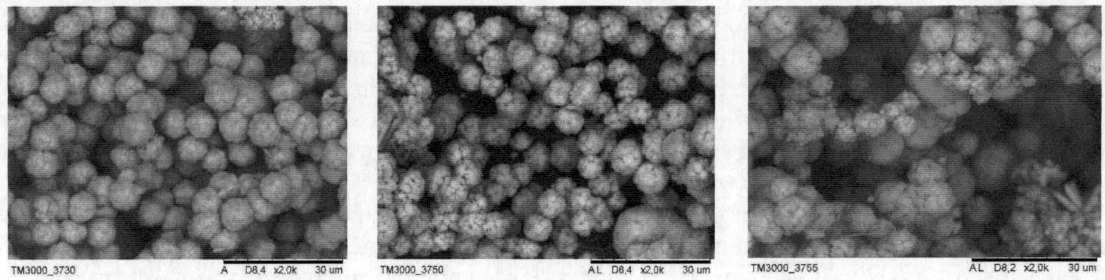

a b c

Figure 2. Electron micrographs of zeolites synthesized without the use of carbon black (a) and with the use of carbon black amounts of 1.0% (b) and 3.5% (c).

Values of specific surface area, area and volume of micropores, as well as the total pore volume of samples obtained on the basis of zeolites synthesized with different contents of carbon black, are presented in Table 1. The largest specific surface area 398 m^2/g is observed for the initial zeolite K1. With an increase in the concentration of carbon black used in the synthesis of zeolite from 0.5 to 5.0%, a decrease in the specific surface area and an increase in the volume of mesopores from 0.02 to 0.08 cm^3/g are observed. At the same time, the volume of micropores slightly varies from 0.15 to 0.18 cm^3/g.

Table 1. Surface properties of zeolites.

Sample	S_{BET} (m^2/g)	S_{micro} (m^2/g)	V_{total} (cm^3/g)	V_{micro} (cm^3/g)	V_{meso} (cm^3/g)
K1	398	322	0.20	0.18	0.02
K3	379	301	0.21	0.15	0.06
K4	378	299	0.22	0.16	0.06
K5	375	290	0.24	0.17	0.07
K6	373	279	0.26	0.18	0.08

Figure 3 shows the dependences of the pore distribution on the size in ZSM-5 zeolites with micro- and mesoporous structures.

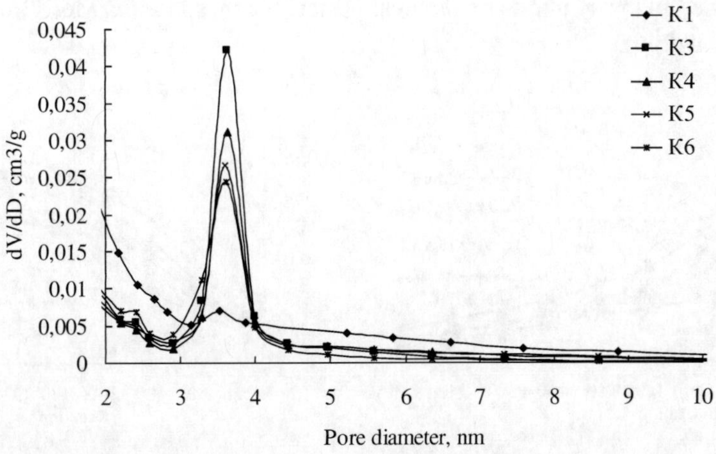

Figure 3. Pore size distribution in ZSM-5 zeolites with micro and mesoporous structure.

It can be seen that the zeolite ZSM-5 with a microporous structure has practically no peak in the range of 3–4 nm, only a small shoulder is observed. When carbon black is added at the stage of zeolite synthesis, mesopores are formed with an average size 3-4 nm, as evidenced by the appearance of a characteristic peak in this region.

The study of the acidic properties of the initial zeolite K1 and zeolites with a mesoporous structure showed the presence of two peaks with clearly pronounced temperature maxima on the thermal desorption curves. This indicating suggests the presence of two types of acid sites. It is evident from the data in Table 2, that the strength and concentration of the acid sites of the initial zeolite K1and the mesoporous zeolite K3 synthesized with the addition of 1.0% carbon black are close.

Table 2. Acidic characteristics of zeolites.

Sample	**Temperature, °C**		**Concentration, µmol/g**		
	T_I	T_{II}	C_I	C_{II}	C_Σ
K1	235	480	753	388	1141
K3	235	475	767	389	1156
K4	230	470	617	364	981
K5	235	470	643	368	1011
K6	245	475	874	372	1246

Note. T_I, T_{II} – temperatures of maximum low- and high-temperature peaks on thermodesorption curves; C_I, C_{II} and C_Σ – concentrations of weak and strong acid sites and their sum, respectively.

An increase in carbon content in the reaction gel from 2.0 to 3.5% results in the formation of mesoporous zeolites K4 and K5 with a lower concentration of weak and strong acid sites. For the sample K6, obtained using 5.0% carbon, a significant increase in the number of low-temperature acid sites (up to 874 µmol/g) and their strength is observed, while the concentration of strong acid centers varies slightly and the total content of acid centers reaches 1246 µmol/g. Thus, the acidic properties of zeolites and their textural and structural characteristics depend on the amount of carbon black used in the synthesis of zeolites.

The test results of 4.0% Mo/ZSM-5 catalysts manufactured on the basis of zeolites synthesized with the addition of different amounts of carbon in the course of non-oxidative conversion of methane to aromatic hydrocarbons are shown in Figure 4a. It can be seen that in all cases the use of carbon in the synthesis of zeolites results in an increase in the activity of the catalysts in comparison with the zeolite K1 obtained without using carbon. The catalytic properties of zeolite catalysts depend on the amount of carbon used in the synthesis of zeolites. Hence, the mesoporous Mo/K3 catalyst exhibits the highest activity and stability.

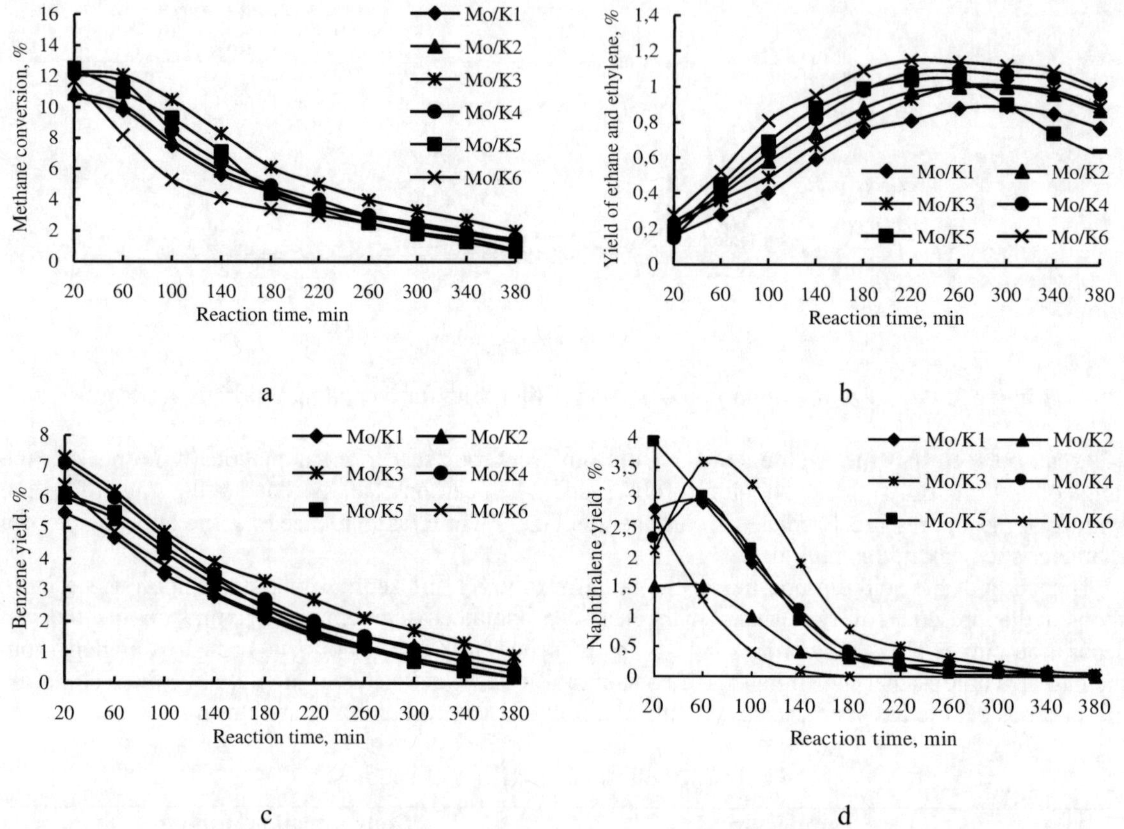

Figure 4. Time dependence of methane conversion (a), total yield of ethane and ethylene (b), and yield of benzene (c) and naphthalene (d) formed due to the use of Mo/K1, Mo/K2, Mo/K3, Mo/K4, Mo/K5; and Mo/K6 catalysts.

An increase in the concentration of carbon black in the reaction alumina silica gel used to synthesize the zeolite up to 3.5% reduces the activity of the Mo-containing zeolite catalyst, but its activity is higher than that of the Mo/K1 catalyst during the first 180 min of reaction. Then the methane conversion approaches to the conversion rates for the Mo/K1 catalyst and remains the same. The increase in the concentration of carbon added in the course of zeolite synthesis to 5.0% results in a noticeable decrease in the activity of the zeolite catalyst.

Analysis of the gaseous products formed in the course of methane conversion shows that they consist mainly of ethane and ethylene, whose yields increase as the process runs, reaching maximum values in the reaction time interval 220-300 minutes (Figure 4b). Then the formation of ethane and ethylene is reduced as a result of the carbonization of the active sites of Mo-containing zeolite catalysts. It is obvious that a higher formation of gaseous products is characteristic of mesoporous catalysts, which, apparently, is due to their higher cracking activity in this process. The highest amount of gaseous products is formed on the catalyst Mo/K6, obtained on the basis of a zeolite synthesized with the highest addition of carbon black and containing the maximum number of acid sites. The liquid products of methane conversion contain aromatic hydrocarbons, mainly benzene and naphthalene (Figure 4c and 4d). The highest amount of benzene is formed in the first 20 minutes of the reaction over all the catalysts under study, after which its yield gradually decreases. The increase in the concentration of carbon black used during the synthesis of zeolites results in an increase in the formation of benzene over the catalysts obtained on their basis. The yield in naphthalene reaches its maximum value in 20-60 min of the reaction, then its formation decreases (Figure 4d). The sharpest decrease in naphthalene yield is observed for the Mo/K6 catalyst.

4. Conclusion

Thus, high silica ZSM-5 zeolites with a hierarchical pore system have been synthesized using carbon black. Investigation of their physicochemical properties has shown that the resulted catalysts differ significantly in textural, structural, and acidic characteristics, which leads to differences in their catalytic properties. The highest activity and stability in the course of non-oxidative conversion of methane into aromatic hydrocarbons have been exhibited by a mesoporous catalyst obtained on the basis of a zeolite, synthezised with carbon used in an amount of 1.0%.

Acknowledgments
The work was performed as part of project No. V.46.2.1 of the Program for Basic Scientific Research of State Academies of Sciences.

References

[1] Zhang C-L, Li S, Yuan Y, Zhang W-X, Wu T H and Lin L 1998 *Catal. Lett.* **56** 207
[2] Ma S, Guo X, Zhao L, Scott S and Bao X 2013 *J. Energy. Chem.* **22** 1
[3] Scmidt I, Christensen C H, Hasselriis P, Kustova M Yu, Brorson M, Dahl S and Johannsen K 2005 *Studies in Surface Science and Catalysis* **154** 1247
[4] Christensen CH, Johannsen K and Schmid I 2003 *Journal of the American Chemical Society* **125** 13370
[5] Dipak B Shukla and Viomesh P Panda J1989 *J. Chem. Tech. Biotech.* **44** 147

IOP Conf. Series: Materials Science and Engineering **597** (2019) 012020 doi:10.1088/1757-899X/597/1/012020

Detoxification of oil-contaminated soils by using humic acids

A V Savelyeva[1,2], E V Linkevich[1], N V Yudina[1] and N A Nebogina[1]

[1]Institute of Petroleum Chemistry, Siberian Branch, Russian Academy of Sciences, 4, Akademichesky Ave., 634055, Tomsk, Russia
[2]National Research Tomsk State University, 36, Lenin Ave. 634050, Tomsk, Russia

E-mail address: anna_savel@mail.ru

Abstract. The paper deals with the study of the effect of peat mechanoactivation on the structure and biostimulating and detoxifying properties of isolated humic acids (HA) in of oil-contaminated soil. It is shown that the mechanical activation of peat leads to an increase in the HA yield by a factor of 2-3. In this case, the changes in the fragment composition of HA are observed. Hence, the percentage of aromatic and oxidized alkyl fragments increases, while the amount of oxygen-containing functional groups decreases. The binding of petroleum organic compounds occurs due to the presence of an aromatic skeleton in the structure of HA. The increase in the proportion of aromatic fragments in the structure of mechanically activated HAs increases their affinity for hydrophobic oil compounds, thereby providing them a detoxifying ability. The maximum detoxifying ability of mechanically activated HA determines their biostimulating properties in the wheat seed germination experiment. The biostimulating effect from the use of mechanically activated HA samples is manifested itself to a greater degree in the increase in the height of the sprout stem and the dry mass of the aerial part of the plant. The processes of biodegradation of oil in the soil in the presence of HA are investigated. It is shown that the amount of bitumoids extracted from the soil in the presence of HA is reduced by 30%. The content of hydrocarbons in bitumoids decreases by 50% due to the biodegradation of low molecular weight alkanes. At the same time, the degree of branching of hydrocarbon chains increases, which suggests a microbiological activity. An increase in alcohol-benzene resins in the composition of bitumoids indicates the stimulation of hydrocarbon-oxidizing bacteria by humic acids. According to the data of IR spectroscopy, the content of paraffin hydrocarbons has decreased during the destruction of oil by soil microflora.

1. Introduction

Oil and petroleum products are widely spread pollutants in nature. Hence, the most important and urgent task in the activities of oil producing and oil refining enterprises is detoxification, cleaning and restoring the properties and fertility of soils polluted by oil and oil products. In the environment, the processes of soil self-purification are extremely slow and need to be stimulated. To remediate the oil-polluted soils, a combination of mechanical, chemical and biological methods is used. When using microbiological methods, complex problems arise in the interaction of populations introduced into the soil with the natural microflora. Microflora includes hydrocarbon-oxidizing microorganisms, which are permanent components of soil biocenoses. It is possible to increase the catabolic activity of microorganisms by introducing peat humic acid-based preparations into contaminated soil.

Humic substances make up 50-80% of the soil organic matter and solid fossil fuels. The presence of a wide spectrum of functional groups in molecules of humic acids (HA) in combination with aromatic fragments determines their ability to enter into ionic and donor-acceptor interactions, to form hydrogen bonds, bind toxic substances (pesticides, polycycloaromatic hydrocarbons, heavy metals, etc.), and actively participate in sorption processes [1-7].

The complex composition of macromolecules of humic acids (HA) and their stochastic nature require the establishment of the relationship of the sorption and detoxifying properties of HA with their structure. One of the methods for the structural modification of HA is the method of mechanical activation of solid caustobioliths. It was previously found that the mechanical activation of solid caustobioliths results in a change in the structural parameters of the isolated HA and increases their efficiency when interacting with ecotoxicants and heavy metals [8-10].

The purpose of this work was to study the detoxifying and biostimulating ability of humic acids in relation to oil.

2. Experimental part

The objects of the study were HA of sphagnum peat, HA1 of the mechanically activated peat without additives, and HA2 of the mechanically activated peat with the addition of solid alkali – 3% NaOH of analytical grade. Mechanical activation of peat was carried out under following experimental conditions: the rotational speed of the drums – 1820 rpm^{-1} and centrifugal acceleration – 600 m/s^2. Grinding bodies were steel balls of 8-10 mm diameter. The weight of balls per one drum loading was 0.2-0.5 kg, sample weight was 15-20 g, and time of treatment was 2 minutes. Humic acids were isolated from peat by treating it with 0.1 n NaOH at a temperature of 90 °C in an amount of 150 ml of solution per 1g of the sample weight for 1 hour. Alkaline extraction was repeated three times. Humic acids in an alkaline solution were precipitated with 4% HCl to pH 2. The brown amorphous HA precipitate was separated by centrifugation and then washed with distilled water to pH 7 and dried in a Petri dish in a vacuum oven to constant weight.

The elemental composition of HA was determined using a Carlo Erba Strumentazione 1106 analyzer (Italy). The analysis of fragment composition was performed via $NMR^{13}S$ spectrometry using a Bruker 300 radiospectrometer (Germany) at an operating frequency of 100 MHz via the fast Fourier transform accumulation method. Spectral sweep width was about 26000 Hz, the time of registration of the free induction decay signal (FID) was 0.6 s, the interval between pulses (Td) was 8 seconds at a pulse width of 90°, and the duration of spectrum accumulation was 24 hours. The quantity of preparation weighed 50-70 mg was dissolved in 0.7 cm^3 of 0.3 M NaOD.

To assess the detoxifying properties of humic preparations, laboratory-based vegetation experiments were conducted on the soil contaminated with a crude oil. A universal, uniform soil from Einheitserde Werkverband E0 (Germany) was used as a model soil. To simulate pollution a crude oil at a concentration of 1.5 wt% was used. The dose of nutrient substrate applied per 100 g of soil was 2 ml of the HA solution with a concentration 0.5 g/l. In this case, the calculated amount of HA was dissolved in a small volume of alkali (0.1 N NaOH), neutralized with 1 M HCl solution to pH 7.0, and made with distilled water up to one liter. The HA solutions were applied to the soil on the 3rd and 5th days of cultivation. Wheat of a genetically unmodified cultivar variety Samurai served as a test object. The duration of the experiment was 20 days at a temperature of 23-27 °C with natural light, daily watering or irrigation with appropriate solutions of HA. The physiological activity of humic preparations was assessed by their effect on the height of the stem of a seedling and its biomass using the standard procedure described in [11].

After the experiment, the content of residual oil in the soil was determined by extraction with chloroform. To determine the group composition of crude and residual oil, the method of column chromatography was used. Destructive changes in oil hydrocarbons were analyzed by IR spectroscopy using a Nicolet 5700 spectrophotometer (FT-IR) in KBr tablets at a ratio of 1:300 within the range from 400 to 4000 cm^{-1}. Based on the results of IR spectroscopy, spectral coefficients were calculated. They represent the normalized optical densities of the following absorption bands: D_{1610}/D_{725} –coefficient of aromatization, D_{720}/D_{1380} – content of paraffin structures, D_{750}/D_{720} – content of condensed aromatics, D_{725}/D_{1465} – content of paraffin structures, and D_{1380}/D_{1465} – branching coefficient.

3. Results and discussion

In Table 1 the results of studies of the effect of peat mechanoactivation (MA) on the yield in HA and their elemental and fragmentary composition are presented. Preliminary mechanical activation of peat promotes an increase in the yield of HA by a factor of 2-3 compared with the initial peat. The highest amount of HA is isolated from peat by MA with 3% NaOH. The increase in the yield of HA is possible due to the destruction of hardly hydrolyzable substances. It is evident from the data of elemental analysis that the H/C and O/C atomic ratios are decreased in HA. This suggests an increase in the degree of aromaticity.

Table 1. Effect of peat mechanoactivation on the yield in HA and their elemental and fragmentary compositions.

Parameter	HA	HA2	HA3
Content, wt%	8.5	15.3	25.4
	Content of elements in HA, wt%		
C	53.4	55.6	57.3
H	6.6	5.6	5.4
N	3.1	2.9	3.0
O	36.9	35.8	34.3
H/C	1.48	1.21	1.13
O/C	0.52	0.48	0.45
	Content of C atoms, rel.%		
C=O 160 ppm	17.8	15.3	14.0
CarO 160-140 ppm	13.2	14.6	13.6
CarCH 140-113 ppm	15.9	16.7	16.8
CalkO 107-54 ppm.	26.8	27.9	28.3
Calk 54-0 ppm	26.3	25.5	27.9

The structural parameters calculated from the results of ^{13}C NMR spectroscopy are shown in Table 1. Data analysis indicates a change in the distribution of carbon atoms by the structural fragments of HA depending on the conditions of mechanochemical activation. The percentage of carbon in aromatic and oxidized aliphatic fragments in the HA1 and HA2 molecules from mechanically activated samples is higher than that in HA of the initial peat. The number of oxygen-containing groups has decreased.

An analysis of the chemical composition of oil-contaminated soil in a month after the introduction of HA from the initial peat into it showed a decrease in the amount of bitumoids under extraction by 30% (Table 2).

Table 2. Changes in the content of bitumoids and n-alkanes extracted from the oil-polluted soil.

Content, wt%	Soil + oil	Soil + oil + HA	Soil + oil + HA1	Soil + oil + HA2
Bitumoids of	55.4	39.5	29.4	32.3
n-alkanes	38.7	30.1	20.5	22.5
Benzene resins	1.5	2.5	2.5	3.5
Alcohol-benzene resins	2.1	4.0	7.5	6.5

The amount of extracted bitumoids decreased by 42-47% after the introduction of mechanically activated samples HA1 and HA2 into the soil. A decrease in the percentage of paraffin-naphthenic fractions, in particular, n-alkanes, was observed in the composition of bitumoids (Table 2). At the same time, the content of resins, in particular, acidic alcohol-benzene resins increased due to the new formation of oxygen-containing compounds. This suggests the destructive oxidative activity of microorganisms, stimulated by the introduction of HAs into the soil.

Based on the results of the analysis of the hydrocarbon composition of bitumoids extracted from the soil, the biodegradation coefficients were calculated (Table 3). A decrease in the n-C_{17}/n-C_{27} ratio in bitumoids, extracted from the soil with the addition of HA suggests a decrease in the percentage of low molecular weight n-alkanes, which are subject of biodegradation in a greater degree. The index i-C_{19}/i-C_{17}, which characterizes the degree of branching of hydrocarbon molecules increases in samples with HA additives and confirms the occurrence of more intensive processes of oil biodegradation. The index $2C_{29}/C_{28}+C_{30}$ reflects the degree of destruction of high molecular weight n-alkanes. The values of this index are similar for hydrocarbons isolated from soils containing HA.

Table 3. Indices of hydrocarbon biodegradation.

Objects	Indices of hydrocarbon biodegradation		
	n-C_{17}/n-C_{27}	i-C_{19}/i-C_{17}	$2C_{29}/C_{28}+C_{30}$
Soil + oil	3.3	0.13	1.1
Soil + oil + HA	2.1	0.19	0.8
Soil + oil + HA1	1.6	0.19	0.8
Soil + oil + HA2	1.9	0.25	0.8

In Table 4 are presented the spectral coefficients characterizing the processes of oil biodegradation by soil microflora and HA additives. As follows from the Table 4, the oil biodegraded by soil microflora, stimulated by the addition of HA2, is characterized by increased coefficients C1, C2, which indicates an increase in the fraction of aromatic structures in oil. Biodegradation of paraffin hydrocarbons is confirmed by a decrease in C3 and C4 coefficients and an increase in the branching degree of C5.

Table 4. Spectral coefficients of oil before and after degradation.

Spectral coefficients	Initial oil	After destruction by indigenous soil microflora	After destruction by indigenous soil microflora + HA2
$C_1=D_{1610}/D_{720}$	0.3	1.1	1.5
$C_2=D_{750}/D_{720}$	0.6	1.3	1.6
$C_3=D_{720}/D_{1380}$	0.5	0.2	0.2
$C_4=D_{725}/D_{1465}$	1.9	0.7	0.03
$C_5=D_{1380}/D_{1465}$	0.9	1.7	4.2

The biostimulating and detoxifying effect of HA solutions was manifested in an increase in the height of the stem of seedling and dry weight of the aerial parts of plants compared to the control (Figure 1). The greatest effect was obtained when treating the soil with solutions of mechanically activated samples HA1 and HA2.

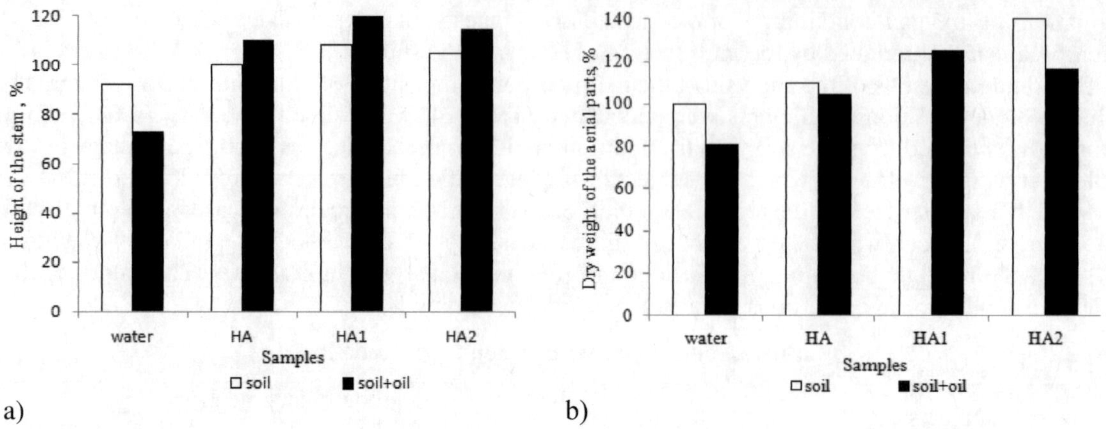

Figure 1. Effect of humic acids on the height of the stem (a) and the dry weight of the wheat seedling (b).

Thus, it has been found that HA exhibit biostimulating and detoxifying properties under conditions of oil-contaminated soil. The greatest effect in this case has been provided by HA of peat, mechanically activated in the presence of alkali. The presence of the aromatic framework provides the ability of HA to bind organic compounds, therefore, as the contribution of the aromatic framework to the HA structure increases, their affinity for hydrophobic organic compounds increases. Maximum aromaticity is characteristic of HA, which determines their high binding and detoxifying ability with respect to oil under the conditions of vegetative experiment.

The processes of oil biodegradation in the presence of HA are accompanied by a decrease in the amount of bitumoids and low-molecular hydrocarbons, and the new formation of high-molecular-weight hydrocarbons. The increase in alcohol-benzene resins in the composition of bitumoids indicates a stimulating effect of humic acids on hydrocarbon-oxidizing microorganisms.

Acknowledgments
The work was performed in the framework of the state assignment of the Institute of Chemical Sciences SB RAS. State registration number AAAA-A17-117030310198-4.

References
[1] Nakashima K, Makia M and Ishikawa F 2007 *Spectrochimica Acta Part A* **67** 930
[2] Mal'tseva E V, Filatov D A, Yudina N V and Chaikovskaya O N 2011 *Solid fuel chemistry* **45** 62
[3] Mal'tseva E V, Yudina N V, Chaikovskaya O N and Nechaev L V 2011 *J. Phys. Chem.* **85** 1558
[4] Perminova I V, Grechishcheva N Yu, Kovalevskii D V, Kudryavtsev A V, Petrosyan V S and Matorin D N 2001 *J. Environ Sci. Tech.* **35** 3841
[5] Wang H B and Zhang Y J 2014 *J. Environ Sci. Health A. Tox. Hazard Subst. Environ Eng.* **49** 78
[6] Plaza C, Brunetti G, Senesi N and Polo A 2006 *Anal. Bioanal. Chem.* **386** 2133
[7] Hernandez D, Plaza C, Senesi N and Polo A 2006 *Environ. Pollut.* **143** 212
[8] Tchaikovskaya O N, Nechaev L V, Yudina N V and Mal'tseva E V 2016 *Luminescence* **31** 1098
[9] Ivanov A A, Yudina N V and Savel'eva A V 2015 *Solid Fuel Chem.* **49** 123
[10] Mal'tseva E V, Yudina N V, Shekhovtsova N S and L P Shilyaeva 2017 *Rus. J. Phys. Chem. A* **91** 1273
[11] Savel'eva A V, Ivanov A A, Yudina N V, Lomovsky O I and Dugarzhav D 2013 *Journal of applied chemistry* **86** 552

Technology for Enhanced Oil Recovery using an Oil-Displacing Acidic System of Prolonged Action

V V Kozlov[1,2], L K Altunina[1,2] , V A Kuvshinov[1], I V Kuvshinov[1] and L A Stasyeva[1]

[1]Institute of Petroleum Chemistry, Siberian Branch of the Russian Academy of Sciences (IPC SB RAS), 4, Akademichesky Avenue, 634055, Tomsk, Russia
[2]National Research Tomsk State University, 634050 Tomsk, Lenin Avenue, 36

kozlov_vv2004@mail.ru

Abstract. An oil-displacing system of prolonged action with adjustable viscosity based on surfactant inorganic acid adduct and polyol is created with the purpose to increase the efficiency of flooding or thermal steam formation treatment and to enhance oil recovery. The laboratory studies of physicochemical, rheological, and filtration characteristics of the system under natural exploitation conditions and thermal-steam treatment are presented. In 2014-2018 the field tests of the technology for enhanced oil recovery were successfully carried out using an acid oil-displacing system at the experimental sites of the Permian-Carboniferous reservoir of the high-viscosity oil of the Usinsk oilfield.

1. Introduction

As oil resources are depleted, technologies aimed at involvement into the development of deposits of hard-to-recover heavy and highly viscous oils have particular relevance [1-5]. The increase in oil recovery of heavy oil deposits can be realized using new physicochemical technologies, in particular, by injection of various compositions of chemical reagents into productive formations. In this regard chemically evolving systems including compositions capable of chemically reacting with the formation rock and fluids show promise. These systems change their composition and properties successively in the course of their motion through the formation. For example, being initially acidic and rock-dissolving, they become neutral or alkaline oil-displacing sol-forming systems after interacting with the rock.

Traditionally, chemically evolving systems develop as the acid composition by the complication with an increase of useful functions. Along with the known acids, complex acids formed by the donor-acceptor interaction of the components of the systems come into use. The composition and properties of complex acids are regulated by the concentrations and ratios of donor and acceptor. The acids have surface-active properties; they can enhance the oil recovery of a low-permeability matrix of the reservoir both due to the dissolution and contra-flow capillary imbibitions of the matrix. Complex acids, much stronger than the initial acid are formed in the 'inorganic polyacid – polyol' systems due to donor-acceptor interaction. The interaction allows increasing the acidity of oil-displacing systems and the duration of their action in the reservoir by increasing the buffer capacity and expanding the range of the buffer action in the acidic pH region. In Figure 1 is shown a scheme for the formation of complex acid and its dissociation into ions by the example of the interaction of boric acid and glycerin.

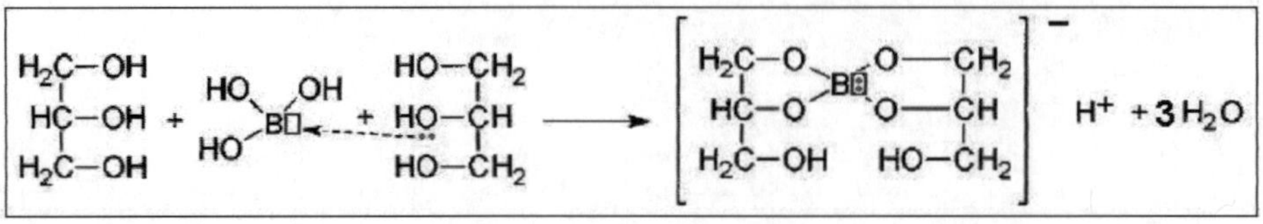

Figure 1. Donor-acceptor interaction of boric acid and glycerol resulted in the formation of complex glycerolboric acid.

The oxygen atom of the hydroxyl group in the molecule of glycerol is the donor; it transfers its lone electron pair to the free orbital of the acceptor that is the boron atom in the boric acid molecule. As a result, one molecule of boric acid and two molecules of glycerol form a molecule of coordination compound, which is glycerol boric acid, four orders of magnitude stronger than boric acid.

To enhance oil recovery in low temperature (20-40 °C) reservoirs, cooled by water injection, and high viscosity oil pools in natural development mode, and in the mode of thermal-steam stimulation, an oil-displacing system based on surfactant, inorganic buffer system, carbamide and polyol with adjustable viscosity or a system of prolonged action based on adduct of inorganic acid was developed at the Institute of Petroleum Chemistry of the Siberian Branch of the Russian Academy of Sciences (IPC SB RAS) in Tomsk. The system has a freezing point of minus 20 ÷ minus 60 °C, low interfacial tension at the boundary with oil, density adjustable in the range from 1.1 to 1.3 kg/m^3, and viscosity from tens to hundreds of mPa·s. The acid system is especially effective in carbonate reservoirs, since it slowly react with carbonate rocks without formation of any insoluble products and promotes the return of the original reservoir permeability. As a result of interaction of the system with the carbonate reservoir, CO_2 is released, which dissolves in oil reducing its viscosity and contributes to an increase in the oil recovery ratio. In addition, as a result of interaction with the carbonate reservoir and hydrolysis of carbamide, which is a part of the composition, its pH rises from 2.8–3.1 to 8.8–10.0 promoting its chemical evolution. The composition is converted into an alkaline oil-displacing system providing prolonged impact on the formation. After thermostatting with the system and carbonate reservoir at 70-120 °C, the viscosity of the oil decreases 1.2-2.7 times.

The paper presents the results of physicochemical studies and laboratory tests of an acid oil-displacing system developed at the Institute of Petroleum Chemistry of the Siberian Branch of the Russian Academy of Sciences.

2. Experimental

The solvent ability relative to carbonate rocks for the oil-displacing acid compositions based on surfactants, inorganic buffer systems (inorganic acid adduct) and polyol with adjustable viscosity was determined by the gravimetric method from the rate of reaction of solutions with marble. To establish the effect of concentration, the mass and surface area of pieces of marble were determined. For this purpose they were placed in bottles with the compositions and kept in an air thermostat at temperatures of 23, 50 and 90 °C for 20 hours. After the experiment, pieces of marble were washed, dried and weighed. The reaction rate of the composition with marble was calculated by the formula:

$$V_p = (m_0 - m)/(S \cdot \tau), \quad (1)$$

where V_p is the reaction rate, g/m^2·h; m_0 is the mass of a piece of marble before the experiment, g; m is the mass of a piece of marble after the experiment, g; S is the area of the piece, m^2; τ is the time of experience, h.

The rheological characteristics were investigated by the method of vibration viscometry using a 'Rheokinetika' vibration viscometer with a tuning fork sensor under heating from 20 to 90 °C and at different shear stresses. The pH values of solutions were obtained by potentiometric method using a glass electrode and a laboratory microprocessor pH meter manufactured by HANNA Instruments. Laboratory filtration tests of an acid system for enhanced oil recovery were carried out using an installation for investigation of filtration characteristics (manufactured by CATACON LLC, Russia), consisting of two parallel columns, each 125 cm^3 in volume.

The filtration characteristics and oil-displacing capacity of the systems were studied using bulk reservoir models made of the disintegrated core material of the Permian-Carboniferous deposit of the Usinsk oilfield, fresh water or a model of reservoir water (synthetic brine) with mineralization 62.1-74.7 g/l of the Usinsk oilfield and degassed oil from the Usinsk oilfield (thermally stabilized oil with the addition of 30 % kerosene). The permeability of the models ranged from 0.893 to 2.300 μm^2; the permeability of parallel columns differed by 1.2-1.9 times. The thermostating time of a heterogeneous reservoir model after treatment with an acid composition was 24 hours and the back pressure was 2 MPa. The effectiveness of the oil-displacing system was studied during the primary displacement of oil and in the course of displacing residual oil with water and steam from two parallel columns with different permeability, as well as under conditions simulating the cyclic steam treatment of producing wells.

The effect of the acid system on the process of oil displacement was studied as follows. First, oil was displaced with water until the products were completely water-flooded from both columns at a given temperature. Every 5-15 minutes, the temperature and the pressure at the inlet and outlet of the columns and the volumes of the displaced oil and water from each column were measured. Based on the data obtained, the gradient of pressure grad P, atm/m, filtration rate V, m/day, fluid mobility k/μ, $\mu m^2/(mPa\cdot s)$, and the coefficient of oil displacement by water C_d, % were calculated. After displacing the oil with water, the bank of the oil-displacing composition was simultaneously injected into both columns, pushed by water for a predetermined distance, and thermostated for a certain time. Then injection of water was continued. In addition, we determined the pH of the fluid at the outlet of the columns and the concentration of urea, which was a part of the composition. Based on the data obtained, the pressure gradient, filtration rate, fluid mobility, and the absolute factor of oil displacement by composition and water were calculated.

3. Results and discussion

As a result of systematic investigations of the effect of pH and the ionic strength of aqueous electrolyte solutions on the interfacial tension of oils and their fractions, an electrocapillary model of the interfacial layer was developed by researchers of IPC SB RAS. The interfacial layer in this model has the properties of an ion-exchange membrane, which is an amphoteric adsorbed layer of the natural oil surfactants [1]. The model explains the extremal dependence of the interfacial tension of oils on the pH of aqueous phase by the existence of an double electrical layer (DEL), resulting from the ionization of the ionogenic groups of natural oil surfactants, which are heteroatomic compounds present in resins and asphaltenes. The model allows correlating the oil component composition with the physicochemical parameters of the interfacial layer. These parameters determine the mechanism of oil displacement by aqueous solutions: interfacial tension, adsorption of natural oil surfactants, composition and concentration of electrolytes of the aqueous phase. The electrocapillary model of the interfacial layer is thermodynamic, i.e. it is free from restrictions on the formation water salinity, in contrast to electrokinetic models of the double layer, e.g. the Gouy-Chapman model and more complex ones. Within this model, the dependence of the interfacial tension of oil σ on the pH of aqueous phase is expressed by the equation [1]:

$$\sigma = \sigma_i - b\cdot\vartheta^2\cdot[-\lg(a_{h+} + K\cdot I)\ ) - pH_i]^2 + c\cdot\vartheta^4\cdot[-\lg(a_{h+} + K\cdot I) - pH_i]^4 - 2.303\cdot X_0\cdot R\cdot T\cdot\lg(1 + K_a\cdot I), \quad (2)$$

where pH_i is the isoion point of the interfacial layer (pH of the maximum of the curve σ – pH); σ_i is the interfacial tension at the isoion point; b is the integral capacitance of DEL; c is the parameter characterizing the dependence of the integral capacity of DEL on its potential; a_{h+} is the activity of hydrogen ions of the aqueous phase; K is the constant ion exchange equilibrium for the interfacial membrane; I is the ionic strength of the aqueous phase; K_a is the adsorption constant for electrolyte ions in the interfacial layer; X_0 is the limiting adsorption of electrolyte ions in the interfacial layer; R is the universal gas constant; T is the temperature; $\vartheta=2.303\cdot R\cdot T/F$; and F is the Faraday number.

Figure 2 shows the calculated (line) and experimental (points) dependencies of interfacial tension on pH for the high-viscosity oil of Permian-Carboniferous deposit of the Usinsk oilfield at the boundary with an aqueous solution of 1-1 electrolyte with constant ionic strength. The calculations were performed for the following values of the equation parameters: σ_i = 37 mN/m; $X_0 = 1.1 \cdot 10^{-6}$ mol/m^2; K_a = 70 kg/mol; I = 0.1 mol/kg; $K = 1 \cdot 10^{-12}$ kg/mol; b = 0.15 F/m^2; c = 0.7 F/(m$^2 \cdot$B^2). Various values of pH of the aqueous phase were set by solutions of acid and alkali, mixed in certain proportions.

The electrocapillary model of the interface predicts the possibility of reducing the interfacial tension in both the alkaline and acidic pH ranges. Hence, in addition to use of alkaline buffer systems, the development and use of acidic buffer systems, such as solutions of complex acids and Lewis acids in coordinating solvents suggest themselves promising. Due to their high buffer capacity, these acid systems will have a prolonged effect and improved surface-active properties, which will allow them to be used as an effective acid agent in carbonate reservoirs of increased permeability of high-viscosity oil deposits.

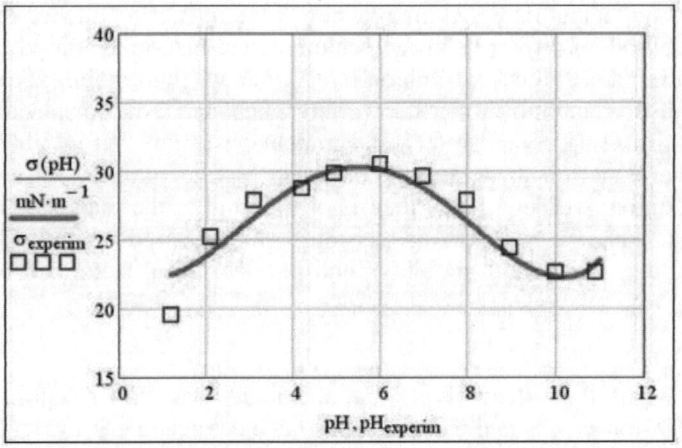

Figure 2. Effect of pH on the interfacial tension of the Permian-Carboniferous deposit of the Usinsk oilfield at the boundary with an aqueous solution of 1-1 electrolyte with an ionic strength 0.1 mol/kg at 80 °C: points – experiment, line – calculation.

The foregoing theoretical models were used for the development of an acid system based on surfactants, an inorganic acid adduct, and a polyhydric alcohol. This system is applicable in a wide range of temperatures, from 10 to 220 °C, in oil fields with terrigenous and carbonate reservoirs, under various geological and physical conditions. The technology of enhanced oil recovery using the composition is aimed at intensifying oil production by increasing the flow rates from producing wells and the injectivity of injection wells by pumping into injection and / or producing wells.

Figure 3 shows the results of the study of the rheological characteristics of oil thermostatted with an acid system. With an increase in the thermostating temperature of oil with solutions of systems based on an inorganic acid adduct and a polyol the pour point of oil decreases from -14 °C to -22 °C at 70 °C of thermostating and from –17 °C to –27 °C at 120 °C.

The dissolution rate of carbonate rock in the acid oil-displacing system decreases with an increase in the amount of urea included in the acid system from 22.1 to 19.2 g /m2·h at 23 °C and from 27.1 to 6.3 g /m^2·h at 90 °C. Lower dissolution rate of carbonate rock in the acid oil-displacing system compared to the dissolution rate of carbonate rock, for example, in hydrochloric acid, allows a long-term impact on the formation rock without destruction of its structure and without the formation of cavities and sediments.

In Table 1 the results of filtration tests of the acid oil-displacing system are presented.

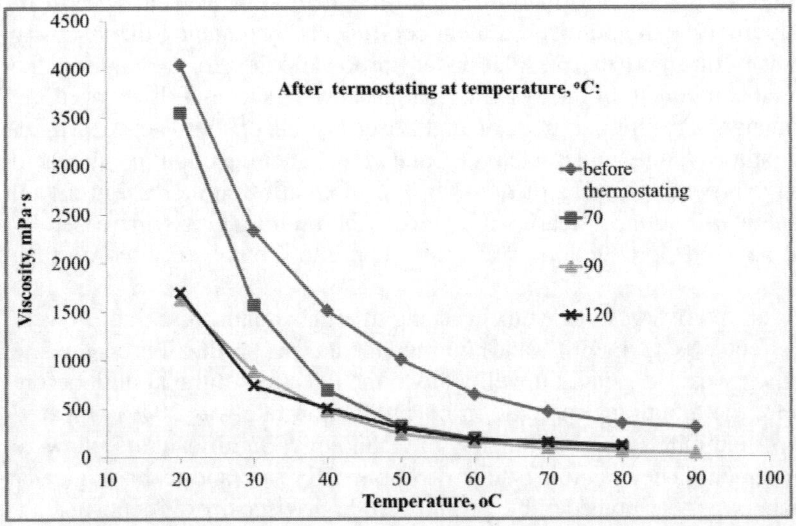

Figure 3. Temperature dependences of viscosity of the oil from the Usinsk oilfield before (initial oil) and after thermostating with the system at various temperatures in the presence of carbonate rock.

Table 1. Filtration characteristics and coefficient of oil displacement under simulated reservoir conditions: natural development conditions (model 1), at low temperatures; natural development conditions followed by cyclic thermal-steam treatment with the acid system (model 2).

Model No	Gas permeability of columns (μm^2)		Mobility ratio		Coefficient of oil-displacement for columns (%)				Increment of the oil-displacement coefficient due to the application of the system (%)	
	1	2	Before injection of the system	After injection of the system	Before injection of the system		After injection of the system for experiment in general			
					1	2	1	2	1	2
1	2.300	1.200	25.5:1	2.6:1	29.6	2.9	45.9	38.1	16.3	35.2
									average 25.8	
2	1.100	0.892	18.5:1	1:1.8	57.6	15.1	69.4	38.1	11.8	23.0
									average 17.4	

As a result of experimental studies, it was found that the treatment of a reservoir with the acid oil-displacing system under conditions of natural development and then under cyclic thermal-steam stimulation contributes to an increase in the oil-displacement ratio of heterogeneous reservoir models. The increment in oil displacement coefficients ranges from 11.8 to 35.2% for the columns. Upon the average, the increment in oil displacement coefficients for heterogeneous reservoir models ranged from 17.4 to 25.8%. The high oil-displacement capacity of the acid system is due to the low interfacial tension at the boundary with the oil, the high washing capacity of surfactants present in the composition, and optimal conditions resulted from the formation of a large-capacity inorganic buffer system. Adduct of an inorganic acid reacts with the polyol to form a stronger acid capable of interacting with the carbonate rock, increasing its injectivity, which also contributes to oil displacement.

Cyclic thermal-steam reservoir treatment, performed after pre-treatment with an acid system of prolonged action, can significantly increase the oil displacement coefficients by reducing the viscosity of oil due to the increase in temperature and the dissolution of carbon dioxide formed during the hydrolysis of carbamide, which is a component of the system. Field tests were successfully carried out in experimental sites of the Permian-Carboniferous pool of high-viscosity oil of the Usinsk oilfield, which were developed in the natural mode under steam stimulation. The injection of the acid composition into 10 low-productive production wells of the Usinsk oilfield allowed increasing the oil production rate by 5.5-14.8 tons/day without any steam stimulation. The liquid rate has increased by 15-25 m^3/day, while incremental oil production was 28 thousand tons, i.e. more than 2000 tons/well [2].

In 2017-2018 the works on oil recovery improvement using thermal-steam stimulation were performed. The duration of the effect was 12 months, and the incremental oil production on the site was 27,000 tons or ~ 3.75 tons/day per each production well. Due to the injection of the system before the cyclic thermal-steam treatment, the additional increase in oil production rate was 20% compared with the previous cycle. Pilot works exhibited a high efficiency of technology. Significant effects were obtained on oil production enhancement. The technology is environmentally safe and technologically efficient. The industrial use of technology has considerable promise for high-viscosity oil deposits.

4. Conclusion

Thus, the acid composition of prolonged action based on surfactant, inorganic acid adduct, and polyol is an effective oil-displacing system having a low interfacial tension at the boundary with oil. The increment of oil displacement coefficients for models of a heterogeneous reservoir using an acid oil-displacing system was 17.4-25.8% on average per model. The dissolution rate of the carbonate rock in the acid oil-displacing system decreases with an increase in the amount of urea included in the composition of acid system from 22.1 to 19.2 $g/m^2 \cdot h$ at 23 °C and from 27.1 to 6.3$g/m^2 \cdot h$ at 90 °C. After thermostating with the system and carbonate reservoir at 70-120 °C, the oil viscosity decreases 1.2-2.7 times.

The acid system of prolonged action could be used as a composition increasing the efficiency of development due to the oil recovery improvement and increase in reservoir coverage and oil displacement, The acid oil-displacing composition could be pumped into injection, steam injection and steam cyclic wells.. It is a low-viscosity, low-pour point, safety liquid, which makes it easy-to-use in the winter period. For the preparation and injection of the acid composition under field conditions, standard oilfield equipment is used. The acid system is applicable both at the early and late stages of development of fields with difficult-to-recover reserves, including deposits of high-viscosity oil.

Acknowledgments

The work was performed within the framework of the Project "V.46.2.3. Physical chemistry and rheology of oil and polydisperse oil-containing systems in the processes of enhanced oil recovery and oil transport (No. 0370-2019-0002)"

References

[1] Altunina L K. and Kuvshinov V A 2007 *Rus. Chem. Rev.* **76 10** 1034-1052
[2] Altunina L K, Kuvshinov V A, Kuvshinov I V, Stasieva L A, Chertenkov M V, Shkrabyuk L S and Andreev D V 2017 *Oil industry* **7** 26-29
[3] Raffan P, Broekhuis A A and Picchioni F 2016 *Journal of Petroleum Science and Engineering* **145** 723-733
[4] Abass A 2014 *Olajire Energy* **77** 963-982
[5] Sorrell S, Speirs J, Bentley R, Brandt A and Miller R 2010 *Energy Policy* **38** 5290–5295

Thermal destruction of the components of high-sulfur vacuum residues

A V Goncharov[1], E B Krivtsov[1,2], N N Sviridenko[1] and A K Golovko[1]

[1]Institute of Petroleum Chemistry, Siberian Branch, Russian Academy of Sciences, 4, Akademichesky Ave., 634055,Tomsk, Russia
[2]National Research Tomsk State University, 36, Lenin Ave., 634050, Tomsk, Russia

E-mail: john@ipc.tsc.ru

Abstract. The problem of processing vacuum residues is relevant due to the enhancement of heavy crude oil recovery. The processing of fuel oil and vacuum residues is complicated due to the high content of resins and asphaltenes present in them. In order to develop new approaches to the processing of vacuum residues, it is necessary to comprehend the mechanisms of thermal degradation of high-molecular-weight oil components. Three samples of high-sulfur vacuum residues are cracked at a temperature of 500 °C and with a various process duration. The characteristic changes in the material and fractional composition of the products of the vacuum residue cracking depending on the process conditions are shown. It is found that the ratio of light to medium distillates in the composition of the products depends on the duration of cracking.

1. Introduction

Due to the wider demands in motor fuels and a simultaneous decrease in light oil reserves, the need for advanced processing of heavy oil residues (vacuum residue, fuel oil) into light fractions (gasoline, diesel) increases [1-4]. Due to the enhancement of heavy crude oil recovery, it is necessary to develop processes based on the destruction of the structures of high-molecular compounds, in particular asphaltenes, with the formation of components of distillate fractions [5, 6]. Asphaltenes, which are contained in significant quantities in oil residues, are mainly responsible for the formation of coke and the deactivation of the catalyst during their processing. The development of thermo-destructive methods for upgrading heavy oil residues will make it possible to reduce the rate of secondary compaction reactions, to achieve a higher degree of feedstock conversion and lower coke formation [7-10]. The aim of the work was to study the patterns of cracking of the components of high-sulfur vacuum residues of different composition.

2. Experimental section

2.1. Objects of study

The objects of study were the vacuum residues of the Novokuibyshevsk refinery, the Omsk refinery, and the vacuum residue obtained from the oil sampled at the Zyuzeyevskoye oil field. Vacuum residues contain a high amount of resin- asphaltene substances; hence, the content of resins reaches 37.8 wt%, while that of asphaltenes – 20.5 wt%. It should also be noted that the vacuum residues are high-sulfur with a low atomic H/C ratio, which characterizes them as unsuitable feed stocks for the production of distillate fractions by the method of catalytic cracking. The main physico-chemical characteristics of the objects of study are presented in Table 1.

Table 1. Physico-chemical characteristics of the objects of study.

Parameter	Novokuibyshevsk vacuum residue (NVR)	Zyuzeyevsk vacuum residue (ZVR)	Omsk vacuum residue (OVR)
Elemental composition (wt%)			
Carbon	82.28	81.56	85.49
Hydrogen	10.73	9.58	10.99
Sulfur	3.04	6.43	1.8
Nitrogen	0.19	0.84	0.56
Oxygen	3.76	1.59	1.16
H/C	1.56	1.41	1.54
Coking behavior	13.59	24.15	14.02
Fractional composition (wt%)			
T_{bp}	343.0	474.0	224.0
200-360	1.4	0.0	0.0
360-500	77.9	0.0	25.9
>500	20.7	100	74.1
Material composition (wt%)			
Oil components	60.7	41.7	65.8
Resins	33.6	37.8	33.1
Asphaltenes	5.7	20.5	1.1
Molecular mass (a.m.u.)			
Resins	600		639
Asphaltenes	718	1552	2142

2.2. Thermal cracking of vacuum residues

Cracking was carried out in 12 cm^3 autoclave reactors at 500 °C for 15-60 minutes. The weight of vacuum residue sample was 7 g. During the experiment, the mass of the empty reactor and the mass of the reactor with the sample were recorded. Upon completion of feedstock heat treatment, the yield in gaseous cracking products was determined from the mass loss of the reactor with the sample after degassing. Then the reactor was washed with chloroform and weighed. The obtained difference between the reactor mass before and after the experiment was determined as coke.

2.3. Determination of the material composition

The material composition of vacuum residues and cracking products was established according to the standard scheme. First, the content of asphaltenes in the sample was determined by the 'cold' Golde method. Next, the content of oils and resins in the resulted maltenes was determined by the adsorption method. For this purpose the sample under analysis was applied on the activated ASC silica gel and the mixture was placed in a Soxhlet extractor for sequential elution of hydrocarbon components. The oil components were extracted with n-hexane, while resins – with ethanol-benzene mixture in the ratio 1:1.

2.4. Determination of fractional composition

The fractional composition of the liquid cracking products was determined by gas-liquid chromatography. Gas-liquid chromatography (GLC) was performed using a Kristall 2000M chromatograph with a quartz capillary column (30 m × 0.22 mm), and a stationary phase SE-54. Helium was used as a carrier gas. The device was equipped with a flame ionization detector (FID).

The linear temperature increase ranged from 80 to 290 °C and the heating rate of the column thermostat was 15 deg/min. The identification of hydrocarbons and the separation of chromatogram segments into gasoline (bp – 200 °C) and diesel (200-360 °C) fractions were performed according to the retention times of n-alkanes (hexane and hexadecane), pristane, and phytane.

2.5. Gas chromatographic analysis
The determination of the hydrocarbon composition of the gaseous products of the cracking of vacuum residues was carried out by the gas chromatographic method using a Kristall 5000 chromatograph according to GOST 31371.3-2008. The detection of hydrogen, oxygen, and nitrogen was carried out using a column filled with NaX molecular sieves (fraction 80-100 mesh., column length 3 m, internal diameter 2 mm). The velocity of the carrier gas (argon) was 30 ml/min. The separation of C_1-C_6 hydrocarbons was performed on a column filled with Porapak R polymer sorbent (fraction 80-100 mesh, Column length 3 m, internal diameter 2 mm). The velocity of the carrier gas (helium) was 30 ml/min. The analysis was carried out in the mode of temperature programming simultaneously on two columns from 35 to 170 °C with heating rate 5 °C/min.

3. Results and discussion
To determine the thermal stability of vacuum residue components, the cracking was carried out at 500 °C. The time of process was 15, 30, 45 and 60 minutes (Table 2). It was found out that during the thermal treatment of the objects of study, the destruction involving mainly resins results in the formation of petroleum coke and gas [11]. At the initial stage of cracking of the vacuum residues from the Novokuibyshevsk refinery (15 min.), the content of asphaltenes in liquid cracking products increases, which is explained by the predominance of the rate of formation of asphaltenes from resins over the rate of condensation of asphaltenes to coke. An increase in the cracking time from 15 to 60 minutes leads to a decrease in the yield of liquid products due to the intensification of gas and coke formation processes. The amount of resin-asphaltene substances in the composition of cracking products decreases: the content of resins by 76 %, and that of asphaltenes – by 86 rel.% (compared with the initial vacuum residue). This is due to the prevalence of reactions of condensation of resin-asphaltene components to coke. The sulfur content in oils increases due to transition of low molecular weight sulfur-containing fragments into the oil fraction [12]. An increase in the cracking time also contributes to a slight increase in the content of sulfur compounds in the composition of liquid products. This is probably due to the predominance of the rate of condensation of sulfur-free components to a solid product and, as a consequence, the concentration of sulfur-containing compounds in the liquid cracking products.

It was found that the initial vacuum residue, extracted from the Zuzeyevsk oil, contains the highest amount of resin-asphaltene substances: resins – 37.8 wt% and asphaltenes – 20.5 wt%, which greatly simplifies the process of their condensation. For the time of cracking 15 minutes, the content of oils in liquid products increases due to the destruction of resins resulted in the formation of low molecular weight components and gaseous products. If the cracking lasts for 60 minutes, the yield in gas and coke exceeds 58 wt%. The content of asphaltenes is reduced by 98% and that of resins – by 82 rel.% compared with the initial vacuum residue as a result of their condensation to solid products of cracking. The amount of sulfur in liquid products decreased by 72 rel.%. It is shown that at the initial stage of cracking (0-15 min) of the vacuum residue from the Omsk refinery the reactions of resin condensation with the formation of significant amounts of newly formed asphaltenes and coke prevail. A further increase in the time of process leads to an intensification of the reactions of destruction of oil components and, as a result, gas formation increases. The content of resins in liquid products is significantly reduced (the residual amount is 24 % relative to their initial content in the vacuum residue).

Table 2. Composition of the products of vacuum residue cracking.

Time	Content, wt%					
	Gas	Liquid (S_{total})	Coke	Oils (S_{total})	Resins	Asphaltenes
			NVR			
Initial	0	100 (3.04)	0	60.70 (1,39)	33.6	5.7
15 min	18.38	81.33 (2.41)	0.29	58.63 (1.36)	16.46	6.24
30 min	19.70	75.06	5.24	56.7	13.5	4.86
45 min	22.48	65.78	11.74	53.69 (1.13)	8.62	3.47
60 min	26.2	57.8 (2.84)	16.0	48.0 (1.86)	9.0	0.8
			ZVR			
Initial	0	100 (6.43)	0	41.7 (1.82)	37.8	20.5
15 min	10.4	72.1 (3.74)	17.5	53.5 (1.80)	9.9	8.7
30 min	13.7	61.8 (2.67)	24.5	52.0 (2.24)	8.3	1.5
45 min	14.7	60.0 (2.34)	25.3	50.2 (2.17)	9.1	0.7
60 min	20.7	41.4 (1.78)	37.9	34.0 (1.54)	6.8	0.6
			OVR			
Initial	0	100 (1.80)	0	64.9 (0.67)	34.0	1.1
15 min	7.9	87.1 (1.05)	5.0	66.9 (0.53)	14.9	5.3
30 min	10.9	78.6	10.5	59.4	15.3	3.9
45 min	11.8	77.1	11.1	59.1	15.1	2.9
60 min	31.6	49.5	18.9	40.7	8.1	0.7

Table 3. Fractional composition of liquid products of vacuum residue cracking.

Time	Content (wt%)	
	B.p. – 200	200 – 360
	NVR	
Initial	0.0	1.4
15 min	13.2	29.4
30 min	27.9	23.4
45 min	27.2	20.3
60 min	30.6	16.9
	ZVR	
Initial	0.0	0.0
15 min	18.8	26.3
30 min	19.2	24.7
45 min	20.1	22.9
60 min	16.0	15.0
	OVR	
Initial	0.0	25.9
15 min	13.6	39.1
30 min	21.3	32.8
45 min	24.6	31.8
60 min	12.5	27.1

The fractional composition of the liquid products of the cracking of vacuum residues is presented in Table 3. It was found that the formation of components of gasoline and diesel fractions occurs at the initial stage of the process. Probably, the fraction bp-200 °C is formed due to cracking of the oil components, while the formation of the fraction 200-360 °C occurs during resin cracking. With an increase in the time of process, the content of resins decreases and, as a result, the rate of formation of components of the fraction 200–360 °C becomes lower than the rate of their cracking. This leads to a decrease in the content of the fraction 200-360 °C in cracking products and an increase in the yield of the fraction bp-200 °C. If cracking of Omsk vacuum residue (OVR) and Zuzeyevsk vacuum residue (ZVR) lasts 45-60 min, the yield in the bp-200 °C fraction decreases. This is probably due to autocatalytic reactions occurring on the surface of coke, which to this time becomes sufficient to initiate the reaction of gasoline cracking on its surface [13].

The analysis of gaseous products showed that the cracking of vacuum residues results in the formation of C_1-C_4 hydrocarbons predominantly (Table 4), while iso- and n-pentanes, carbon dioxide are also present. The main gaseous cracking products are methane and ethane (~ 85 rel.% of the total amount of all gases). An increase in time of the process leads to an increase in the content of C_1–C_3 hydrocarbons in the composition of cracking gases of all the objects of study. The content of C_4–C_5 hydrocarbons remains the same, being less than 1wt%.

Table 4. Composition of gaseous products of cracking of vacuum residues.

Gases	Content, wt%											
	OVR 15	OVR 30	OVR 45	OVR 60	ZVR 15	ZVR 30	ZVR 45	ZVR 60	NVR 15	NVR 30	NVR 45	NVR 60
H_2	0.23	0.39	0.67	0.90	0.49	0.70	0.57	0.45	1.07	0.79	0.53	0.82
CH_4	4.52	7.26	8.48	21.42	6.50	8.58	10.04	14.46	12.11	14.28	17.28	17.31
CO_2	0.17	0.06	0.08	0.24	0.29	0.37	0.28	0.36	0.58	0.34	0.42	0.26
C_2H_4	0.02	0.01	0.01	0.01	0.01	0.01	0.01	0.0	0.04	0.01	0.01	0.26
C_2H_6	1.50	2.05	1.68	6.49	1.96	2.52	2.52	3.61	2.97	2.88	4.54	4.88
C_3H_8	0.96	0.82	0.61	1.36	0.79	1.06	0.91	1.32	1.19	1.02	1.78	1.92
i-C_4H_{10}	0.12	0.08	0.06	0.26	0.09	0.12	0.10	0.14	0.12	0.11	0.19	0.19
n-C_4H_{10}	0.27	0.17	0.14	0.66	0.19	0.24	0.19	0.28	0.23	0.20	0.52	0.41
i-C_5H_{12}	0.05	0.03	0.03	0.11	0.04	0.05	0.04	0.04	0.04	0.03	0.09	0.07
n-C_5H_{12}	0.06	0.03	0.04	0.15	0.04	0.05	0.04	0.04	0.03	0.04	0.12	0.08
$\sum$gas	7.90	10.90	11.80	31.60	10.40	13.70	14.70	20.70	18.38	19.70	25.48	26.20

4. Conclusion

It has been found out that resins prevail in the composition of initial resins (33-38 wt%). The dependence of the degree of conversion of the components of vacuum residues on the duration of cracking has been revealed. It has been shown that the reactions of condensation of resins to asphaltenes and coke readily run at the initial stage of the process (0-15 min). With an increase in the duration of cracking, reactions of destruction of oil and resin components with the formation of distillate fractions begin to proceed, and gas and coke formation also increases.

The content of distillate fractions in liquid products of vacuum residue cracking can reach 50 wt%. At the initial stage of cracking, accumulation of the fraction 200-360 °C occurs in the composition of products. The increase in the time of process results in cracking of this fraction and the accumulation of distillate (boiling point 200 °C) in the liquid cracking products.

Acknowledgements

The work was carried out as a part of project No. V.46.2.2 of the Program for Basic Scientific Research of State Academies of Sciences.

References

[1] Okunev A G, Parkhomchuk E V, Lysikov A I, Parunin P D, Semeykina V S and Parmon V N 2015 *Russ. Chem. Rev.* **84** 987-999
[2] Sadrameli S M 2016 *Fuel* **173** 285-297
[3] Lababidi H, Sabti H and Humaidan F 2014 *Fuel* **117** 59-67
[4] Khalikova D A, Petrov S M and Bashkirtseva N Y 2013 *Bull. KSTU.* **16** 217-221
[5] Eletsky P M 2017 *J .SibFU Chem.* **4** 545-572
[6] Humaidan F, Hauser A, Rana M, Lababidi H and Behbehani M 2015 *Fuel* **150** 558-564

IOP Conf. Series: Materials Science and Engineering **597** (2019) 012022 doi:10.1088/1757-899X/597/1/012022

[7] Xin S, Liu Q, Wang K, Chen Y, Yuan P, Cheng Z and Yuan W 2016 *J. Chem. Eng. Sci.* **146** 115-125
[8] Kaminski T and Husein M M 2018 *Fuel Process. Technol.* **181** 331-339
[9] Mousavi M, Adbollahi T, Pahlavan F and Fini E H 2016 *Fuel* **183** 262-271
[10] Kannaiyan S, Madhavan V R, Rajagopal S and Jayabalan A 2016 *Appl. Therm. Eng.* **97** 13-21
[11] Karpov Y, Krivonosova A, Krivtsov E, Golovko A and Krivtsova N 2014 *Proceed. Chem.* **10** 424-429
[12] Shah A, Fishwick R, Wood J and Lake G A 2010 *J. Energy Environ*. Sci. **3** 700-714
[13] Menoufy M F, Ahmed H S, Betiha M A and Sayed M A 2014 *Fuel* **119** 106-110

IOP Conf. Series: Materials Science and Engineering **597** (2019) 012023 doi:10.1088/1757-899X/597/1/012023

Differential potentiometric precipitation titration of zinc(II) and copper(II) using carbon composite electrode

J A Poluboyarinova[1] and S V Shumar[2]

[1]National Research Tomsk Polytechnic University, 30 Lenina Avenue, Tomsk 634050 Russian Federation
[2]National Research Tomsk State University, 36 Lenina Avenue, Tomsk 634050 Russian Federation

E-mail: [1]ju-li-a.kz@mail.ru; [2]shumar.svetlana@yandex.ru

Abstract. The article presents the results of investigation of potentiometric titration of the working solution containing Zn(II) and Cu(II) by sodium diethyldithiocarbamate (NaDEDC) using carbon composite electrode as the indicator electrode. The proposed method for the separate determination of copper ions (II) and zinc (II) under the joint presence was tested on data from titration model of binary mixtures and the real object. The titration curves were processed by the linearization method based on the transformation of a titration curve into the multiphase linear regression, whose parameters determine the equivalence point with high accuracy.

1. Introduction

Zinc and copper are metals which are widely used in various areas of industry, science and technology. The control of amount of these metals in different objects is performed by physical and physicochemical methods of analysis, many of which demand the expensive equipment, reagents and additional sample preparation step [1–4]. Electrochemical methods of analysis are advantageous because of their polyfunctional performance, high accuracy, low cost and simplicity. The efficiency of an electrochemical method is determined by the use of appropriate detection system (i.e. indicating electrode or sensor). Therefore, development of new electrode systems and their implementation in the routine analysis are prospective areas of analytical chemistry. Usually potentiometric determination of copper and zinc is carried out using metal, ion-selective, carbon and other indicator electrodes [5-9]. A new electrochemical sensor, carbon composite electrode (CCE), was previously shown to use in potentiometric determination of some heavy metals [10]. The sensor demonstrated long lifetime and ease working surface renovation; it can be modified (bulk and surface) with a conducting or insolating phase; also, it has low cost. The CCE may be used in various electrochemical techniques (e.g. coulometry, voltammetry, potentiometry, etc.) due to its polyfunctional properties.

In this work, the CCE has been proposed as a sensor for the potentiometric titration of Zn(II) and Cu(II). The conditions for the separate determination of zinc(II) and copper(II) were studied and have been optimized. The developed technique has been applied for the determination of zinc and copper in simulation solutions and in a jewelry alloy containing copper and zinc.

2. Experimental part

The potentiometric measurements were performed with the pH meter/ionomer Ethan (Russia). The working electrode was CCE. The CCE was made of polyethylene concentrate containing 30% carbon with uniform distribution of particles in the bulk of electrode [11]. The solution of NaDEDC was added to analyt with a pneumatic microburette. The acidity of a solution was measured with pH-meter.

The potentiometric titration of Zn(II) and Cu(II) was performed according to the following procedure. The 0.5 ml portions of a titrant were added to the solutions. The portions were reduced to 0.2 ml near the equivalence point. The electromotive force was recorded after each addition of a titrant portion and solution stirring until a stable potential is reached. All the titration curves were processing by the fragmentary linear approximation method in order to determine the equivalence point. This method was developed by Maryanov. This method helps to determine the equivalence point at any part of the titration curve. The principle of the method is to use the mathematical model to transform a titration curve into a linear regression [12].

2.1 Titrant for potentiometric titration of Zn (II) and Cu (II)

Dithiocarbamates have gained widespread use in potentiometry among chelating agents for titrimetric determination of metal ions. NaDEDC in particular is widely used in analysis due to its reactivity, simple synthesis and other essential analytical properties.

It can form a low-soluble precipitate with Cu(II) - Zn(II) ions in the ratio 1:2 in a wide pH range according to Eqs. 1 and 2:

$$Cu^{2+} + 2DEDC^{-} \rightarrow Cu\,(DEDC)_2\downarrow \quad (1)$$

$$Zn^{2+} + 2DEDC^{-} \rightarrow Zn\,(DEDC)_2\downarrow \quad (2)$$

The sensitivity of the CCE and silver electrode toward $DEDC^{-}$ ion may be explained by the redox properties of the titrant. The redox pair of oxidized and reductive forms of dithiocarbamate – thiuram disulfide reagent arises in NaDEDC solution (Eq. 3) and the indicator electrode exhibits its redox behavior.

$$2(C_2H_5)_2NCS\text{–}S^{-} \rightleftarrows (C_2H_5)_2NS\text{–}S\text{–}S\text{–}SN(C_2H_5)_2 \quad (3)$$

2.2 The analytical response of Zn (II) and Cu (II) on CCE

The typical potentiometric curves for Zn(II) and Cu(II) with NaDEDC using CCE are shown in Figures 1 and 2. The titration curves were transformed into a linear plot by the linearization method in order to determine the equivalence point. Since there are pronounced jumps on the titration curves, the AB regions were processed for this purpose.

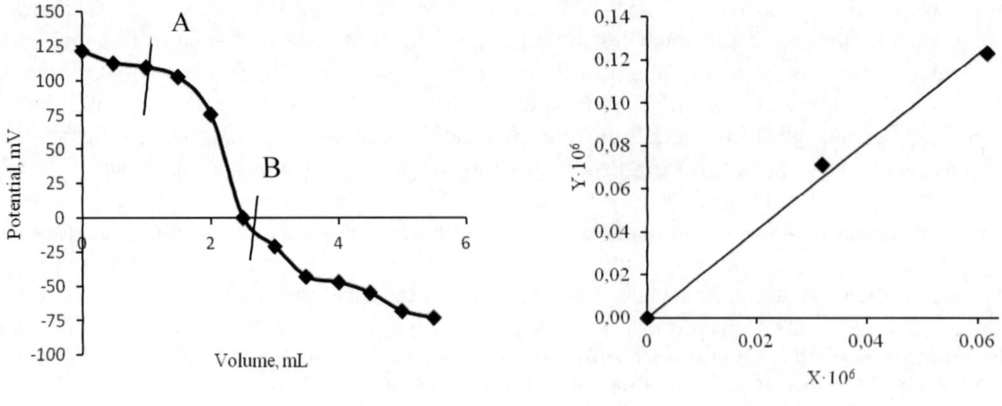

a b

Figure 1. a – Potentiometric titration curve for the simulation solution of $2{\cdot}10^{-3}$ M Zn(II) with $2{\cdot}10^{-2}$ NaDEDC using CCE; the aliquot volume is 25 ml; b – the linear plot for AB fragment of the titration curve.

Figure 1b shows the linear plot for AB region, which is approximated by a straight line with a good correlation coefficient r = 0.9999; the slope V_{eq} (Zn) is 2.314 ± 0.043. This demonstrates good agreement between the titration model and the experimental data. The precipitate solubility K_{sp} value of a precipitate formed has a sufficient impact on the results of potentiometric titration The experimental solubility product of $Zn(DEDC)_2$ was calculated to be 4.034×10^{-15} using a regression intercept term y_0. Figure 2b shows the linear plot for AB region, which is approximated by a straight line with a good correlation coefficient r = 0.9999; the slope V_{eq} (Cu) is 3.322 ± 0.001. This also demonstrates a good agreement between the titration model and the experimental data. The experimental solubility product of $Cu(DEDC)_2$ was calculated to be 1.729×10^{-19} using a regression intercept term y_0.

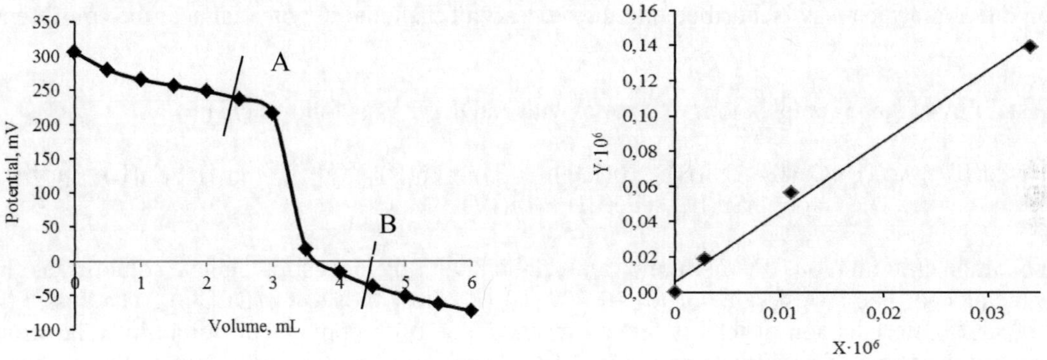

a b

Figure 2. a – Potentiometric titration curve for the simulation solution of $2.314\cdot10^{-3}$ M Cu(II) with $2\cdot10^{-2}$ NaDEDC using CCE; the aliquot volume is 25 ml; b – the linear plot for AB fragment of the titration curve.

2.2.1 Effect of supporting Electrolyte

Supporting electrolytes and buffer solutions were used in order to stabilize the ionic strength and to keep pH constant for better reproducibility of analysis results. It is especially necessary to add the supporting electrolyte to very dilute solutions with low conductivity in order to reach a stable electrode potential.

A decrease of the jump in potential is observed for Zn(II), with the pH increase up to 8 in ammonia buffer solution. It is impossible to conduct the potentiometric titration in strong acidic solutions at pH3 because of the decomposition of NaDEDC in acidic solutions. The curves with the most pronounced jumps in potential were obtained at the pH range from 4 to 6 using H_3BO_3 supporting electrolyte. The solution of H_3BO_3 can stabilize the ionic strength and prevent the Zn(II) surface adsorption. Unlike zinc, copper can be titrated in an alkaline medium using ammonia buffer solution pH8. The curves with the most pronounced jump in potential were obtained at the pH range from 7 to 8 using NH_4OH supporting electrolyte.

Tables 1 and 2 present the results of titration of simulation solutions of zinc and copper with NaDEDC.

Table 1. The results of titration of simulation solutions of zinc.

	Added (mg)	Found (mg)	Δ (%)	S_r
$1\cdot10^{-2}$	8.1	8.0±0.1	1.6	0.007
$2\cdot10^{-3}$	1.6	1.57±0.03	3.6	0.007
$2\cdot10^{-4}$	0.162	0.181±0.005	11.6	0.011
$2\cdot10^{-5}$	0.016	0.010±0.006	15.6	0.005

Table 2. The results of titration of simulation solutions of copper.

	Added (mg)	Found (mg)	Δ (%)	S_r
$2\cdot10^{-2}$	18.6	19.1±0.1	2.6	0.0027
$2\cdot10^{-3}$	1.86	1.83±0.03	2.0	0.0068
$2\cdot10^{-4}$	0.186	0.179±0.002	3.1	0.0044
$2\cdot10^{-5}$	0.0186	0.0203±0.0004	9.4	0.0098

An error of Zn(II) and Cu(II) potentiometric determination using CCE is increased with the solution dilution, which may be attributed to the decrease of the jump in potential near the equivalence point.

The stability of some complex ions of a metal with NaDEDC is as follows [13]:

Hg(II) > Pd(II) > Ag(I) > Cu(II) > Ni(II) > Bi(III) >Co(II) > Pb(II) > Cd(II) > Tl(I) > Zn(II) > In(III) > Sb(III) > Fe(III) > Te(IV) > Mn(II).

In potentiometric titration of a mixture of ions, the ion with the lowest precipitate solubility is first being precipitated. The lowest value of K_{sp} =1.729×10^{-19} corresponds to Cu(DEDC)$_2$. The possibility of the separate determination of ions is limited by the value of the precipitate solubility. The strong difference in the value of precipitate solubility of the ions increases the possibility of their separate determination.

2.3 Simulation solutions of Zn (II) and Cu (II)
The titration of the simulating solutions containing zinc(II) and copper(II) with various rations with NaDEDC has been done. The titration curve (Figure 3a) has a jump in potential corresponding to the simultaneous precipitation of Zn(DEDC)$_2$ and Cu(DEDC)$_2$, so the separate determination of copper and zinc ions is impossible under these conditions.

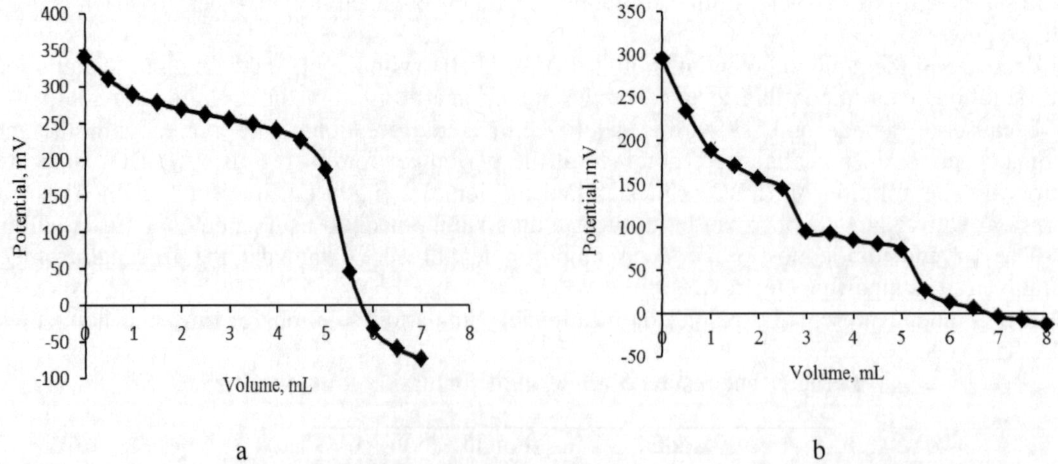

Figure 3. - Potentiometric titration curves of 25 ml solution containing $2\cdot10^{-3}$ M of zinc(II) and $2.3\cdot10^{-3}$ M copper(II) with 0.02 M NaDEDC using CCE; a - without addition of the supporting electrolyte, b - with an addition of the supporting electrolyte.

It has been shown that the separate determination of copper(II) and zinc(II) ions by potentiometric titration with CCE is possible using EDTA as a background electrolyte. The medium alkaline pH7-8 was maintained with ammonia buffer solution. This helped to avoid ion hydrolysis and significantly affect the solubility constant for the copper and zinc complexes obtained. A sufficient difference in the conditional constants makes it possible to carry out the differential titration. The titration curve (Figure 3b) contains two jumps in potential. The first jump correlates with copper(II) titration whereas the second one correlates with the total precipitation of zinc(II) and copper(II) ions.

Table 3 presents the results of the titration of simulation solutions containing Zn(II) and Cu(II) with various ratios with NaDEDC titrant using CCE.

Table 3. The results of the titration of simulation solutions containing Zn(II) and Cu(II) with various ratios with NaDEDC titrant using CCE (P = 0.95, n = 3).

	Ratio of the components	Added (mg)	Found (mg)	Δ (%)	S_r
Cu	1	1.86	1.89±0.05	1.86	0.011
Zn	1	1.58	1.5±0.1	4.13	0.026
Cu	2.5	4.64	4.82±0.02	3.86	0.0017
Zn	1	1.58	1.46±0.15	7.96	0.043
Cu	5	1.86	1.78±0.01	3.94	0.0024
Zn	1	0.32	0.347±0.002	9.5	0.0275

Table 3 shows that the increase of the difference in the ratio of the components leads to the increase of error which may reach 10%. A larger error for zinc ion determination compared to the one for copper ions can be explained by a partial co-precipitation of the second component of a mixture onto the first component precipitate. The method of differential determination of copper and zinc ions has been developed and tested in the analysis of a jewelry alloy containing copper and zinc. The results were in a good agreement with those obtained by X-ray fluorescence within random error, and the total content of zinc and copper correlated with the sample weight. Relative standard deviation for the copper and zinc determination was typical for potentiometric titration analysis.

3. Summary

The basic characteristics of a new sensor for potentiometric analysis have been studying. The conditions for Zn(II) and Cu(II) titration in working solutions (composition of supporting solution and pH) have been optimized, and the procedure for Zn(II) and Cu(II) determination by potentiometric precipitation titration has been developed. The conditions of differentiated determination of copper ion (II) and zinc (II) from a single solution by potentiometric titration with computer processing results of fragmentary linearization method were studied. The random errors for the titration of $1\cdot10^{-2}$-$1\cdot10^{-4}$ M solutions are the range of 2-10%, S_r is 0.003–0.01. The developed procedure has been successfully tested in the analysis of a jewelry alloy, which has demonstrated that the results meet the expected values within the confidence interval and are confirmed by precipitation titration and X-ray fluorescence.

References

[1] Uddin M N and Salam A M 2013 *Chemosphere* 366-73
[2] Sardans J, Montes F and Penuelas J 2011 *Soil and Sediment Contamination* **20** 447-91
[3] Jen-Fon J and Chih-Shih C 1992 *Analytica Chimica Acta* **270** 55-61
[4] Krawczynski , Krawczyk T, Trojanowicz M, Hulanicki A 1988 *Anal. Chem.* **33** 69-73

IOP Conf. Series: Materials Science and Engineering **597** (2019) 012023 doi:10.1088/1757-899X/597/1/012023

[5] Mohammadzadeh K R 2013 *Russian Journal of Electrochemistry* **5** 515
[6] Vlasov Y G, Ermolenko Y E, Legin A V, Rudnitskaya A M and Kolodnikov V V 2010 *Anal Chem* **9** 900-19
[7] Honeychurch K C and Rymansaib Z 2018 *Sensors and actuators* **267** 476-82
[8] Shumar S V, Kuzminskaya E A and Gavrilenko M A 2015 *Bulletin of the Tomsk Polytechnic University* **8** 71-8
[9] Shumar S V, Kuzminskaya E A and Lapova T V 2015 *Yale Review of Education and Science* **16** 496-504
[10] Shumar S V, Kuzminskaya E A and Petrova M A 2016 *AIP Conf. Proc.* **1772** 050010-1-050010-7
[11] Noskova G N, Zakharova E A, Chernov V I, Zaichenko A V, Elesova E E and Kabakaev A S 2011 *Anal. Methods* **5** 1130-5
[12] Marianov B M, Zarubin A G and Shumar S V 2003 *Journal of Anal Chem* **58** 342-6
[13] Chebotarev V K 1998 *Talanta* **47** 1043-51

IOP Conf. Series: Materials Science and Engineering **597** (2019) 012024 doi:10.1088/1757-899X/597/1/012024

Synthesis of aluminum-oxide adsorbents using alkalis at the hydration stage and study of their physicochemical and adsorptive properties

A V Livanova[1], Zh B Budaev[1], E P Meshcheryakov[1], S I Reshetnikov[2], L A Isupova[2] and I A Kurzina[1]

[1]Tomsk State University, Russia, Tomsk, Lenin ave., 36, 634050
[2]Boreskov Institute of Catalysis SB RAS, Russia, Novosibirsk, 5 Lavrentieva ave., 630090

E-mail: truelivanova@mail.ru

Abstract. The paper presents the research results of physicochemical characteristics, adsorption kinetics and value of adsorption capacity with respect to water vapors of the aluminum-oxide adsorbent modified with potassium and sodium cations. It was shown that modification of desiccants by potassium and sodium cations at the hydration stage of products of hydrargillite thermal activation during synthesis allowed obtaining adsorbents-desiccants with high values of static (more than 20 g/100g) and dynamic capacitance (more than 5.0 g/cm^3) by water vapors.

1. Introduction

One of the key priorities in perfection of the process of sorption air drying is the right choice of commercially produced adsorbents depending on the required degree of air drying [1], dynamic capacitance and regeneration conditions, mechanical strength, resistance to condensed moisture [2], abrasion and etc. or development of new sorbing materials possessing high sorptive and strength characteristics [3]. Generally, silica gels, zeolites, active aluminum oxide are used in the systems of air drying. It is common knowledge that aluminum oxide in its characteristics surpasses silica gels at low humidity and zeolites at high humidity. Besides, an important positive feature of aluminum oxide is its water resistance. This is an indicator that often predetermines the choice of aluminum oxide as an adsorbent for drying and reprocessing of media, which contain condensed moisture [1]. In addition, aluminum-oxide adsorbents have such advantages over other desiccants as high mechanical strength, stability of adsorption capacity during multicycle work, completeness and simplicity of regeneration, resistance to wear, resource availability [4, 5]. The purpose of the work is to obtain samples of aluminum-oxide adsorbents – desiccants – modified by potassium and sodium cations at the hydration stage during synthesis and to study their physicochemical characteristics.

2. Materials and methods of research

To obtain adsorbents based on aluminum oxide, a method of centrifugal thermal activation (CTA) of hydrargillite (HG) with subsequent hydration of the obtained product in mild conditions was used, which allowed synthesizing adsorbents-desiccants with a high specific surface [6, 7].

Hydration of the product by CTA of HG was carried out in the temperature range of 97-98 °C at atmospheric pressure during 6 hours in the alkaline medium in the reactor during constant stirring. A ratio of solid and liquid phases was =1:4; pH of the medium at the moment of loading the CTA powder was equal to 10-12. An alkaline module (alkaline mole / Al_2O_3 moleKOH medium), and during synthesis of samples 5A, 8A – 0.06 (in the NaOH medium). Hydration products, for the purpose of cost-cutting of the technological process, were not washed free from admixtures; they were discharged into a stainless steel container and placed in the drying cabinet with forced convection at temperature of 120 °C during 24 hours. After drying, the sample was discharged and ground in the ball mill for 8 hours. As a result of it, fine-dispersed aluminum hydrate with the powder particle size of 5-25 μm was obtained. Later ground aluminum hydrate in the amount of 250 g was poured into the mixer with Z-shaped blades. Then the aqueous solution of electrolyte-peptizer was added. In case of necessity, an additional amount of water was added drop-wise to obtain plastic mixture suitable for forming by extrusion. During peptization, nitric acid was used to synthesize all the samples. An acid module (acid mole/Al_2O_3 mole) for samples 1A, 2A, 5A was 0.05, and for sample 8A – 0.08. The finished plastic mixture was pressed through the die obtaining granules of cylinder shape (diameter – 3.5 ÷ 3.6 mm, length – 5.0 ÷ 6.0 mm). The obtained extrudates were dried in the air during 24 hours, whereupon they were exposed to baking in the flow of dry air in the pipe furnace with a volume of 5 litres at 450 °C for 4 hours with an air flow rate of 5000 h^{1}. The duration of the baking after acceleration was 4 hours. The baking rate was no more than 50 °C/hour. After the baking, the samples were cooled and tested by measuring static and dynamic capacitance by water vapors, and other characteristics were studied, and other characteristics were studied, and other characteristics were studied.

The phase composition and structural parameters of the samples were researched using the diffractormeter XRD-6000 based on CuK_{α}-radiation. Analysis of the phase composition was conducted using data bases PDF 4+, as well as programs of full-profile analysis POWDER CELL 2.4. The content of the mass fraction of Na and K was determined by atomic emission spectrometry with inductively-coupled plasma (AES-ICP). Analysis was conducted using the atomic emission spectrometer with inductively-coupled plasma OPTIMA 4300DV produced by PERKINELMER company with the wavelength operation range from 170 to 800 μm. The specific surface and porous structure of initial products of CTA and obtained adsorbents were determined using the apparatus produced by Quantachrome Corporation by adsorption and desorption of nitrogen. To calculate the values of the specific area of the BET surface, pore volume and pore size distribution, the program "Gas Sorpsion Report Autosob for Windows for AS-3 and AS-6" Version 1.23, was used. The differential thermal analysis (DTA) was conducted using the apparatus NETZSCH STA 449C in the temperature range from 20 to 1000 °C with a heating rate of 10 deg./min in the air; the sample weight was 0.2 g, the accuracy of determination of mass losses was ± 0.5 %. The static capacitance of adsorbents with regard to water vapors was determined by the gravimetric method by the mass fraction of water absorbed by the mass unit of the adsorbent by the time of reaching adsorption equilibrium at temperature of 25.0 °C in the atmosphere with relative humidity of 60.0 %. Dynamic adsorption capacitance of adsorbents with regard to water vapors was determined by measuring the amount of water vapors adsorbed from their mixture with the air, passing through the layer of adsorbent granules from adsorber by the time of reaching the dewpoint temperature of gas flow on exit from the adsorber (minus 40.0 °C). The resulted value was referred to 100.0 cm^3 of the tested sample. The kinetics of adsorption-desorption of water vapors at pressure close to atmospheric pressure was studied on the adsorption plant using the quartz balance described in work [8]. The experiments were made at room temperature (25 °C) and relative humidity of ~ 100 % using a fraction of 0.5-1.0 mm.

3. Results

According to the thermal analysis, the performed calculation of the content of hydroxides in the hydration products, proceeding from the temperatures of decomposition and weight losses accepted for them, showed that in the products of alkaline hydration there were up to 50 % of trihydroxides as well as pseudoboehmite and amorphous hydroxide. The content of the modifying admixtures and physicochemical characteristics of the synthesized samples after heat treatment are given in Table 1.

Table 1. Physicochemical properties of adsorbent samples.

Sample	S_{sp} (m^2/g)	V_{por} (cm^3/g)	Content (mass. %)		D (Å)	Mechanical strength (MPa)	Dynamic capacitance (g/100 cm^3)	Static capacitance (g/100g) with 60 % humidity
			Na^+	K^+				
1A	354	0.38	0.10	1.85	42.6	5.8	5.0	20.0
2A	346	0.47	0.10	2.20	42.8	3.7	5.5	20.1
5A	314	0.31	2.20	0.01	39	5.3	6.5	20.2
8A	317	0.30	2.30	0.01	38	8.4	7.5	20.2

According to the results of the X-ray phase analysis, the major phase of the studied samples is a low-temperature modification η - of aluminum oxide. A degree of crystallinity of the samples varies within the range from 30 to 40 %.

Table 1 shows that for the samples, modified by potassium, the specific surface value of the adsorbent accompanied by the increase in potassium content in the sample (1A, 2A) practically did not change. However, there was a tendency to an increase in the volume and average diameter of pores during reduction of mechanical strength and static capacitance with respect to water vapors. At the same time, there was no dependence in the change of dynamic capacitance. When replacing KOH by NaOH at the hydration stage, the specific surface value, the volume and the average diameter of pores of the obtained adsorbents reduce with a commensurable value of static capacitance (samples 5A and 8A, Table 1). It is possible to note that with a similar content of the alkaline metal in the adsorbent, mechanical strength and dynamic capacitance increase during transition from sample 2A to sample 5A. An increase in the concentration of peptizer - HNO_3 during synthesis (sample 8A) allowed a greater growth in the value of mechanical strength and dynamic capacitance of the adsorbent-desiccant. Kinetic curves of adsorption-desorption of water vapors on the samples are shown in figure 1. An irreversibility effect turned out to be typical of the adsorbents under study: after desorption of water vapors, curves did not regain their initial state (the difference was ~ 8 mass. %). This can be explained by the presence on the surface of adsorbents of Lewis sites capable to thoroughly adsorb water. At the initial stage of adsorption of water vapors, the fast growth of adsorption capacity (Figure 1) was conditioned by formation of a monomolecular layer on the adsorbent surface – chemisorption. Further multi-molecular adsorption was conditioned by means of molecular interaction. The energy of these two processes differs significantly: for chemisorption - 6-9 kcal/mole of the adsorbed substance, and for physical adsorption - ~0.5 kcal/mole. It is possible that ~ 8 mass. % of residual water is a result of chemical interaction of water vapors with the adsorbent surface.

An increase in the alkaline content at the hydration stage during synthesis of sample 2A as compared to sample 1A led to the decrease in the adsorption rate (Figure 1, a; Table 2). This could be explained by an increase in the amount of potassium admixture in the near-surface layer, which led to a slowdown of mass transfer and diffusion reduction [9]. Nevertheless, the value of the ultimate adsorption capacity for these samples coincided. As Figure 1 b) shows, kinetic curves for these samples practically coincide.

The data obtained during analysis of kinetic curves of water vapors adsorption on samples 1A, 2A, 5A, 8A, are given in Table 2. The kinetic curves for all samples are described by the following equations: $-\ln(a_m - a) = \beta t - \ln a_m$, $a = k^*\ln\tau$, $a = A^*\tau^{1/2}$.

Approximation of the experimental data by the above-mentioned mathematical functions shows that equation $a=A*\tau^{1/2}$ is best suited to describe kinetic curves on all experimental samples, since in this case the linear correlation coefficient is closest to one. The adsorption rate of water vapors for the samples under study is similar.

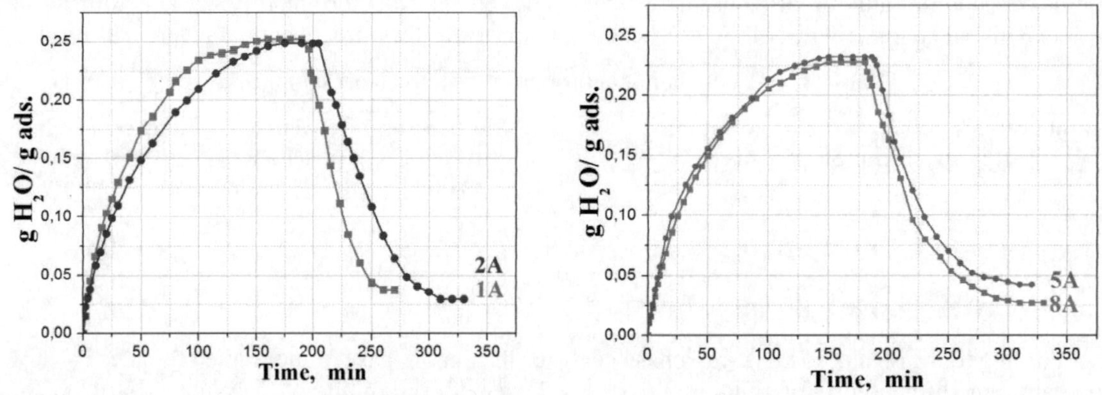

a) samples 1A and 2A b) samples 5A and 8A

Figure 1. Kinetics of water vapor adsorption on the fraction of 0.5-1.0 mm of adsorbent granulate based on aluminum oxide, modified by cations of alkaline metals (velocity of the gas – carrier during adsorption – 30 l/h, during desorption – 10 l/h at T = 25 °C).

It is common knowledge that [10] for conventional porous desiccants (silica gel, zeolite, aluminum oxide), a limiting adsorption stage is water vapor diffusion in the adsorbent pores. This process is usually realized by the volume, more rarely – by the surface. The analysis of the kinetic curves of water vapor adsorption on the experimental samples of aluminum-oxide adsorbents, conducted in this work, confirms the appropriateness of using diffusion models to describe the process of water absorption by adsorbents.

Table 2. Characteristics of experimental samples.

№ Sample	a_{max}, g_{water}/g_{ads}	$-\ln(a_m - a)= \beta t - \ln a_m$		$a=k*\ln\tau$		$a=A*\tau^{1/2}$	
		β, min⁻¹ → β, min^{-1}	R_0	k	R_2	A	R_1
1A	0.25	0.0279	0.97	0.0619	0.97	0.0220	0.98
2A	0.25	0.0234	0.95	0.0524	0.91	0.0205	0.99
5A	0.23	0.0298	0.95	0.0517	0.95	0.0209	0.99
8A	0.23	0.0258	0.98	0.0517	0.93	0.0208	0.99

4. Conclusions

As a result of the undertaken studies it was shown that modification of desiccants by potassium and sodium cations at the hydration stage of products of hydrargillite thermal activation during synthesis allowed obtaining adsorbents-desiccants with high values of static (more than 20 g/100g) and dynamic capacitance (more than 5.0 g/cm³) by water vapors. To obtain the adsorbent with a more optimal set of characteristics, it was preferable to use sodium cations as a modifier and maintain the acid module at the level of 0.06 at the peptization stage.

Acknowledgements
The research was carried out with the financial support of the Ministry of Education and Science of the Russian Federation. Agreement No. 14.575.21.0139, identifier RFMEFI57517X0139.

References
[1] Keltsev N V 1984 *Basics of adsorption technology* (Moscow: Chemistry) p 512
[2] Ulianova M A et al 2006 *Waterproof silica gels and areas of their application* (Tambov: Tambov State Technical University vol 12, No.1A) p 83–91
[3] Maiorov V V et al 2012 *Economic justification of the choice of the adsorbent in the air drying process* (Angarsk: Angarsk State Technical University vol 1, No.1) p 14-20
[4] Stiles E 1991 *Carriers and supported catalysts* (Moscow: Chemistry) p 240
[5] Sergunin A S et al 2012 *Study of the dynamics of adsorption and desorption of water vapors by active aluminum oxide and NaX zeolite* (Tambov: Tambov State Technical University vol 18 No. 3) p 664-670
[6] Danilevich V V et al 2012 *Highly effective adsorbents-desiccants based on aluminum oxide Kinetics and catalysis* **53 5** 673–680
[7] Zotov R A, Isupova L A, Danilevich V V, Babina A A, Sinelnikov A N, Meshcheryakov E P and Kurzina I A 2016 Dynamic capacity at high pressure of dehumidifiers based on modified aluminum oxide *Catalysis in industry* **16** 6–13
[8] Reshetnikov S I, Livanova A V, Meshcheryakov E P, Kurzina I A and Isupova L A 2017 Kinetic laws of adsorption on aluminum-oxide desiccants doped with alkali metal cations *Journal of Applied Chemistry* **90 11** 1451-57
[9] Kivshinnikov I M 1987 *Mineral fertilizers and salts: Properties and techniques of their improvement* (Moscow: Chemistry) p 256
[10] Ostrovski N M et al 2007 Simulation of the limiting stage of the process of water sorption by composite sorbents "calcium chloride in a porous matrix" *Theoretical Foundations of Chemical Technology* **41 1** 88-94

IOP Conf. Series: Materials Science and Engineering **597** (2019) 012025 doi:10.1088/1757-899X/597/1/012025

Influence of the composition of the adsorbent "hygroscopic salt/aluminum oxide" on its physicochemical properties

Zh B Budaev[1], A V Livanova[1], E P Meshcheryakov[1], L A Isupova[2] and O V Magaev[1]
[1]Tomsk State University, Russia, Tomsk, Lenin ave., 36, 634050
[2]Boreskov Institute of Catalysis SB RAS, Russia, Novosibirsk, 5 Lavrentieva ave., 630090

E-mail: truelivanova@mail.ru

Abstract. The paper presents the research results of physicochemical characteristics, adsorption kinetics and value of adsorption capacity with respect to water vapors of the composite adsorbent "$CaCl_2$ – aluminum oxide". Sorbent synthesized by centrifugal thermal activation of hydrargillite and its subsequent hydration in mild conditions was used as a carrier. It was shown that, in the studied range of change in the $CaCl_2$ content of the adsorbent (8.6-15.5% mass.), a sample containing 13.8% mass. of $CaCl_2$ was the most promising when used as a desiccant. This sample was characterized by static capacitance that was higher than 20 g/100 g, by quite high dynamic capacitance of 5.8 g/100 cm^3 and good mechanical strength of 4.9 MPa.

1. Introduction

It is common knowledge that classic adsorbents (activated carbon, silica gel, aluminum oxide, zeolite) can act as a matrix for different hygroscopic inorganic salts [1]. Dehumidifying effect of such composite materials, known as sorbents of the "salt in the porous matrix" type, is based on a combination of adsorption principles owing to developed specific surface and volume absorption of water by the solution of hygroscopic salt. During gas dewatering, these sorbents have advantages over traditional desiccants since they possess higher sorption capacity and lower (150-200 °C) regeneration temperature [2]. The porous matrix is not only able to absorb adsorbate, but also acts as a dispersion medium, forming the required particle size of salt and its surface, can change the salt state and its properties and provides heat input through the solid phase and gas transport through the system of pores. Development of energy saving and environmentally safe technology of obtaining effective sorbent based on aluminum oxide applying centrifugal thermal activation (CTA) of hydrargillite and its subsequent hydration in mild conditions has allowed obtaining aluminum oxide which has a more developed surface and is characterized by the presence of meso- and micropores, as well as high strength [3-7]. In this connection, the study of a possibility of using this aluminum oxide adsorbent as a matrix is of interest. For instance, $CaCl_2$ can act as hygroscopic salt. Calcium chloride is a strongly hygroscopic salt; it absorbs energetically water vapors, thus forming a number of crystalline hydrates. Transitions of crystalline hydrates into each other are characterized by different temperatures: $CaCl_2 \cdot 6H_2O$ at 30.1 °C melts and transforms into $CaCl_2 \cdot 4H_2O$, then into $CaCl_2 \cdot 2H_2O$ (at 45.1 °C) and into $CaCl_2 \cdot H_2O$ (at 175.5 °C). Calcium chloride is completely dehydrated only at temperatures above 250 °C [8]. However, practical application of massive salt is limited by the fact that at low relative pressure of water vapors (about 0.2), the solution possessing corrosion properties starts to form, which can be dangerous for reactors (adsorbers). To eliminate this effect, the salt is introduced into the adsorbent matrix pores [9]. In this paper, a series of adsorbents $CaCl_2/Al_2O_3$, differing by the content of hygroscopic salt in the samples, has been obtained and considered. Their physicochemical and adsorption characteristics have been studied in detail.

2. Materials and methods of research

A γ- Al_2O_3 sample in the form of granules of cylinder shape (5.0÷6.0 mm in length, 3.5÷3.6 mm in diameter), synthesized by centrifugal thermal activation of hydrargillite with its subsequent hydration in mild conditions, was used as a matrix. Conditions of synthesis of this sample were described in paper [10]. The mass fraction of Ca in synthesized sorbents was determined by atomic emission spectroscopy of microwave plasma using the atomic emission spectrometer of microwave plasma "Agilent 4100" (Agilent -Technologies, USA) with preliminary "disclosure" of acids in the mixture in the system of microwave sample preparation "Speedwave four SW-4" (Berghof, USA). Then, proceeding from the metal content, the salt content in the adsorbent was recalculated. Textural characteristics of the adsorbents were determined by isotherms of nitrogen adsorption-desorption at 77 K using the 3Flex sorptometer (Micromeritics, USA). The specific surface area was measured using the BET method. The mesopore volume was determined by analyzing the integral pore volume distribution curve depending on the radius (along the desorption branch); the average pore diameter (in nm) was determined by equation $d_{ave} = 4000V_{pore}/A$, where d_{ave} – average pore diameter; V_{pore} – volume pore; A was the pellet surface area [11]. The surface area occupied by micropores was evaluated by treating isotherms of nitrogen adsorption using Dubinin-Astakhov equation. To perform thermal analysis of aluminum oxides in argon atmosphere, the synchronous thermal analysis instrument "STA 449 F3, Jupter", made by NETZSCH (Germany), was used. The device simultaneously captured the differential scanning calorimetry (DSC) curves and the weight loss (WL). The temperature measurement range was 30-350 °C; the heating rate of the samples was 10 deg/min.

The mechanical strength of granules was assessed on the device "IPG -100" (Federal State Unitary Enterprise "UNIHIM with OZ", Russia). The measurements were taken for 30 granules and the average value was found.

The dynamic adsorption capacity (DC) of adsorbents with reference to water vapors was determined by means of measuring the amount of water vapors adsorbed from their mixture with the air, passing through the layer of adsorbent granules in the adsorber, at the moment of reaching the dew point temperature of the gas flow on exit from the adsorber (minus 40.0 °C). The resulting value was related to 100.0 cm^3 of the test sample. The layer height of the test sample (with the granule sizes of 2.0-4.0 mm in diameter and 2.0-4.0 mm in length) was 28.0-29.0 cm.

Static adsorption capacity of the adsorbents with respect to water vapors was measured using the gravimetric method in the exsiccator. The static adsorption capacitance (SC) of the samples was determined by the mass fraction of water absorbed by a mass unit of the adsorbent by the time of reaching adsorption equilibrium at temperature of 25.0 °C in the atmosphere with a relative humidity of 60.0 %.

Experiments on the study of adsorption kinetics of water vapors on the modified samples were conducted using the adsorption plant and the quartz balance at temperature of 25.0 °C and relative humidity of 100 % according to the procedure, described in work [10].

IOP Conf. Series: Materials Science and Engineering **597** (2019) 012025 doi:10.1088/1757-899X/597/1/012025

3. Results

Before impregnation of the aluminum oxide, the salt used for impregnation was baked at temperature of 200 °C for 4 hours in the muffle electric furnace "EKPS 10" (Smolenskoe SKTB-SPU, Russia). The impregnated $CaCl_2$ samples of aluminum oxide adsorbents were obtained by impregnation from the solution excess (equilibrium deposition filtration) [12]. The composition and textural characteristics of the initial sample of aluminum oxide (sample No. 1) and samples of modified adsorbents synthesized on its basis (samples 2-5) are given in Table 1.

Table 1. Textural characteristics, crushing strength and modifying additives content in the alumina samples.

Sample	Composition of the sample	Content of hygroscopic salt (wt. %)	S_{BET} (m²/g)	$V_{mesopores}$ (cm³/g)	$d_{mesopores}$ (nm)	Micropore surface area (m²/g)	Limiting micropore volume (cm³/g)	Crushing strength (MPa)
1	Al_2O_3	0	290	0.358	4.8	226	0.104	7.1
2	Al_2O_3 +	8.6	225	0.314	5.2	177	0.080	4.1
3	$CaCl_2$	12.7	182	0.301	6.0	143	0.065	4.1
4		13.8	144	0.245	6.1	119	0.055	4.9
5		15.5	96	0.180	6.6	79	0.036	6.8

As a result of studying the adsorption process, it was established that isotherms of nitrogen adsorption at relative pressures of 0.05-0.2 on all the samples were described by the BET equation. These isotherms belonged to isotherms of the IV type, according to the IUPAC classification with an initially sudden rise at a low relative pressure, which was conditioned by the presence of micropores and a gradual increase in the adsorption value with a pressure increase. Isotherms for the initial sample and sample No. 5 are shown in Figure 1 as an example.

In the range of relative pressures over 40 %, adsorption-desorption isotherms exhibited pronounced hysteresis. This type of the isotherm pointed to the presence of mesopores on the surface of the samples under study, as well as the presence of reversible capillary condensation in adsorbent mesopores. The initial sample of aluminum oxide possesses maximum nitrogen adsorption capacity, specific surface area, volume of micro- and mesopores, mechanical strength. When the salt content in aluminum oxide increases, the nitrogen adsorption capacity, volume of meso- and micropores, specific surface area decrease along with the increase in the diameter of mesopores (Table 1, Figure 2). The mechanical strength of adsorbent granules containing 8.6 % mass. of the salt is significantly lower than that of the initial sample granules of aluminum oxide. However, when the salt content in the aluminum oxide rises, the strength of the modified sample granules increases (Table 1).

According to literature data [13], the process of water precipitation from Al_2O_3 was stepwise. At the first stage (up to 100 °C) there was precipitation of physically adsorbed water. At the second (100-350 °C) and third (350-550 °C) stages there was precipitation of chemically combined water. According to the data of work [14], the water from the sample, similar in composition to sample No 1, started to exude already at 52 °C, and the maximum of exudation fell to 136-153 °C. Peaks in the high-temperature region on the thermogram were absent for this sample.

The results of the study of the samples by thermogravimetric analysis (TGA) and differential scanning calorimetry (DSC) are given in Figure 3 and in Table 2.

As it turned out, there were two peaks in the temperature region up to 350 °C for the obtained samples on the DSC curves (Table 2). It is possible to note that when the salt content in the sample grew, there was a tendency towards an increase in the peak maximum temperature being in a higher temperature region (100-300 °C). At the same time, sample mass losses in the temperature range of 30 - 350 °C, determined by the TG-curve (Table 2), increased naturally as the content of hygroscopic salt in the

IOP Conf. Series: Materials Science and Engineering **597** (2019) 012025 doi:10.1088/1757-899X/597/1/012025

sample grew from 4.46 % mass. for sample No. 2 to 7.19 % mass. for sample No. 5. The water was removed from the samples more intensively in the temperature region of 100-350 °C.

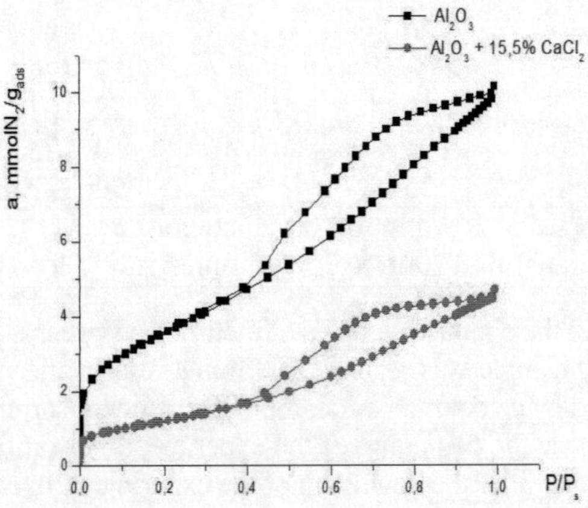

Figure 1. Isotherms of nitrogen adsorption desorption on the initial Al_2O_3 sample and the sample containing 15% mass. of salt.

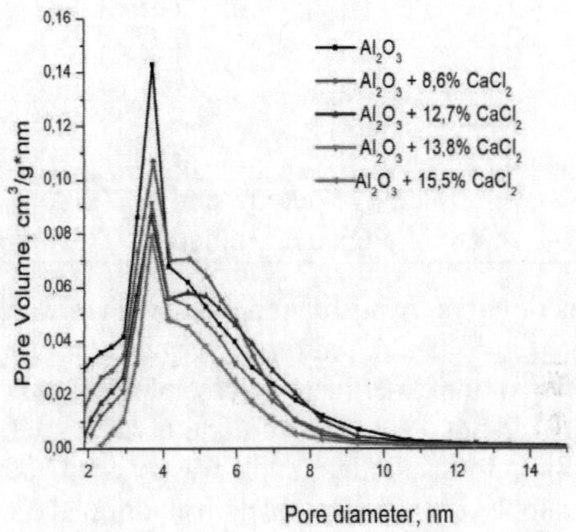

Figure 2. Pore size distribution of samples.

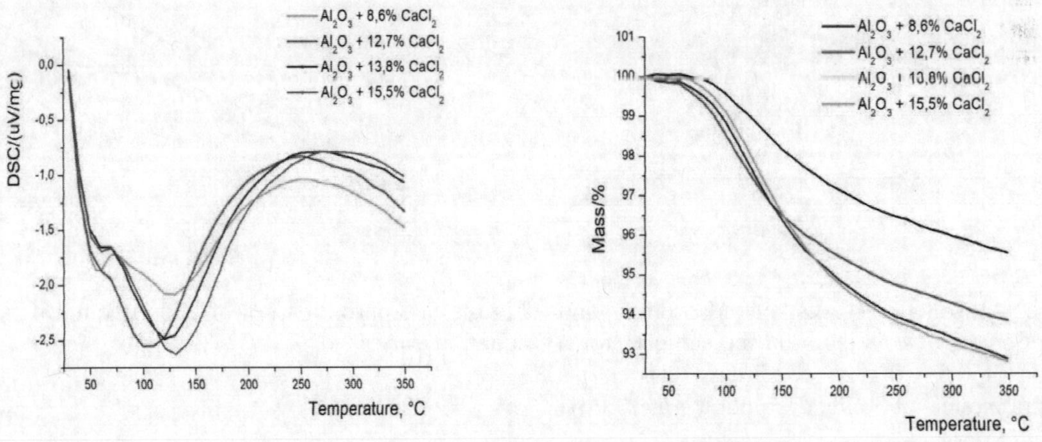

a) b)

Figure 3. DSC (a) and TG (b) curves for modified samples.

Table 2. DTA and TGA data of the alumina samples.

Sample	**The weight loss on the TG curve (%)**

	to 100°C (Δm_1[a])	from 100 °C to 350°C (Δm_2[b])	T_{max} (DSC) (°C)
1	0.20	3.30	140
2	0.42	4.04	58; 126
3	1.45	4.70	58; 103
4	1.15	5.95	60; 123
5	0.94	6.25	59; 129

[a]Δm_1 – weight change determined by TGA at a heating rate of 10°/min to 100 C integration
[b]Δm_2 – weight change determined by TGA at a heating rate of 10°/min to 350° C and holding for 1h

Kinetic curves of adsorption-desorption of water vapors on the obtained samples of modified adsorbents are presented in Figure 4. Adsorption equilibrium for the initial sample was reached 200 minutes later after the beginning of the experiment. In case of the samples modified by calcium chloride, the water vapor adsorption capacity did not reach the equilibrium value after 5 hours since the beginning of the experiment. Adsorption equilibrium for the initial sample was reached 200 minutes later after the beginning of the experiment. In case of the samples modified by calcium chloride, the water vapor adsorption capacity did not reach the equilibrium value after 5 hours since the beginning of the experiment.

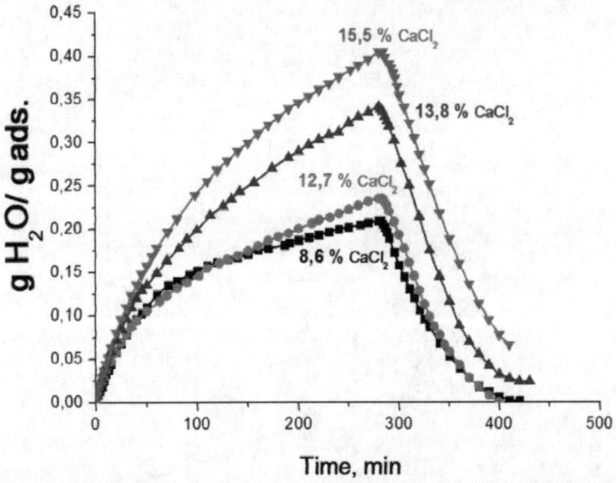

Figure 4. Kinetic curves of adsorption and desorption of water vapors on the samples for a fraction of 0.5–1.0 mm (conditions: carrier gas adsorption rate – at 30 l/h, desorption – at 10 l/h).

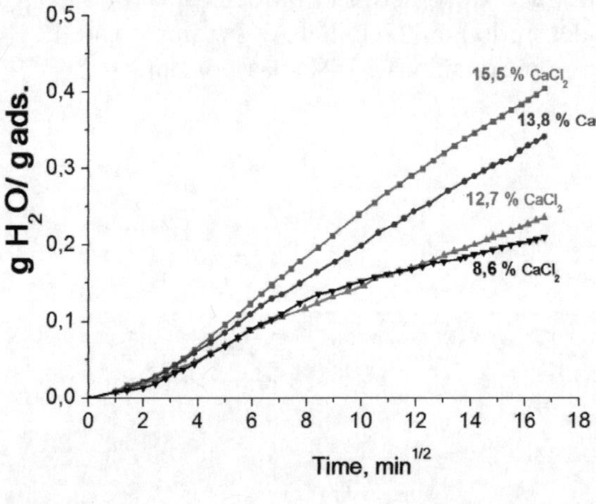

Figure 5. Approximation of experimental data by equation $a=At^{1/2}$.

Approximation of the experimental data by different kinetic equations showed that the rate of the adsorption process was described best of all by equation $a=A\ t^{1/2}$ (Figure 5), where a - the amount of water adsorbed at time t; A - the rate constant, g/min$^{-1/2}$; t - time, min. Values of rate constants A determined in this equation for the samples under study are given in Table 3. The equation of linear dependence of the adsorption value on was earlier used specifically in the description of water vapor adsorption kinetics on aluminum oxide [15] and the composite material – "calcium chloride/silica gel" [16]. The equation was evidence of the fact that the limiting stage of the process is water diffusion in the adsorbent pores.

Table 3. Characteristics of the kinetic curves of water vapor adsorption on alumina samples and the values of static and dynamic capacitance.

Sample	a, t=280min (g/g $_{ads}$)	A (g /min$^{-1/2}$)	SC (g/100g $_{ads}$) at 60% humidity	DC (g/100 cm^3)
1	0.246[a]	0.0199	23.4	4.6
2	0.210	0.0134	21.4	7.0
3	0.235	0.0146	21.0	6.8
4	0.341	0.0217	23.2	5.8
5	0.404	0.0263	29.5	4.3

[a]for the initial sample, reaching ultimate water adsorption capacity for 200 minutes was registered.

The adsorption capacity of the modified samples with respect to water vapors and rate constants (Table 3), determined from kinetic data, increase when the salt content grows. When the content of aluminum oxide is at the level of 13.8 mass. % of $CaCl_2$ and higher, these characteristics exceed the values determined for the initial sample. Values of static capacitance also change symbatically for these samples. Dynamic capacitance with respect to water vapors, when introduced into aluminum oxide, at first increases, but the salt content higher than 12.7% mass. gradually decreases.

4. Conclusions

Introduction of hygroscopic salt into aluminum oxide makes major changes in properties of hygroscopic salts and initial adsorbent used as a matrix. This is manifested through the change in textural, adsorption and strength characteristics, in the form of curves on thermograms. When the salt content increases, the value of specific surface decreases; the volume of meso- and micropores decreases. The porous matrix, in this case aluminum oxide, is primarily a medium facilitating dispersion of salt. It is noted in work [1] that on such carriers as γ - Al_2O_3, having 6-9 nm pores, there is stronger dispersion of salt, which significantly influences the sorption behavior of impregnated desiccant. In the pores of the mentioned systems, stoichiometric crystalline hydrates does not form at all. This fact can explain the absence of peaks on DSC curves, corresponding to their decomposition. The water vapor adsorption capacity of impregnated aluminum oxide desiccants, determined by kinetic curves, grows with the increase in the percentage content of salt in the samples. The same tendency is observed in case of static capacitance. This bears evidence of predominant contribution of hygroscopic salt in the adsorption process. Since just adsorption properties of hygroscopic salt determine the properties of the entire composite, a low value of specific surface of obtained adsorbents does not play a pivotal role. The kinetic curves of aluminum oxide adsorbents with addition of hygroscopic salts are best described by the equation used for description of processes the speed of which is limited by diffusion in adsorbents pores. Adsorption rate constant A increases along with the amount of the introduced modifier. Among the experimental samples impregnated by hygroscopic salts throughout the entire set of characteristics, sample No. 4 is the most promising for the use as a desiccant, of which the following is typical: the value of static capacitance is higher than 20 g/100g, the high value of dynamic water capacitance is at the level of 5.8 g/100 cm^3, and good mechanical strength is 4.9 MPa.

Acknowledgments

The study was financially supported by project RFMEFI57517X0139 of the Ministry of Education and Science.

References

[1] Buluchevskij E A, Lavrenov A V and Duplyakin V K 2007 *Russian Journal of General Chemistry* **77** 2284-91
[2] Buluchevskij E A, Lavrenov A V and Duplyakin V K 2009 *Protection of Metals and Physical Chemistry of Surfaces* **45** 605-09

IOP Conf. Series: Materials Science and Engineering **597** (2019) 012025 doi:10.1088/1757-899X/597/1/012025

[3] Tanashev Yu.Yu. *et al.* 2007 *Kinetics and Catalysis* **48** 153-61

[4] Kharina I V, Isupova L A., Litvak G S, Moroz E M, Kryukova G N, Rudina N A, Tanashev Yu Yu and Parmon V N 2007 *Kinetics and Catalysis* **48** 327-35

[5] Danilevich V V, Isupova L A, Danilova I G, Zotov R A and Ushakov V A 2016 *Russian Journal of Applied Chemistry* **89** 343-53

[6] Zotov R, Isupova L, Danilevich V, Babina A, Sinelnikov A, Meshcheryakov E and Kurzina I 2016 *Catalysis in industry* **16** 6–13

[7] Livanova A, Meshcheryakov E, Reshetnikov S and Kurzina I 2019 *AIP Conf. Proc* 1145 012041. DOI:10.1088/1742-6596/1145/1/012041

[8] https://cniga.com.ua/index.files/calciumchloride.htm

[9] Aristov Yu I, Gordeeva L G and Tokarev M M 2008 *Composite sorbents "salt in a porous matrix": synthesis, properties, application* (Publ. House of the Siberian Branch of the Russian Academy of Science, Novosibirsk) p 362

[10] Reshetnikov S I, Livanova A V, Meshcheryakov E P, Kurzina I A and Isupova L A 2017 *Russian journal of applied chemistry* **90** 1760-65

[11] Greg S and Sing K *Adsorption, specific surface area, porosity* (Moscow: Mir) p 310

[12] Bourikas K, Kordulis C and Lycourghiotis A 2006 *Catal. Rev.* **48** 363-444

[13] Alex T C 2014 J. *Therm. Anal. Calorim.* DOI 10.1007/s10973-014-3670-1

[14] Zotov R, Meshcheryakov E, Livanova A, Minakova T, Magaev O, Isupova L and Kurzina I. 2018 *Materials* **11** 132 DOI: 10.3390/ma11010132

[15] Marcussen L 1974 *Chem. Eng. Sci.* **29** 206

[16] Ovoshchnikov D S, Glaznev I S and Aristov Yu I 2011 *Kinetics and Catalysis* **52** 620-28

A study on the oxidative coupling of methane over $La_{2-x}Mg_xTiO_4$ catalysts

Yu Ivanova, R Petrov, M Arhipova, S Reshetnikov and L Isupova

Boreskov Institute of Catalysis SB RAS, 5 Lavrentieva Avenue, Novosibirsk 630090, Russian Federation

E-mail: ivanova@catalysis.ru

Abstract. $La_{2-x}Mg_xTiO_{5-\delta}$ samples with different content of Mg and La were synthesized by the precipitation from a mixed solution of $Mg(NO_3)_2$ and $La(NO_3)_3$ in the presence of TiO_2 suspension. The samples were characterized using X-ray diffractometry, mercury porosimetry, and BET techniques. Specific surface area of the samples ranges from 0.1 to 4.17 m^2/g, and pore volume, from 0.005 to 0.023 cm^3/g. Activity and selectivity of the samples in the oxidative condensation of methane (OCM) were measured using a fixed catalyst bed flow quartz reactor. It was demonstrated that samples with $x \leq 0.6$ showed the highest activity toward OCM reaction. This can be explained by a high content of the main active phase $(La,Mg)_2TiO_{5-\delta}$ with the layered perovskite structure and a sufficient specific surface area.

1. Introduction

Oxidative condensation of methane (OCM) is one of the promising methods for obtaining the ethane-ethylene mixture directly from natural gas without a complicated step of syngas production. Practical implementation of the process is hindered by the low yield of target products that does not exceed 26% at conversion above 35%, according to experimental data, which is related to the low activity/selectivity of the catalysts.

In this connection, titanium-containing materials with the perovskite structure deserve special attention. Owing to their crystal structure (the closest cubic packing of oxygen ions and metal cations), they possess some unique properties, such as the ability to ionic exchange and intercalation of oxygen ions, high thermal stability, etc. [1, 2] and are promising catalysts for heterogeneous catalysis [3], particularly for the OCM process [4]. The binding energy of oxygen in the perovskite structure plays an important role. Activation of methane starts from cleavage of the C-H bond upon interaction with the oxide surface [5]. Active sites with the strongly bound oxygen decrease the contribution of undesirable processes of complete oxidation of methane, products and semiproducts of oxidative condensation of methane [6].

Of particular interest are oxide systems with the perovskite-like structure containing Mg and La. Individual oxides MgO and La_2O_3 are difficultly reducible and require a large amount of energy to form oxygen vacancies [4]. Systems containing La_2O_3 were successfully employed as catalysts for the OCM process [7-9]. Many researchers noted that the addition of MgO to various oxide systems exerted a beneficial effect on the performance of OCM reaction [10-13]. In [14], active oxygen species in MgO-La_2O_3 mixed oxides of different composition were studied. Strong interactions between MgO and La_2O_3 oxides led to the formation of oxygen vacancies; as a result, an additional type of adsorbed oxygen was produced. This type of adsorbed oxygen is very active toward the low-temperature oxidation of H_2 and can be used to activate CH_4 at high temperatures under relatively mild conditions of the reaction at 600°C. In [15], La_2TiO_5 oxide with the layered perovskite structure was shown to be highly active; its activity increased upon substitution of lanthanum by strontium.

In our work, the effect of La substitution by Mg in the structure of perovskite-like titanates ($La_{2-x}Mg_xTiO_{5-\delta}$) on their physicochemical and catalytic properties in the oxidative condensation of methane was investigated.

2. Experimental

Precipitation from a mixed solution of $Mg(NO_3)_2$ and $La(NO_3)_3$ in the presence of TiO_2 suspension was used to synthesize $La_{2-x}Mg_xTiO_{5-\delta}$ samples with a step x = 0.2, where x is the degree of lanthanum substitution by magnesium. A 0.5 M solution of $(NH_4)_2CO_3$ served as a precipitating agent. The precipitation was performed at room temperature until pH became equal to 8–8.5; after that, the precipitate was filtered, washed to pH 7, and dried at 120°C. All the synthesized samples were calcined at 1100°C for 4 hours. The concentration of lanthanum (molar fraction) C_{La} = La/(La+Mg) in the oxides is related with the substitution degree as $x = 2\cdot(1 - C_{La})$.

Activity and selectivity of the samples in the oxidative condensation of methane were measured using a flow quartz reactor with the diameter 5 mm and a fixed catalyst bed [15]. Conditions of experiments were as follows: temperature T = 700–800°C; pressure P = 1 atm; volume of the catalyst 0.5 ml; catalyst fraction with the grain size 0.25–0.5 mm diluted with 2 g quartz having the grain size 0.25–0.5 mm; and feed space velocity of the reaction mixture ($CH_4/O_2/N_2$ = 45/11.3/43.7 % (mol.)) 30 l/h. The reaction mixture was sampled and analyzed on LHM-80 and Tsvet-500 chromatographs with thermal conductivity detectors connected in series. The data obtained were used to calculate methane conversion (X_{CH4}), selectivity (S_{C2}), and yield (Y_{C2}) of C2 products using the following formulas:

$$X_{CH4} = 1 - \frac{C_{CH4}}{C^0_{CH4}} \cdot U \tag{1}$$

$$S_{C2} = \frac{2\cdot(C_{C2H4} + C_{C2H6})}{C_{CO} + C_{CO2} + 2\cdot(C_{C2H4} + C_{C2H6})} \tag{2}$$

$$Y_{C2} = X_{CH4} \cdot S_{C2} \tag{3}$$

where C^0 is the inlet concentration of the corresponding reactant; C is the outlet concentration of the corresponding substance; and U is the ratio of output to input streams.

The specific surface area (S_{BET}, m^2/g) was estimated according to BET from thermal desorption of argon. The pore structure was examined by nitrogen porosimetry under pressure using a QUADRASORB evo™ (Quantachrome, USA) instrument.

The phase composition was determined by X-ray diffraction analysis (XRD). Diffraction patterns were obtained on a Bruker D8 (Germany) diffractometer using a CuK_α source ($\lambda = 1.5418$). Each sample was scanned point-by-point with a 0.05° step in the 2θ range from 10 to 70°. The phase composition was revealed by comparing the experimental diffraction patterns with the JCPDS database. The size of coherent scattering region (CSR) was found using the Scherrer formula.

3. Results and Discussion

3.1. Materials characterization

Phase composition and textural characteristics of the samples are listed in Table 1.

According to XRD data, all the synthesized samples contain perovskite-like phases, some samples (at $x < 0.8$) have titania impurities, but single-phase samples are only those with x = 0 and 0.8.

A sample with the composition x = 0 has a well crystallized La_2TiO_5 structure of layered perovskite. At $x < 0.8$ ($C_{La} > 0.6$), a mixture of the main $(La,Mg)_2TiO_{5-\delta}$ and $(Mg,La)TiO_3$ and impurity TiO_2 and $MgTiO_3$ phases is formed. A sample with the substitution degree x = 0.8 (C_{La} = 0.6) has reflections of

IOP Conf. Series: Materials Science and Engineering **597** (2019) 012026 doi:10.1088/1757-899X/597/1/012026

only one well crystallized phase with the $LaTiO_3$ perovskite structure [16]. This indicates that at this substitution degree lanthanum and magnesium intercalate into the perovskite-like structure and form a $(La,Mg)TiO_3$ compound. A sample with x = 2 (Figure 1) contains well crystallized $MgTiO_3$ and Mg_2TiO_4 phases. The SCR sizes of the observed main perovskite-like phases vary in the range of 190-650 Å; single-phase samples have greater CSR sizes.

Table 1. Phase composition, CSR size, specific surface area and pore volume of the synthesized $La_{2-x}Mg_xTiO_{5-\delta}$ samples.

Composition of synthesized $La_{2-x}Mg_xTiO_{5-\delta}$ samples	S_{BET} ($m^2\ g^{-1}$)	V_{pore} ($cm^3\ g^{-1}$)	Main phase	CSR (Å)	Impurity phase
La_2TiO_5	1.05	0.018	La_2TiO_5	560	
$La_{1.8}Mg_{0.2}TiO_{5-\delta}$	0.98	0.016	$(La,Mg)_2TiO_{5-\delta}$ $(La,Mg)TiO_3$	320 400	TiO_2, $MgTiO_3$
$La_{1.4}Mg_{0.6}TiO_{5-\delta}$	4.17	0.023	$(La,Mg)_2TiO_{5-\delta}$ $(La,Mg)TiO_3$	190 370	TiO_2, $MgTiO_3$
$La_{1.2}Mg_{0.8}TiO_{5-\delta}$	0.10	0.007	$(La,Mg)TiO_3$	500	
Mg_2TiO_4	1.59	0.005	$MgTiO_3$ Mg_2TiO_4	650 410	

Specific surface area of the samples ranges from 0.1 to 4.17 m^2/g, and pore volume, from 0.005 to 0.023 cm^3/g. A sample with x = 0.6 has the highest values of S_{BET} and V_{pore}, whereas the single-phase sample with x = 0.8 has the lowest S_{BET} and V_{pore} (Table 1). Overall, non-single-phase samples are characterized by higher specific surface area and pore volume due to their insufficient sintering.

Thus, it was demonstrated that variation of the elemental composition – the La to Mg ratio in $La_{2-x}Mg_xTiO_{5-\delta}$ – makes it possible to synthesize a double oxide with the perovskite structure at x = 0.8, or at a molar concentration of lanthanum C_{La} = La/(La+Mg) = 0.6.

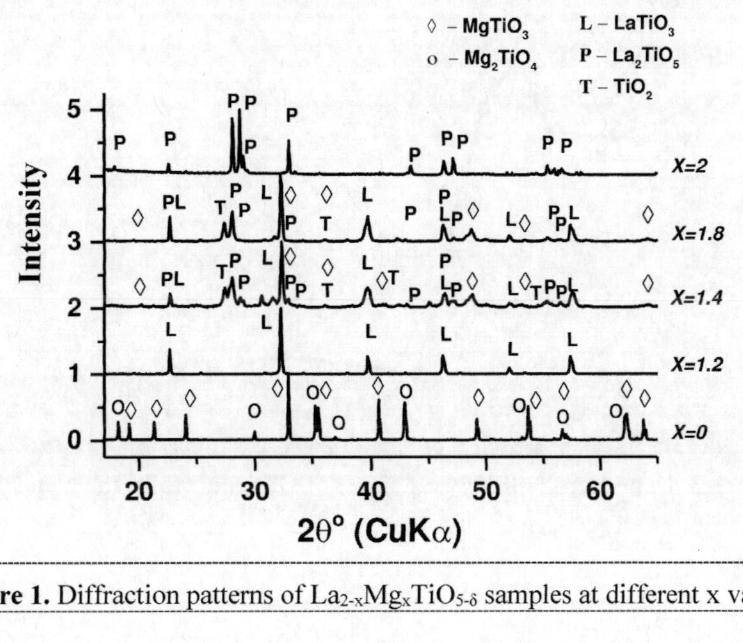

Figure 1. Diffraction patterns of $La_{2-x}Mg_xTiO_{5-\delta}$ samples at different x values.

3.2. Catalytic tests

OCM is a selective process. Products of selective oxidation (ethane and ethylene) are formed along with the deep oxidation products (CO and CO_2).

Dependences of methane conversion (X_{CH4}), selectivity (S_{C2}) and yield (Y_{C2}) of C2 products on the content of magnesium (x) are displayed in Figures 2 and 3. High activity of the samples was observed for the compositions with $x \leq 0.6$ containing the $(La,Mg)_2TiO_{5-\delta}$ phase. Selectivity of the indicated samples (with $x \leq 0.6$) for C2 products was 55-60%. The highest activity was observed for the samples with x = 0.4-0.6.

Overall, an increase in x above 0.6 decreased the methane conversion and C_2 yield; therewith, the $(La,Mg)_2TiO_{5-\delta}$ phase was not detected in the composition of the samples. A single-phase $La_{1.2}Mg_{0.8}TiO_{5-\delta}$ sample with the $LaTiO_3$ structure is low active and has a low specific surface area.

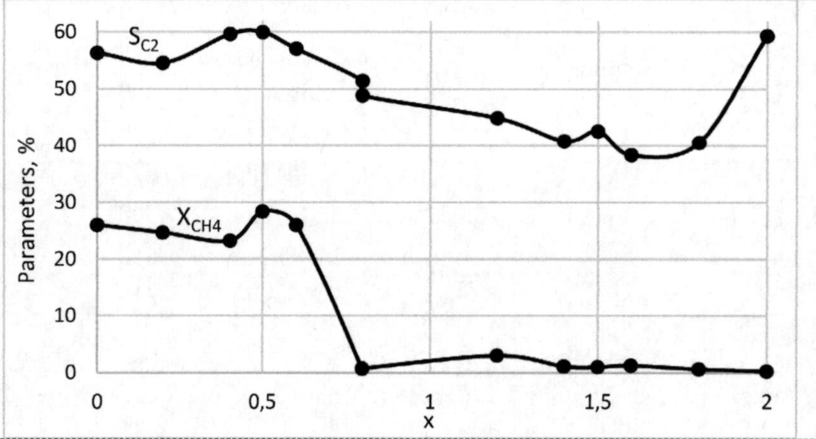

Figure 2. Dependences of methane conversion (C_{CH4}) and selectivity (S_{C2}) of C2 products on the content of magnesium (x) in $La_{2-x}Mg_xTiO_{5-\delta}$ oxides. Conditions: gas composition $CH_4/O_2/N_2$ = 45/11.3/43.7 % (mol.), P = 1 atm.

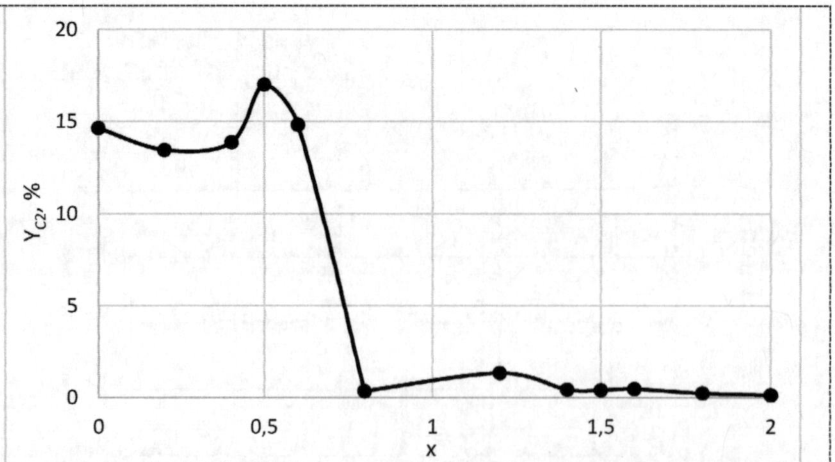

Figure 3. Dependence of the yield (Y_{C2}) of C2 products on the content of magnesium (x).

4. Conclusion

$La_{2-x}Mg_xTiO_{5-\delta}$ samples with different content of Mg and La were synthesized. It was shown that the introduction of magnesium (x) leads to the appearance of the perovskite $(La,Mg)TiO_3$ phase, which content increases with x. The possibility to synthesize a single-phase double oxide system with the $(La,Mg)TiO_3$ perovskite structure at x = 0.8 was demonstrated.

Among the tested samples, those with $x \leq 0.6$ showed the highest activity toward oxidative condensation of methane. This can be attributed to a high content of the main active phase $(La,Mg)_2TiO_{5-\delta}$ with the layered perovskite structure and a sufficient specific surface area.

Acknowledgments
This work was conducted within the framework of the budget project AAAA-A17-117041710090-3 for the Boreskov Institute of Catalysis.

References

[1] Pena M A and Fierro J L G 2001 *Chem. Rev.* **101** 1981
[2] Vereshchagin V I, Pletnev P M, Surzhikov A P and Fedorov V E 2009 *Modified ceramics with perovskite and spinel phases* (Moscow: Nauka) p 223 (in Russian)
[3] Royer S, Duprez D, Can F, Courtois X, Batiot-Dupeyrat C, Laassiri S and Alamdari H 2014 *Chem. Rev.* **114** 10292
[4] Kumar G, Lau S L J, Krcha M D and Janik M J 2016 *ACS Catal.* **6** 1812
[5] Fung V, Polo-Garzon F, Wu Z and Jiang D 2018 *Catal. Sci. Technol.* **8** 702
[6] Krylov O V 2002 *Heterogeneous catalysis* (Novosibirsk: Novosibirsk State University Press) p 260 (in Russian)
[7] Chu C, Zhao Y, Li S and Sun Y 2014 *J. Phys. Chem. C* **118** 27954
[8] Conway S J, Greig J A and Thomas G M 1992 *Appl. Catal. A: General* **66** 199
[9] Huang S-J, Walters A B and Vannice M A 1998 *Appl. Catal. B: Environmental* **17** 183
[10] Ivanov D V, Isupova L A, Gerasimov E Yu, Dovlitova L S, Glazneva T S and Prosvirin I P 2014 *Appl. Catal. A: General* **485** 10
[11] Wu M C, Truong C M, Coulter K and Goodman D W 1992 *J. Am. Chem. Soc.* **114** 7565
[12] Zavyalova U, Holena M, Schlogle R and Baerns M 2011 *Chem. Cat. Chem.* **3** 1935
[13] Schwach P, Schlögl R and Trunschke A 2013 *11-th European Cong. on Catalysis – EuropaCat-XI (Lyon)*
[14] Wu X, Fang Z, Pan H, Zheng Y, Jiang D, Ni J and Li X 2017 *Catal. Sci. Technol.* **7** 797
[15] Ivanova Y, Petrov R, Arhipova M, Reshetnikov S and Isupova L. 2019 *J. Phys.: Conf. Ser.* **1145** 012024
[16] Eitel M and Greedan J E 1986 *J. Less. Common Met.* **116** 95

IOP Conf. Series: Materials Science and Engineering **597** (2019) 012027 doi:10.1088/1757-899X/597/1/012027

Polyvinyl Alcohol-Based Cryogels for the Oil Industry

E S Kashlach[1], L K Altunina[1,2], V N Manzhai[1,2] and M S Fufaeva[2]
[1]National Research Tomsk State University, 634050 Tomsk, Lenin Avenue, 36
[2]Institute of Petroleum Chemistry, Siberian Branch of the Russian Academy of Sciences (IPC SB RAS), 4, Akademichesky Avenue, 634055, Tomsk, Russia

E-mail: maria81@ipc.tsc.ru

Abstract. The paper presents the results of studies of rheological properties of the initial viscous solutions of polyvinyl alcohol and 'oil-in-water' (O/W) emulsions. It is found that aqueous solutions of polymer and polymer-based colloidal systems have typical non-Newtonian properties. After the freeze-thaw cycle, the viscous liquid systems pass into a solid state of aggregation to form multicomponent cryogels with rubber-like structure. The elastic properties of two-component cryogels consisting of polyvinyl alcohol and water only, three-component compositions formed on the basis of an 'aqueous PVA solution-mineral oil' emulsion, and cryogels containing loose dispersed fillers are investigated. The physicochemical properties of the obtained elastic samples are studied. The prospects of application of cryogels for the oil producing well and road construction are outlined. A comparative analysis of physicomechanical properties of cryogels depending on the nature of the filler is carried out. In order to simulate the real conditions of oil field construction in the permafrost zone, samples of cryogels with solid dispersed fillers pre-wetted with oil are prepared and their mechanical and hydrophobic properties are investigated. Hydrophobic properties of cryogels make it possible to use them as waterproofing layer (membrane) in the construction of the asphalt pavement and insulation of the bottoms and walls of hydraulic structures. Cryogels can be recommended for use as polymer 'pigs' for removal of undesirable deposits in the pipeline during its cleaning or for pipeline pressure testing. In this case, 'batching pigs' are introduced to the boundary between hydrocarbon liquid and water.

1. Introduction

Nowadays, cryotropic nanostructures based on polyvinyl alcohol (cryogels) have become the object of increased interest. Freeze of viscous aqueous solutions of polyvinyl alcohol (PVA), keeping in crystalline state at temperature $T < 0$ °C for several hours and subsequent thaw within the range of positive temperatures (at $T > 0$ °C) result in the formation of elastic cryogels (rubber-like bodies). In this case, a transition of two-component solutions occurs from the liquid state of aggregation to the solid state without the use of 'cross-linking' chemical reagents, i.e. due to the enhancement of the intermolecular interaction of the PVA polymer chains [1, 2]. Cryogels formed under conditions of solvent crystallization are thermally reversible but in contrast to low-molecular compounds, they have no fixed phase transition point, hence they are melted at a temperature ($T_{melt.}$), which is several tens of degrees above the temperature of structurization of the initial aqueous solutions ($T = 0$ °C). Cryogels are non-toxic, environmentally friendly materials used in biotechnologies and medicine [3]. The structural elements of cryogels are polyhedral cells with wall thickness from units to tens of nanometers, formed from macromolecules of polyvinyl alcohol, which give a unique complex of

IOP Conf. Series: Materials Science and Engineering **597** (2019) 012027 doi:10.1088/1757-899X/597/1/012027

physicomechanical properties to cryostructures. The internal cavities of the cells are filled with water. Two-component cryogels scatter light due to the presence of excessive fluctuations in the density of the polymeric substance in the cell walls. Hence, in contrast to transparent PVA solutions, cryogels are opaline bodies. The mechanical properties of such a system are determined by the nanoscale thickness of the cryogel network. Deformation due to the application of external stress, for example, bending of the wall of polymer network, leads to an elastic (entropic) change in the shape of macromolecules, since the wall thickness is commensurable with the values of the moving segments of the macromolecules.

In climatic zones of permafrost, cryogels are used in plugging operations to form blocking screens [4] that prevent undesirable (destructive) filtration of water through dams and bottoms of hydraulic structures. Cryogels are promising materials for the development of new technologies of the construction and repair of producing wells, where, as a result of warm oil seepage, wellhead funnels (sinkholes) are formed, and for cleaning of pipelines. The introduction of modifiers and fillers into the polymer matrix of a cryogel at the stage of its formation changes the mechanical and physicochemical properties of the material; therefore, the study of such properties is of practical interest [5, 6].

2. Experimental section

In order to prepare aqueous polymer solutions, samples of polyvinyl alcohol were used, which contained no more than 1% of residual acetate groups in their structure. Emulsions with hydrophobic properties were prepared by adding predetermined portions of mineral oil to an aqueous PVA solution under vigorous stirring. Mineral oil is a complex multicomponent system consisting mainly of naphthenic and paraffinic hydrocarbons, and aromatic compounds. To form more rigid multicomponent cryogels, fine additives were introduced into an aqueous solution of polyvinyl alcohol: quartz sand or clay (bentonite). To simulate the real conditions of the oil fields, samples of cryogels with bulk fillers pre-wetted with oil were also prepared. Dynamic viscosity (η, Pa•s) of individual liquids (aqueous solutions of polyvinyl alcohol or mineral oil) and their emulsions were measured on a "Rheotest-2' rotational viscometer at different temperatures (T, °C) in a wide range of shear rates (j, s^{-1}).

To manufacture two-component cryogels (polyvinyl alcohol–water), polymer solutions of different concentrations were poured into cylindrical cells and frozen at negative temperature -20 °C. Then solid samples were thawed at room temperature +20 °C. After the freeze–thaw cycle, elastic cryogels were ready for use. Cryogels from emulsions and suspensions were manufactured in cells in a similar way. Figure 1 shows the stages of formation of cryogels of different compositions.

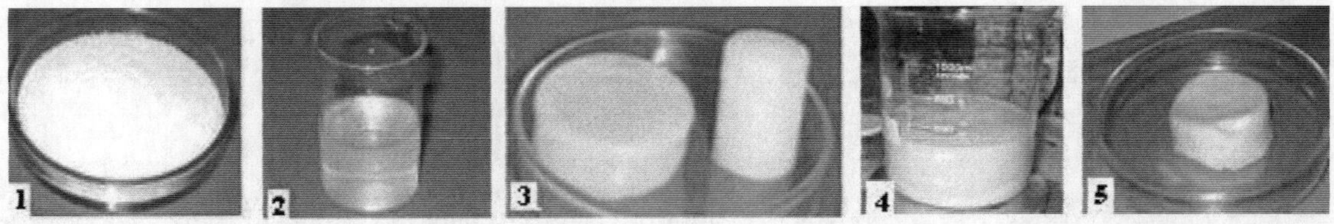

Figure 1. Stages of preparation of filled cryogels: 1 – dry powder of polyvinyl alcohol, 2 – aqueous solution of polyvinyl alcohol, 3 – two-component cryogel, 4 – 'mineral oil in an aqueous solution of polyvinyl alcohol' emulsion, 5 – elastic cryogel containing water, polyvinyl alcohol, and mineral oil.

The degree of hydrophobicity of the surface of oil-filled cryogels was determined by computer video scanning. Droplets of water or oil were applied to the surface of the obtained cryogels and the images of changes in the size of droplets were registered using a microscope. Using the computer program of image processing, the area that a droplet of water takes after acertain time was determined. The formed elastic cryogel samples were deformed (relative strain γ), the elastic stress (P, Pa) arising in the material was measured, and then the modulus of elasticity (G, Pa) using Hooke's formula was calculated. The melting point of cryogels ($T_{melt.}$, °C) was determined by the falling-ball method, detailed in [2]. The thermal insulating properties of cryogels were estimated by the value of the thermal conductivity coefficient (λ, W/K•m), which was determined using the installation with the main working unit consisting of two steel coaxial cylinders. The medium under study was placed into the gap between the cylinders. The value of the thermal conductivity coefficient was calculated using the formula

$$\lambda = Q \ln \langle R_1 | R_2 \rangle \left[2\pi Lt(T_0 - T) \right]^{-1}$$

where R_1 is the inner radius of the large cylinder; R_2 is the outer radius of the small cylinder; Q is the amount of heat transferred from the heated water of the thermostat to the water of inner cylinder; L is the height of the small cylinder; T is the current water temperature in the inner cylinder at some moment in time (t); and T_0 is the temperature of the heat carrier in the thermostat.

3. Results and discussion

The method of rotational viscometry was used to conduct rheological studies of solutions of polyvinyl alcohol and emulsions at different shear rates (j). It was found that these liquids are typical non-Newtonian systems (Figure 2).

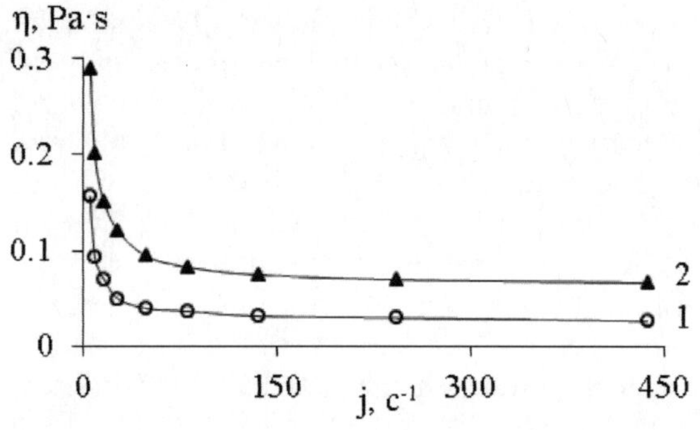

Figure 2. Aqueous solution of polyvinyl alcohol (5 wt%); 2 – 'Oil-in-water'emulsion obtained from an aqueous solution of polyvinyl alcohol (5 wt%) and mineral oil (30 wt%).

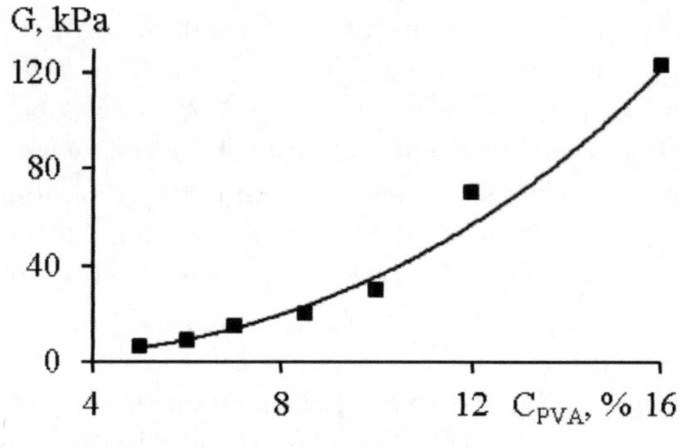

Figure 3. Elastic modulus of cryogels (G) vs the concentration of polyvinyl alcohol (C_{PVA}).

It was experimentally established that the viscosity exponential dependence on temperature (T), described by the Arrhenius–Frenkel–Eyring equation $\eta = A \cdot \exp(E / RT)$, and the viscosity power-law dependence on concentration (C) of the dissolved polymer

$$\eta_P = \eta_S (1 + [\eta]C + k[\eta]^2 C^2 + ...)$$

are characteristic features for aqueous PVA solutions and PVA-based emulsions, as well as for Newtonian liquids. The equation also suggests that a high molecular weight polymer should be used to achieve the maximum viscosity of the solution at the minimum concentration [7].

When studying the viscous properties of PVA solutions in the concentration range of 1–10%, it was revealed that the Weissenberg effect was observed only at a polymer concentration of 5% and higher. It has also been experimentally found that solutions are capable of cryogel forming at a polymer concentration of at least 5%. These experimental facts indicate that a continuous fluctuation network of mutually entangled macromolecules exists already in the initial solutions with such a concentration. The presence of this network is confirmed by the manifestation of the Weissenberg effect.

IOP Conf. Series: Materials Science and Engineering **597** (2019) 012027 doi:10.1088/1757-899X/597/1/012027

By giving experimentally the strain (γ) to formed cryogel samples and by measuring the elastic stress (P) arising in the material, the elastic modulus (G) was calculated using the Hooke formula $P = G \cdot \gamma$. It follows from Figure 3 that the modulus of elasticity of polymer bodies increases noticeably from 10 to 120 kPa with an increase in the PVA concentration of the samples from 5 to 16%.

Studies of the effect of the number of freeze-thaw cycles (n) on the elastic properties of two- and multicomponent cryogels revealed an increase in the elastic modulus after each subsequent cycle. The most elastic properties were observed for samples containing mineral oil in the polymer matrix of the cryogel (Figure 4).

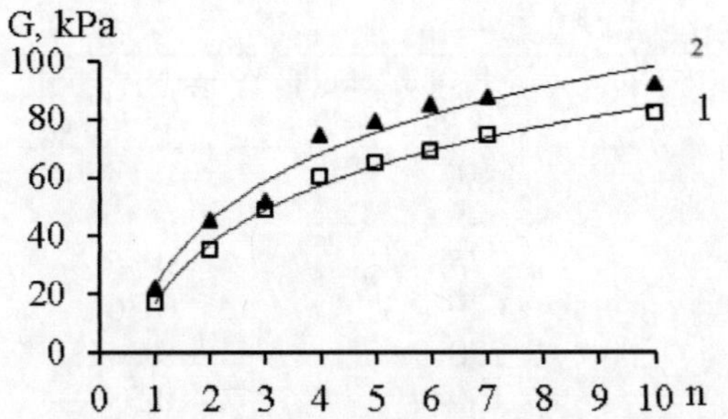

Figure 4. Elastic modulus of the cryogel (G) vs the number of freeze-thaw cycles (n). The composition of systems: 1. PVA – 10%; 2. PVA – 10%, and mineral oil – 30%.

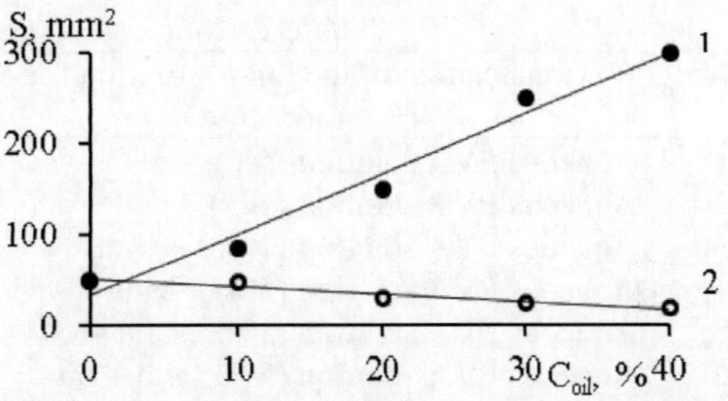

Figure 5. Areas of droplets of oil (1) and water (2) on the surface of cryogels vs the content of mineral oil present in them.

It was found that due to the spreading of a drop of oil, the degree of hydrophobicity of cryogels filled with mineral oil increases significantly with an increase in the content of hydrocarbon filler, while the area of a droplet of water decreases (Figure 5).

The samples with the lowest content of polyvinyl alcohol, which is the main structure-forming component, have the greatest use or economic value for application of cryogels as structural materials. Therefore, in subsequent experiments, the use was made of a 5% aqueous PVA solution doped with various additives increasing the rigidity of cryogels: inorganic salts, glycerin, quartz sand, clay (bentonite), and cement. All these ingredients significantly increase their elastic moduli of the samples (Figures 6-8 and Table 1). To simulate the real conditions of the oil fields, samples of cryogels with finely dispersed fillers previously soaked in oil were prepared from suspensions. Then, the elastic moduli (G, kPa) and thermal conductivity coefficients (λ, Wt / Km) of the samples were measured.

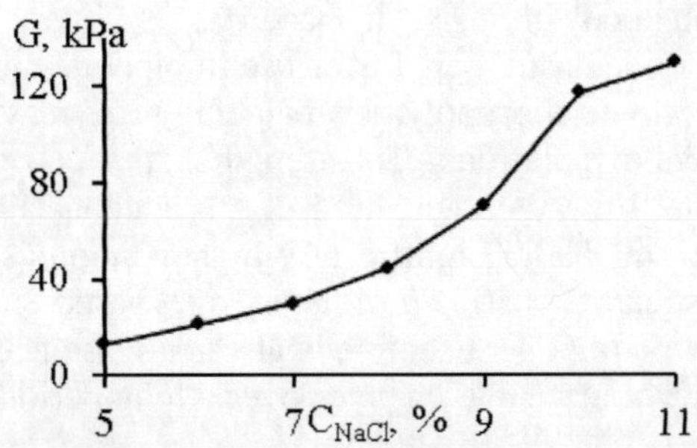

Figure 6. Elastic modulus of a cryogel (G) containing 5% PVA vs the concentration of sodium chloride (C_{NaCl})

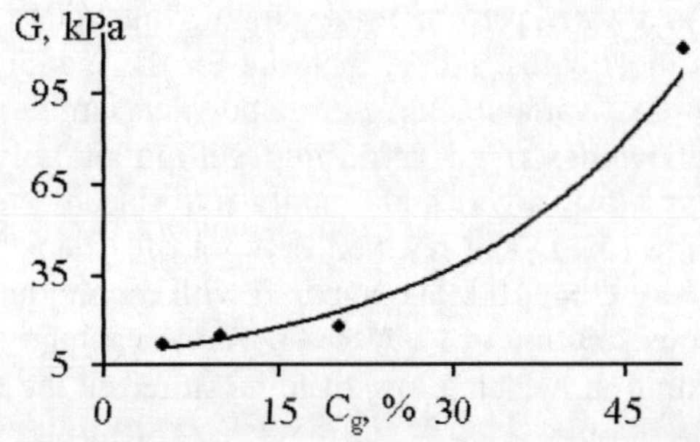

Figure 7. Elastic modulus of a cryogel (G) containing 5% PVA on the concentration of glycerol (C_g, %)

It is evident from Table 1 that the elasticity of these samples substantially exceeds the elasticity of simple two-component cryogels (PVA-water) even in the presence of hydrophobic fillers (sand, bentonite, and cement, pretreated with oil). It is known that the coefficient of thermal conductivity of water is equal to $\lambda_B = 0.62$ Wt/Km. The coefficients of thermal conductivity of wet soil (sand, clay, cement-sand solution, etc.) have even higher values in the range (1-2) Wt/Km. The layers of the polymer matrix between the fine particles of the filler perform a 'damping' function and reduce the thermal conductivity of the cryogel by half as compared with water. Thus, the excellent mechanical and thermal insulation properties of cryogels make them advisable to use for the construction of producing oil wells, in the mouth of which paraffin deposits are often observed when cooling the oil coming to the surface.

Table 1. Composition and properties of filled cryogels.

No.	Composition of the initial solution for cryogel formation, %	Properties of cryogels	
		G, kPa	λ, Wt/Km
1	Aqueous PVA solution (5%)	10	0.33
2	Aqueous PVA solution (10%)	40	0.31
3	Aqueous PVA solution (10%) + sand	775	0.35
4	Aqueous PVA solution (5%) + sand	629	0.36
5	Aqueous PVA solution (10%) sand + oil	270	0.34
6	Aqueous PVA solution (5%) sand + oil	150	0.35
7	Aqueous PVA solution (10%) + bentonite	587	0.34
8	Aqueous PVA (5%) + bentonite	465	0.35
9	Aqueous PVA solution (10%) bentonite + oil	589	0.33
10	Aqueous PVA solution (5%) bentonite + oil	345	0.34
11	Aqueous PVA solution (10%) + cement	963	0.37
12	Aqueous PVA solution (5%) + cement	885	0.38
13	Aqueous PVA solution (10%) cement + oil	677	0.34
14	Aqueous PVA solution (5%) cement + oil	416	0.35

Transportation of oil through pipelines is accompanied by the deposition of the mechanical micro impurities and water, as well as asphaltenes, resins, and paraffins. These deposits accumulated in low-lying pipeline sections reduce the effective cross-sectional area of the pipeline and increase the energy consumption of the pumping. To remove undesirable deposits, mechanical scrapers are usually used, which have a significant drawback of inability to pass through local pipe resistances (line bends, restrictions/expansions, etc.). It is possible to solve this problem by means of viscoelastic polymer 'pigs', which are introduced into pipelines when their periodic flow testing for the purpose of separate transfer of dissimilar liquids. These 'batching pigs' introduced at the boundary between the hydrocarbon fluid and water, perform the separating function of the working body during such operations.

In most cases, source material for the manufacture of viscoelastic gels fit for use in pipelines with sections of variable diameter, is polyacrylamide. Another way to form polymer 'batching pigs' may be the above-described cryotropic gelling of polyvinyl alcohol solutions. To strengthen the cryogel, glycerol (50%) was additionally introduced into the initial binary composition of the aqueous PVA solution (5%). The resulted cryogel of a density 1200 kg/m^3 had a significantly higher modulus of elasticity G = 110 kPa compared with the modulus of elasticity G = 10 kPa of the cryogel based on an aqueous solution of PVA (5%). The advantage of the three-component sample is also its resistance to dehydration, which allows it to be stored in the open air without visible changes over a long period of time.

To test the technological properties of the cryogels obtained, a laboratory setup was assembled, which included the following units connected in series: a pressure pump, a pressure gauge for recording differential pressure ΔP, a polymer 'pig' injecting chamber, and a pipeline model consisting of three segments: an U-shaped tube, spiral-shaped and straight sections. Similar diameters of the injecting chamber and the segments of pipeline were equal to $4 \cdot 10^{-3}$ m. In technologies for cleaning of pipelines and separate pumping of liquids using a 'batching pig', the L/R ratio between the length L of the pig and its radius R is of great importance. Therefore, we formed gels in two chambers having the same radius $R = 2 \cdot 10^{-3}$ m and different lengths $L_1 = 8 \cdot 10^{-2}$ m and $L_2 = 18 \cdot 10^{-2}$ m. The spiral-shaped and straight sections were filled with gasoline. Metal filings, simulating mechanical deposits in pipelines were put into the U-shaped tube. During the experiment, water was pumped into the injecting chamber. Due to the immobility of the 'batching pig', the injection of the first portions of water was accompanied by an increase in pressure up to a certain threshold limit value ΔP. Upon reaching this threshold value, the polymer 'pig' was separated from the pipe wall and its further uniform motion along the pipeline segments occurred under the influence of some constant pressure drop ΔP_{pipe}.

The shear stress for the separation from the wall (τ_1) and shear stress for uniform motion of cylindrical cryogel 'pigs' having binary and ternary compositions (τ_2) along the pipe were calculated by the formula $\tau = (R/2L) \cdot \Delta P$. The results of the experiments are presented in Table 2.

Table 2. Hydrodynamic parameters of the polymer 'pig' motion in pipes of various geometries.

No. of gel pig injecting chamber	L/R	Binary composition		Ternary composition	
		τ_{pig}, Pa	τ_{pipe}, Pa	τ_{pig}, Pa	τ_{pipe}, Pa
1	40	380	200	1250	290
2	90	1100	450	2160	870

When passing of cryogel 'pigs' through the model pipeline system, a complete removal of mechanical impurities from the pipeline body and the absence of a zone of mixing of water and hydrocarbon fluid were observed. After analyzing the results of laboratory experiments (Table 2), we tried to predict the potential industrial use of PVA-based cryogels. It can be expected that the a polymer 'batching pig', for example, 10 m long and 1.2 m in diameter, will separate from the pipe wall at an excess pressure not exceeding $1 \cdot 10^5$ Pa ($\Delta P < 1$ kg/cm^2). Hence, low values of the pressure drop, which is necessary for the separation of the polymer 'pig' from the pipe wall exclude the possibility of blocking the tube side and creating an emergency situation.

4. Conclusions

Introduction of a hydrophobic ingredient to the polymer matrix of cryogel at the stage of its cryostructuring is accompanied by the appearance of water-repellent properties of the material.

Cryostructuring of a suspension of loose inorganic material (sand or clay) in a solution of polyvinyl alcohol could be used in regions of permafrost in road and oil well construction.

Cryogels can be recommended as elastic 'batching pigs' for cleaning of pipelines from undesirable impurities.

Acknowledgments

The work was performed within the framework of project V.46.2.3. of the Program for Basic Scientific Research of State Academies of Sciences (No. 0370-2019-0002).

References

[1] Lozinsky V I 2014 *Advances in Polymer Science* **263** 48
[2] Hassan C M, Peppas N A 2000 *Macromolecules* **33** **7** 2472
[3] Lozinsky V I, Damshkaln L G, Kurochkin I N, Kurochkin I I 2016 *European Polymer Journal* **53** **1** 189
[4] Altunina L K, Kuvshinov V A, Dolgikh S N 2006 NATO Science Series. IV. Earth and Environmental Sciences (Springer) **65** 103
[5] Podorozhko E A, Ul'yabaeva G R, Tikhonov V E, Grachev A V, Vladimirov L V, Antonov Y A, Kil'deeva N R, Lozinsky V I. 2017 *Colloid Journal* **79** **1** 94
[6] Lozinsky V I, Podorozhko E A, Nikitina Y B, Klabukova L F, Vasil'ev V G, Burmistrov A A, Kondrashov Y G, Vasiliev N K 2017 *Colloid Journal* **79** **4** 497
[7] Manzhai V N and Fufaeva M S 2015 *Chemistry and Technology of Fuels and Oils* **5** 38

IOP Conf. Series: Materials Science and Engineering **597** (2019) 012028 doi:10.1088/1757-899X/597/1/012028

Sol-gel synthesis of spherical biomaterials of TiO_2–SiO_2–P_2O_5/MgO composition and study of their properties

V Kozik, L Borilo, E Lyutova, E Izosimova and A Brichkov

Department of Inorganic Chemistry, National Research Tomsk State University, 36 Lenina Avenue, Tomsk 634050, Russian Federation

E-mail: lyutova.tsu@mail.ru

Abstract. The authors obtained spherical biomaterials of TiO_2 – SiO_2 – P_2O_5/MgO composition by sol-gel method. TOKEM-200, a weakly acid porous cation exchanger based on acryl-divinylbenzene was used as organic matrices. The cation exchangers were saturated with Mg^{2+} ions. Sol based on titanium butoxide, tetraethoxysilane and phosphoric acid was deposited on them. The temperature condition for producing spherical materials TiO_2–SiO_2–P_2O_5 / MgO were proposed. To obtain a spherical material, the treated cation exchangers were subjected to stepwise heat treatment in the temperature range from 60 to 600 °C. To study the biologically active properties of the material, the samples were immersed in simulation body fluid (SBF).

1. Introduction

The discovery and development of new technologies and innovations, as well as the development and synthesis of new biomaterials are necessary to meet the requirements in the fields of tissue engineering and bone regeneration [1–3]. The restoration and replacement of bone tissue lost as a result of trauma, osteoporosis and other diseases caused by aging or an accident are key factors in orthopedic tissue engineering [4, 5]. Metallic prostheses, materials based on metal alloys, polymers, calcium phosphate ceramics, glass ceramics and bone grafts are widely used as implants [6, 7]. Phosphate materials are used in various forms, which differ in the method of preparation and particle size [8]. There are many methods for producing phosphates. One of the common methods of synthesis is the sol-gel method, which replaces the traditional melt cooling, is carried out at room temperature and improves the homogeneity and purity of the synthesis products [9, 10].

One of the promising areas recently is the production of layered spherical materials with a hierarchical structure. Such materials are designed to fill the volume of complex forms, when restoring bone tissue. Material is proposed for the development of additive technologies, such as 3D printing of biomedical materials [11, 12]. To improve the functional characteristics of a promising material for implants, compositions based on calcium phosphate or magnesium modified with silicon and titanium can be considered [13, 14]. It is proved that the presence of silicon and titanium in the volume of phosphate-calcium or magnesium material and on its surface accelerates the connection of the implant with the bone (osteointegration) [5, 8, 13].

There are many studies describing the different effects of elements, such as strontium, zinc, copper, silver, lithium and magnesium in the composition of phosphate bioceramics to achieve certain characteristics, such as osteoconductiveness, angiogenicity, and antibacterial properties [5, 6]. Magnesium is the second most common intracellular cation and the fourth most common metal cation in humans [15]. About 50–60% of total amount of Mg in the body is contained in the bones [16, 17]. In addition, Mg plays an important role in the development of bones, increases the activity of osteoblasts and inhibition of osteoclasts [17]. Mg is an element that can be absorbed by the human body, it is vital for the assimilation and fixation of calcium in the bones [16, 17].

But still not resolved two issues. Firstly, it is the effect of the Mg2+content in the composite on the rate of hydroxyapatite (HA) formation on the surface of the biomaterial. Secondly, it is the optimal amount of substituted MgO in materials, which enhances cell proliferation.

2. Experimental part

It was obtained spherical biomaterials, the outer frame of which was TiO_2–SiO_2–P_2O_5 and the inner part was filled with MgO oxide (sample TiO_2 – SiO_2 – P_2O_5/MgO). TOKEM-200, a weakly acid porous cation exchanger based on acryl-divinylbenzene (PO "Tokem" LLC, Kemerovo, Russia), was used as organic matrices for the preparation of TiO_2 – SiO_2 – P_2O_5/MgO oxide composites. This cation exchanger had a spherical shape with a granule size from 0.315 to 1.600 mm and a high selectivity to Mg^{2+}.

The study of the total exchange and sorption capacity of ion exchangers was performed using sorption methods [18], the moisture content was investigated by gravimetry. The content of Mg^{2+} ions in solutions with a concentration of ~ 0.001 M and above was determined by the method of complexometric titration with the indicator murexide [18].

Spherical materials with the composition TiO_2 – SiO_2 – P_2O_5/MgO were obtained by thermal treatment of a cation exchanger (TOKEM-200) with a supported sol based on butanol, titanium butoxide (Acros, USA) and tetraethoxysilane (Russia). For this, cation exchangers were placed in a saturated aqueous solution of magnesium nitrate (pure 99%, Yugreaktiv, Russia) and kept for 1 day at room temperature. The cation exchangers were washed with a small amount of water, filtered and dried at a temperature of 60 °C to constant weight after ion exchange. Then, dried cation exchangers saturated with Mg^{2+} ions were placed in a sol. For the formation of a framework of a bioactive material of spherical shape, an aggregally stable sol was prepared. Butanol (99.9%, Russia) was used as a solvent, and phosphoric acid (5 mol.%, Russia) was used as a polycondensation catalyst. After the onset of chemical equilibrium in a solution of C_4H_9OH – H_3PO_4, a mixture of titanium butoxide (35 mol.%, Russia) and tetraethoxysilane (60 mol.%. Russia) was added.The ripening of sols was carried out at room temperature for 3 days. The method of preparation of the sol is presented in [18]. After applying the sol to cation exchangers, the samples were dried at 60°C to constant weight. To obtain a spherical material, the treated cation exchangers were subjected to stepwise heat treatment in the temperature range from 200 to 600 °C. The heating rate of the muffle furnace was 14 °/min.

The kinematic viscosity of the solutions was determined in a HPLC-2 capillary viscometer. The thermo-gravimetric analysis was performed on a NETSCH Jupiter STA 449 F1 thermal analyzer in the temperature range from 25 to 1000°C with a heating rate 30°/min in air atmosphere. IR spectra of dried solutions were reordered on a Agilent Cary 630 FTIR IR spectrometer in the frequency range from 400 to 4000 cm^{-1}. Phase composition of the dispersed materials was determined by X-ray diffraction on Rigaku MiniFlex 600 diffractometer (CuKα radiation). Surface morphology of the obtained materials was studied using a scanning electron microscope (SEM) HITACHI TM-3000. Elemental composition was determined by X-ray microanalysis by using a Quantax-70 instrument. Evaluation of the bioactivity of the obtained materials was studied in vitro by keeping the samples in a cell-free simulation of SBF for 14 days. Samples were placed in SBF at a temperature 37 ± 0.5°C. Value of pH was 7.4. The solution was changed every day for 14 days. The composition of the SBF solution is described in [19].

3. Results and discussion

The results of studies of the physicochemical properties and selectivity of sorbents to the Mg^{2+} ion were used as the basis for the choice of cation exchanger for the production of spherical materials. During the study, quantitative characteristics of the physicochemical properties of the TOKEM-200

IOP Conf. Series: Materials Science and Engineering **597** (2019) 012028 doi:10.1088/1757-899X/597/1/012028

cation exchanger were obtained. It was established that the total exchange value was 10.25 ± 0.27 mmol/g, and the sorption capacity was 6.95 ± 0.13 mmol-eq/g, which was 63% of the total exchange capacity. The moisture content of the cation exchanger was 54.0 ± 0.5%. It follows the most of the functional groups are involved in the sorption process.

The aggregatively stable sol was prepared to form a frame of a bioactive material of spherical shape. As a criterion of the film-forming ability of solutions, their viscosity was measured (Figure 1). It was found that the viscosity of the solutions after their preparation, increases sharply during two days. Films were obtained after the viscosity reached a value of 3.2 mm2/s. Solutions were stable for 12 days.

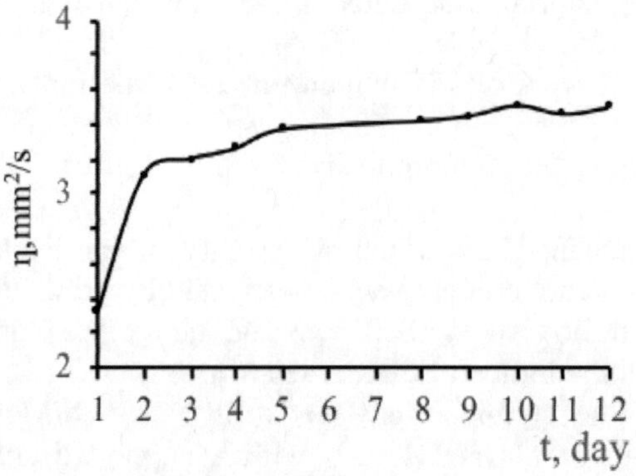

Figure 1. Viscosity values for prepared solutions in dependence on time.

The film-forming ability of solutions was achieved due to hydrolysis and polycondensation processes. They are accompanied by an increase in the viscosity of the solutions. Electrostatic interactions between dissolved ions and solvent molecules also lead to an increase in viscosity. Phosphoric acid increases the acidity of the medium and leads to acceleration of the processes of hydrolysis and polycondensation.

The results of thermal analysis showed that the decomposition of cation exchangers had two exothermic effects (Figure 2).

Exothermic effects with a maximum at a temperature of 285°C correspond to the removal of physically adsorbed H_2O. The complete combustion of hydrocarbon fragments to CO_2 occured above 450°C, the product yield was 37.6%.

Despite the fact that the decomposition of the sample was completed at about 500°C, an annealing temperature of 600°C was chosen. This is related to the literature data [38], according to which the crystalline phase of silicon oxide (IV) is formed at a temperature of 600°C.

For a more complete removal of the organic component of the cation exchanger, the dried samples were subjected to stepwise heat treatment at 200°C, 300°C, 350°C, and 450°C. The heat treatment time at each temperature was 60 minutes. After that, the samples were annealed at 600°C for 4 hours.

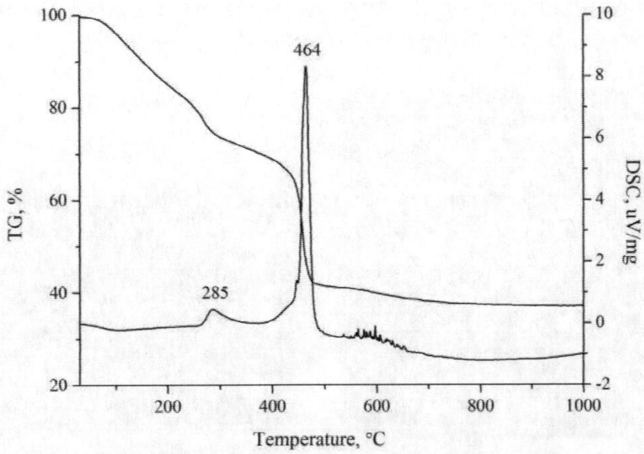

Figure 2. Thermal analysis data

According to the results of scanning electron microscopy (Figure 3), the sample has a spherical structure and the $TiO_2 - SiO_2 - P_2O_5$ framework is uniformly fixed on the cation exchanger.

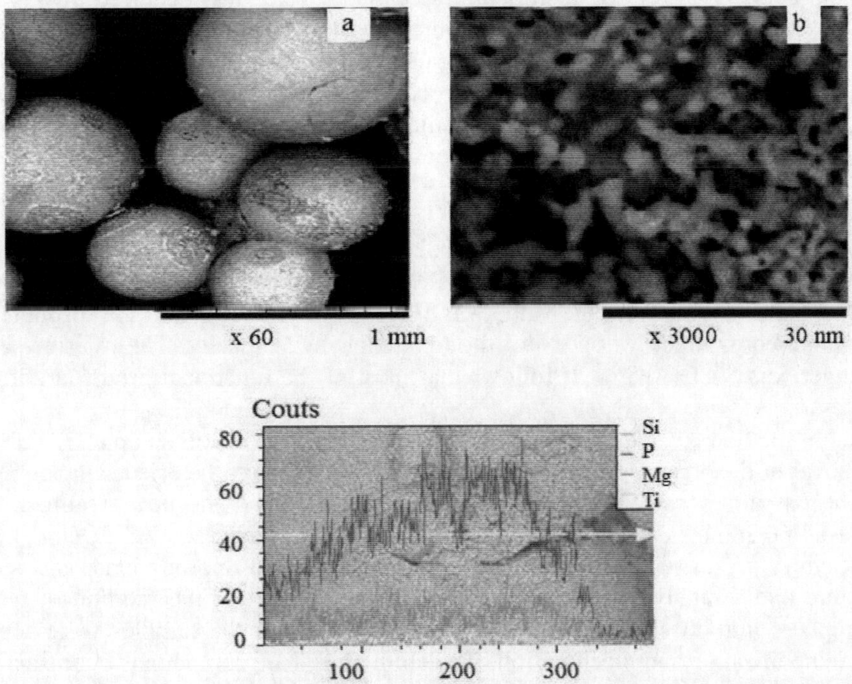

Figure 3. Micrographs of the samples at different magnifications and distribution of elements along the line, where: a – magnification of 60, b – magnification of 3000.

Heat treatment of hybrid mesostructures leads to the formation of mesoporous materials with a specific regular structure and a highly developed surface. It is necessary for the fixation of biological cells when the sample is introduced into the biological medium. Elemental analysis (Figure 4) showed a uniform distribution of elements throughout the sample.

The evaluation of the bioactivity of the obtained materials was studied in vitro by keeping the samples in cell-free imitation of SBF blood plasma for 14 days.

When studying the morphology of the coatings obtained after immersion in the SBF solution, there are large, loose and porous particles on the surface. Their presence is favorable for the formation of a strong connection of the implant with the bone. (Figure 4).

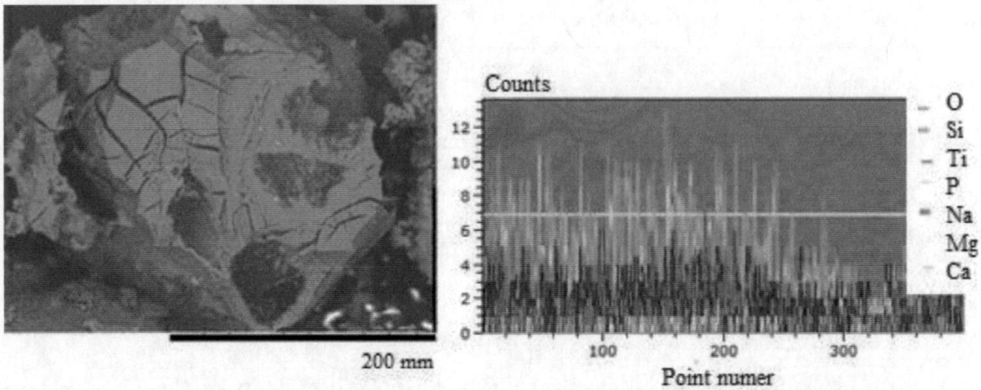

Figure 4. Micrographs of the samples and the distribution of elements along the line after immersion in SBF.

After the samples were kept in the SBF solution, the mineralization and precipitation of calcium phosphate compounds on the active centers of the coating surface was occurred. It was evidenced by the presence of Ca and P ions on the data X-ray microanalysis (Figure 4). After immersion in the SBF solution, sodium ions were fixed on the sample surface. It indicates the precipitation of the solution components. Ion Na+ helps to accelerate the process of bone tissue recovery [12-20]. Silicon ions improve the process of bone recovery, because silicon-oxygen bridges help to build in ions from the surface of materials into the intercellular fluid, promoting adhesion to bone tissue [18].

4. Conclusions

Thus, biomaterials of spherical shape based on Tokem-200 were synthesized by sol-gel method from FFS. The framework of the material was TiO_2–SiO_2–P_2O_5, and the inner part was filled with MgO. It was established that the total exchange value was 10.25 ± 0.27 mmol/g, and the sorption capacity was 6.95 ± 0.13 mmol-eq/g, which was 63% of the total exchange capacity. The moisture content of the cation exchanger was 54.0 ± 0.5%. It follows that most of the functional groups are involved in the sorption process.

It was established that the solutions were suitable for obtaining materials up to 12 days. For a more complete removal of the organic component of the cation exchanger, the dried samples were subjected to stepwise heat treatment at 200°C, 300°C, 350°C, and 450°C. The heat treatment time at each temperature was 60 minutes. After that, the samples were annealed at 600 °C for 4 hours. Heat treatment of hybrid mesostructures leads to the formation of mesoporous materials with a specific regular structure and a highly developed surface. It is necessary for the fixation of biological cells when the sample is introduced into the biological medium. After the samples were kept in the SBF solution, the mineralization and precipitation of calcium phosphate compounds on the active centers of the coating surface was occurred. It proves biologically active properties of the material.

Acknowledgments
The results were obtained within the framework of the state task of the Ministry of Education and Science of the Russian Federation, project No.10.2281.2017/4.6.

References

[1] Barinov S M 2010 Calcium phosphate-based ceramic and composite materials for medicine *Russ. Chem. Rev.* **79** 13
[2] Dorozhkin S V 2016 Calcium Orthophosphate-Based Bioceramics and Biocomposites New

York: Wiley 115

[3] Kalantari E, Naghib S M, Naimi-Jamal M, Mozafari M 2017 Green solvent-based sol–gel synthesis of monticellite nanoparticles: a rapid and efficient approach *J. Sol-Gel Sci.Technol.,* **84** 87

[4] Barinov S M, Komlev V S 2014 Calcium phosphate bone cements (a review): I. Binder systems *Materialovedenie* **1** 33

[5] Jmal N and Bouaziz J 2017 Synthesis, characterization and bioactivity of a calcium-phosphate glass-ceramics obtained by the sol-gel processing method *Mat. Sci. Eng.* **71** 279

[6] Borilo L P, Lyutova E S and Spivakova L N 2016 Study of biological properties of thin-film materials on the basis of the SiO_2–P_2O_5–CaO system *Key Eng. Mater.* **683** 427

[7] Yamaguchi S, Nath S, Matsushita T, Kokubo T 2014 Controlled release of strontium ions from a bioactive Ti metal with a Ca-enriched surface layer *Acta Biomaterialia* **10** 2282

[8] Popa A C, Stan G E, Husanu M A, Mercioniu I, Santos L F, Fernandes H R 2017 Bioglass implant-coating interactions in synthetic physiological fluids with varying degrees of biomimicry. *International Journal of Nanomedicine* **12** 683

[9] Moghanian A, Sedghi A, Ghorbanoghli A and Salari E 2018 The effect of magnesium content on in vitro bioactivity, biological behavior and antibacterial activity of sol–gel derived 58S bioactive glass *Ceramics Int.* **44** 9422

[10] Kukueva E V, Putlyaev V I, Tikhonov A A and Safronova T V 2017 Octacalcium phosphate as a precursor for thr fabrication of composite bioceramics *Inorg. Mater.* **53,2** 212

[11] Li X, Wang M, Deng Y, Xiao Y, Zhang X 2017 Fabrication and Properties of Ca-P Bioceramic Spherical Granules with Interconnected Porous Structure *ACS BIOMATER SCI ENG* **3** (8) 1557

[12] Bjornoy S H, Bassett D C, Ucar S, Andreassen J P, Sikorski P 2016 A correlative spatiotemporal microscale study of calcium phosphate formation and transformation within an alginate hydrogel matrix. *Acta Biomaterialia* **44** 254

[13] Oliveira W F, Arruda I R S, Silva G M M, Machado G, Coelho L C, Correia M T S 2017 Functionalization of titanium dioxide nanotubes with biomolecules for biomedical Applications *Mater. Sci. Eng. C.* **81** 597

[14] Evdokimov P V, Putlyaev V I, Ivanov V K, Garshev A P, Shatalova T B, Orlov N K, Klimashina E S, Safronova T V 2014 Phase equilibria in the tricalcium phosphate-mixed calcium sodium (potassium) phosphate systems *Russ. J. Inorg. Chem.* **59** 1219

[15] Dorozhkin S V 2010 Bioceramics of calcium orthophosphates *Biomaterials Biomaterials* **31** 1465

[16] Bagherpour I, Naghib S M, Yaghtin A H 2018 Synthesis and characterisation of nanostructured hardystonite coating on stainless steel for biomedical application *IET nanobiotechnology* **12** 895

[17] Dezfuli S N, Huan Z, Mol A, Leeflang S, Chang J, Zhou J 2017 Advanced bredigite-containing magnesium-matrix composites for biodegradable bone implant applications *Mater. Sci. Eng. C.* **79** 647

[18] Shamsutdinova A N, Kozik V V 2016 Obtaining and properties of thin films based on titanium, silicon and nickel oxides *Chemistry for Sustainable Development* **24 (5)** 699

[19] Kokubo T, Kushitani H and Sakka S 1990 Solutions able to reproduce in vivo surface – structure changes in bioactive glass – ceramic *Biomaterials* **24** 721

Technology and Manufacture of Reinforced Concrete Structures with Application of Silica Fume for Multi-storey House-building in the Republic of Khakassia

L P Nagruzova, G N Shibaeva, K V Saznov and A Kubanychbek kyzy

Construction Department, Khakass Technical Institute, Siberian Federal University, 27 Shchetinkin str., Abakan, Russia

E-mail: aitbu.kubany4bekkyzy@yandex.ru

Abstract: Silica fume is a processing waste of ferroalloy factories. The application of silica fume instead of cement allows producing high strength concrete, as well as concrete types of the middle and low classes providing a saving in cement of up to 50 %. The authors have developed a technology for preparing a concrete mixture for reinforced concrete structures with silica fume. The technology is simple and does not require additional expensive equipment having low power intensity. The technological parameters have been investigated. They are related to slump, maturing conditions during steam curing and monolithic casting. The economic efficiency of applying concrete mixtures containing silica fume has been determined.

1. Introduction

One of the main objectives in the construction industry is cutting of construction cost value by utilizing the latest technologies and modern materials. Construction in Khakassia is considered a dynamically developing industry. Most of the building structures related to residential construction are made of monolithic and prefabricated monolithic reinforced concrete. One of the ways to reduce the cost of materials for multi-storey residential buildings is to lower the amount of reinforcement used in the structure by contracting the construction section, which is possible using high strength concrete. The development of production technology of high strength concrete and study of its raw materials sources are of great importance for manufacturing the concrete type. In recent years in all developed regions the high strength concrete containing silica fume has been rather widely spread, which enables to reduce significantly material consumption and to increase durability, frost resistance, water resistance. In Khakassia, the Department of Urban Development made a decision on the design and construction of high-rise residential buildings (17-, 19-, 23-storey buildings). In this regard, the use of high strength concrete containing silica fume is an urgent issue, since applying of silica fume allows producing concrete class B80 and higher using cement of brand 400.

Batrakov V.G., Kaprielov V.S., Schonfeld A.V., Gamaliy E.A., Pertsev V.T., Mkrtchyan A.M. studied the problem of making high strength concrete or concrete types of middle classes having a reduced cement consumption while preserving simultaneously strength and other indicators of concrete containing silica fume [2, 3, 4, 5, 6, 8].

2. Purpose

The purpose of the work consists in selecting concrete compositions of different classes subjected to steam curing in natural conditions that are made of local materials extracted and produced in the

IOP Conf. Series: Materials Science and Engineering **597** (2019) 012029 doi:10.1088/1757-899X/597/1/012029

Republic of Khakassia and in studying physical and mechanical properties and technology of manufacturing concrete containing silica fume.

3. Experimental part

Silica fume is a superdispersed material captured by bag filters of gas-cleaning installations at ferroalloy plants. The main component of superdispersedwaste is silicon dioxide having amorphous modification SiO2.

The authors conducted a research to investigate the impact of silica fume on cement stone. The average density of silica fume is 2.1-2.3 g/cm³ (cement 3.1 g/cm³), bulk density is 0.15-0.20 g/cm³, grain size is less than 0.1 μm (50 times smaller than the particle size of cement) specific surface is 200,000-250,000 cm²/g (cement is about 4,600 cm²/g). The silicon oxide content in silica fume reaches 90.7% and sometimes up to 96 %. Oxides of calcium, magnesium and iron predominate among other components.

It is important to take into account a complex of factors and indicators that are reflected in technological and technical characteristics of cement stone: required amount of water, strength, and consumption of SPs (superplasticizers), since silica fume due to its large specific surface area has higher water demand a superplasticizer is applied. The efficiency evaluation of silica fume is shown in Figure 1.

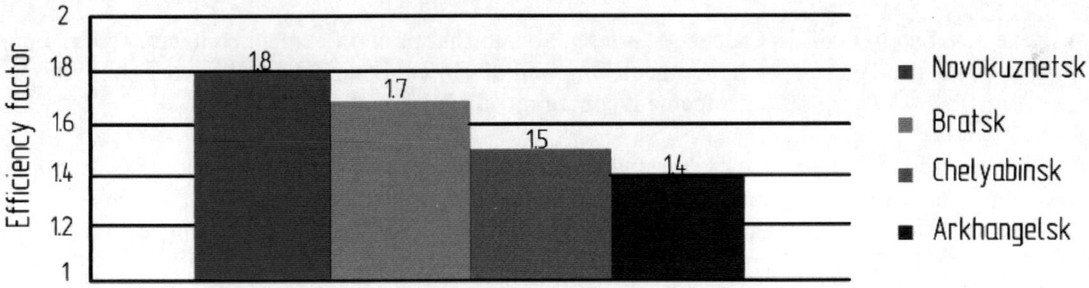

Figure 1. Efficiency of silica fume produced at different plants.

According to Figure 1 it is seen that the most effective waste of ferroalloy production is silica fume made in the city of Novokuznetsk. Namely, we study this substance.

The authors conducted studies of the effect of silica fume on cement stone applying cement brand 400 (produced at Achinsk and Topkinsk plants). To estimate the strength gain at the age of 28 days, an empirical formula (2.1) was proposed, which allows determining this parameter depending on thedosage of silica fume, its pozzolanic activity, degree of hydration, capillary porosity of the system.

The formula, which is quite exact at different dosages of silica fume depending on the weight of cement, is derived on the basis of a number of prerequisites:

- the value of pozzolanic silica fume activity depends on the content of amorphous SiO2;
- the change in the density structure of cement stone when adding silica fume is connected with the volume of gel and capillary pores and to a lesser extent with the volume of macropores;
- gel porosity is determined by the degree of hydration in the cement stone which is affected by SiO2 / Ca(OH)2 ratio. Capillary porosity is determined by water-cement ratio, degree of hydration and, silica fume content.

$$R = K\left(\frac{S}{SI}\right)^2\left(\frac{B-0{,}5\cdot\alpha\cdot P}{10\cdot C}\right)^2 \%, \qquad (1)$$

K is a coefficient considering the difference in molecular masses of SiO_2 and Ca $(OH)_2$;
S is an absolute content of SiO_2 in the composition of mixed binder, %;
SI is a content of Portland cement in cement stone without silica fume, it is assumed to be 15 %;
α is a degree of hydration of Portland cement, %;
C is a dosage of silica fume, %;

IOP Conf. Series: Materials Science and Engineering **597** (2019) 012029 doi:10.1088/1757-899X/597/1/012029

P is a percentage of cement

Figure 2 shows the dependences of concrete strengths on the consumption of cement containing silica fume. The results of Figure 2 indicate that class B 55 concrete produced has a cement consumption of 340-350 kg/m³. Figure 2b shows the saving of cement containing silica fume when comparing it with the control factory strengths of concrete.

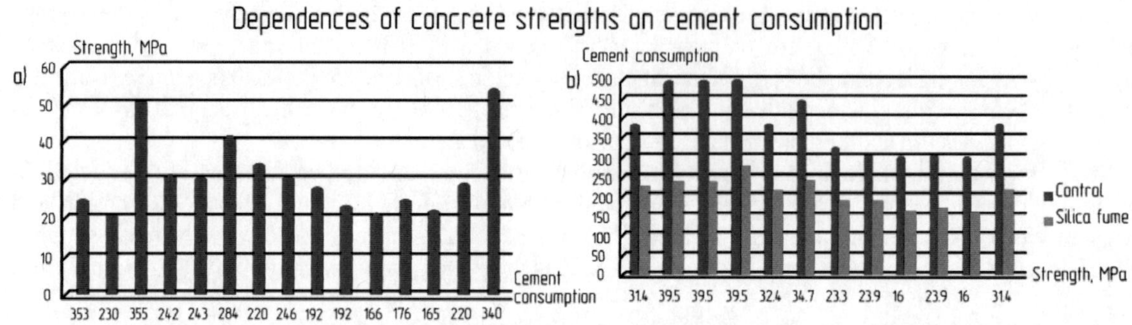

Figure 2. a) Diagram of dependence of strength on consumption of cement containing silica fume; b) diagram of dependence of strength on consumption of cement having control factory strengths and of cement containing silica fume.

The conditions of hardening have been studied. When increasing in dosage of silica fume, the amount of free lime in the cement stone samples decreases. Figure 3a presents the results of determining CaO in samples of cement stone, depending on the conditions of hardening and dosing of silica fume. Practical dosage above 30% of cement weight leads to the fact that this dosage is an efficiency threshold of silica fume. The use of silica fume in combination with the plasticizer SP1 or S3 is an effective means of saving cement in concretes having both low and high cement content. In addition, the received results indicate that along with considerable economy of cement, there is a significant increase in the strength of the material, which provides certain opportunities for obtaining high-strength concrete (Figure 3b).

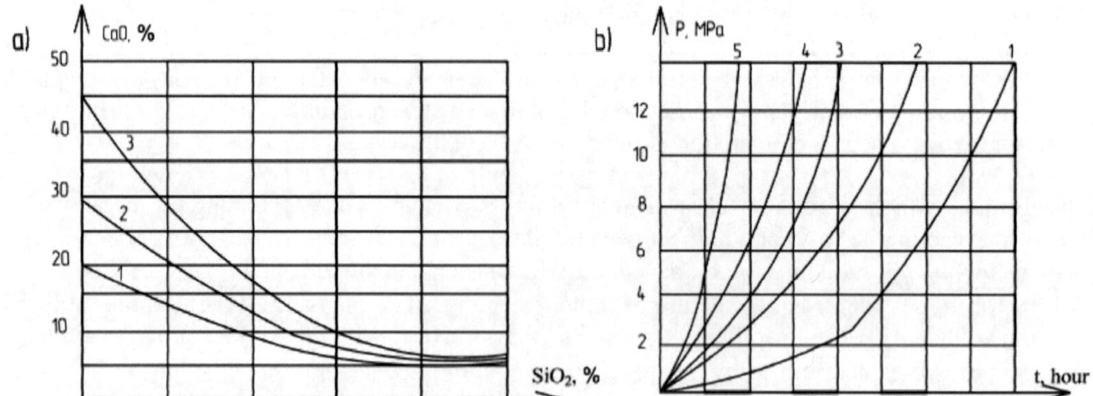

Figure 3. a) Content of free lime in the samples of cement stone with different dosages of silica fume; b) cement without additives, 2-cement +10% of silica fume, 3- cement +20% of silica fume, 4 – cement +30% of silica fume, 5 – cement+40% of silica fume.

In order to evaluate the properties of concrete mixtures and determine the effect of technological sequence of mixture preparing on the properties of concrete the samples made in a forced mixer were tested in two ways. In the first case, fillers were introduced in the mixture of cement and S-3, then after

IOP Conf. Series: Materials Science and Engineering **597** (2019) 012029 doi:10.1088/1757-899X/597/1/012029

pre-mixing, water was added with S-3 additive. In the second case, the mixture was prepared in the usual sequence, and silica fume was introduced in the concrete mixer in the form of a suspension of 40% concentration with a solution of S-3 additive and water (survivability time of the suspension is three days according to the specification). The mixing time of weighed materials is 3 minutes.

In the course of the experiment the change of the flowability of the mixtures in time and the dependence of placeability characterized by rigidity according to GOST 10181.1-81 on the flowability measured by slump were determined. The strength was defined by testing the samples having edge size of 10 cm which were hardening in normal conditions and during steam curing that was carried out in two modes. The first mode: 3+3+6+2 at the isotherm temperature of 900°C, the second one had the same duration but the isotherm temperature was 60°C.

As a result of the experiment, it has been established that the mixtures prepared by introducing silica fume in the dry form into the mixer lose rapidly their flowability in 10 minutes which creates certain inconveniences by performing works. This is connected with kinetics of measuring the water demand of mixtures until saturation of ultradispersed material with water, which occurs depending on the silica fume dosage in 6-10 minutes after mixing all components of the concrete mixture (see Figure 4).

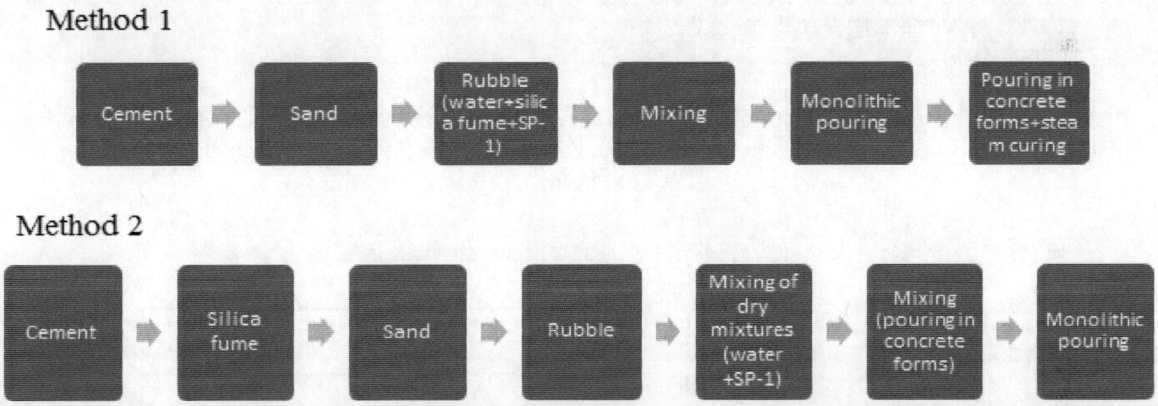

Figure 4. Process flowsheet of producing concrete of different brands.

By preparing mixtures with introduction of silica fume in the form of suspension in the mixer, the flowability in time varies relatively evenly. The consistency of concrete mixtures in time proved to be one of the decisive factors that determined the choice of technology with feeding silica fume to a mixer in the form of suspension.

The data presented in Figure 5 indicate that concrete mixtures containing ultradisperse filler have better placeability, that is, stronger thixotropic properties than ordinary mixtures of the same placeability. It is to mention the placeability increases with an increase in the proportion of cement substituted by silica fume in the mixture composition.

The conditions of hardening have a significant effect on the strength properties of concrete: when hardening in normal conditions an intensive increase in strength occurs within 7-20 days; when hardening occurs in steam curing conditions, the strength increases with increasing isotherm temperature up to 900°C (see Figure 6).

The above stated results lead to the conclusion that the technology of manufacturing a concrete mixture of different brands is simple and does not require special equipment.

Additionally, an autonomous loading of silica fume is provided which depending on the production conditions can be supplied in a dry state or in the form of suspensions (see Figure 6 b).

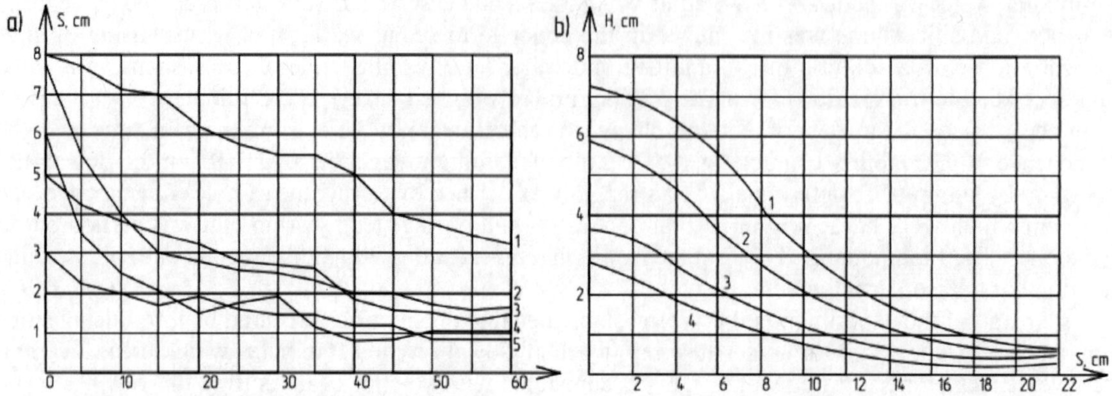

1- mixture without addition of silica fume; 2, 4- mixtures with addition of silica fume in the volume of 30 and 20 % produced by introducing silica fume in dry form; 3, 5 - the same mixtures produced by introducing silica fume in the form of suspension

1 -mixtures without addition of silica fume and S-3; 2 - mixtures without addition of silica fume containing S-3; 3 - mixtures with S-3 and substitution of 20% of cement by silica fume

Figure 5. a) Change in the flowability of concrete mixtures in the course of time; b) Changes in the placeability of concrete mixtures using silica fume depending on flowability (Binder consumption is 300 kg/m³).

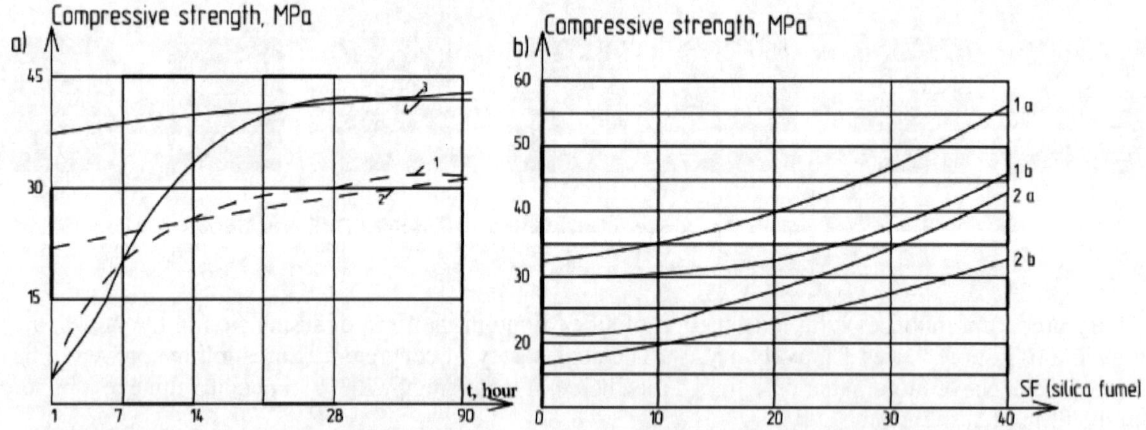

Check sample with cement consumption of 300 kg/mi when standard hardening; 2 - the same when steam curing occurs according to the mode 2+3+5+2 when temperature isotherm is 90°C; 3 - sample with cement consumption of 230 kg/mi and feeding silica fume in the volume of 85 kg /mi by standard hardening; 4 - the same, when steam curing occurs according to the mode 2+3+5+2 at temperature isotherm of 90°C

1a, 1b - samples produced by using Portland cement M500 of Topkinsk plant at isotherm temperature of 85° C and 60°C;
2a, 2b - samples produced by using Portland cement M400 of Topkinsk plant at isotherm temperature of 90° C and 60°C;

Figure 6. a) Kinetics of hardening concrete samples with addition of silica fume; b) Concrete strength depending on the dosage of silica fume and steam curing mode.

The time variation of flowability of concrete mixtures with silica fume additive depends on the method of their preparation: when preparing mixtures with feeding silica fume into the concrete mixer in the initial state (dry), there is a sharp decrease in flowability of the mixtures; flowability in time changes evenly when supplying silica fume in the form of suspension. From the point of view of

preserving the consistency in time, it is preferable to prepare concrete mixtures supplying silica fume in the form of suspension in the mixer, this applies to the silica fume noncompacted.
Concrete mixtures containing silica fume have pronounced thixotropic properties. This is manifested in increased placeability of low-slump mixtures. Supplying silica fume in concrete compositions at constant cement consumption allows to obtain an increase in the strength of concrete from 40 to 70 %. The positive effect of silica fume additive on the strength properties of concrete can be used to significantly reduce the cost of cement.
The strength of concrete containing silica fume is considerable affected by the mode of steam curing; hardening of concrete occurs more intensively when increasing the isotherm temperature.

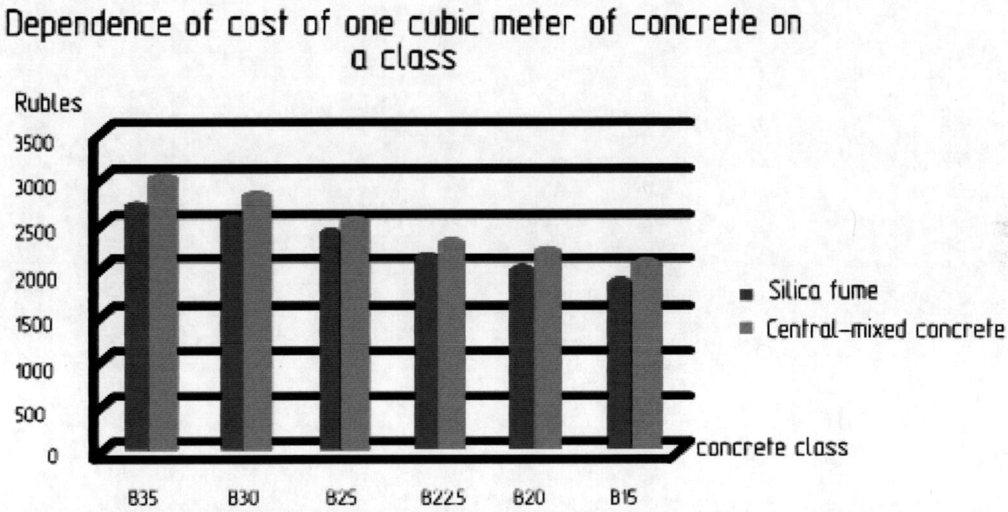

Figure 7. Dependence of cost of one cubic meter of concrete on a class.

4. Conclusion

The study of concretes having middle activity has shown that supplying silica fume in concrete mixture will allow: producing high-strength concrete (or cement saving up to 50%); making concretes that are not steam-cured and have specified delivery strength B30 within 24 hours; reducing the duration of heat and moisture treatment for 3-4 hours; increasing frost résistance F and water resistance w of concrete; as well as increasing sulphate-resistance of concrete made using ordinary Portland cement; and improving bonding of cast concrete mixtures. The safety of reinforcing steel in concrete is ensured at a dosage of silica fume that should be not more than 20% of cement weight (if it is more than 20% steel corrosion inhibitors are to be fed – sodium nitrate). The environmental situation will be also improved while providing a significant effect in the construction industry.

Refrences

[1] Nagruzova L and Saznov K 2015 Selection of compositions of concrete mixture: M 250, M 300, M 400 containing silica fume made of local materials (Abakan: Khakass Technical Institute) p 86
[2] Batrakov V, Kapriyelov S and Sheynfeld A 1989 Effectiveness of applying ultradispersed waste of ferroalloy production *Con. and rei. con.* (Moscow) **8** 24-25
[3] Kapriyelov S1992 Silica fume: classification: main factors of action mechanism: effects on the structure and properties of cement stone Chemical additives and their application in the production technology of precast concrete (Moscow: Central Russian House of Knowledge) pp 62-68
[4] Gamaly E and Trofimov B 2009 Structure and properties of cement stone containingsilica fume

additives and softener polikarboxylat *Bul. of Sou. Ura.Sta. Uni.Con.* and Arc. **16** 29-35
[5] Ye G and van Breugel K 2009 Simulation of connectivity of capillary porosity in hardening cement-based systems made of blended materials *Heron* **54 (2)** 163-184
[6] Pertsev V, Ledenev A and Halilbekov Y 2017 Complex organomineral layer additives for concreteInt. *Sci. Mag.* **2** 89-91
[7] Nagruzova L, Kiselman A, Strelnikov D and Tankov E 2018 Improvement of concrete and production of reinforced concrete structures with savings of up to of 40-50% of cement using microsilica *Sci. and Bus.: Way of Dev.* **10** 61-66
[8] Mkrtchyan A 2013 Reinforced concrete columns from high-strength concrete on materials of the Republic of Armenia p 24 (Rostov-on-Don)

IOP Conf. Series: Materials Science and Engineering **597** (2019) 012030 doi:10.1088/1757-899X/597/1/012030

Wave mechanism of structure formation in cement compositions

N P Gorlenko[1], Yu S Sarkisov[1], V I Syryamkin[2], L B Naumova[2], A N Pavlova[1] and B I Laptev[3]

[1]Tomsk State University of Architecture and Building, 2, Solyanaya Sq., Tomsk, 634003, Russia

[2]National Research Tomsk State University, 36, Lenin Ave., Tomsk, 634050, Russia

[3]Nove Technologije, D. O. O. Ljbljana 1000, Sloveniya

E-mail: gorlen52@mail.ru

Abstract. The development of new methods of physicochemical analysis opens vast opportunities for studying the kinetics and mechanisms of the structure formation in cement systems. A study of the structure formation and cement hardening processes will allow widening a range of methods of their non-destructive testing and developing production techniques for advanced composite materials. The X-ray microtomography system is used to analyze the formation of alternating dark and bright concentric rings both on the surface and in the bulk of the specimens. It is supposed that the formation of concentric rings is caused by the generation of acoustic vibrations in the cement-water system at early stages of the structure formation.

1. Introduction

Today, digital technologies are applied in investigations of physicochemical and service properties of construction materials allowing users to estimate their non uniformity, detect structural defects, pore size and pore distribution in the bulk of the material [1-4]. As is known, the inner structure of the cement-water system depends on many factors and is determined mostly by the formation of the capillary-porous, three-phase system solid phase-water-entrained air. The volume of unreacted cement particles, gel and capillary pores significantly changes during its curing. Control for their development and, as a consequence, the properties of construction materials can be performed only using new experimental data and detailed analysis of this process development. This will provide the digital technology-based online research in the field cement compositions [5].

2. Materials and Methods

PC-500 cement bricks 2×2×2 cm in size were used in the experiment. The water-cement ratio was 0.34 for humid curing conditions.

The X-ray microtomography (XMT) system used in this work has the following specifications: 1–13 μm image segmentation; soft X-ray source with energy range from 20 to 160 kV; 0–250 mA anode current; < 5 μm focal spot size (at 4 W). Image reconstruction time is 10 min/cm^3. The X-ray detector has a resolution of 4872×3248 pixels, with the pixel size of not over 7.4×7.4 μm. The time of three-dimensional image reconstruction is 10 min. The system allows studying objects at any angle with ±1 μm accuracy of location without destroying the original object.

IOP Conf. Series: Materials Science and Engineering **597** (2019) 012030 doi:10.1088/1757-899X/597/1/012030

3. Results and Discussion

Figures 1 and 2 present a chain of the XMT images of the cement brick morphology with the increasing scanning depth.

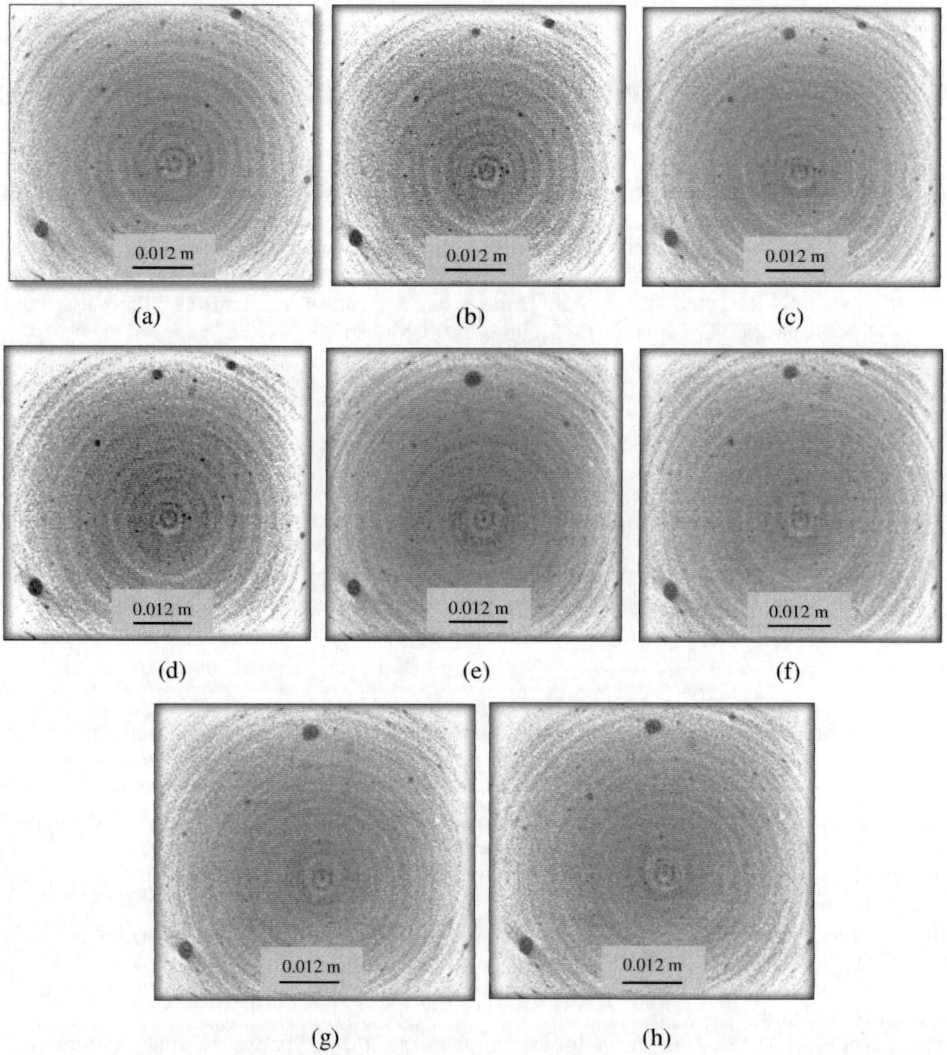

Figure 1. XMT scans of cement brick layers (a, b, c, d, e, f, g, h) at a 2 μm scan depth. Magnification: ×5.

The XMT scans reveal the detailed morphology of the cement specimen, which contains dark formations (pores) and alternating dark and bright concentric rings. The following conclusions can be made after the analysis of the XMT images.

1. The pore distribution over the area and the bulk of specimens is random, and the total areal porosity at different depth is not a constant value.
2. The pore size ranges mostly from 10^{-2} to 10^{-3} μm at a depth varying from several units to hundreds of micrometers.

IOP Conf. Series: Materials Science and Engineering **597** (2019) 012030 doi:10.1088/1757-899X/597/1/012030

3. Entrapped air is both of symmetrical and asymmetrical configurations, which may be caused by the alternating density gradients at the cement-air interface at the end of the cement setting time.
4. The configuration of the observed alternating dark and bright concentric rings changes with the scanning depth. At the same time, the distance between two minima of the holes (dark rings) or between two maxima of the shoulders (bright rings) is constant and amounts on average to 0.006 m.

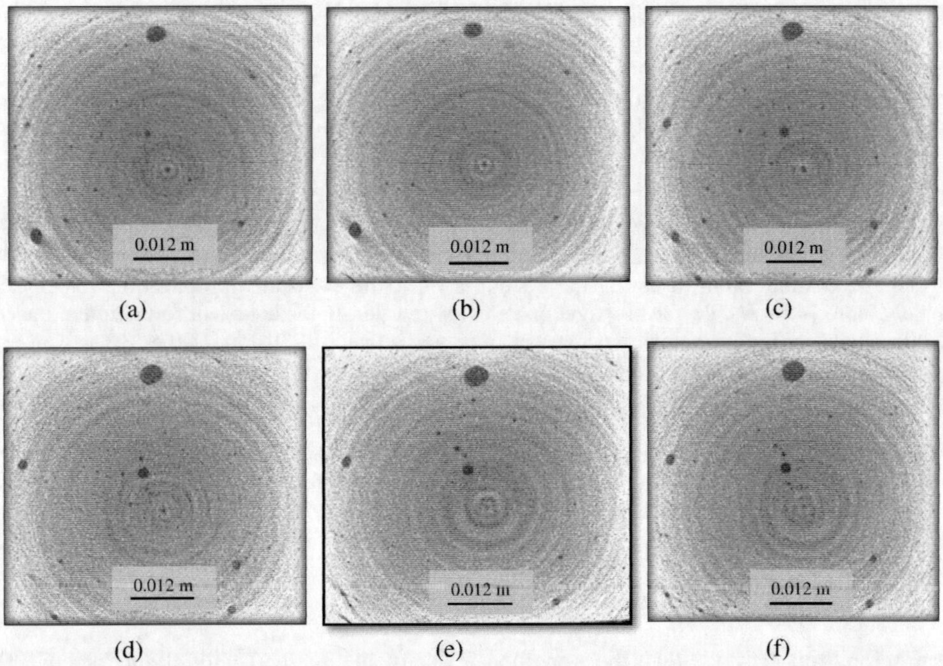

Figure 2. XMT scans of cement brick layers (a, b, c, d, e, f) at a 4 μm scan step. Magnification: ×5.

The formation of these concentric rings is probably caused by the process development of the electromagnetic wave generation in the bulk of the hardening cement-water system. It is well known that acoustic vibrations accompany the processes of cement hydration and curing with the frequency ranging from 100 Hz to 20 kHz and intensity of 3–10 dB observed at the dispersion of cement particles [6]. It is possible to assume that the higher frequency vibrations, including ultrasonic, occur in the cement-water system especially at a time of cement mixing. Probably, ultrasonic vibrations is one of driving forces of the cement particle dispersion, whereas the concentric rings result from the formation of standing waves in the cement-water system (see Figure 3).

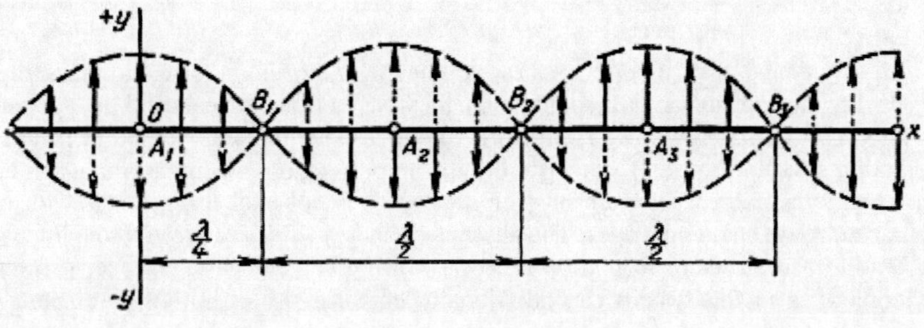

Figure 3. Schematic of standing wave formation: B – wave nodes; A – vibrational amplitude; λ – wavelength.

It is known that standing waves appear when two waves with the similar amplitude and length travel in opposite directions. This can occur if the electromagnetic vibrations are generated by particles or their new formations constituting the bulk of particles with similar size and properties. In other words, the system represents the totality of particles with more or less evenly distributed parameters, such as, for example, weight, elasticity, size close to that respective parameters in each elementary volume of the system. It is also known that due to the ultrasonic waves, the smaller particles and the liquid phase shift toward the standing wave node, whereas coarse particles shift toward the larger amplitude of vibration. As a result, the cement paste compacts nearby the contacting area of the electromagnetic source (dark rings). Outside this source, the formation of the lower density area is observed (bright rings). At the same time, the wavelength for a standing wave is determined by the double distance between the dark and bright rings and amounts on average to 0.012 m as presented in Figure 2.

The vibration frequency of ultrasonic waves in the cement-water system can be found from

$$f = c/\lambda,$$

where c is the velocity of the ultrasound propagation in the curing system; λ is the wavelength.

Given that at the initial point of curing the system is fluid, the ultrasound propagation velocity can be conventionally set to 1490 m/s, i.e. the propagation velocity in water. In the course of formation of the cement brick the ultrasound propagation velocity increases and ranges between 1930 and 3000 m/s.

Let us calculate the frequency of standing waves according the given above equation:

$$f = (1490 - 3000)/0.012 = (124.2 - 250.0)\ \text{kHz}.$$

Actually, as follows from Figure 1, these standing waves are characterized by a range of frequencies which include multiple harmonics due to which a wider frequency band can be expected, namely from acoustic to megahertz range observed in experiments.

In our view, the electromagnetic source is the eigen-frequency vibrations of cement particles which can be calculated as

$$f = \frac{1}{2\pi}\sqrt{k/m},$$

where f is the eigen-frequency; $k = F/l$ is the proportionality constant representing the elastic force F acting on the bond length unit l; m is the particle mass.

At the initial point of curing, the cement paste is a colloidal system with the particle size ranging between 10^{-6} and 10^{-8} m. Assuming that the density of the cement paste is 1400 kg/m^3, let us find the mass of the individual particles:

$$m = V\rho = (10^{-6}–10^{-8})^3 \times 1400 \approx (10^{-15}–10^{-21})\ \text{kg}.$$

For particle which manifest the dipole-dipole interaction, $k = (0.9–1.8)\times 10^{-7}$ kg/m [7]. Therefore,

$$f = 0.16\sqrt{\frac{(0.9-1.8)\times 10^{-7}}{(10^{-15}-10^{-21})}} = (2\times 10^{3} - 1.5\times 10^{6})Hz.$$

Accordingly, if the system contains a population of particles with the similar size and mass, the formation of standing waves having the acoustic and ultrasonic frequency bands is possible during the generation of free electromagnetic vibrations. The energy conversion from kinetic to potential and *vice versa* occurs within the section of the oscillating system (Figure 3), namely from the amplitude to the neighboring wave node. In this case, there are no energy losses because each of such section does not exchange energy with neighboring sections that is characteristic to standing waves. The absence of energy losses can lead to the intensification of the dispersion, hydration and structure formation processes in the cement paste. Hence, preconditions for maintaining standing waves in the system can additionally enhance the operating parameters of cement compositions.

Circumstantial evidence that ultrasound exerts an effect on the intensification of the dispersion, hydration and structure formation processes, is the literature data [8, 9]. One can suppose that along, for example, with a thermal effect, the generation of the internal electromagnetic waves in the cement-water system at the initial point of curing is the additional factor that affects the curing process.

It is important to note that the formation of concentric rings can be also observed at a travelling spherical wave. This is conditioned by different propagation rates of transverse and longitudinal waves. The ratio between these waves is determined by the Poisson ratio:

$$\frac{V_1}{V_2} = \sqrt{\frac{1-2\,\nu}{2-2\,\nu}}\ ,$$

where V_1, V_2 are the propagation velocities of the longitudinal and transverse waves, respectively; ν is the Poisson ratio.

Given that $\nu = 0.2$ for concrete, we get:

$$\frac{V_1}{V_2} = \sqrt{\frac{1-2\times 0{,}2}{2-2\times 0{,}2}} = 0.375.$$

Taking all these calculations into account, the curing cement-water system can be considered as a medium with the electromagnetic field impedance, where $V_1 < V_2$, that results in the formation of visible concentric rings.

4. Conclusions

The X-ray microtomography-based analysis showed that the alternating dark and bright concentric rings both on the surface and in the bulk of the specimens were formed at a distance of about 0.006 m between them.

Based on calculations, it was assumed that one of the possible mechanisms of the observed phenomenon could be the generation of internal electromagnetic waves having the acoustic and ultrasonic frequency bands. The obtained experimental data and theoretical calculations indicated to a possible wave mechanism of the structure formation in cement compositions.

References

[1] Peng R, Yan Y, Ju Y, Mao L and Yang Y 2011 Computation of fractal dimension of rock pores based on gray CT images *Chinese Sci. Bull.* **56**(31) 3346 3357

[2] Withers P J and Preuss M 2012 Fatigue and damage in structural materials studied by X-ray tomography *Annu. Rev. Mater. Res.* **42** 81–103

[3] Hussain A and Akhtar S 2017 Review of Non-Destructive Tests for Evaluation of Historic Masonry and Concrete Structures *Arabian Journal for Science and Engineering* **42**(3) 925-940

[4] Bogomolov E N and Syryamkin V I 2013 Digital X-ray tomograph. Basic physics of tomographs *Izv. Vyssh. Uchebn. Zaved., Fiz.* **56**(10/2) 48–54

[5] Ksenev N I, Syryamkin V I and Shidlovskii S V 2009 Reconstructed structures in the problems of control and investigation of surface images processing algorithms for evaluating solid-body deformations *Telecommunications and Radio Engineering (English translation of Elektrosvyaz and Radiotekhnika)*, **68**(9) 813–820

[6] Lotov V A 2018 Cement particles interaction with water or cement hydration and curing mechanisms *Bulletin TPU. Geo Assets Eng.* **329**(1) 99–110

[7] Akhverdov I N 1961 *Basic physics of concrete* (Moscow: Stroiizdat) 404 p

[8] Prša M A, Kasaš-Lažetić K.K.2018_Electromagnetic fields and their impacts *IOP Conference Series: Materials Science and Engineering* **294**(1) 012001

[9] Zhong B, Zhang Y, Zhou J, Chen L, Li D and Tan J 2018 UV irradiation improves the bond strength of resin cement to fiber posts *Dental Materials Journal* **30**(4) 455–460

Technology of foam concrete production with ultradispersed quartz waste

I N Kuznetsova, M A Darulis and A F Kosach
Yugra State University, Russia, Khanty-Mansiysk, 628012, Russia

E-mail: inkuznetsova-gsh@mail.ru

Abstract. The article presents a technology for producing foam concrete using activated quartz waste. The waste is an ultrafine filler in the cement stone that builds pore-forming walls and generates a structure with minimal pores (with a diameter down to 50 microns) and walls with similar cross-section. The average density of such foam concrete is 500 kg/m^3 and the total porosity equals 69-76%. The specimens demonstrate increased strength and reduced thermal conductivity. The use of the ultrafine filler in foam concrete reduces cement consumption by 10%, increases the density of the cement stone by 4%, augments the strength of the foam concrete by 10% and reduces its thermal conductivity by 8%. The results of the tests approve increased operational, physical and mechanical properties of such foam concrete.

1. Introduction

Foam concrete is a solid material from cement stone forming walls of evenly distributed pores (macropores). The capillary and gel pores (micropores) are evenly distributed in the cement stone. The macroporosity and microporosity of foam concrete depend on the type and properties of the foam-generating agent, granulometric composition of the filler, chemical additives, preparation technology, solid-to-liquid ratio, water-to-cement ratio and temperature of the mixture.

The subject of the study is the effect of activated quartz waste as ultrafine filler on the structure of the pore walls in foamed concrete. The study shows that the strength properties of foamed concrete increase due to the discrete reinforcement of the pore walls and, as a consequence, reduces cement consumption.

Hydration of cement determines the properties of cement-stone pore walls. Various physical and chemical processes are responsible for the formation of high-strength cement stone. A crucially important characteristic is strong adhesion of quartz particles to cement formations conditioned by secondary bonding with quartz grains formed from hydrates recrystallized from plates. These particles give rise to zones of crystalized inclusions. The filler should actively react with $Ca(OH)_2$ and other clinker hydration products. Its surface should have maximum possible compatibility with the structure of crystallizing hydrates, serving as a substrate for them [1–4].

The aim of the study was to improve a technology of foam concrete production using activated quartz waste as an ultrafine filler of the cement stone structure forming pore walls of the foam. Such technology can enable an evenly distributed pore structure with the same cross-section of pore walls, increase strength and reduce thermal conductivity of foam concrete [5].

2. Research methods

The structure of the specimens was determined using a REM-100U scanning electron microscope (Russia). The specific surface and average particle size were determined on PSKh-12 instrument

(Russia). The heat conductivity of the specimens was measured by the method of stationary heat flux using ITP-MG4 device (Russia). The porosity was measured by mercury porosimetry on a Porosimeter 2000 pore analyzer (Fision Instruments, Spain).

3. Experimental

To prepare the mortar for foam concrete production, quartz waste from production of ultrapure quartz concentrate (LLC "Polar Quartz", Russia) was used. It consisted of secondary waste after grit magnetic separation (25–26 μm) and powder classification (6–7 μm), and dust from an on-site suction system (3–6 μm).

To prepare the experimental concrete specimens, the following technological scheme was used.

1. The quartz waste was fed into a centrifugal disk mill for activation.

2. The ultrafine quartz filler (10 wt%) and cement (TsEM II / A-Sh 42.5N) with common granulated blast furnace slag were dosed and mixed, and then fed into a feed hopper by a pneumatic transporter.

3. A frothing agent and water at a temperature of 20–25 °C were fed to a foam generator. The frothing time was 5–6 minutes.

4. The foam was mixed with the mortar in a mixer for 3–4 minutes;

5. Then, the foam was poured into molds, and cured for 3–4 hours at a temperature of 20 °C.

6. Then, the specimens were subjected to hygrothermal treatment in an autoclave at 12 atm as per the following heating regime: gradual temperature increase up to 80 °C over 4 hours; isothermal soaking at 80 °C for 6 hours; and gradual temperature decrease down to 20 °C over 4hours.

The flowchart of foam concrete specimen preparation is depicted in Figure 1. The values of specific surface area (S_{sp}) and the average size (d) of quartz waste particles after dry activation are presented in Table 1.

Table 1. Specific surface area and average quartz particle size.

Waste type	Specific surface area [cm²/g]	Particle size [μm]
Grit magnetic separation waste	1575–1951	17–20
Powder classification waste	3321–4815	4–7
Dust from on-site suction system	5579–9547	2–4

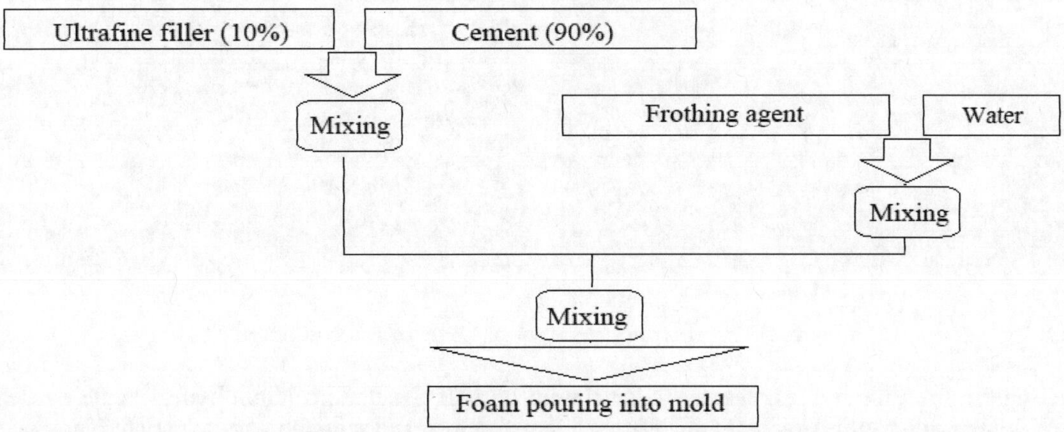

Figure 1. Flowchart of foam concrete production.

The proposed technology of foam concrete provides an evenly distributed closed-pore structure with pore walls having similar cross-section. The walls are reinforced by ultrafine quartz filler: needle-like crystals penetrate the pore space and strengthen the structure of the foam concrete. The properties

IOP Conf. Series: Materials Science and Engineering **597** (2019) 012031 doi:10.1088/1757-899X/597/1/012031

of the specimen are affected by pore volume and pore dimensions: the total pore volume in the pore walls of the specimens was 0.04 cm^3/g with a pore diameter of 0.061 μm, while the total pore volume of the specimens was 0.75 cm^3/g with a pore diameter of 1–50 μm.

Figures 2 and 3 illustrate the microstructure of pore space and pore walls of the specimens, respectively.

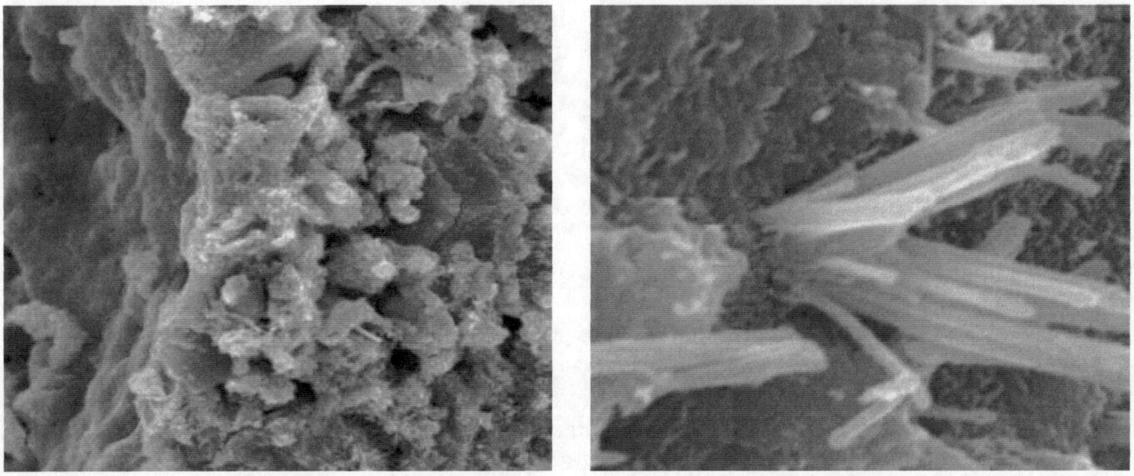

10μm 1mm

Figure 2. Microstructure of pore space of foam concrete.

5μm 5μm

Figure 3. Microstructures of pore walls of foam concrete.

Evidently from the figures, needle-like ettringite crystals penetrate calcium hydrosilicate particles, which contributes to the strengthening of the microstructure and augments the characteristics of the foam concrete specimens as a whole.

The main physical and technological properties of foam concrete specimens with addition of activated ultrafine quartz filler are given in Table 2.

IOP Conf. Series: Materials Science and Engineering **597** (2019) 012031 doi:10.1088/1757-899X/597/1/012031

Table 2. Physical and technological characteristics of experimental foam concrete specimens.

Characteristic	Value
Concrete grade in terms of average density	D400
Heat conductivity coefficient [W/(m K)]:	
• grit magnetic separation waste	0.113
• powder classification waste	0.119
• dust from on-site suction system	0.122
Water vapor transmission rate [kg/(m h Pa]	0.18
Sorption humidity [%] (at relative humidity of 75%)	less than 7–9
Compression strength [MPa]:	
• grit magnetic separation waste	1.57
• powder classification waste	1.21
• dust from on-site suction system	0.98

4. Conclusion

Following the results of the research, the addition of an ultrafine filler (activated secondary quartz waste after grit magnetic separation and powder classification, and dust from on-site suction system) to the foam cement mortar leads to the quartz particles penetrating the structure of pore walls, which saves up to 10% of cement, increases the cement stone density by 4%, improves the foamed concrete of by 10% and lowers the thermal conductivity coefficient by 8%.

The technology of foam concrete with the use of activated quartz waste allows creating a structure with a pore diameter of up to 50 microns, average density of foamed concrete of 500 kg/m^3 and total porosity of the foam concrete specimens of 69-76%. It increases strength and reduces thermal conductivity of the concrete. The share of the pore volume in the cement stone pore walls is less than 0.4% of the total pore volume and varies with the cement hydration degree. The increased pore wall strength is due to discrete reinforcement by ultrafine quartz waste filler.

References

[1] Particle Sciences 2012 An Overview of the Zeta Potential Part. Sci. 2 1–4

[2] E P Kearsley, P J Wainwright. The Effect of Porosity on the Strength of Foamed Concrete *J Cem. Concr. Res.*, 2002, **32**:233–239

[3] Luo X, Dong J, Zhang L, Du J, Wang H and Gao W 2017 Preparation of silica micro spheres via a semibatch sol–gel method *J. Sol-Gel Sci. Technol.* **81** 669-77

[4] Verba, Circe, Montross, Scott, Spaulding, Richard, Dalton, Laura, Crandall, Dustin, Moore, Johnathan, Glosser, Deborah, Huerta, Nik, and Kutchko, Barb. *Foamed Cement Interactions with CO2*. United States: N. p., 2017. Web. doi:10.2172/1373376

[5] Haljasmaa, Igor, Kutchko, Barbara, Spaulding, Richard, Crandall, Dustin, Moore, Johnathan, Gill, Magdalena, Benge, Glen, Hawthorne, Steven B., Dockter, Bruce A., Beddoe, Chris, Sorensen, Jim, Fazio, James, and Gieger, Connor. *Foamed Cement: Correlation of Foam Quality with Mechanical and Physical Properties*. United States: N. p., 2019. Web. doi:10.18141/1506981

IOP Conf. Series: Materials Science and Engineering **597** (2019) 012032 doi:10.1088/1757-899X/597/1/012032

Structure and properties of the cement stone modified by ultradispersed quartz waste

M A Darulis, I N Kuznetsova and A F Kosach
Yugra State University, Russia, Khanty-Mansiysk, 628012, Russia

E-mail: mpermitina@mail.ru

Abstract. The paper scrutinizes the application of an ultradispersed filler from activated quartz waste as a modifier of cement stone and resulting physicomechanical characteristics of the latter. The deliverables justify the addition of the ultradispersed filler to cement stone in production of cement concrete. The physicomechanical characteristics of quartz-cement stone after dry and wet activation are given. Goal: the goal of the present work is to study the physicomechanical properties of the cement stone modified by ultradispersed quartz filler. Methodology: mechanical milling is used to get ultrafine particles. Research results: the research establishes that the effective ratio of ultradispersed activated quartz to cement is 30:70. The application of the ultrafine filler based on activated quartz waste allows increasing the density by 7%, strength by 35% and lowering the heat conductivity by 16%. Practical use of results: the work proposes a method for producing cement stone with augmented physicomechanical properties using quartz waste in the amount of 30% in total cement volume. Value: the work scientifically justifies the possibility to produce cement stone under hydraulic activation with denser structure with the major volumetric fraction of the filler represented by quartz waste particles with dimensions less than 20 μm.

1. Introduction

One of the most abundant minerals in the earth crust is quartz which is also a top-priority object of study by geologists for industrial application.

Unfortunately, the information on the world's stock and production rates of quartz is not published, which is particularly connected with its strategic importance. It is believed that the quartz stock is fairly large. The main producers of natural quartz are known to be Brazil, Russian Federation and PRC.

There are 13 quartz fields in Russia with quite uneven distribution of the reserves along the state territory: 77.6% are in Ural Federal District, 16.5% in Siberian Federal District and 5.9% in Northwestern Federal District. Interestingly, a lot of fields are not accounted by the state register of mineral resources.

There is a plant producing ultrapure quartz concentrate—Polyarny kvarts, OJSC (Polar Quartz)—in Nyagan city of Khanty-Mansiysk Autonomous Okrug – Yugra. The project is of tremendous socio-economic value and develops two municipal formations of the Beryozovsky District: Ust-Puiva settlement (intermediate terminal with primary enrichment facility for quartz ore) and Nyagan city (plant for producing ultrapure quartz concentrate).

Quartz waste is a side product of quartz feedstock processing and amounts to 30% of total pure quartz concentrate production. One of the demanded solutions for utilization of a non-demanded product of the plant is production of high-strength small-grain concrete on the basis of ultradispersed quartz waste after production of ultrapure quartz concentrate

2. Research methods

The work considers application of waste after ultrapure quartz concentrate production as an ultradispersed filler for production of nanostructured concrete with high physicomechanical properties. The physical model of nanostructured cement stone is given in Figure 1.

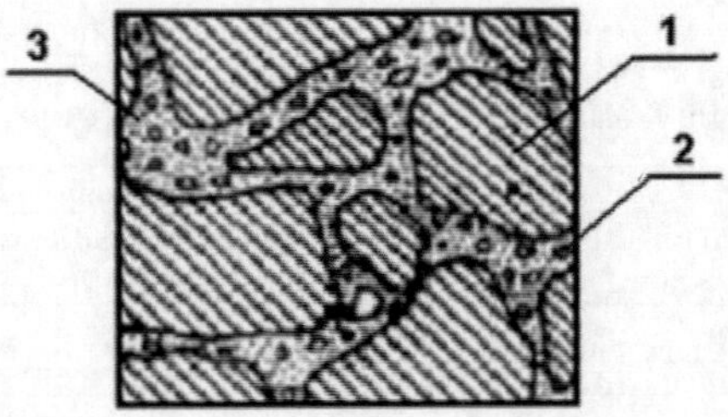

Figure 1. Nanostructurization of cement stone: 1 - cement particles; 2 - ultradispersed quartz waste; 3-hydration products of clinker components.

The main goal of the research was to investigate the effect of the filler based on ultradispersed quartz waste on the physicomechanical properties of cement stone.

To reach the goal, a structural scheme of the work was elaborated (Figure 2) which envisaged the investigation of secondary quartz waste application as an ultradispersed filler to produce concrete from a nanostructured cement gel. The gel was obtained using various ultradispersed quartz waste: 1) dust from on-site suction system (3–6 μm); 2) powder classification residue (6–7 μm); 3) grit magnetic separation waste (25–26 μm).

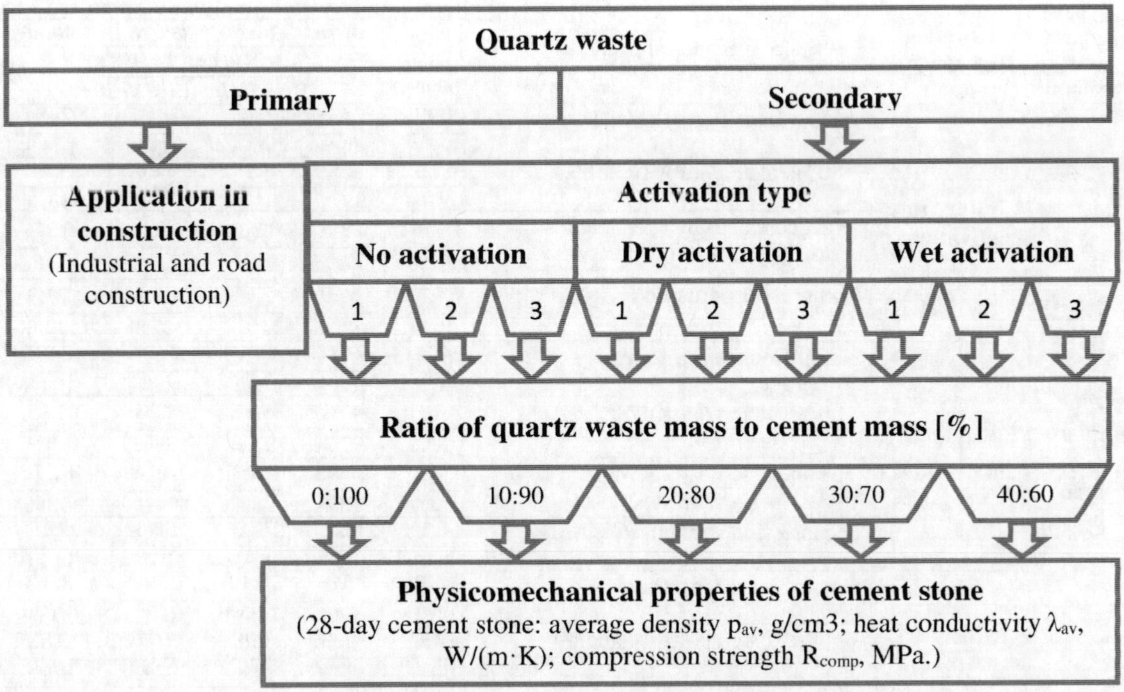

Figure 2. Structural scheme of investigation.

The following materials were used in the study:

1. Binder (PTs-400-D20 cement).

2. Quartz waste after production of ultrapure quartz concentrate (Polyarny kvartz, OJSC). Secondary waste: Residues of grit magnetic separation, powder classification and dust collection (3–26 μm).
3. Tap water.

The specific surface and average size of the particles was determined by PSKh-12 device (Russia). The data is given in Table 1.

Table 1. Physicomechanical properties of quartz waste.

Property to be determined	Waste name								
	1) Dust from on-site suction system			2) Powder classification residue			3) Grit magnetic separation residue		
Activation technology	No activation	Dry	Wet	No activation	Dry	Wet	No activation	Dry	Wet
D (particle size) (μm)	3–6	2–4	1-3	7–11	4–7	3-5	25–29	17–20	13-15
Uc (specific surface) (cm2/g)	8017–4052	9547–5579	10860–8017	3249–1817	4815–3249	8115-4317	858–501	1951–1575	2360-2235
ρ (bulk density), (g / cm^3)	1.59–1.78	1.68–1.86	1.93–2.08	1.55–1.69	1.61–1.75	1.81–2.07	1.73–1.83	1.55–1.77	1.87-1.99

The specimens were produced from cement slurry with normal viscosity and water consumption of 100 ml per 400 grams of cement. The strength of the quartz-cement stone was determined using nine specimens with dimensions of 2x2x2 cm for each composition, according to the structural investigation scheme. The parameters of the tests are presented in Table 2.

Table 2. Physicomechanical properties of specimens

Type of quartz waste (Secondary)	Activation technology	Physico-mechanical properties	Ratio of quartz waste mass to cement mass (%)				
			0: 100 Absolute value	10:90	20:80	30:70	40:60
				Obtained results / *% of absolute value*			
1) Dust from on-site suction system (Figure 4a)	Initial (no activation)	Density ρ_{av} (kg/m^3)	2397	2435	2451	2470	2491
			100 %	102 %	102 %	103 %	104 %
		Calculated porosity Pc (%)	4.76	4.63	4.45	4.26	4.17
			100 %	98 %	94 %	90 %	88 %
		Heat conductivity coefficient λ_{av} (W/m·K)	0.406	0.396	0.388	0.381	0.377
			100 %	97 %	96 %	94 %	93 %
		Compression strength R_{comp} (MPa)	64.7	67.9	77.1	82.1	71.9
			100 %	105 %	119 %	127 %	111 %
	Dry activation	Density ρav (kg/m^3)	2401	2437	2455	2474	2495
			100%	101%	102%	102%	104%
		Calculated porosity Pc (%)	4.73	4.61	4.41	4.23	4.15
			100%	97%	93%	89%	88%
		Heat conductivity coefficient λav (W/m·K)	0.405	0.395	0.386	0.379	0.376
			100%	98%	95%	94%	93%
		Compression strength R_{comp} (MPa)	65.1	68.1	77.6	82.6	72.4
			100%	105%	119%	127%	111%
	Wet activation	Density ρ_{av} (kg/m^3)	2419	2516	2540	2588	2564
			100%	104%	105%	107 %	106%
		Calculated porosity Pc (%)	4.45	4.23	4.05	3.87	3.83
			100%	95%	91%	87%	86 %
		Heat conductivity coefficient λav (W/m·K)	0.393	0.366	0.354	0.342	0.334
			100%	93%	90%	87%	85%
		Compression strength R_{comp} (MPa)	71.1	76.1	83.9	94.6	81.8
			100%	107%	118%	133%	115%

Continuation of Table 2

Physico-mechanical properties	Ratio of quartz waste mass to cement mass (%)

Type of quartz waste (Secondary)	Activation technology		0: 100 Absolute value	10:90	20:80	30:70	40:60
				Obtained results % *of absolute value*			
2) Powder classification residue (Figure 4b)	Initial (no activation)	Density ρ_{av} (kg/m^3)	2387	2426	2443	2461	2481
			100%	101%	102%	103%	104%
		Calculated porosity Pc (%)	4.86	4.71	4.55	4.35	4.26
			100%	97%	94%	90%	88%
		Heat conductivity coefficient λav (W/m·K)	0.404	0.394	0.386	0.379	0.375
			100%	98%	96%	94%	93%
		Compression strength R_{comp} (MPa)	64.1	67.2	76.3	81.4	71.0
			100%	105%	119%	127%	111%
	Dry activation	Density ρ_{av} (kg/m^3)	2394	2432	2447	2467	2488
			100%	101%	102%	103%	104%
		Calculated porosity Pc (%)	4.80	4.66	4.50	4.31	4.21
			100%	97%	94%	90%	88%
		Heat conductivity coefficient λav (W/m·K)	0.402	0.392	0.383	0.376	0.373
			100%	98%	95%	94%	93%
		Compression strength R_{comp} (MPa)	64.7	67.9	77.1	82.1	71.9
			100%	105%	119%	127%	111%
	Wet activation	Density ρ_{av} (kg/m^3)	2421	2494	2518	2566	2542
			100%	103%	104%	106%	105%
		Calculated porosity Pc (%)	4.55	4.28	4.05	3.96	3.91
			100%	94%	89%	87%	86%
		Heat conductivity coefficient λ_{av} (W/m·K)	0.379	0.345	0.330	0.322	0.318
			100%	91%	87%	85%	84%
		Compression strength R_{comp} (MPa)	81.3	88.6	98,4	109.7	97.6
			100%	109%	121%	135 %	120%
3) Grit magnetic separation residue (Figure 4c)	Initial (no activation)	Density ρ_{av} (kg/m^3)	2378	2417	2433	2451	2473
			100%	102%	102%	103%	104%
		Calculated porosity Pc (%)	4.91	4.75	4.59	4.41	4.3
			100%	97%	93%	90%	88%
		Heat conductivity coefficient λ_{av} (W/m·K)	0.401	0.391	0.383	0.374	0.370
			100%	98%	96%	93%	92%
		Compression strength R_{comp} (MPa)	57.4	63.1	71.1	75.3	66.0
			100%	110%	124%	131%	115%
	Dry activation	Density ρ_{av} (kg/m^3)	2385	2423	2439	2459	2478
			100%	102%	102%	103%	104%
		Calculated porosity Pc (%)	4.85	4.7	4.55	4.37	4.25
			100%	97%	94%	90%	88%
		Heat conductivity coefficient λav (W/m·K)	0.399	0.389	0.382	0.371	0.369
			100%	97%	96%	93%	92%
		Compression strength R_{comp} (MPa)	57.9	63.6	71.6	75.9	66.4
			100%	110%	124%	131%	115%
	Wet activation	Density ρ_{av} (kg/m^3)	2433	2482	2530	2555	2532
			100%	102%	104%	105%	104%
		Calculated porosity Pc (%)	4.73	4.45	4.21	4.12	4.07
			100%	94%	89%	87%	86%
		Heat conductivity coefficient λ_{av} (W/m·K)	0.375	0.341	0.326	0.319	0.315
			100%	91%	87%	85%	84 %
		Compression strength R_{comp} (MPa)	70.5	79.7	90.9	95.2	93.1
			100%	113%	129%	135%	132%

The quartz waste was activated on a Viyuga-3 continuous rotary mill (Russia). The mill subjects the ultradispersed quartz materials to mechanical activation during dry milling, and to mechanical and hydrodynamic activation during wet milling (Figure 3).

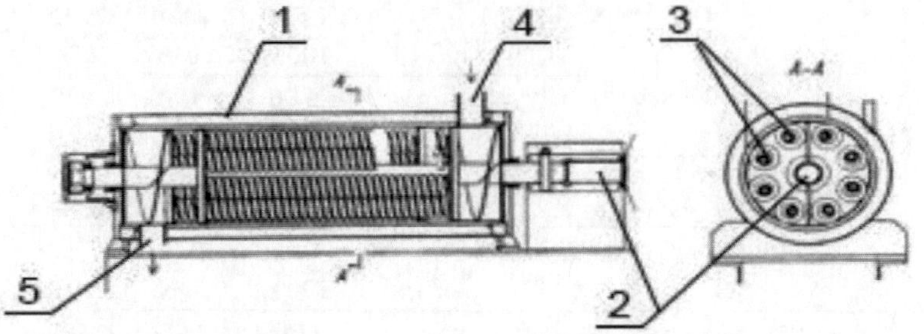

Figure 3. Continuous rotary mill Viyuga-3: 1) cylindrical case; 2) cylindrical finishing case (eight-element cylindrical spirals); 3) rod elements with vertical plates supporting eight cylindrical spirals (mounted in the case with bearings); 4) loading bay; 5) unloading bay for activated material.

3. Experimental

The specific surface of the filler is determined by the size, shape and roughness of its grains: the smaller they are, the larger the surface roughness and the specific surface of the filler grain are.
The weight distribution of secondary quartz waste particle size before and after activation was determined by a Microsizer 201 particle analyzer (Scientific Instruments, Russia). Below is the figure illustrating the weight distribution of particle size depending on the quartz waste type: Figure 4a depicts data for the dust from on-site suction system, Figure 4b, for the powder classification residue and Figure 4c, for the grit magnetic separation residue.

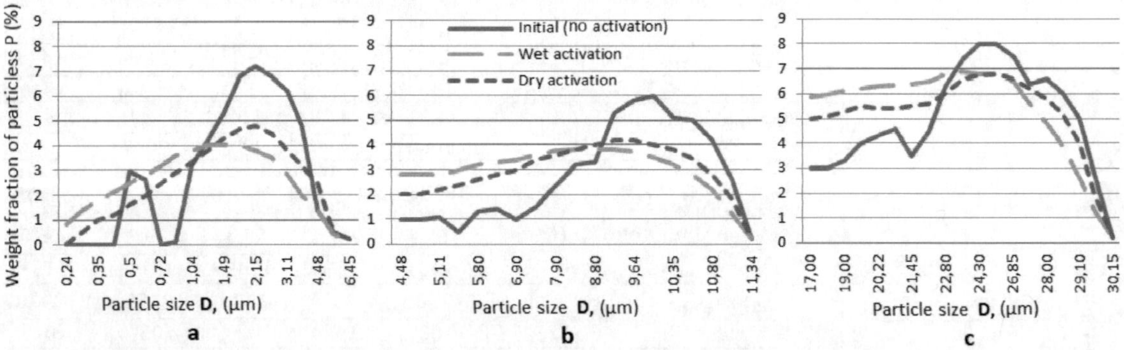

Figure 4. Weight distribution of particle size depending on the quartz waste type: a) dust from on-site suction system; b) powder classification residue; c) grit magnetic separation residue.

The structure of the specimens was studied by a REM 100U scanning electron microscope (Russia). The microphotographs (Figure 5) clearly depict needle-like crystals and their agglomerates that are typical for ettringite (calcium hydrosulfosilicate). The prismatic crystals evidence the presence of alite, while rounded crystals indicate belite.

The images of the pore volume and pore walls clearly show the aggregates of well crystallized long-fiber hydrosilicates forming both at the interface of a pore and pore wall, and inside a pore. These crystallization contacts make up a rigid skeleton from fibrous (needle-like) crystals penetrating the porous volume of concrete, which improves its hardness and compression strength.

The analysis has detected calcium hydroxides crystallizing as elongated crystals and massifs (Figure 5c). Such massifs enhance heat conductivity.

The studies allow concluding that the component of the cement stone based on quartz waste finely milled down to 20 μm demonstrates its activity by building bridges binding it with the cement matrix and playing the role of discrete reinforcement. The presence of such needle-like build-ups may evidence increased strength of the material, because they play a reinforcing role in the concrete structure.

a b c

Figure 5. Microstructure of cement stone using activated secondary waste of ultrapure quartz concentrate production a) grit magnetic separation residue; b) powder classification residue; c) dust from on-site suction system.

The received results show that the higher the specific area of the micro-filler produced from the waste of ultrapure quartz concentrate production in Nyagan city is, the more actively it interacts with other hydration products via faster reaction forming finely dispersed hydrates.

4. Conclusion

The deliverables testify that the application of activated waste of ultrapure quartz concentrate production of the plant in Nyagan city allows saving up to 30% of cement.

The application of the ultradispersed filler based on activated quartz waste allows improving the physicomechanical properties of cement stone: increase the density by 7% and strength by 35%, decrease the heat conductivity coefficient by 16%.

Wet activation increases the strength of the specimens 1.5 times as compared to non-activated specimens. In contrast to dry milling, the strength of the specimens increased 1.3 times.

The study of the implementation of the activated quartz waste as an ultradispersed quartz-cement binder yielded the effective ratio of quartz to cement (30:70) applicable in both industrial and civil construction. Assumably, the industrial implementation of received results will attract considerable amount of investments in Khanty-Mansiysk Autonomous Okrug, which will positively affect economic and social situation in the region and solve local ecological problems.

This will reduce expenses for production of pure quartz and basic cost of quartz concentrate, improve the competitive ability of the products and reduce the environmental footprint of the plant, because the production waste requires disposal activities, expenses for packaging, temporary storage, transportation to disposal field, field maintenance and land remediation.

5. References

[1] Particle Sciences 2012 An Overview of the Zeta Potential Part. Sci. 2 pp 1–4

[2] E P Kearsley and P J Wainwright 2002 The Effect of Porosity on the Strength of Foamed Concrete *Cem. Concr. Res.* 32 pp 233–239

[3] Luo X, Dong J, Zhang L, Du J, Wang H and Gao W 2017 Preparation of silica micro spheres via a semibatch sol–gel method *J. Sol-Gel Sci. Technol.* 81 pp 669-77

[4] Verb C, Montross S, Spaulding R, Dalton L, Crandall D, Moor J, Glosser D, Huerta N, and Kutchko B 2017 Foamed Cement Interactions with CO_2 United States doi:10.2172/1373376

[5] Haljasmaa I, Kutchko B, Spaulding R, Crandall Du, Moore J, Gill M, Benge G, Hawthorne S B, Dockter B A, Beddoe C, Sorensen J, Fazio J, and Gieger C 2019 Foamed Cement: Correlation

of Foam Quality with Mechanical and Physical Properties United States doi:10.18141/1506981

IOP Conf. Series: Materials Science and Engineering **597** (2019) 012033 doi:10.1088/1757-899X/597/1/012033

Application of sea sand for 3D concrete printing

A V Sharanova, A O Tovpinets and M A Dmitrieva
Immanuel Kant Baltic Federal University, 14 Nevskogo St, Kaliningrad 236016, Russia

E-mail: a.sharanova@gmail.com

Abstract. The paper shows the promising use of sea sand for additive technologies in construction. The use of sea sand reduces the setting time of the concrete mix and increases the initial strength compared to conventional quarry construction sand. A comparative analysis of the characteristics of sea and quarry sands was performed. It has been experimentally proven that the use of sea sand instead of quarry sand in fine-grained concrete mix leads to a decrease in the setting time of the fresh mix and an increase in the strength of concrete samples. The flexural strength and compressive strength of samples containing sea sand at the age of 28 days increases by 35.7% and 16.7%, respectively, comparing to samples containing quarry sand.

1. Introduction

Construction using the additive technologies, or concrete 3D printing, is a new and promising method in the construction industry. The main difference of this method compared to traditional construction is layer-by-layer growing of structural elements without the use of formwork and the use of fine-grained concrete mix as a building material. Concrete for 3D printing should quickly set and gain sufficient initial strength to maintain the weight of the layers printed above. Concrete mixture must have good fluidity for free extrusion through the extruder nozzle, and at the same time sufficient viscosity to maintain the printed form. Fine additives based on industrial waste (fly ash, silica fume), or additives of thermally treated clay (metakaolin) give to concrete the necessary properties [1].

It is known that the use of sea sand as a fine aggregate for concrete leads to a decrease in setting time and an increase in the initial strength of concrete [2-7]. But at the same time, due to the presence of chloride ions in the sea sand, the durability of concrete with sea sand is lower than that of ordinary concrete, since chlorides cause corrosion of steel reinforcement and generally reduce the service life of concrete structures [4]. It becomes possible to assume that the use of sea sand as a fine aggregate can be effective in constructions, where reinforcement with polypropylene, basalt or other types of fibers is used as an alternative to steel reinforcement. In addition, the use of mineral additives, such as fly ash, silica fume, metakaolin leads to the blocking of chloride ions. Due to this fact, there is an increase in long-term strength and service life of concrete structures [2, 5, 6]. In [7], it is noted that in metakaolin-modified concrete, chloride ions introduced from seawater can be immobilized through the formation of hydrocalumite (Friedel's salt), which improves the corrosion resistance of concrete and makes the addition of metakaolin to sea concrete more beneficial. Therefore, sea sand is promising for use as fine aggregate to concrete for construction using additive technologies.

2. Materials and methods

The paper compares the effect of two types of sand on the characteristics of fine-grained concrete. According to the results of previous work on the selection of the composition for additive technologies in construction, the optimal composition was determined that meets the requirements of concrete 3D printing. Two mixtures of fine-grained concrete are made of this composition; in the first case, sea sand is used as fine aggregate, in the second case, quarry sand is used.

2.1. Materials

The following materials were used for the preparation of samples:

- Portland cement Eurocem super 500, CEM I 42.5H, Petersburg Cement;
- Sand: for the composition No. 1: sea sand from the coast of the urban-type settlement Yantarny, Kaliningrad region; for the composition number 2: sand from the quarry near village Sinyavino, Kaliningrad region;
- Polypropylene fiber: length of 12 mm, thickness of 35 μm, country of origin: Russia;
- Silica fume (SF): wastes of the metallurgical industry, country of origin: Poland;
- Highly active metakaolin (MK): white, country of origin: Russia, Chelyabinsk region, Synergo;
- STACHEMENT 1267 hyperplasticizer based on polycarboxylates for the production of ready-mixed concrete: country of origin: Poland.

Characteristics of sand are presented in Table 1.

Table 1. Sand characteristics.

Location of origin	Size group	Fineness modulus	Content of dust and clay particles (%)	True density (kg/m^3)	Bulk density (kg/m^3)	Voidness (%)
Coast	Medium	2.1	-	2612	1474	43.6
Quarry	Fine	2	2.2	2642.5	1330	49.7

All the characteristics were determined according to Russian State standards GOST 8736-2014 "Sand for construction work. Specifications" and GOST 8735-88 "Sand for construction work. Testing methods".

The appearance of the sand is shown in Figures 1 and Figure 2. The images were obtained using an optical electron microscope.

Figure 1. Sea sand.

Figure 2. Quarry sand.

The consumption of materials per 1 m^3 of the concrete mixture is given in Table 2.

Table 2. Materials per 1 m^3.

W/C ratio	Cement (kg/m^3)	Sand (kg/m^3)	Water (l/m^3)	Silica fume (SF) (kg/m^3)	Metakaolin (MK) (kg/m^3)	Fiber (kg/m^3)	Hyperplast. (l/m^3)
0.66	545	1168	360	156	78	1.2	11.7

2.2. *Methods*

Two compositions of fine-grained concrete were made. In the first case, sea sand was used as fine aggregate, in the second case, quarry sand was used. For each of the compositions were determined: the setting time using Vicat apparatus with a needle (GOST 30744-2001); the consistency of the cement paste by the Vicat apparatus with a plunger (GOST 30744-2001); the consistency of the fresh mixture by the flow table test (GOST 310.4-81); the workability by the cone penetrometer test (GOST 5802-86); the flexural strength and compressive strength of samples 40x40x160 mm (GOST 310.4-81) at the age of 1, 3, 7 and 28 days.

3. Results

The test results for the comparison of compositions containing different sands are given in Table 3. The strengths of the samples are presented in graphs (Figures 3, 4) and in Table 4.

Table 3. Compositions containing different sands.

Composition No.	Initial setting time	Final setting time	Consistency by the Vicat apparatus (mm)	Flow table test (mm)	Cone penetrometer test (mm)
1	1h 40 min	2h 30 min	5	148.5	47
2	2h	2h 40 min	37	145.3	46

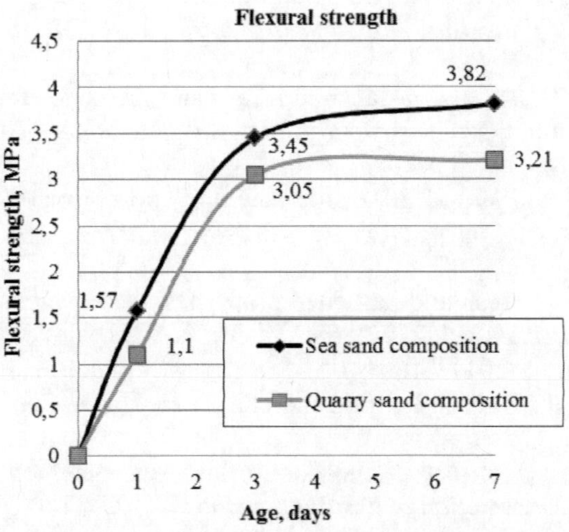

Figure 3. Flexural strength at 1, 3 and 7 days.

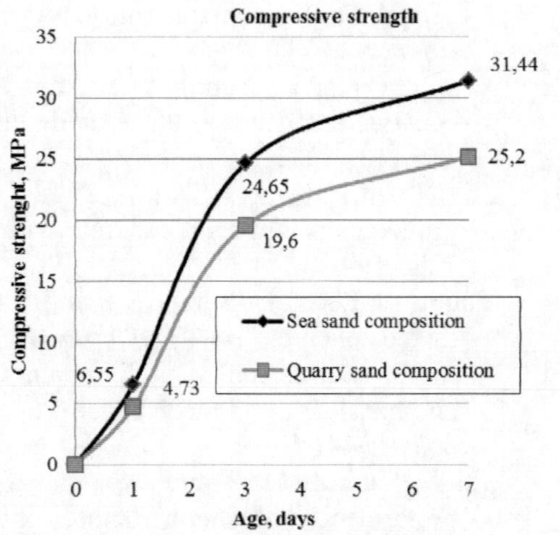

Figure 4. Compressive strength at 1, 3 and 7 days.

Table 4. The strengths of the samples.

Composition No.	Sand type	Flexural strength (MPa)				Compressive strength (MPa)			
		1 day	3 days	7 days	28 days	1 day	3 days	7 days	28 days
1	Sea sand	1.57	3.45	3.82	5.09	6.55	24.65	31.44	36.66
2	Quarry sand	1.10	3.05	3.21	3,75	4.73	19.60	25.20	31.40

The flexural strength and compressive strength of samples with sea sand at the age of 28 days increases by 35.7% and 16.7%, respectively, comparing to the samples with quarry sand.

4. Conclusion

As a result of the research, it was proved that the use of sea sand as an aggregate in fine-grained concretes leads to a decrease in the setting time and an increase in the initial strength. It can be assumed that the use of sea sand as an aggregate in concrete is promising for additive construction methods and is relevant for coastal areas. It should be noted that the use of sea sand in compositions for 3D printing requires further, more thorough research. There is a need to define additional characteristics of sea sand, such as salt content, mineralogical composition, etc. The investigation of joint work of mineral additives and salt-containing sand components, the appearance of salt on the surface of the finished concrete, its influence on the concrete structure are insufficient and need further study.

Acknowledgement

The study is supported by the Russian Foundation for Basic Research grant No. 19-48-390001 "Multicomponent nano-modifiers of concrete mixes". The project was supported by the results of a competitive selection of research projects in the Regional Competition for Fundamental Research Projects.

References

[1] Vatin N, Chumadova L, Goncharov I, Zykova V, Karpenya A, Kim A and Finashenkov E 2017 3D printing in construction *Construction of Unique Buildings and Structures* **1(52)** 27–46 (rus)
[2] Xiao J, Qiang C, Nanni A and Zhang K 2017 Use of sea-sand and seawater in concrete construction: Current status and future opportunities *Construction and Building Materials* **155** 1101-1111
[3] Zhu N, Jin F, Kong X, Xu Y, Zhou J, Wang B and Wu H 2018 Interface and anti-corrosion properties of sea-sand concrete with fumed silica *Construction and Building Materials* **188** 1085-1091
[4] Younis A, Ebead U, Suraneni P and Nanni A 2018 Fresh and hardened properties of seawater-mixed concrete *Construction and Building Materials* **190** 276-286
[5] Cheng S, Shui Z, Sun T, Yu R, Zhang G and Ding S 2017 Effects of fly ash, blast furnace slag and metakaolin on mechanical properties and durability of coral sand concrete *Applied Clay Science* **141** 111-117
[6] Shi Z, Shui Z, Li Q and Geng H 2015 Combined effect of metakaolin and sea water on performance and microstructures of concrete *Construction and Building Materials* **74** 57-64
[7] Li Q, Geng H, Shui Z and Huang Y 2015 Effect of metakaolin addition and seawater mixing on the properties and hydration of concrete *Applied Clay Science* **115** 51-60

IOP Conf. Series: Materials Science and Engineering **597** (2019) 012034 doi:10.1088/1757-899X/597/1/012034

Study of the kinetics of hardening in "activated cement-water" system

A Rubanov, Y Sarkisov, D Mokshin, K Gauss and R Mokshin
Tomsk State University of Architecture and Building, Russia, Tomsk

E-mail: rubal06@yandex.ru

Abstract. The paper considers alternative methods of activation of cement systems based on the use of internal properties of cement. Their peculiarity is the use of activated cement and changing the sequence of mixing of cement mortar. Mixing of the main volume of Portland cement with cement slurry and its prior curing for 15 minutes allowed increasing the strength characteristics of hardened cement paste. As shown by experimental studies, the strength of the samples increases by 10–30% depending on the method of preparation and curing time of the cement slurry. Further experiments will be aimed at researching "activated cement–activated water" system.

1. Introduction

Cement compositions (hardened cement paste, building mortars, concretes, dry mixtures, etc.) represent the most typical building materials widely applied in construction. Improvement of the main properties of cement compositions include qualitative selection of their compositions based on various components and modification of the technology of their production. Initially, the increase of strength parameters was provided due to appropriate selection of concrete and mortar composition by adding chemical admixtures, whereas in recent times their properties are improved by modifying the technology of producing building materials based on Portland cement, including activation of their components.

The previous studies [1, 2] show that activation of mixing water for cement compositions using combined material and field research methods is the most available and economically feasible method for improving strength and other operational characteristics of hardened cement paste.

Activation of Portland cement is usually aimed at increasing its reactivity to water. This task can be performed in different ways [3, 4]; however they are not always successful.

The study of V. Solomatov [5] demonstrates the principal possibility of the so-called intense separate technology of concrete mixtures production. It provides separation of producing concrete and mortar mixtures and transition from their slow mixing in concrete mixers to brief intensification of this process.

At the same time, it is known that size distribution of Portland cement particles has the shape of a classical Gaussian curve, according to which circa 50–55% of particles are 8–20 μm, 25–30% are less than 8 μm, and 15–20% are larger than 20 μm. When using mechanical activation by dispersing Portland cement, the number of smaller particles grows, which leads to increase of its hydraulic activity. However, Portland cement where smaller particles prevail is unable to develop increased long-time strength, as obviously the most part of it is hydrated at early age.

Activation of all Portland cement required for producing mortar and concrete mixtures is also economically inefficient.

The purpose of this study is development of economically and technically feasible technology for activation of Portland cement due to actualization of its inner properties.

The following key tasks were set to achieve the above purpose:

1. develop alternative technology for cement systems activation;
2. perform physical and chemical study of curing products;
3. define physico-mechanical characteristics of hardened cement paste.

To perform the listed tasks the study offers several options of preparing cement mortar with the use of activated cement:

Option 1 – reference samples (traditional technology of samples production) are prepared both with M500 Portland cement and with admixture of activated cement with larger specific surface;

Option 2 – part of Portland cement (10% of total volume) is subjected to mechanical activation, mixed with the total amount of water, right after which the obtained suspension is added to the rest part of non-activated cement;

Option 3 – part of Portland cement (10% of total volume) is subjected to mechanical activation, mixed with the total amount of water, cured for 15 minutes, and then the obtained suspension is added to rest part of non-activated Portland cement.

Options 2 and 3 were repeated in both M500 Portland cement and in cement with activated Portland cement admixture.

2. Experiment

The following equipment was used while doing the research:

– Planetary mill MP 4/1, in which Portland cement was activated;
– PSH 10 setup for defining specific surface of Portland cement;
– PSU-10 hydraulic press for defining strength of the samples;
– Instron 3382 setup for testing for compression and/or tension;
– DRON-6 setup for X-ray diffraction analysis.

In order to define compressive strength of samples cubes with the size $(2\times2\times2)\times10^{-2}$ m were prepared from M500 Portland cement mortar according to the 3 offered options. Water-cement ratio in all variants was permanent and equal to 0.34. The samples were produced in metal molds, cured in moist cabinet and tested for compressive strength within the settled period and in compliance with the standard methods.

3. Results

After 15 minutes of mechanical activation in a planetary mill the specific surface of Portland cement increased from 4,422 cm^2/g to 4,976 cm^2/g, which makes about 13%.

Compressive strength of the samples was defined with the use of PSU -10 hydraulic press and Instron 3382 setup.

The results of compressive strength testing of the sample using a hydraulic press are given in Tables 1 and 2.

Table 1. Strength of hardened cement paste based on Portland cement.

Type of production	Samples strength, MPa / % R_{28}, during the hardening process, days			
	1	3	7	28
Option 1	34.2/56	43.0/71	49.3/81	60.8/100
Option 2	38.1/63	50.3/83	56.3/93	61.3/101
Option 3	45.1/74	55.3/91	62.3/102	66.8/110

Table 2. Strength of hardened cement paste with the addition of activated cement.

Type of production	Samples strength, MPa / %R_{28}, during the hardening process, days			
	1	**3**	**7**	**28**
Option 1	39.5/57	51.1/74	64.5/92	69.5/100
Option 2	41.8/60	52.5/75	66.5/95	77.5/112
Option 3	45.4/65	56.8/82	65.1/94	91.3/131

It is seen from the Tables that adding 10% of activated Portland cement results in the increase of compressive strength in samples by 15–30 % depending on the curing time. One might suggest that highly dispersed Portland cement particles more actively react with water forming the coalescence contacts of colloid and crystal nature, which leads to strength increase in cement samples.

When mixing Portland cement with suspension strength of hardened cement paste grows by 10–15% with Portland cement and by 5–10% with activated cement admixture. Additional curing of cement suspension for 15 minutes and subsequent mixing it with Portland cement results in the average increase of samples strength by 10–30% depending on the curing time.

The results of a set of tests with samples using Instron 3382 setup after 7 days' curing are presented in Figure 1.

The obtained data confirm the test results for PSU -10 hydraulic press. Also, the graphs analysis shows that variation of sequence of mixing the cement samples causes both increase of strength compared to the reference samples and improvement of their mechanical properties. The difference of strains in reference samples and the ones produced according to the offered options makes 30–50%, which indicates a more integrated structure of hardened cement paste and a less amount of pores. Besides, the limit of elastic strain was increased, which demonstrated the ability of structures to adapt to variation of operation loads. The average increase of elastic modulus by 20–25% is observed.

The results of X-ray diffraction analysis are given in Figure 2.

X-ray diffraction analysis of hardened cement paste showed that the intensity of diffraction peaks in portlandite $Ca(OH)_2$ prepared according to the Option 3 increased by 25–30%, it may indicate a higher cement hydration degree.

It is known that hydrated lime $Ca(OH)_2$ is formed due to alite hydrolysis in the following reaction

$$2(3CaO \cdot SiO_2) + 6H_2O = 3CaO \cdot 2SiO_2 \cdot 3H_2O + 3Ca(OH)_2 + 502\ J/g,$$

which proceeds with heat generation. One the one hand, more intense production of hydrated lime $Ca(OH)_2$ enables production of ettringite which forms films on reaction products within first minutes of hydration; the system grows in volume and silicate structure gradually replaces initial aluminum phase, due to which hardening of cement paste occurs. At the same time, excess of hydrated lime may lead to the so-called leaching corrosion, i.e. lime washout from the hardened cement paste. However, further structure solidification and the ability of hardened cement paste to grow hard for a long period of time within decades are impossible without the sufficient amount of $Ca(OH)_2$. In this regard, the effect obtained after cement composition activation can be considered an efficient method of increasing strength and other operational characteristics of cement compositions.

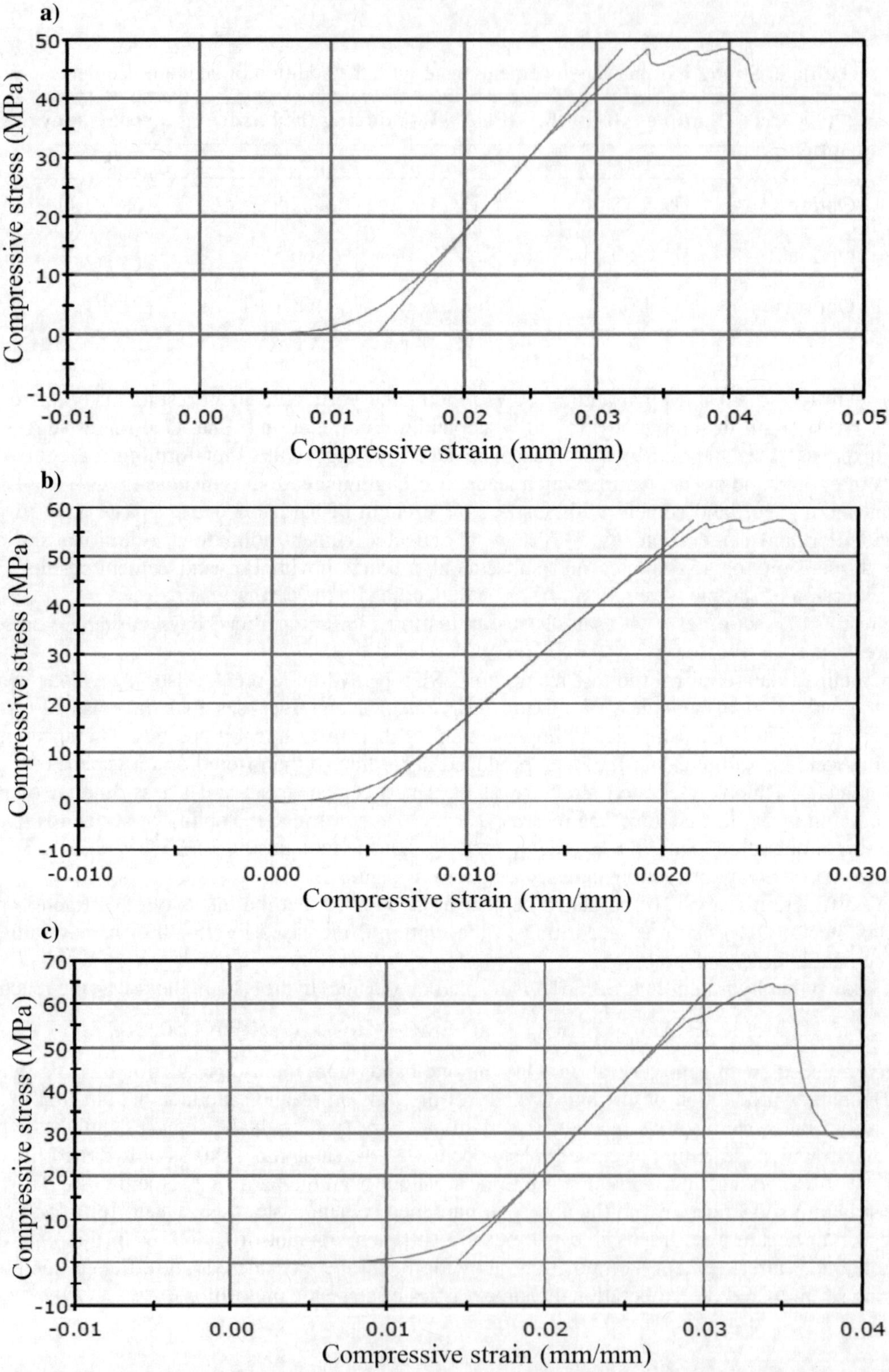

Figure 1. Results of testing hardened cement paste samples produced according to Option 1 (a), 2 (b), 3 (c)

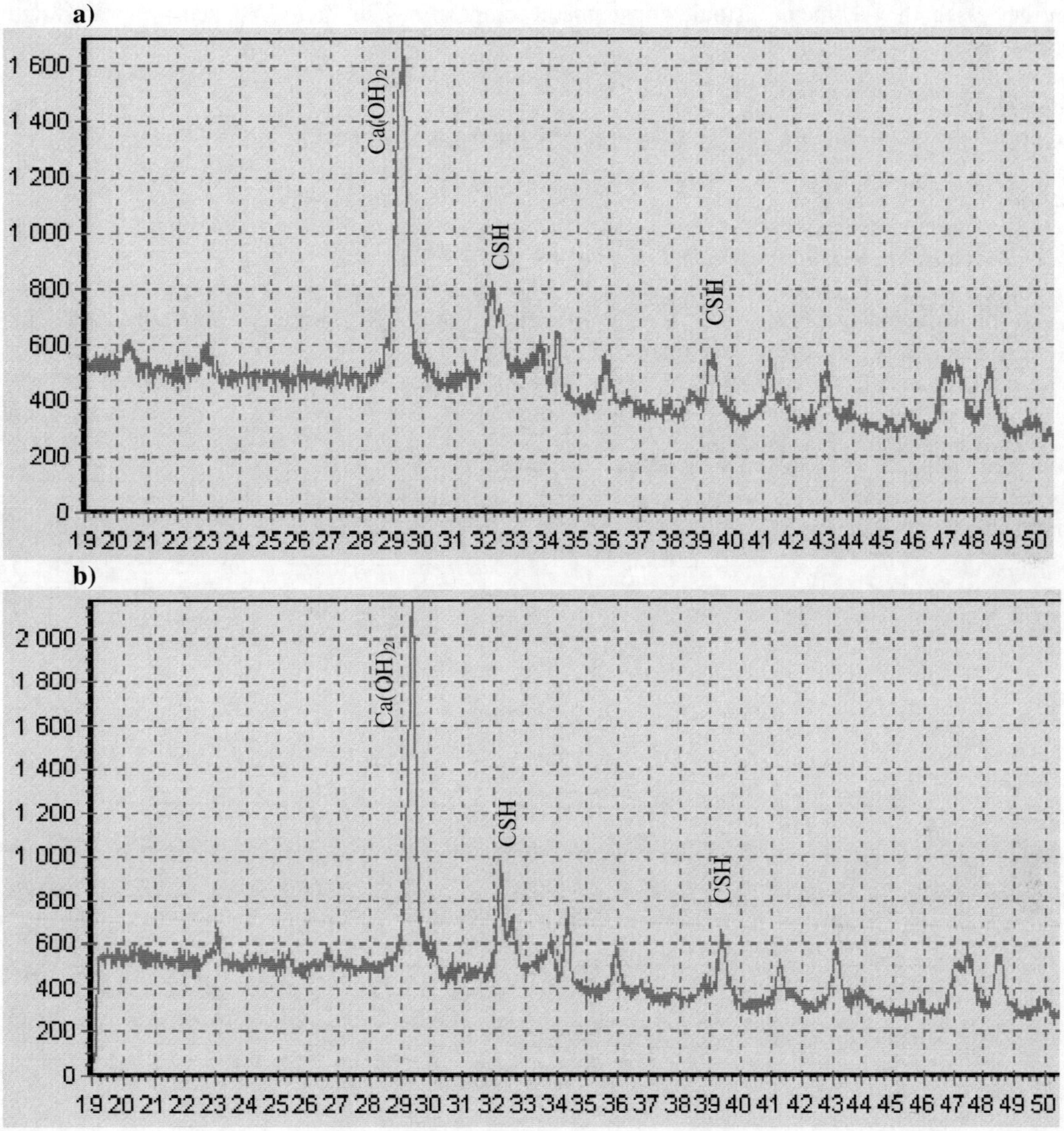

Figure 2. X-ray diffraction analysis of hardened cement paste produced according to Option 1 (a) and 3 (b) with activated cement admixture

The next stage of the research includes studying the behavior of "activated cement – water" system in concrete samples.

4. Conclusion.

The research outcomes include techniques of preparation of cement compositions with improved strength parameters. Mixing of Portland cement with suspension increases strength of the hardened cement paste by 5–15 %. Besides, additional curing of cement suspension for 15 minutes with further mixing it with Portland cement results in 10-30% growth of strength depending on the curing time. It is found that due to adding 10% of activated Portland cement strength of hardened cement paste increases by 15–30 % depending on the curing time.

Further research will include studies on cement compositions in "activated cement – activated water" system.

Acknowledgments
The reported study was funded by RFBR according to the research project № 18-38-00674.

References

[1] Allahverdi A, Maleki F, Mahinroosta M 2018 Chemical activation of slag-blended Portland cement *Journal of Building Engineering* vol 18, pp 76–83
[2] Rubanov A, Gorlenko N, Vergasov V 2018 Material and Field Impacts on Cement Compositions MATEC Web 4th International Young Researchers Conference "Youth, Science, Solutions: Ideas and Prospects" p 143
[3] Zhou W, Li L., Liu S-H, Vinh T N D, Liu X-H 2017 Hydration properties and thermal analysis of cement-based materials containing limestone powder *Journal of Central South University* 24 (12) pp 2932–39
[4] Liu S-H, Kong Y-N, Wang L 2014 Hydration mechanism of low quality fly ash in cement-based materials *Journal of Central South University*, 21 (11), pp 4360–67
[5] Solomatov V 1989 Problems intensive separate technology *Concrete and reinforced concrete* 7 p 4

Simulation of Curing Processes in Cast-in-place Thin-walled Structure under Different Climatic Conditions

D Mokshin, K Gauss, A Rubanov and R Mokshin
Tomsk State University of Architecture and Building, Russia, Tomsk

E-mail: i2ii@bk.ru

Abstract. The paper presents the study on the effect of heating wire pitch on the strength of concrete in cast-in-place thin-walled wall structure when heated with the use of heating wire in the mold made of 40 mm thick wood depending on the wind speed and outdoor air temperature using ELCUT software. The conclusions include recommendations on cold weather concreting of a cast-in-place reinforced concrete wall with the surface modulus $S_m = 10$.

1. Introduction

Concrete is an inherent element of modern construction. It is one of the most widespread materials which largely defined the level of development of our civilization. At the same time, concrete is a complex artificial composite material that may have completely unique properties. Strength is the key parameter that ensures quality of concrete structures and constructions. Development of strength occurs due to hydration, i.e. interaction between cement and water. Such chemical reaction requires conditions close to normal ones.

Considering the geography of Russia, especially location of oil and gas producing regions, specific methods shall be used for creating and keeping the required temperature in the concrete body for it to reach the ultimate strength, as freezing early age concrete has negative effect on its strength characteristics. At negative temperatures, water transits into solid state and hydration reaction stops. Internal pressure forces simultaneously occur in concrete due to water volume increase during solidification. At early stage, immature concrete structure is unable to withstand these forces. When thawing the hydration process resumes, however strength characteristics irreversibly decrease.

The science progresses and there are multiple methods (both heating and non-heating ones) of cold weather concrete production [1–4]. According to the standard regulations SP 70.13330.2012 “Load-bearing and separating constructions” the most economically viable curing method for cold weather concreting of cast-in-place structures is selected depending on the type of structure (surface modulus) and outdoor air temperature.

2. Experiment

The paper presents the study on the effect of heating wire pitch (HWP = 100, 150, and 200 mm) on the strength of concrete mixture in cast-in-place structure under various conditions: wind speed (v = 0, 1, 5, 10, and 15 m/s; convective heat transfer coefficient of the environment $\alpha_{o.c}$= 3.78; 6.87; 18.14; 28.58 and 37.29 W/(m^2·K), respectively) and outdoor air temperature ($t_{o.a}$ = – 5, – 15, – 25, and – 35 °C).

The structure under study was a wall sized 0.2×3.0×6.0 m. Surface modulus S_m = 10, id est it is a thin-walled structure with minor exothermic processes that were neglected in this study.

Table 1 presents the mold type used in the calculations.

Table 1. Mold parameters.

Mold type	Material	Layer thickness, mm	Heat transfer resistance coefficient (W/ m²·°C)		
			Wind speed, m/s		
			0	5	15
II	Board	40	2.03	3.6	3.94

Three design models (DM) with finite element meshes (FEM) with different heating wire pitch (HWP = 100, 150, and 200 mm) were built within the research. The pitch was defined as a set of peaks at the depth of 30 mm with heat production capacity of q = 35 W/m. Heat losses occur through the mold side surfaces. B30 concrete with initial temperature of concrete mixture $t_{c.m}$ = 10 °C was used.

The value $\alpha_{o.c.}$can be calculated with the help of Frank's equation (1):

$$\alpha_{o.c} = 6.31\nu^{0,656} + 3,78e^{-1,91\nu}, \quad (1)$$

whereν is wind speed, m/s; e = 2.718 is the base of the natural logarithms.

The first term of equation (1) considers forced convection, and the second term refers to natural convection. Figure 1 presents DM No. 2 with FEM that has 3 blocks and 7,649 nodes.

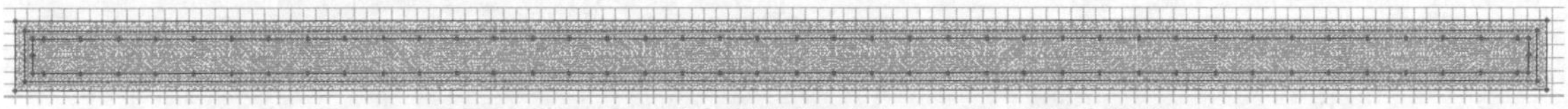

Figure 1. DM No. 2 with FEM (mold type II, HWP 150 mm).

Thermophysical properties of the mold were taken from the software reference book.

Table 2. Thermophysical properties of materials.

Material	Thermal conductivity (W/m·K)	Specific heat capacity (J/K·kg)	Density (kg/m³)
Metal	58	470	7.850
Concrete	2.07	1.050	2.500
Pine	0.18	2.300	500

Heating parameters and time (3 days) were selected. Then follows task solving which can be called a "rapid calculation". Such calculation is done using ELCUT software. The interval of solution is selected by the user. In this research the interval was set for 2 hours. The used software complex produces task solution with the account of heat capacity and thermal conductivity of all materials of the structure in the design scheme with regard to space and time. Convective heat exchange with the environment is also taken into account.

3. Results

Figure 2 demonstrates dependency graphs of the impact of heating wire pitch on the concrete body temperature of the thin-walled wall structure (t_c) under various environmental conditions. Figure 2 also contains initial temperature of concrete mixture and the limits of heating temperature specified in the construction standard SP 70.13330.2012.

It is clear from Figure 2 that heating wire pitch is the key parameter having the most effect on the body temperature of the concrete structure. After studying one of the cases at constant wind speed v = 5 m/s and outdoor air temperature $t_{o.a}$ = – 25 °C it was found that the concrete body temperature decreases by 44.14 % as the heating wire pitch grows from 100 mm to 150 mm, and by 73.5 % with HWP increase from 100 mm to 200 mm.

While studying the case of HWP = 50 mm the concrete body temperatures exceeding those specified by SP 70.13330.2012 were observed.

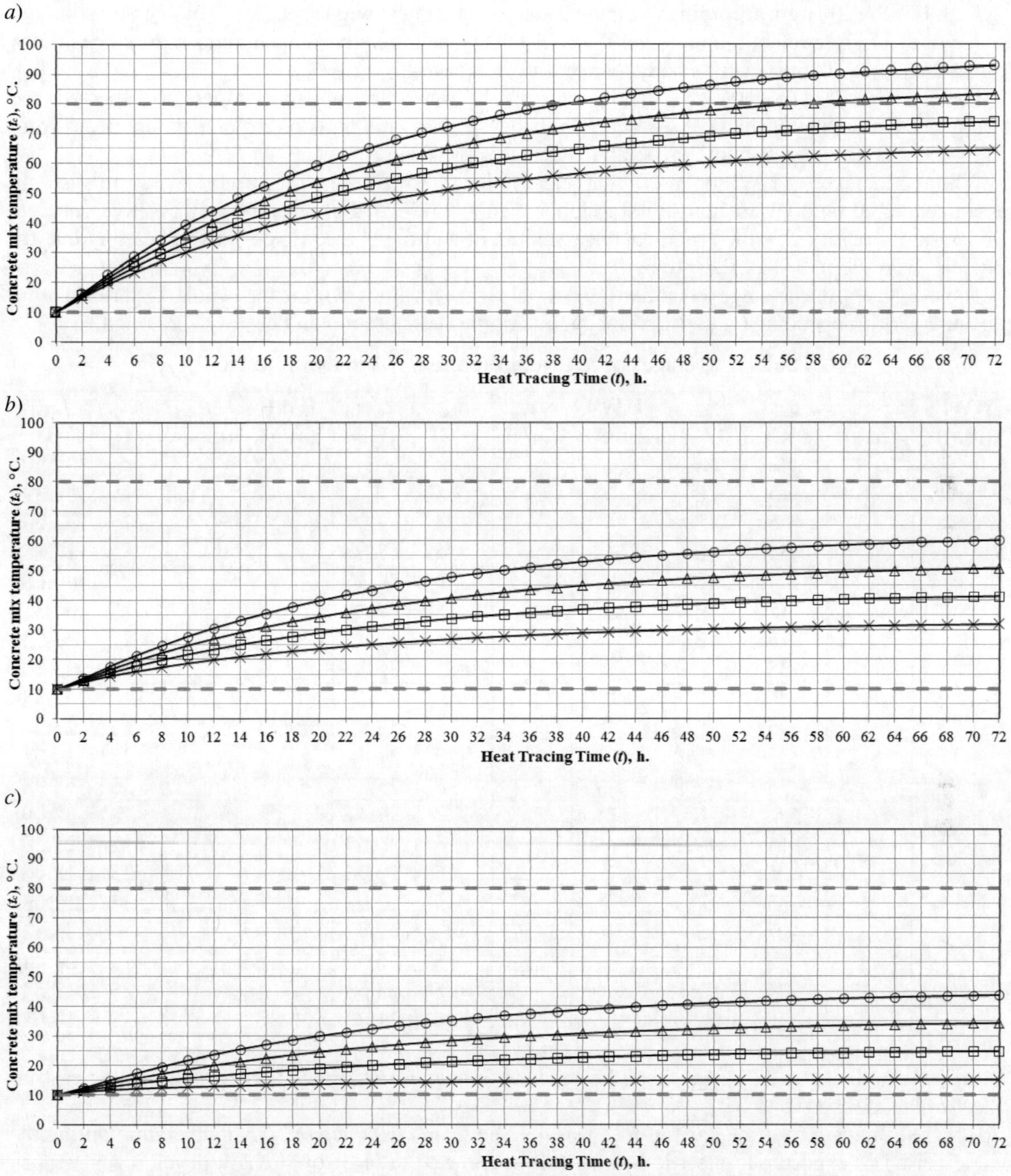

Figure 2. Dependency graph of the impact of heating wire pitch on the concrete body temperature of the thin-walled wall structure (t_c) at various values of outdoor air temperature and wind speed v= 5 m/s subjected to electric heating for the period from 0 to 72 hours:
a – HWP = 100 mm; *b* – HWP = 150 mm; *c* – HWP = 200 mm;
× –– 5 °C; □ –– 15 °C; Δ –– 25 °C; ○ –– 35 °C;
— – temperature limit up to + 80 °C specified by SP 70.13330.2012;
— – initial temperature of concrete mixture $t_{c.i}$ = 10 °C.

With HWP = 100 mm, according to Figure 2 electric heating at wind speed v = 5 m/s is only admitted for the outdoor air temperature no lower than – 15 °C. When the environment temperature goes down from – 25 to – 35 °C the concrete body temperature decreases by 12.9 %.

The most appropriate is HWP = 150 mm, as no overheating or low positive temperatures are observed during electric heating of the wall structure. In this case, when the outdoor air temperature drops from – 5 to – 35 °C the concrete body temperature decreases by 47.2 %.

With the increase of HWP up to 200 mm electric heating with the wind speed of v = 5 m/s is only recommended at outdoor air temperature higher than – 25 °C. Otherwise, too low positive concrete temperature is observed, which requires huge amount of additional electric power and heating time for concrete to develop 70 % of *R*28 strength.

Table 3 gives the average concrete temperature ($t_{c.av.}$) values within the whole electric heating exposure period depending on the air velocity, outdoor air temperature and heating wire pitch.

Table 3. Average concrete temperature ($t_{c.av.}$) in the structure.

HWP (mm)	***v* (m/s)**	**$t_{o.a.}$ (°C)**	**$t_{c.av.}$ (°C)**	**HWP (mm)**	***v* (m/s)**	**$t_{o.a.}$ (°C)**	**$t_{c.av.}$ (°C)**	**HWP (mm)**	***v* (m/s)**	**$t_{o.a.}$ (°C)**	**$t_{c.av.}$ (°C)**
100	0	- 5	E	150	0	- 5	E	200	0	- 5	47.30
		- 15	E			- 15	E			- 15	42.39
		- 25	E			- 25	52.68			- 25	37.20
		- 35	E			- 35	47.65			- 35	32.15
	1	- 5	E		1	- 5	54.35		1	- 5	40.68
		- 15	E			- 15	48.5			- 15	34.81
		- 25	E			- 25	42.66			- 25	28.96
		- 35	E			- 35	36.79			- 35	23.10
	5	- 5	E		5	- 5	45.7		5	- 5	33.88
		- 15	E			- 15	39.05			- 15	27.26
		- 25	55.62			- 25	32,44			- 25	20.63
		- 35	48.99			- 35	25.83			- 35	13.98
	10	- 5	E		10	- 5	43.41		10	- 5	32.09
		- 15	58.91			- 15	36.58			- 15	25.28
		- 25	51.95			- 25	29.76			- 25	18.46
		- 35	45.11			- 35	22.93			- 35	11.63
	15	- 5	E		15	- 5	42.42		15	- 5	31.33
		- 15	57.3			- 15	35.52			- 15	24.44
		- 25	50,4			- 25	28.63			- 25	17.54
		- 35	43,49			- 35	21.73			- 35	10.63

In Table 3, the letter “E” means that the concrete temperature $t_{c.}$ in the body of thin-walled cast-in-place wall exceeds +80 °C.

Concrete strength in percentage of *R*28 was defined from the concrete body temperature during the whole period of exposure to electric heating (72 hours) on the basis of strength development graphs from MDS 12-48.2009 “Cold weather concreting with the use of heating wires”. For B30 (M400) concrete design compressive strength is 38.35 MPa according to the standard document GOST 26633-2015 “Heavy-weight and sand concretes. Specifications”.

The research outcomes are presented in Table 4 using which is convenient for defining strength depending on the heating wire pitch (HWP = 100, 150, and 200 mm), outdoor air temperature ($t_{o.a}$ = – 5, 15, 25, and 35 °C) and wind speed (v = 0, 1, 5, 10, and 15 m/s) after 3 days’ exposure to electric heating.

Strength for intermediate wind speed values can be obtained by using interpolation.

Table 4. Concrete strength in thin-walled wall structure depending on temperature, wind speed and heating wire pitch.

HWP (mm)	v (m/s)	$t_{o.a}$ (°C)	$t_{c.av}$ (°C)	HWP (mm)	v (m/s)	$t_{o.a}$ (°C)	$t_{c.av}$ (°C)	HWP (mm)	v (m/s)	$t_{o.a}$ (°C)	$t_{c.av}$ (°C)
100	0	- 5	E	150	0	- 5	E	200	0	- 5	68.4
		- 15	E			- 15	E			- 15	65.4
		- 25	E			- 25	72.2			- 25	62.0
		- 35	E			- 35	68.6			- 35	58.5
	1	- 5	E		1	- 5	73.5		1	- 5	64.4
		- 15	E			- 15	69.1			- 15	60.4
		- 25	E			- 25	65.6			- 25	56.3
		- 35	E			- 35	61.8			- 35	52.2
	5	- 5	E		5	- 5	67.4		5	- 5	59.7
		- 15	E			- 15	63.3			- 15	55.1
		- 25	74.5			- 25	58.7			- 25	50.4
		- 35	69.4			- 35	54.1			- 35	45.1
	10	- 5	E		10	- 5	66.1		10	- 5	58.5
		- 15	77.13			- 15	61.6			- 15	53.7
		- 25	71.6			- 25	56.8			- 25	48.8
		- 35	67.1			- 35	51.4			- 35	43.3
	15	- 5	E		15	- 5	65.5		15	- 5	57.9
		- 15	75.8			- 15	60.9			- 15	53.1
		- 25	70.3			- 25	56.0			- 25	48.0
		- 35	66.1			- 35	51.2			- 35	42.5

4. Conclusion

Mold of 40 mm thick wood is very susceptible to external effects of the environment, especially to wind speed. At constant outdoor air temperature $t_{o.a} = -25$ °C and HWP = 150 mm and with the the increase of wind speed v from 0 to 15 m/s the concrete body temperature drops by 54,3 %.

In isolated cases 70 % of *R*28 strength is developed after 3 days of electric heating (Table 4), otherwise, additional time is required, which causes extra energy costs. Based on this one may conclude that this type of mold shows little efficiency.

Using the obtained research outcomes the following recommendations on concreting cast-in-place thin-walled wall structure in 40 mm thick wooden mold are developed:

- HWT = 50 mm are not recommended for use.
- HWT = 100 mm should be used in open and well ventilated plots or objects with wind speed higher than 1 m/s and outdoor air temperature lower than – 15 °C.
- HWP = 150 mm is recommeded to use for all the studied wind speed (v = 0, 1, 5, 10, and 15 m/s) and outdoor air temperature (to.a = – 5, – 15, – 25, and – 35 °C) values. However, it should be mentioned that 3 days of exposure to electric heating will not suffice for developing 70 % of R28 strength.
- With HWP = 200 mm electric heating shows lower efficiency. It is recommended for use in closed or poorly ventilated premises with wind speed less than 5 m/s and outdoor temperature higher than – 25 °C. For other environment-related parameters low electric heating temperature is observed. In all cases, additional time of electric heating is required for concrete to develop 70 % of *R*28 strength.
- With HWP = 100, 150, and 200 mm development of ultimate concrete strength of 30% of *R*28 is provided in all cases.
- HWP = 250 mm is not recommended for use because of low electric heating temperature, and in cases of high wind speed and low temperature the concrete body temperature decreases to 0 °C and lower.

Acknowledgments

The reported study was funded by RFBR according to the research project № 18-38-00674.

References

[1] Zheng L 2009 Discussion and Control of High Strength Self-compact Negative Temperature Concrete Structure Damage *Underground Engineering and Tunnels* **1**

[2] Yang S-w, Yang Y-z, Deng H-w and Ba H-j 2004 Early demage degree of construction in minus temperature concrete (Institute of Materials Science and Engineering, Harbin University of Industry, Harbin 15001, Heilongjiang, China)

[3] Roberts-Wollman C L, Breen J E and Cawrse J 2002 Measurements of Thermal Gradients and their Effects on Segmental Concrete Bridge *Journal of Bridge Engineering* **7** 3

[4] Poon C-S, Kou S C and Chan D 2006 Influence of steam curing on hardened properties of recycled aggregate concrete *Magazine of Concrete Research* **58** **5** 289-299

Thermal protection of buildings from sandwich panels for Southern Siberia conditions

G Shibaeva, E Ibe, D Portnyagin and E Afanasyeva
Khakass Technical Institute, Siberian Federal University, Abakan, Russia

E-mail: Katerina.ibe@mail.ru

Abstract. Article represents recommendations on decrease in heat losses through heat-conducting inclusions of sandwich panels. Recommendations are based on the analysis of calculations by a finite element method in software package Elcut and thermal imaging. The purpose of this study is to increase the thermal protection of buildings built with sandwich panels for the South Siberia conditions. The way of decrease in the factor of cold bridges is studied by adding thermal insulation of panel node connections. The authors analyzed constructive joints: roof parapet, roof cornice and junction of wall panels.

1. Introduction

External fencing sandwich panels are known for a long time. [1] At present, the construction of frame public buildings using sandwich panels as enclosing structures has become very popular in Abakan. Mineral wool and polystyrene foam are used as insulation layers in such panels.

In papers [2–4] the authors note that the sandwich panels, due to the peculiarities of the design, have several disadvantages, including various types of deformations. In the cold climate of Southern Siberia, temperature distortions are especially important.

The disadvantages of panels with mineral wool and metal frame is the possibility of precipitation of insulation and low efficiency of thin insulating gaskets, which impairs the thermal properties of the fence. Although mineral wool is impregnated with phenol-formaldehyde resins, which makes mineral wool a hygroscopic material, it is still the case that in the absence of ventilation the middle layer of the panel accumulates moisture.

The disadvantage of sandwich panels with a middle layer of polyurethane foam is a low fire resistance due to the flammability of the insulation.

An important problem in the operation of sandwich panels is significant stresses and deformations during temperature and humidity effects of the environment. In winter, the temperature difference between the inner layer of a three-layer sandwich panel and the outer one causes bending stresses in the sandwich panels. Especially for roofing panels, where inevitably accumulated snow enters into account, depending on the slope, respectively, and its amount.

The crumpling of the sandwich panel on the supports also affects the self-tapping screws, which eventually begin to "crawl out" of the panel. Because of this, moisture will penetrate under the gasket of the screw, accumulating in the core of the three-layer panel; the process of corrosion of the screw will begin to develop.

However, due to the low cost and speed of construction, this design is very common. The main problems are the thermal diversities that thermal bridges pass through. The main part of heat losses and the thermal bridges meets in the structural sites of exterior walls. [5–9]

IOP Conf. Series: Materials Science and Engineering **597** (2019) 012036 doi:10.1088/1757-899X/597/1/012036

Insufficient knowledge of the issues of behavior of sandwich panels on the effects of temperature effects, which requires solving these issues, which are interesting from a scientific point of view and important from a practical point of view.

The purpose of this study is to increase the thermal protection of buildings built with sandwich panels for the South of Siberia conditions.

2. Methods

In order to analyze the joints of enclosing structures made of sandwich panels, the objects were monitored. As the objects of study were selected buildings in operation in the city of Abakan, the Republic of Khakassia (table 1).

Thermal insulation control of the above objects was carried out in December 2018. At the time of the survey objects were heated. The survey was carried out in the absence of wind, precipitation, fog and smoke. In the process of measurements, the enclosing structures of the buildings being examined were not exposed to direct and reflected solar radiation.

The following types of work were performed during thermal imaging:

- inspection of objects using a thermal imager for the formation of a general characteristic of objects and the identification of areas to be further thermographed;
- overview thermography of the external surfaces of the enclosing structures to identify temperature anomalies;
- detailed thermography of selected areas of external surfaces of walling to clarify the temperature anomalies.

Table 1. Objects of study

Photo	Thermogram
Object 1: Swimming pool on Torosova str. There is a stone wool insulation 150 mm.	
	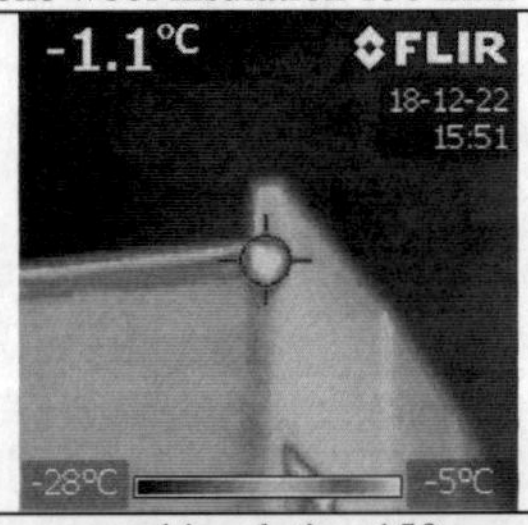
Object 2: Shopping centre on Nekrasova str. There is a stone wool insulation 150 mm.	
	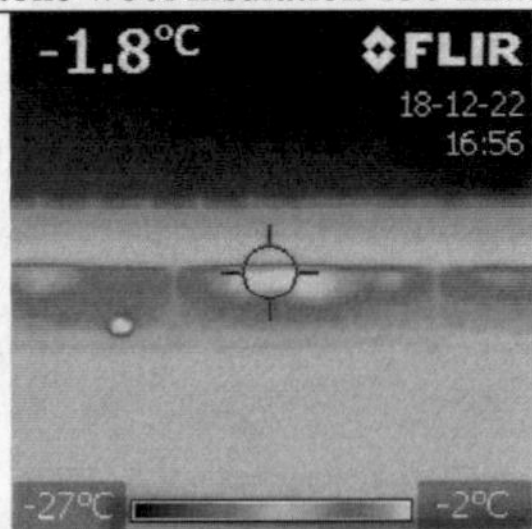
	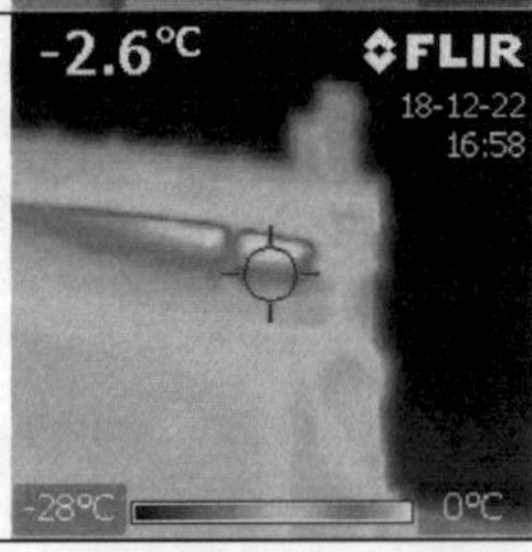

IOP Conf. Series: Materials Science and Engineering **597** (2019) 012036 doi:10.1088/1757-899X/597/1/012036

Object 3: Shopping centre on Igarskaya str. There is a stone wool insulation 120 mm.	
	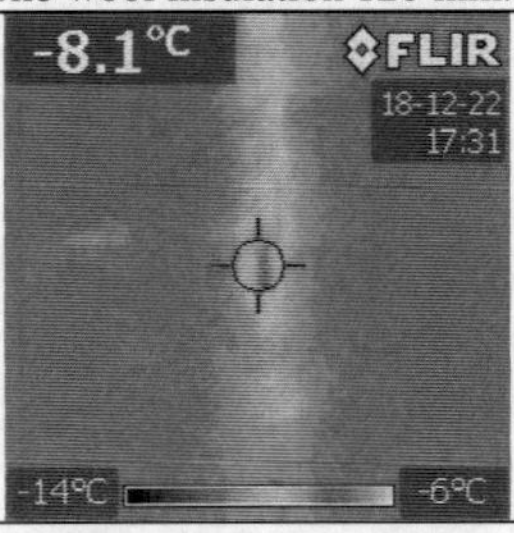

The second part is a computation of structural sites using the soft package Elcut to study the thermal field of the external wall, comparison of results depending on the type of insulation.

Mathematical modeling of the thermal regime of the structure in the cold season was carried out under the following boundary conditions:

- inside air temperature in the building tint = 20 °C (Russian standard GOST 30494–2011);
- outside air temperature text = –37 °C (Russian construction norm SP 50.13330.2012);
- heat transfer coefficient at the internal surface of the envelope α_{si} = 8.7 W/(m^{2}°C) (Russian construction norm SP 50.13330.2012);
- heat transfer coefficient at the external surface of the envelope α_{se} = 23 W/(m^{2}°C) (Russian construction norm SP 50.13330.2012).

3. Results and discussions

The performed thermal imaging surveys show that there are large heat losses in all the joints of the sandwich panels.

Thermal imaging of objects 1 and 2 shows the temperature difference between the outer surface of the wall and the corner joint, which is more than 25 C.

The temperature difference in the front butt joint is 8 C.

For cold climate conditions, such a drop will inevitably lead to the formation of condensate and ice on the inner surface of the fence.

It should be noted that the lifetime of buildings ranges from 3 to 15 years. In this case, the defects of the thermal shell of buildings are almost identical. In the process of inspection of the joints, it was established that the assembly foam completely lost its original properties, has a large shrinkage and detachment from the panel surface. In this regard, it can be said that the design joint of sandwich panels made of assembly foam is ineffective in cold climates and the manufacturer should develop various design solutions for different climatic conditions.

Figures 1–3 show temperature fields in considered joints according to design solution.

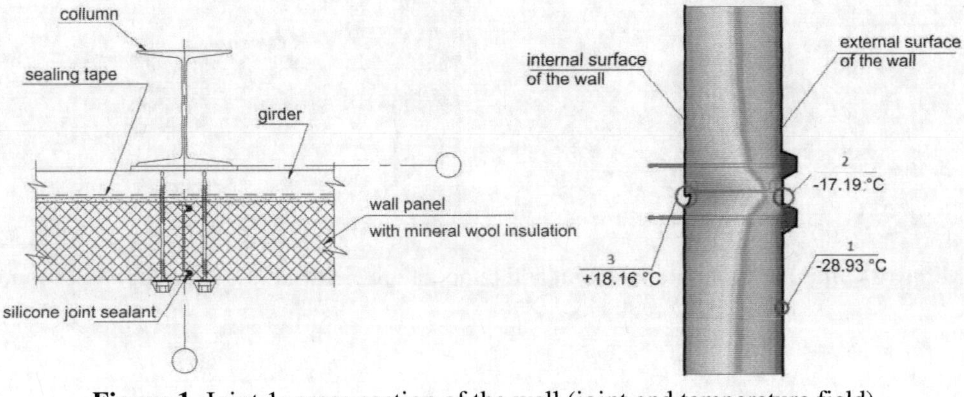

Figure 1. Joint 1: cross section of the wall (joint and temperature field)

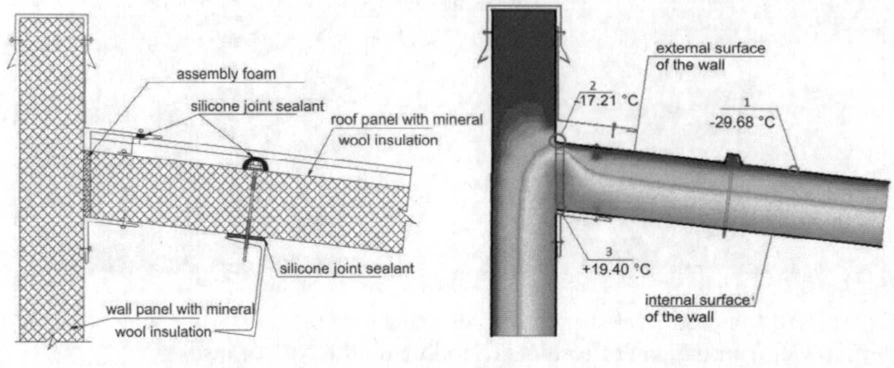

Figure 2. Joint 2: roof parapet (joint and temperature field)

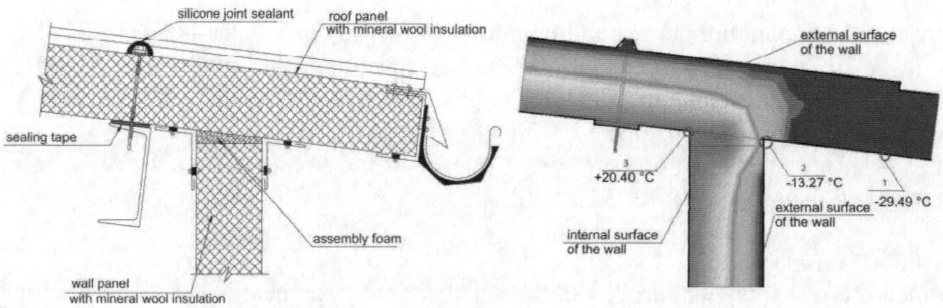

Figure 3. Joint 3: roof cornice (joint and temperature field)

Analysis of figures 1-3 shows that in all corner joints there is a large exfiltration of warm air, due to which the temperature of the outer contour of the panel is 10–15 °C higher than the temperature of the main part of the panel. In this regard, large heat losses are generated in buildings.

In this regard, for the temperature conditions of the south of Siberia, horizontal and angular joints of the sandwich panels are proposed to be sealed not with assembly foam, but with polyurethane foam, which is elastic and resists significant temperature deformations. The results of calculations of joints with additional insulation are presented in fig. 4-5, table 2.

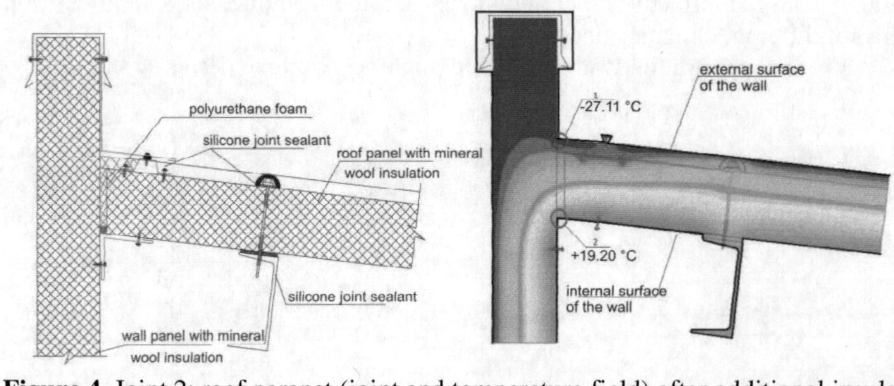

Figure 4. Joint 2: roof parapet (joint and temperature field) after additional insulation

IOP Conf. Series: Materials Science and Engineering **597** (2019) 012036 doi:10.1088/1757-899X/597/1/012036

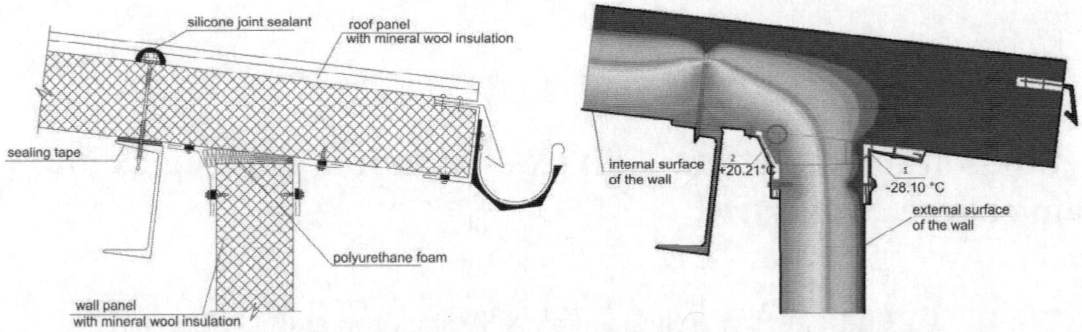

Figure 5. Joint 3: roof cornice (joint and temperature field) after additional insulation

Table 2. Temperature on external and internal contour before and after additional insulation

Number of joint	Before additional insulation		After additional insulation	
	External surface temperature T, °C	Internal surface temperature T, °C	External surface temperature T, °C	Internal surface temperature T, °C
1	-17.2	+18.2	-28.3	+25.7
2	-17.2	+19.4	-27.1	+19.2
3	-12.4	+15.1	-26.7	+19.2

Analysis of temperature fields after additional insulation shows that these nodes are more efficient from the standpoint of energy saving, there is no exfiltration. With this solution, the heat loss is reduced by 3 times

Analyzing the temperature change in the nodes of the enclosing structure in the annual cycle of operation, we can assume in which node connections the processes of precipitation and accumulation of condensation moisture are possible, as well as roughly determine the periods of "wetting" and "drying" of enclosing structures.

4. Conclusions

Based on the research we can draw the following conclusions:

1. In the conditions of Southern Siberia, it is impossible to use a standard constructive solution of sandwich panels without additional thermal insulation.
2. The use of polyurethane foam in butt joints of panels leads to the fact that after 5 years of operation, the properties of the insulating gasket are reduced to zero, which leads to large heat losses and reduces the durability of the outer fence.

References

[1] Chamberlain M 1963. *U.S. Patent No. 3,111,787*. Washington, DC: U.S. Patent and Trademark Office.
[2] Li D, Deng Z, Chen G, Xiao H and Zhu L. 2017 *Composite Structures* **169** 29-41
[3] Frostig Y and Thomsen O 2008 *Composites Part B: Engineering* **39 (1)** 165-184
[4] Li D, Deng Z, Xiao H and Jin P 2018 *Thin-Walled Structures* **122** 8-16
[5] Vedishcheva I, Ananin M, Al Ali and Vatin N 2018 *Magazine of Civil Engineering* **78 2**
[6] Korniyenko S 2014 In *Applied Mechanics and Materials* **618** 509-513
[7] Kornilov T and Nikiforov A 2018 *Magazine of Civil Engineering* **84(8)** 140–149
[8] Pukhkal V and Mottaeva A 2018 *Magazine of Civil Engineering* **81(5)** 202–211
[9] Khalimov OZ, Shibaeva GN, Portnyagin DG and Ibe EE 201) *IOP Conference Series: Materials Science and Engineering* **365** 042071

Methods of introduction of glyoxal-containing additives into foam concrete mixture

A I Kudyakov[1], A B Steshenko[1], A S Simakova[1] and A D Latypov[2]
[1]Tomsk State University of Architecture and Building, Department "Building materials" 2, Solyanaya Sq., 634003, Tomsk, Russian Federation
[2]Tomsk State University, 36 Lenina Avenue, 634050, Tomsk, Russian Federation

E-mail: steshenko.alexey@gmail.com

Abstract. The present paper investigates a cement based foam concrete with glyoxal-containing additives which are introduced into the mixture in different ways. The relevance of the given study is conditioned by the necessity to provide the required quality parameters of mixture for transportation and laying the formwork, as well as providing strength and physical characteristics of wall structures for housing construction. Four ways of introducing additives into the foam concrete mixture have been proposed: spraying with 40% glyoxal aqueous solution; joint grinding of sand with crystalline glyoxal in the amount of 0.01% by weight of cement; introduction of crystalline glyoxal into the foam concrete mixture; introduction of a 40% aqueous solution of glyoxal to the foam concrete mixture. The study was conducted in the Laboratory of Tomsk State University of Architecture and Building which is accredited in accordance with the national standards requirements. Glyoxal-containing additives applied within the technology of cement foam concrete of natural hardening contributes to ensuring increased aggregative stability of mixtures and compressive strength at the age of 28 days. The elaborated composition and technology of foam concrete production with glyoxal-containing additives is intended for use in the walls structures as well in monolithic construction as in pre-cast production.

1. Introduction

In accordance with the concept of Russia's development up to 2020, great attention is paid to building materials based on local raw materials with the required performance characteristics, low energy intensity and cost. Such materials include porous wall materials based on cement binders. Porous concrete (foam concrete) can be effectively used for the manufacture of wall blocks, constructing monolithic wall structures in low-rise construction, mainly individual housing. The construction of low-rise housing with the use of foam concrete in enclosing structures is a favorable object for small businesses.

Great attention is paid to reducing thermal conductivity and improving the strength, uniformity of quality parameters, and the rational and economical use of raw materials in the production of foam concrete of natural hardening. Effective modifying additives for binders and concretes [1-3] are widely used for managing the quality of cement based concretes. In the Russian building sector mainly used imported additives. Therefore, the development and use of domestic modifying additives is relevant. This is possible with the development of innovative technologies in the chemical industry. The Tomsk region (Russia) developed a technology for the synthesis of oxalic dialdehyde, glyoxal, which has a

IOP Conf. Series: Materials Science and Engineering **597** (2019) 012037 doi:10.1088/1757-899X/597/1/012037

number of unique properties, which has a number of unique properties that can be effectively used in the pharmaceutical industry, construction, refining, textile, leather, paint and varnish, glue and metallurgical industries, etc.

Glyoxal can be attributed to surface-active substances. The surface activity value of 0.045 $N{\bullet}m^2/kg$ is provided by the concentration of glyoxal in an aqueous solution of 10 g/l. The dependence of surface activity on the concentration of glyoxal in solution is close to logarithmic. The maximum value of surface activity is observed at low concentrations. The manifestation of the surface-active properties of glyoxal in aqueous solution allows us to predict the plasticizing effect when it is use as additive to composite cement based building materials.

The pH value of the cement paste reaches 12.5 after cement interacts with water. Conditions are created for the Cannizzaro reaction to proceed after glyoxal is introduced into cement paste. Assuming the pH of the cement paste is 12, the introduction of glyoxal into the cement paste leads to the Cannizzaro reaction (Figure 1). Presumably, the interaction of glyoxal with calcium hydroxide (the product of the hydrolysis of alita in the cement paste), as a result of the Cannizzaro reaction, calcium glycolate is formed [4-7]. An increased heat release is observed in the cement paste after adding water to the mixture. The cement paste expands due to the intensive increase in the volume of newly formations ($(HO\ CH_2\text{-}COO)_2Ca$ during the cement hydration. In addition, it was found that the resulting calcium glycolate has great strength.

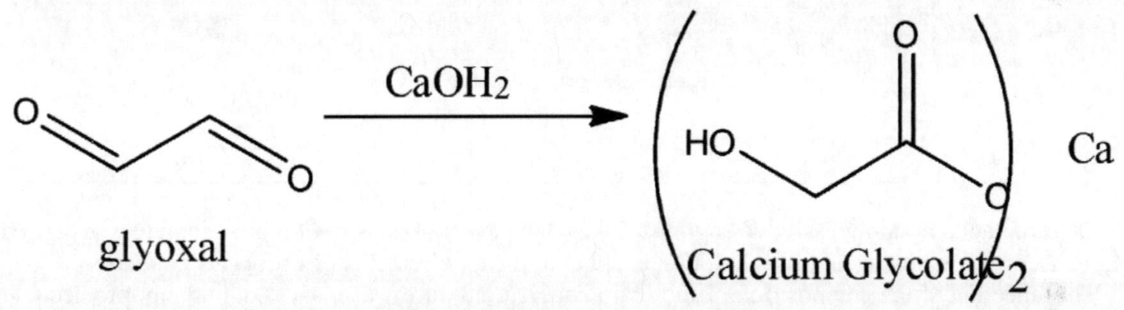

Figure 1. Cannizzaro reaction

The heme-diol and dimeric forms of glyoxal remain for some time resistant to the alkaline environment of concrete, since aldehyde groups are involved in the Cannizzaro disproportionation of aldehydes, and not heme-diol groups, it takes time to shift the balance to the aldehyde form. Since water is weakly bound to heme-diol forms, the amount of water decreases during the hydration process, and it starts to be released from these complexes. This provides prerequisites for the scientific substantiation of the choice of glyoxal as a modifying additive in the production of building materials. The ability to plasticize, retain water and give it over time into the volume of the composition allows the forming of a reserve of structure formation in cement compositions [8,9]. Glyoxal is capable of forming structures cross-linked with silicates during the hardening process of the hardened cement paste.

The above properties of glyoxal in cement systems allow us to recommend it to control the structure formation of cement based building compositions of various functional purposes. The use of glyoxal in the technology of building materials has recently been investigated by A.S. Knyazev, V.V. Strokova, Yu.S. Sarkisov, N.P. Gorlenko, A.I. Kudyakov. Based on the results of the analysis of security documents and published articles about the use of glyoxal in the building sector, the development of research and the acquisition of new knowledge in order to develop new building materials using Russian-made glyoxal is relevant.

2. Materials and methods

In carrying out experimental research were used raw materials that meet the requirements of national standards of Russia. As binders Portland cement of Topki Cement Factory (Kemerovo Region, Russia)

CEM I 42.5H (Russian State Standard GOST 30515-2013) was used, sand of Kudrovskoe deposit of Tomsk Region (Russia) with fineness modulus 1.86 (Russian State Standards GOST 8736-2014 and GOST 26633-2012), water (Russian State Standard GOST 23732-2011) and foam agent PB-2000.

A 40% aqueous solution of glyoxal and crystalline glyoxal (TU 2633-003-67017122-2011), manufactured by TD Novokhim (Tomsk, Russia) were used as a structure formation accelerator for foam concrete mixtures. Technical characteristics of glyoxal-containing additives are presented in table 1.

Table 1. Technical characteristics of glyoxal-containing additives

Indicator	Crystalline glyoxal	40% aqueous solution of glyoxal
Appearance	white to yellowish beige powder	colorless / pale yellow liquid
Mass fraction of the main substance,%	80	40±1
The rate of activity of hydrogen ions H * (pH)	-	2-3.5
Density, g /cm^3	1.0	1.20-1.35
Mix	water, ethanol, methanol, isopropanol	ethyl alcohol, acetone
Volatility	-	low

Determination of compressive strength, shrinkage and average density of foam concrete was carried out in accordance with standard techniques, adapted to the requirements of European standards.

The selection of the composition of the foam concrete mixture was carried out taking into account the requirements of SP 277-80 “Instructions for the production of cellular concrete products”. Taking into account the results of optimization of technological modes of preparation of foam concrete mixture in a laboratory mixer, the basic composition of foam concrete was developed, which is given in table 2.

Table 2. Basic composition (control) of foam concrete D500 per 1 m^3

Cement, kg	Sand, kg	Water, kg	Foaming agent, l
270	135	202	1.3

When establishing the regularities of the influence of additives on the properties of the foam concrete mixture and concrete foam, additives were introduced into the basic composition of the foam concrete without changing the number of components and then adjusting the recommended composition according to the values of the actual average density of the concrete mix. The preparation of the foam concrete mixture was carried out using a single-stage technology with laboratory foam concrete mixer.

The prospects of mechanical and chemical methods of activation are substantiated according to the results of the analysis of various methods of activation of the components of foam concrete, which are aimed at improving the contact interaction of the hardened cement paste with the aggregate.

The positive effect is achieved by controlling the structure of the foam concrete. Improvement of the performance characteristics of foam concrete is possible through the development of new technological methods regulating the physicochemical phenomena at the contact of the interacting components of the hardened cement paste-aggregate section [10]. Regardless of their state of aggregation, all raw materials are usually in an inactive position and with complex processing may acquire chemical activity as a result of interaction at the interface. Activation of the components of the foam concrete mixture can significantly improve the basic technical properties of foam concrete. And the aggregate should be modified to obtain the greatest effect from the action of additives. It has been established that surface treatment of the aggregate with aqueous solutions of acids leads to a significant increase in the microhardness of the hardened cement paste on contact with the aggregate grains and to an increase in the strength of the samples. The process of contact formation begins with the moment of contacting the surface of the grains of the aggregate, mineral particles and binder with liquid phase of cement paste. By controlling the surface properties of the contacting particles, it is possible to influence the processes of structure formation, primarily in the zone of contact of cement and sand.

We have proposed four ways of introducing glyoxal-containing additives into the foam concrete mixture:

- spraying of 40% glyoxal aqueous solution onto the aggregate surface;
- joint grinding of sand with crystalline glyoxal in the amount of 0.01% by weight of cement;
- introduction of crystalline glyoxal into the foam concrete mixture in the amount of 0.01% by weight of cement;
- introduction of a 40% aqueous solution of glyoxal to the foam concrete mixture in the amount of 0.01% by weight of cement.

Spraying with 40% glyoxal aqueous solution

The ways of chemical activation of the aggregate are considered, which leads to the enhancement of their contact interaction with the products of hydration of binders. In conducting studies quartz sand was treated with 40% aqueous glyoxal solution by spraying with a spray gun. Next, the foam concrete mixture was prepared using a single-stage method in a laboratory foam concrete mixer. The finished foam concrete mixture was placed in metal forms 15x15x15 cm. The foam concrete samples were kept at standard conditions at a temperature of 20±2°C for 24 hours. Next, the samples were solid cubes in a natural hardening chamber, at a temperature of 20±2°C and a relative air humidity of at least 90% for 27 days prior to testing.

Introduction of glyoxal into the foam concrete mixture

Pre-dosed water with a foaming agent was added to the foam concrete mixer and mixed for 1 minute. Then sand and cement were loaded into the mixer and the mixture was mixed for 2 minutes until a homogeneous state was obtained. It is very important to distribute the cement well in the sand. Then, 40% aqueous solution of glyoxal was added to the mixture and all components were mixed for another 4.5 minutes. Crystalline glyoxal was introduced using the same technology.

Joint grinding of sand with crystalline glyoxal

Sand sieved on a sieve before grinding. The dosed components are loaded into a laboratory mill and subjected to joint grinding. Activation occurs due to abrasion – simultaneous deformation of compression and shear. Preparation of a foam concrete mixture with activated aggregate was carried out by a single-stage method.

3. Results and discussion

The results of studies of the influence of glyoxal-containing additives and the method of their introduction on the rheological properties of the foam concrete mixture and the physicmechanical properties of foam concrete are given in table 3.

Shrinkage deformations of foam concrete, which lead to deformations and cracking, are of great importance for foam concrete building technologies throughout the entire life cycle of building management. Plastic shrinkage occurs in freshly molded products. Plastic shrinkage is a consequence of the rapid loss of water from the surface of the concrete, leading to the formation of negative capillary pressure in the micropores.

It was established that the introduction of additives into the foam concrete mixture leads to reduce plastic shrinkage by 29-69% (2.86-1.24 mm/m). The obtained data confirm the results of a study on the effect of modifying additives on the process of early structure formation of foam concrete [11]. Glyoxal additive accelerates the process of early structure formation, which allows stabilizing the porous structure of the foam concrete mixture and thereby reducing shrinkage deformation of the foam concrete mixture. The aggregative stability of the foam concrete mixes increases from the beginning to the end of the cement paste setting [11].

Table 3. Properties of foam concrete mixtures and foam concrete

Properties	The compositions				
	Control	Crystalline glyoxal 0.01%	40% glyoxal aqueous solution 0.01%	Sand treatment with 40% glyoxal water solution	Co-grinding sands with crystal glyoxal
Plastic shrinkage, mm/m	4.02	2.86	2.56	2.20	1.24
Drying shrinkage, mm/m	3.00	1.75	2.8	1.63	1.37
Spread, cm	11	11.5	14	14.4	11.9
Compressive strength, MPa	0.70	0.83	0.70	1.28	1.50
Average density in dry condition, kg/m^3	455	478	490	480	480

According to the experimental results presented in table 3, it was found that the introduction of glyoxal-containing additives into the mixture leads to reduce drying shrinkage of natural hardening foam concrete by 7-54%. The introduction of glyoxal contributes to the reduction of chemically unbound water in foam concrete, as a result, a decrease in shrinkage is observed during the drying of foam concrete [1].

During natural hardening of foam concrete, a significant decrease in shrinkage deformations allows predicting a reduced level of stress formation during structure formation in the volume of foam concrete and, as a result, increased operational reliability of products (structures) made of this material [13]. Processing sand by spraying with a 40% aqueous solution of glyoxal leads to increase in the compressive strength of foam concrete by 30%. The introduction of crystalline glyoxal in the amount of 0.01% by weight of cement, leads to a slight increase in the average density of foam concrete, while the compressive strength increases by 56% (0.83 MPa).

Joint grinding of sand with crystalline glyoxal in the amount of 0.01% by weight of cement leads to enhancing the compressive strength of foam concrete D500 up to 1.50 MPa. The positive effect of grinding sand with a glyoxal additive on the strength of foam concrete is explained by the mechanochemical activation of the surface of quartz sand, which leads to a decrease in plastic shrinkage and an increase in the contact strength of foam concrete . At joint grinding of quartz sand with crystalline glyoxal the class of foam concrete increases to B1, while maintaining the average density D500.

4. Conclusion

The introduction of modifying glyoxal-containing additives into the concrete mix during its preparation leads to reduce shrinkage deformations, to accelerate the process of structure formation and to increase the compressive strength of foam concrete.

1. Optimal content of 40% aqueous solution of glyoxal or crystalline glyoxal is 0.01% by weight of cement in foam concrete mixture based on Portland cement, sand, water and a foaming agent. Foam concrete mixture with 40% aqueous solution of glyoxal has better flowability of mixture from 11 to 14 cm and reduced plastic shrinkage of natural hardening foam concrete by 45 %.

2. Joint grinding of sand with crystalline glyoxal in the amount of 0.01% by weight of cement leads to enhancing the compressive strength of foam concrete D500 up to 1.50 MPa and also leads to reduce drying shrinkage by 54%.

IOP Conf. Series: Materials Science and Engineering **597** (2019) 012037 doi:10.1088/1757-899X/597/1/012037

3. Good agreement of the experimental results obtained by the authors was established during the elaboration of scientifically justified compositions of foam concrete with application of glyoxal-containing additives, and that agrees with the other private research results of other authors obtained during application of structure formation accelerator for foam concrete mixtures and given in independent sources. Thus, glyoxal-containing additives are an interesting object for obtaining new knowledge and developing innovative technologies for hardening cement compositions in construction.

Acknowledgement
The research was conducted with the financial support of Russian Foundation for Basic Research and administration of Tomsk Region within the research project No. 18-48-703042.

References

[1] Kudyakov A I and Steshenko A B 2015 Investigation of the influence of the crystalline glyoxal on properties of air hardened cement based foam concrete *Lett. Mater* **5** **(1)** 3-6
[2] Kudyakov A I and Steshenko A B 2016 Study of hardened cement paste with crystalline glyoxal *Key Eng. Mater.* **683** 113-117
[3] Simakova A, Kudyakov A, Efremova V, Latypov A 2017 The effects of complex glyoxal based modifiers on properties of cement paste and hardened cement paste *AIP Conf. Proc.* **18000** 020006
[4] Kudyakov A I, Simakova A S, Kondratenko V A, Steshenko A B, Latypov A D 2018 Cement paste and brick properties modified by organic additives *Journal of Construction and Architecture* **6** 138-147
[5] Kirsten W L, Charles A K, Nichole M P, David O D 2006 Oligomer Formation in Evaporating Aqueous Glyoxal and Methyl Glyoxal Solutions *Environ. Sci. Technol.* **40** 6318-23
[6] Maruful M and Jeffrey A J 2000 Temperature dependent near-UV molar absorptivities of glyoxal and gluteraldehyde in aqueous solution *Spectrochim. Acta, Part A* **56** 2653–58
[7] Salomaa P 1956 The kinetics of the Cannizzaro reaction of glyoxal *Acta Chem. Scand.* **10** 2 311–319
[8] Gorlenko N P, Rubanov A V, Sarkisov Y S 2015Antifreeze agent based on sodium salt of glyoxalic acid *Journal of Construction and Architecture* **5** 110-116
[9] Gorlenko N P, Sarkisov Y S, Volkov V A, Kul'chenko A K 2014 Structurization Processes in the System Cement-Water with Chemical Addition of Glyoxal *Russian Physics Journal* **57** 2 278-284.
[10] Luk'yanchikov S A and Luk'yanchikova A S 2016 Enrichment of low-grade local aggregates for the production of concrete by the method of chemical activation *Actual problems of the modernity* **2** 181-186
[11] Steshenko A B and Kudyakov A I 2015 Early structure formation of foam concrete mix containing modifying admixture *Magazine of Civil Engineering* **2** 56-62
[12] Kudyakov A I and Steshenko A B 2015 Cement Foam Concrete with Low Shrinkage *Adv. Mater. Res.* **1085** 245-249
[13] Kudyakov A I and Steshenko A B 2015 Shrinkage deformation of cement foam concrete. *IOP Conf. Ser.: Mater. Sci. Eng.* **71** 012019

IOP Conf. Series: Materials Science and Engineering **597** (2019) 012038 doi:10.1088/1757-899X/597/1/012038

In-situ curing mechanism of Portland cement

Yu A Abzaev, A I Gnyrya, S V Korobkov, D O Dudov and D A Mihailov
Tomsk State University of Architecture and Building, 2, Solyanaya Sq., 634003, Tomsk, Russia

E-mail: korobkov_1973@mail.ru

Abstract. The paper presents the results of predictive modelling of the mechanism of Portland cement curing using the VCCTL tool. The curing is studied at the water-cement (W/C) ratio of 0.30 and 0.35, during 28 days at room temperature. It is shown that the hydration degree correlates with the heat generation and the accumulation rate of hydration products. At W/C=0.30, the content of clinker phases and hydration products is higher than at W/C=0.35, but the rate of the indicated processes is almost similar. The growth in the water-cement ratio reduces the pore space filled with free moisture as well as the moduli of rupture. The correlation is detected between the Portland cement strength and Young's modulus in the form of quadratic and cubic polynomials. These dependencies are in good agreement with the literature data and testify to the suitability of the VCCTL microstructural model of Portland cement curing.

1. Introduction

One of the most efficient tools of predictive modelling of the Portland cement concrete hydration and gain strength is the Virtual Cement and Concrete Testing Laboratory (VCCTL) [1–6]. It seems to be very interesting to study the hydration process regularities and search for the correlation between the experimental parameters of stress-strain curves and elastic moduli of curing Portland cement in ordinary conditions as well as the influence of the water-cement ratio on this correlation.

The aim of this work is to explore the regularities of the cement hydration process and predict the strength properties of the type CEM I 42.5B cement using the VCCTL computer modeling tool. The mechanical tests of Portland cement were carried out after 4 hours and 1, 3, 7, 14 and 28 days of curing. The analysis is performed for the curing mechanism of Portland cement having different cement-water ratio and its effect on the correlation between the strength properties.

2. Materials and methods

The type CEM I 42.5B Portland cement was used in this experiment to investigate its mechanical properties. The Bogue calculations were used for this cement composition of 61.06 g C_3S, 13.62 g C_2S, 12.79 g C_4AF and 6.50 g C_3A [7]. The cement paste with the water-cement ratio of 0.30 and 0.35 was prepared in a common form 20×20×20 mm in size and was then compacted on a vibrating table [8, 9]. Isothermal hardening of Portland cement was provided by a climate chamber at 20 °C and 100% humidity. A test machine INSTRON was used to measure the ultimate compressive strength at room temperature and a 0.5 mm/min loading rate. The mechanical tests of Portland cement were carried out after 4 hours and 1, 3, 7, 14 and 28 days of curing. The stress-strain curves were obtained. Due to the brittle fracture of Portland cement, its strength property was determined by the yield stress. Its values were averaged over three points. These experiments constitute a part of our previous research [10, 11].

IOP Conf. Series: Materials Science and Engineering **597** (2019) 012038 doi:10.1088/1757-899X/597/1/012038

Prediction of the concrete hardening is performed using the Virtual Cement and Concrete Testing Laboratory (VCCTL) [1, 2]. The VCCTL is used to model concrete at the microstructural level based on two-dimensional images of the initial phases and hydration products obtained by a scanning electron microscopy. A plane, random distribution of single elements obtained by Voronoi decomposition [3] is converted to a three-dimensional model. Each unit volume (voxel) of the microstructure is assigned with a phase element. The interaction between neighbouring voxels of the microstructure is based on the cellular automata rule. According to stoichiometry of the current states, autocorrelation of neighbouring voxels, development of phases and water-filled pores, the composition of Portland cement can be determined by the dissolution, growth, the intermediate phase suppression and the formation of final products (hydrates). The main hydration products, such as low-crystallized, almost amorphous, calcium silicate hydrate having the properties of gel, and calcium hydroxide (C–H), with reasonable certainty, form on the grain surface contacting with water, and appear in available pore space. The initial random three-dimensional model of the microstructure is recovered by means of autocorrelation functions and contains four mineral clinker phases and different forms of calcium sulphate. Calcium silicate hydrate (amorphous or low-crystallized) forming during the C_2S and C_3S hydration possesses a wide spectrum of compositions which combine general concept of calcium silicate hydrate or C-S-H gel. The alkalinity of water in pores and its saturation have an effect on the kinetics of chemical reactions. At the initial stages of modeling, we utilize a quadratic dependence of the normalized volume of C-S-H gel, while at later stages the probability of the partial surface dissolution relative to the surface of the initial cement is determined. The qualitative phase composition of the solid substance allows us to estimate the moduli of rupture of Portland cement during its curing using the moduli of the individual phases in the initial and final products [12]. The volume of the hydration products strongly depends on the water-cement ratio. The hydration process is modeled using the VCCTL tool [1–6]. And the initial data are used as the input parameters for the VCCTL modeling of Portland cement hardening.

3. Results and discussion

Figures 1–4 show the main modeling results and mechanical test results of Portland cement curing with the water-cement (W/C) ratio of 0.3 and 0.35. The curing lasts for 28 days at 20 °C.

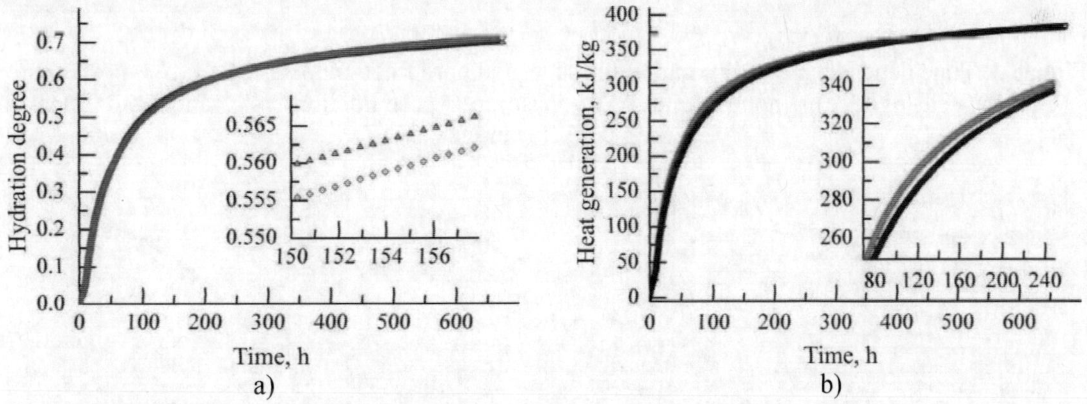

Figure 1. Time dependences of hydration (a) and heat generation (b). W/C ratio: 0.30 (upper curve); 0.35 (lower curve).

IOP Conf. Series: Materials Science and Engineering **597** (2019) 012038 doi:10.1088/1757-899X/597/1/012038

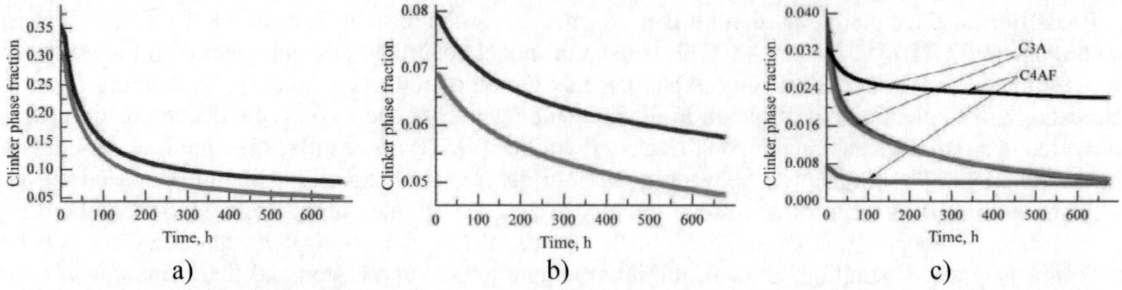

Figure 2. Time dependencies of clinker phase fraction: a) – alite (C_3S); b) – belite (C_2S); c) – aluminate (C_3A), alumoferrite (C_4AF). W/C ratio: 0.30 (upper curve); 0.35 (lower curve).

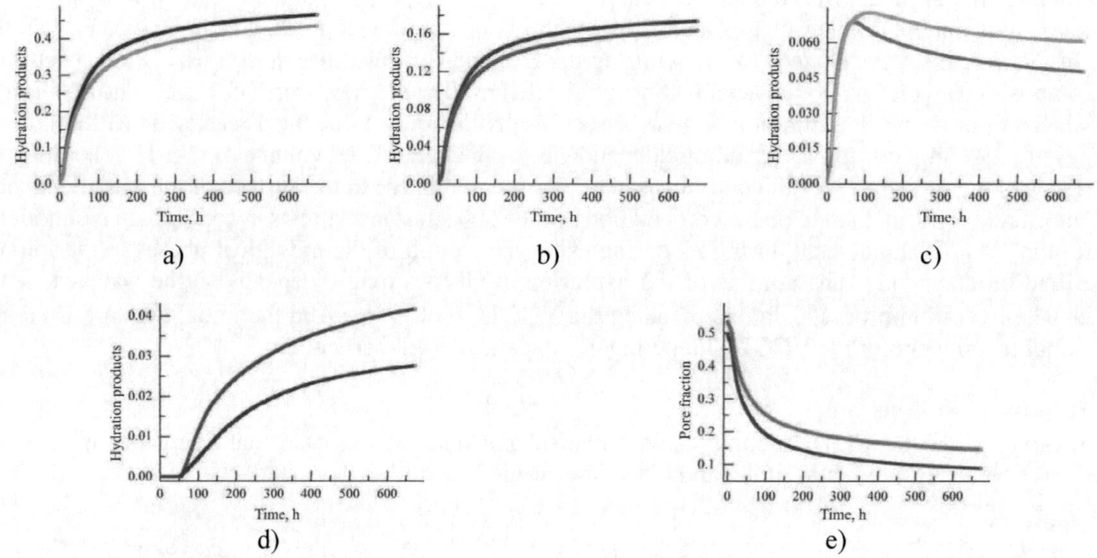

Figure 3. Time dependencies of hydration products and pore fraction: a) – C-S-H; b) – portlandite (CH); c) – ettringite, d) – monosulfate (AFm-phase); e) – pore fraction, W/C ratio: 0.30 (upper curve); 0.35 (lower curve).

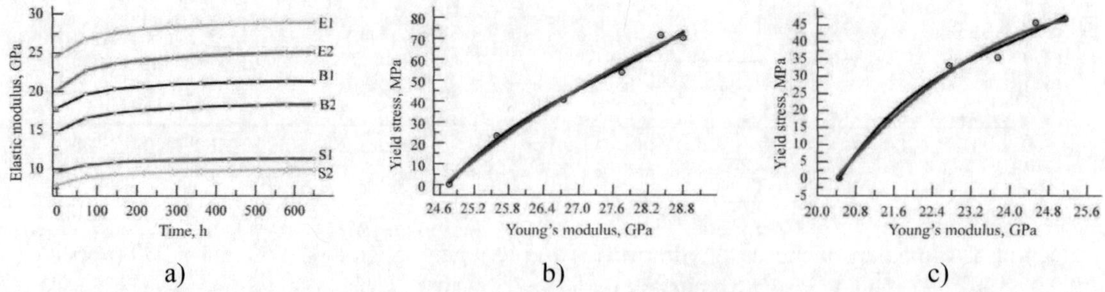

Figure 4. Time dependences of strength properties of Portland cement: a) – shear modulus, W/C = 0.30 (curve $E1$), W/C = 0.35 (curve $E2$); bulk modulus, W/C = 0.30 (curve $B1$), W/C = 0.35 ($B2$); shear modulus, W/C = 0.30 (curve $S1$), W/C = 0.35 (curve $S2$). Yield stress: b) – W/C = 0.30; c) – 0.35. Approximations use quadratic (red curve) and cubic (black curve) polynomials.

These figures show that the main hydration process develops within the time period not exceeding 100 hours. Within this period, we observe the intensive cement hardening and heat generation which correlate with each other. The water-cement ratio insignificantly affects these processes (figure 1). Inserts in figure 1 indicate the effect from the water-cement ratio. After 100 hours, the hydration intensity monotonely decreases. The clinker phase fraction substantially lowers during 100 hours. However, after 20-day curing, the initial clinker phases do not completely vanish (figure 2). It is found that the water-cement ratio affects the process of clinker dissolution. Since the cement paste was prepared, the amount of alite, belite and particularly alumoferrite is higher at W/C = 0.30 than at W/C = 0.35 (figure 2). As can be seen from figure 3, the main hydration products are C–S–H, portlandite, ettringite and to a lesser extent, monosulfate (AFm-phase). With the curing time, the amount of the indicated products increases, except for ettringite which changes non-monotonely. The highest rate of the hydration product accumulation (see figure 3) is observed at the initial stage of 100 h, and then it monotonely lowers. The degree of hydration correlates with the heat generation and the accumulation rate of hydration products (see figs 1–3). In modeling, it is assumed that the pore space is filled with water. According to figure 3*e,* the amount of free moisture in pores is higher at a 0.30 water-cement ratio. These results indicate that at W/C = 0.35, the hydration process in Portland cement occurs under conditions close to saturation. As presented in figure 4, the moduli of rupture are obtained for Young's modulus E, the bulk modulus B and the shear modulus S, depending on the time of curing and water-cement ratio. The latter significantly affects the elastic properties of the material, although the behaviour of $E1$, $E2$ and $B1$, $B2$ and also $S1$ and $S2$ curves is almost the same during the curing process. The lower water cement ratio enhances the strength properties of cement despite the large pore space. It can be supposed that the elastic moduli are strongly affected by the greater amount of clinker phases. The analysis presented in [12] for the correlation between the models of hardening and strength properties of cements shows that the most suitable are models based on the dependencies in the form of polynomials of ~ (2.5–3) degrees. Figure 4 contains the approximated dependences between the yield stress and Young's modulus. With a high degree of reliability (≈0.97), the experimental values of the yield stress can be approximated using either quadratic or cubic polynomials. This is proved by the results obtained in [12]. A comparison of the parameters of cubic polynomials shows that the rate of the strength gain at a 0.30 water-cement ratio is higher than at 0.35 (see table 1). The parameters of quadratic polynomials are less sensitive to the water-cement ratio and differ insignificantly.

Table 1. Parameters of quadratic and cubic polynomials.

Parameters	W/C	A[a]	B	C	D
Cubic	0.30	–15827.170	1712.59335	–62.06803	0.75599
	0.35	–5924.28961	735.06165	–30.4467	0.4239
Quadratic	0.30	–1356.23386	87.17665	–1.30545	–
	0.35	–932.66324	74.94812	–1.43216	–

[a] A, B, C, D are cubic and quadratic polynomial approximation parameters

4. Conclusions

The analysis of the predicted phase compositions in curing Portland cement under normal conditions showed that the water-cement ratio had a notable effect on the parameters investigated. At W/C = 0.30, the curing process was slower than at W/C = 0.35. Thus, the amount of clinker phases was rather higher, while the development of hydration products was insignificant. Also, free moisture in the pore space was higher. The high values of elastic moduli were provided by the higher content of clinker phases. The results of the VCCTL modeling of Portland cement curing were supported by the exponential function of the yield stress and Young's modulus with a high degree of reliability.

Acknowledgment
Authors wish to acknowledge the financial support from the Russian Foundation for Basic Research (Grant N 18-08-01025).

References
[1] Bullard J, Garboczi E and Stutzman P 2013 *VCCTL Software* In: Overview and Opportunities (Gaithersburg: National Institute of Standards and Technology, USA) pp 1–17
[2] Bentz D P 1995 *A Three-Dimensional Cement Hydration and Microstructures Program* Pt I Hydratation Rate, Heat of Hydration and Chemical Shrinkage, Report NISTIR 5756. (Gaithersburg: U.S. Department of Commerce) pp 1–47
[3] Bentz D P, Coveneyt P V, Garboczit E J, Kleynf M F and Stutzmant P E 1994 Cellular automaton simulations of cement hydration and microstructure development *Modelling Simul Mater. Sci. Eng.* **2** 783-808
[4] Bentz D P 2000 *CEMHYD3D: A Three-Dimensional Cement Hydration and Microstructure Development Modelling Package Version 2.0,* Report NISTIR 6485, U.S. Department of Commerce
[5] Bentz D P, Haecker C J, Feng X P and Stutzman P E 2003 Process Technology of Cement Manufacturing *Proc. 5th Int. VDZ Cong. on Process Technology of Cement Manufacturing* (Düsseldorf: Germany) pp 53–63
[6] Stutzman P E, Bullard J W and Feng P 2015 *Quantitative Imaging of Clinker and Cement Microstructure*, Report NISTIR 1877, U.S. Department of Commerce
[7] Taylor H F W 1996 Cement Chemistry (Moscow: Mir) 560 p
[8] SNiP 310.3-76 'Cements. Definition of normal density, setting time and uniform volume distribution' (in Russian)
[9] SNiP 22685-89 'Molds for concrete test samples' (in Russian)
[10] Gnyrya A I, Abzaev Yu A, Korobkov S V and Gauss K S 2018 *Proc. 5th All-Russ. Conf. on Polar Mechanics (Novosibirsk, Russia)* In: IOP Conf. Series: Earth and Environmental Science pp 1–5, doi: 10/1088/1755-1315/193/1/012010/
[11] Abzaev Yu, Gnyrya A, Korobkov S, Gauss K, Boyarintsev A and Tomrachev S 2019 *Proc. 15th Int. Conf. of Students and Young Scientists Prospects of Fundamental Sciences Development* In: Journal of Physics: Conf. Series **1145** 012016 pp 1-8, doi:10.1088/1742-6596/1145/1/012016/
[12] Erdogan S T, Garboszi E J and Bullard J W 2008 *Proc. Int. Conf. on Durability of Building Materials and Components* (Istanbul – Turkey) pp 1–8

Deformability criteria of metal at uniaxial tension

A M Zharmukhambetova[1] **and S A Barannikova**[1,2]

[1] Institute of Strength Physics and Materials Science SB RAS, Tomsk, Russian
[2]Tomsk State University of Architecture and Building, Tomsk, Russian Federation

E-mail: zharmukhambetova@gmail.com

Abstract. This paper is devoted to study of the laws of metals deformation behavior determined by carrying out standardized uniaxial tensile tests. The analysis of deformability of materials was carried out. The metals vary in yield limit from 40 to 1050 MPa and in ultimate strength from 460 to 1240 MPa. The ratio of the deformation work of uniform plastic deformation (from the yield point to the ultimate strength of materials) to the volume of the material was determined to assess the mechanical deformability. An inverse relationship between uniform deformation work and ratio of yield limit to ultimate strength was established. It was also found that the normalized specific work of deformation linearly increases with strength. The deformability criterion can be considered as the universal criterion for ferrous and non-ferrous metals and its magnitude can be estimated from known quantity of the ultimate strength. In practice, the numerical values of compliance criterion can be used to predict the behavior of materials with different strength under mechanical perturbations in the course of operation and machining (metal forming and cutting).

1. Introduction

One of the burning issues is development of reliable methods for assessing the carrying capacity of constructions and for defining the remaining life of technical systems, which have been in service for long. However, problem solution for precluding accidents is never 'cut and dried matter'; it calls for a fresh approach to this urgent problem which would combine coordination and systematization of studies in this sphere.

Material fracture problem is a complex of scientific and technical issues. The beginning of active research in this sphere dates back to A. Griffiths whose followers were E. Orowan, J. Irwin and others [1-3]. Due to a joint effort of the workers G.I. Barenblatt [4], three basic sets of fracture criteria were developed in strength physics and fracture mechanics: these are energy, force and deformation criteria. Traditionally, the threshold force or energy characteristics of material response to the external action are used as probable criteria of strength and fracture of solids. Among these are the following characteristics, which can be determined experimentally: yield limit, ultimate strength, impact ductility, etc., as well as critical values of crack tip stress intensity factor for normal cleavage and in-plane and out-of plane shear. There are also other criteria of crack propagation, which take into account the structural characteristics of a solid in combined stressed state [5, 6].

The application of new materials in machine building, modernization of technologies of material treatment and enhancement of the performance of mechanical equipment – all this is directly related to the advancement of science in the sphere of material strength and endurance of constructions. The structure and substructure of metallic materials and the kinetics of structural changes are investigated employing a wide range of methods, i.e. X-ray diffraction methods developed by Berg-Barret, Shultz,

Fujivar and Lang; method of moving X-ray detector, diffractometry of residual stresses, etc. Thus X-ray methods were used to examine test samples under special loading conditions. The data obtained on the dislocation and block-type structure of metallic materials is of basic importance; it enables gaining an insight into the nature of plastic deformation in metallic materials. Moreover, X-ray methods are found to be an effective tool for assessing the state and endurance of construction parts. Of the many challenges facing researchers is the development of an innovative approach to the problem of strength and fracture on the base of dependable criteria and trustworthy coefficients.

The basic assumption that crack occurrence in the loaded solid is implicit has to be reexamined, since it puts limitations on this approach by precluding the issue of crack nucleation, which was studied in sufficient detail over the last few decades by fracture physics [7]. In accordance with the conventional concepts of physics and mechanics of fracture, it is implicitly assumed that the stress-strain state of material would remain homogeneous on the macro-scale up to the point of main crack development or the onset of necking in plastic materials. However, the experimental evidence obtained by a number of workers strongly suggests that the plastic flow in metals would exhibit an inhomogeneous behavior at different stages of loading [8].

A comprehensive research of plastic deformation macrolocalization was performed for a range of metals and alloys in [9-12]. It is found that after the attainment of yield limit, the plastic flow would exhibit an inhomogeneous behavior; the inhomogeneities have characteristic size comparable to the deforming solid dimensions. The macroscopic inhomogeneity is a specific feature of plastic deformation development, which has been observed for all the studied metals and alloys by all types of loading. Of particular note is the space-time periodicity of localization patterns observed for most deforming solids. All the types of macrolocalization patterns have been examined in sufficient detail. It is found that within the localized plasticity nuclei defect fracture development would occur at a higher rate relative to the rest of material volumes, so that the likelihood of crack nucleation in the same regions is high. This finding suggests that plastic flow inhomogeneities are directly related to the processes involved in crack nucleation and development at the pre-fracture stage.

Computational evaluation methods of the stress-strain state of materials and products are often complex in dealing with problems of stress and strain distribution in samples or real products of complex shape especially when the material has an internal structure. This structure can change during the deformation due to successive processes of formation and relaxation of stress concentrators [13]. As a result of mechanical tests, stress-strain curves are obtained, which qualitatively and quantitatively characterize the material's response to loading, and determine the strength and ductility of materials. This work is devoted to the study of the laws of the deformation behavior of metals, determined in standard uniaxial tension tests. The basis for evaluation of the deformability criteria will be deformation work, determined by the area of the tension diagram.

2. Experimental part

The uniaxial tension tests were performed on universal test machine (Instron-1185) at a constant speed of gripping $V_m = 0.2$ mm / min. A wide range of metals and alloys (Cu; γ-Fe; Ni; Sn; Nb; Zn; Mg; V; α-Fe; Zr; Al; Ti; Mo) with various characteristics of durability and plasticity were studied and tested. The load curves of the studied materials are presented in Figure 1. In the thermodynamic terms the macroscopic result of deformation (changes in shape and size) is a dissipative process, that is the transition of a part of the kinetic energy to the internal energy of the deformable metal. It is realized in the formation of a certain internal structure [14 - 17]. Furthermore, the amount of energy absorbed during deformation is determined by the work of deformation, and absorption rate is defined as the deformation resistance [18]. In the case of tension test, the deformation work is numerically equal to the surface area under stress-strain curve, and is the sum of the work of elastic and plastic deformation.

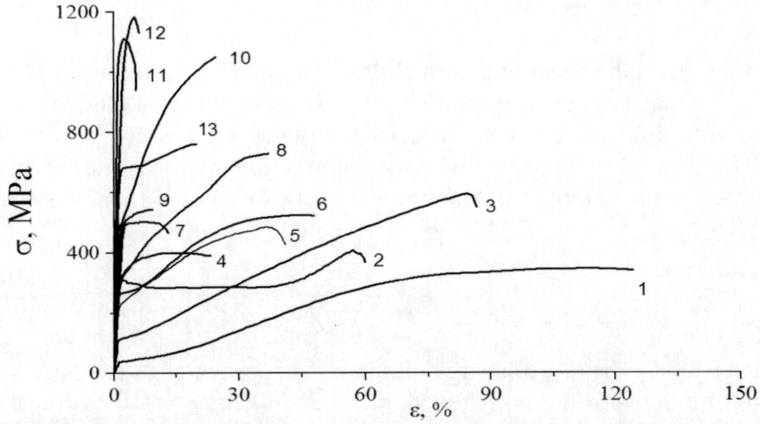

Figure 1. Deformation curves of the test materials: *1 – Cu; 2 – γ-Fe; 3 – Ni; 4 – Sn; 5 – Nb; 6 – Zn; 7 – Mg; 8 – V; 9 – α-Fe; 10 – Zr; 11 – Al; 12 – Ti; 13 – Mo.*

3. Results and discussions

The stress-strain curves (Figure 1) obtained as a result of mechanical tests were fitted in the form of fourth-order polynomials by using software package Table Curve:

$$\sigma_i = a_0 + a_1 \cdot \Delta\varepsilon_i + a_2 \cdot \Delta\varepsilon_i^2 + a_3 \cdot \Delta\varepsilon_i^3 + a_4 \cdot \Delta\varepsilon_i^4 \quad (1)$$

At least 15 points were used. The polynomial was used as an integrand to calculate the surface area under stress-strain curve, i.e. the deformation work. An inverse proportion between the deformation work on a field of uniform plastic deformation and the ratio of the yield limit to the ultimate strength of test materials was established and the correlation coefficient is 0.83 (Figure 2a).

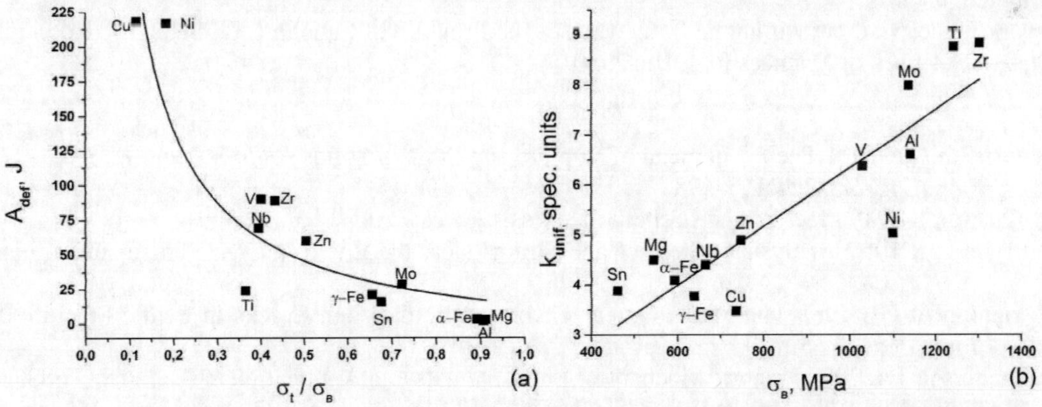

Figure 2. The dependence of the deformation work on the ratio of yield limit σ_t to ultimate strength σ_B (a) and the dependence of the coefficient k_{unif} on the ultimate strength (b).

To estimate the mechanical deformability, the ratio of deformation work (from yield limit to ultimate strength of materials) to the deformable volume calculated as the product of the initial area and working length of tested sample was determined. Thus, the specific work of uniform plastic deformation was determined as:

IOP Conf. Series: Materials Science and Engineering **597** (2019) 012039 doi:10.1088/1757-899X/597/1/012039

$$\alpha_{unif} = {A_{unif}}/{V_{def}} \tag{2}$$

The specific work reflects the material deformability in terms of the limiting work that is irreversible for shaping of the given metal in the individual stages [17], as such it quantitatively reflects the energy dissipation. Another reflection of the material deformability may be the "compliance" proposed in [18] as a resistance to deformation under the condition the deformation increases due to external factors. This quality of metal is defined as the ratio of the specific work to the deformation corresponding to the ultimate strength:

$$k_{unif} = {\alpha_{unif}}/{\varepsilon_B} \tag{3}$$

It is actually characterized by the rate of energy dissipation under uniaxial tension.

A linear relationship between the normalized specific work k_{unif} and the ultimate strength of the studied metals σ_B was established with correlation coefficient of 0.9 (Figure 2 b). The quantity of specific energy, in units mJ/mm^3, required to increase the true strain ε by 1% is taken as the conventional units of the coefficient k_{unif}.

4. Conclusions

The analysis of deformability of materials was carried out. Noted that the yield limit is 40–1050 MPa, and the ultimate strength is 460–1240 MPa for all studied materials used for construction products. It was established that the specific work of concentrated deformation normalized to the corresponding deformation k_{unif} increases linearly with increasing strength. In the applied aspect, the numerical values of k_{unif} can be used to predict the behavior of materials with different strength under mechanical perturbations in the course of operation and machining (metal forming and cutting).Taking into account the wide range of material's strength, the deformability criterion (Figure 2b) can be considered as the universal criterion for ferrous and non-ferrous metals and the magnitude of k_{unif} can be estimated from known quantity of the ultimate strength.

Acknowledgements

The work was carried out within the framework of the Program of Fundamental Scientific Research of the State Academies of Sciences for 2013-2020.

References

[1] Griffith A A 1921 The Phenomena of Rupture and Flow in Solids *Philos. Trans. R. Soc. A Math. Phys. Eng. Sci.* **221** 163–198

[2] Orowan E 1949 Fracture and strength of solids *Reports Prog. Phys.* **12** 309

[3] Irwin G R 1957 Analysis of Stresses and Strains Near the End of a Crack Traversing a Plate *J. Appl. Mech.* **24** 361–364

[4] Barenblatt G I 1962 The Mathematical Theory of Equilibrium Cracks in Brittle Fracture *Adv. Appl. Mech.* **7** 55–129

[5] Anderson T L 2007 Fracture Mechanics: Fundamentals and Applications (CRC Press) 688

[6] Rabotnov Y N, Vasilchenko G S, Koshelev P F and Merinov G N 1978 Brittle fracture design of structures *Applications and Non-metals* (Elsevier) pp 765–772

[7] Indenbom V L 1992 The Microscopic Theory of Cracks *Mod. Probl. Condens. Matter Sci.* **31** 253–268

[8] Walgraef D and Aifantis E C 2012 On certain problems of deformation-induced material instabilities *Int. J. Eng. Sci.* **59** 140–155

[9] Barannikova S A, Ponomareva A V, Zuev L B, Vekilov Y K and Abrikosov I A 2012 Significant correlation between macroscopic and microscopic parameters for the description of localized plastic flow auto-waves in deforming alloys *Solid State Commun.* **152** 784–787

[10] Zuev L and Barannikova S A 2011 Plastic Flow Localization Viewed as an Auto-Wave Process Generated in Deforming Metals *Solid State Phenom.* **172–174** 1279–83

[11] Barannikova S A, Bochkareva A V., Lunev A G, Shlyakhova G V. and Zuev L B 2016 Changes in ultrasound velocity in the plastic deformation of high-chromium steel *Steel Transl.* **46** 552–557

[12] Zuev L B, Barannikova S A, Lunev A G, Kolosov S V. and Zharmukhambetova A M 2019 Basic Relationships of the Autowave Model of a Plastic Flow *Russ. Phys. J.* **61** 1709–17

[13] Erdogan F 2000 Fracture mechanics *Int. J. Solids Struct.* **37** 171–183

[14] Kostina A A, Bayandin Y V. and Plekhov O A 2014 Model of energy accumulation and dissipation in plastically deformed metals *Phys. Mesomech.* **17** 43–49 in Russian

[15] Pashinskaya E H, Tkachenko V M and Zavdoveev A V 2016 Energy dissipation during the tension of the st3 steel produced by severe plastic deformation by the rolling with shear *High Press. Phys. Technol.* 20–26 in Russian

[16] Bugrov Y V 2012 Determination of the specific work of plastic deformation under tension of metals *Zavod. Lab., Diagnost. Mater* **78** 66–68 in Russian

[17] Gliner R E, Dubinskii V N, Katyukhin E B and Pryanichnikov V A 2017 Assessment of steel deformability in terms of energy dissipation *Steel Transl.* **47** 589–593

[18] Smirnov-Alyaev G A 1978 *Resistance of materials to plastic deformation* (Leningrad: Mashinostroenie) 54 in Russian

IOP Conf. Series: Materials Science and Engineering **597** (2019) 012040 doi:10.1088/1757-899X/597/1/012040

Rotational analysis of the inversion-vibrational spectra of the $^{15}NHD_2$ molecule: ν_6 fundamental band

A L Fomchenko, A S Belova and A V Kuznetsov

Research School of High-Energy Physics, National Research Tomsk Polytechnic University, Tomsk, 30 Lenin av., 634050, Russia

E-mail: fomchenko@tpu.ru

Abstract. This paper considers an assignment of the high-resolution Fourier spectra of $^{15}NHD_2$ molecule in the region of ν_6 band. The analysis is based on the combination differences method. As a result of the analysis, the energy structure of inversion-vibrational states (v_6=1, *s*) and (v_6=1, *a*) was obtained. More than 550 vibration-rotation transitions were found and values of about 140 upper vibrational energy levels were determined in total.

1. Introduction

High-resolution infrared and microwave spectra of molecules are a source of important information about the internal structure and physical properties of molecules. One of the most common methods for analyzing experimental rotational spectra is based on using the effective Hamiltonian model. The use of such a model makes it possible to determine many spectroscopic constants, primarily rotational constants, which contain information about the structure of the molecule. In addition, solving an inverse spectroscopic problem with an effective Hamiltonian makes it possible to obtain centrifugal distortion parameters. A complete analysis of the rotational spectrum of the ground and lower excited vibrational states gives accurate information about the energies of the interacting vibrational states, as well as more detailed information about the corresponding intramolecular interactions occurred because of the resonances.

Present study is devoted to the high resolution study of di-deuterated ^{15}N-ammonia species. During the decades, gaseous ammonia is known to have played a key role in atmospheric chemical processes and in the bio-, geo-chemical processes that occur in sensitive ecosystems (forests, soils, streams, and coastal waters) after the deposition. Astrophysical monitoring of ammonia allows one to recognize both main mother molecule and its isotopically substituted species as an important gas tracers for characterizing different evolutionary stages before and after the formation of stars [1-4]. Thus, high resolution study of ammonia and its isotopologues gives possibility to supplement the line lists needed and used for astrophysical applications. From the theoretical point of view, the inversion at the pyramidal nitrogen atom is among the most interesting theoretical problems in intramolecular dynamics and, in particular, tunneling dynamics, including the possibility of molecular chirality with appropriate substitution [5].

At present, there are already data on studies of the ammonia molecule and its isotopologues (for example, [6-9]), some of which are presented in the HITRAN database. However, for a complete and correct description of the intramolecular structure of ammonia, it is necessary to have data on all its isotopologues. In this article, we continue the analysis of the ammonia molecule.

2. Experimental part

The high-resolution absorption spectra of the $^{15}NHD_2$ molecule were recorded using the Bruker IFS 125HR Fourier transform spectrometer at LISA in Créteil (France). Experimental setup one can see in Table 1 for both spectra. Due to the rapid exchange of H / D, spectra have a significant number of lines belonging to the isotopologues of $^{15}NH_2D$, $^{15}NHD_2$, $^{15}NH_3$ and $^{15}ND_3$. The spectra were recorded at a stabilized room temperature of (297 ±1) K and calibrated with CO_2 and H_2O lines. The resulting accuracy of recorded spectra is ±0.0003 cm^{-1} for well isolated lines.

Table 1. Experimental setup for the spectral region 1000 – 1800 cm^{-1}.

Spectrum	Resolution, cm^{-1}	Optical path length, m	Pressure, Pa	Temperature, K	Measuring time, h	No of scans	Calibration gas
I	0.004	5.649	237.5	297	8	320	CO_2, H_2O
II	0.004	5.649	52.1	297	10.5	420	CO_2, H_2O

3. Theoretical part

The $^{15}NHD_2$ molecule is an asymmetric top with values of the three rotational parameters close to 7.45, 5.37, and 3.80 cm^{-1}. $^{15}NHD_2$ is the deuterated isotopologue of the NH_3 molecule, which is very well studied in ro-vibrational spectroscopy. For that reason, the most of the effects and features, which are peculiar to the NH_3 molecule, can also be found in the $^{15}NHD_2$ one. It is, first of all, inversion motion in the ammonia molecule which, in combination with the isotopically substituted configuration of $^{15}NHD_2$, leads to a very complicated picture of the ro-vibrational spectra.

Figure 1 shows equilibrium configuration of mother molecule NH_3 and isotopologue $^{15}NHD_2$. Taking into account the numbering of nuclei and orientation of axes in the $^{15}NH_2D$ molecule one can easily estimate the values of the equilibrium inertia momenta of the $^{15}NH_2D$ molecule and to show that the following relation is fulfilled: $I^e_{zz} > I^e_{xx} > I^e_{yy}$.

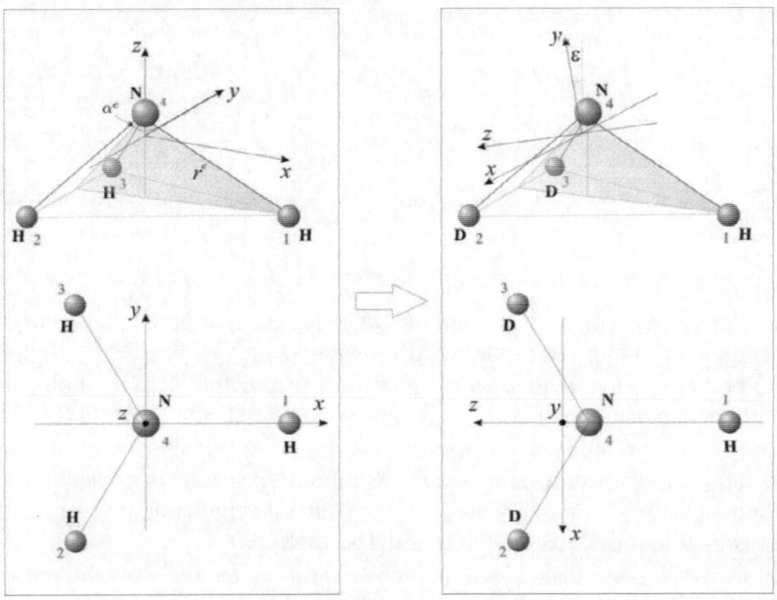

Figure 1. Equilibrium structure of ammonia molecule NH_3 and isotopologue $^{15}NHD_2$.

At the same time, the mostly efficient and the widely used model of the Hamiltonian of asymmetric top molecule (the *A*-reduction and I^r-representation) assumes that for the rotational parameters the following relations are fulfilled: $I^e_{yy} > I^e_{xx} > I^e_{zz}$. To satisfy these conditions in our case, it is necessary to replace the coordinate axis of the molecular fixed coordinate system: z axis by the y axis, and y axis by the z one.

To provide the assignment of spectra method of combination differences was used. Such an efficient method requires some theoretical calculations. First, one needs to know the energy structure of ground state of the studied molecule with high accuracy. As there is inversion motion in ammonia molecule, the ground state of $^{15}NHD_2$ consists of two sub-states, (GS, *s*) and (GS, *a*). The parameters of ground state were taken from [10].

Secondly, it is important to know the selection rules specific to the studied band. In accordance with the general principles of quantum mechanics (see, for example, [11]), the possibility of the transition from the quantum state | φ> to another quantum state | ϕ> is determined by the inequality to zero of the square of the matrix element of the dipole moment P_Z (here P_Z is the component dipole moment operator in the laboratory coordinate system) on the functions | φ> and | ϕ>. Due to the symmetric properties of the molecule and considering that operator of the dipole moment P_Z of a molecule is transformed in accordance with the irreducible representation A_2 of the C_{2v} group one can show that the ν_6 band consists of two sub-bands of the *b*-type: ν_6 $(s\leftarrow s)$ and ν_6 $(a\leftarrow a)$. So selection rules are: $\Delta J=0,\pm1$; ΔK_a=odd; ΔK_c=odd $(s\leftarrow s)$ or $(a\leftarrow a)$.

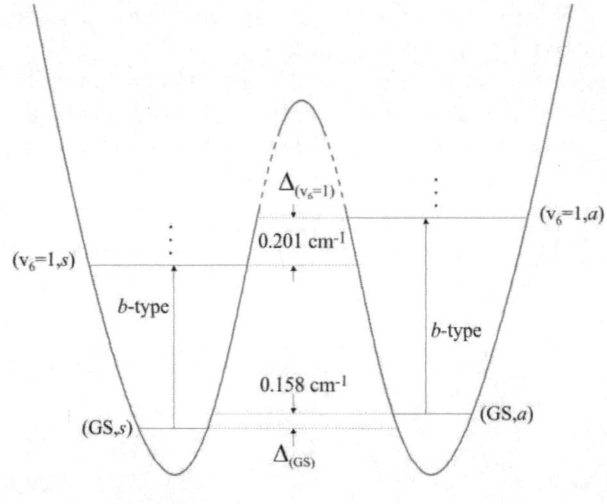

Figure 2. Scheme of possible transitions in the ν_6 band of the $^{15}NH_2D$ molecule. Types of possible rotation-inversion-vibration transitions are marked.

4. Results of analysis

To solve the inverse spectroscopic problem and obtain spectroscopic parameters, one needs to provide assignment of the inversion-vibrational-rotational transitions. In the part (a) of Figure 3 one can see the central part (Q-branch) of the ν_6 inversion–vibration band of the $^{15}NHD_2$ molecule. The presence of two strongly interacting sub-bands ν_6 $(s\leftarrow s)$ and ν_6 $(a\leftarrow a)$ leads to complicated picture of the spectrum with the absence of regular structure of transitions. One can see two closely located sub-band centres near 1459 cm^{-1}. A small part of the high-resolution spectrum is presented in the part (b) of Figure 3. The assignment is also complicated by the fact that the sample is consists of mixture of different ammonia isotopologues ($^{15}NH_3$, $^{15}ND_3$, $^{15}NH_2D$, $^{15}NHD_2$).

As the result of analysis, more than 550 transitions belonging to the two sub-bands ν_6 $(s\leftarrow s)$ and ν_6 $(a\leftarrow a)$ were assigned in the experimental spectrum. Correctness of assignment was confirmed by the presence of numerous combination differences. Small part of assigned transitions and energies of upper state is presented in Table 2. We obtained the values of about 140 upper energy levels of (ν_6=1, *s*) and (ν_6=1, *a*) states with the values of quantum numbers $J^{max.}$ / $K_a^{max.}$ equal to 11/6.

The obtained data will be used as an input in a weighted least square fit with the aim to determine rotational, centrifugal distortion, and resonance interaction parameters of the effective Hamiltonian. As an outcome the set of spectroscopic parameters will be obtained.

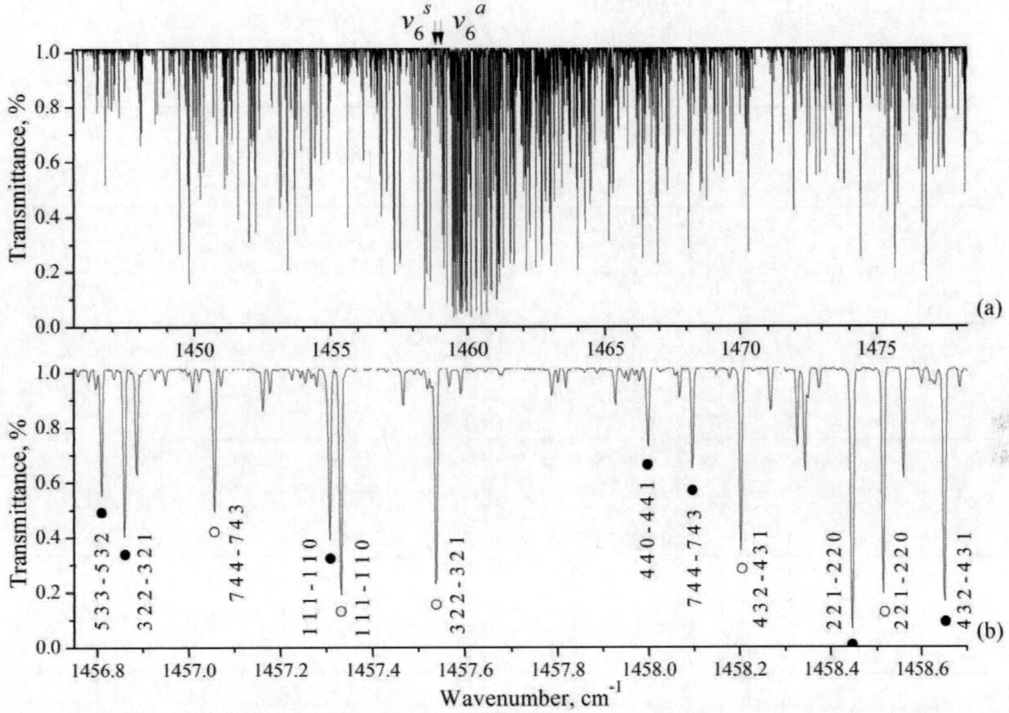

Figure 3. The central part (Q-branch) of the ν_6 inversion-vibration band of the $^{15}NHD_2$ molecule of the spectrum II (part a). Fragment of the high-resolution spectrum II in the 1456.75–1458.70 cm^{-1} region (part b). Lines belonging to the ν_6^s and ν_6^a bands are marked by dark and open circles, respectively. Quantum numbers J K_a K_c of upper and lower states are indicated.

Table 2. Small part of assigned transitions and energies of (v_6=1, s) and (v_6=1, a) states.

Quantum numbers of upper state			Quantum numbers of groung state			Wavenumber	Transmittance	Energy	Spectrum	State
J	K_a	K_c	J	K_a	K_c	cm^{-1}	%	cm^{-1}		
2	2	1	2	2	0	1458.4450	6.2	1497.7789	II	(v_6=1, s)
			2	0	2	1471.1693	57.9			
			3	2	2	1431.8411	19.7			
3	2	2	2	2	1	1486.7880	46.6	1525.4895	II	(v_6=1, s)
			3	2	1	1456.8601	38.9			
			3	0	3	1473.7093	86.6			
			4	2	3	1423.7597	23.7			
4	3	2	3	2	2	1495.9239	12.2	1561.8615	II	(v_6=1, s)
			4	2	2	1453.6901	33.2			
			4	0	4	1477.5766	73.8			
			5	2	4	1416.1519	8.6			
5	2	4	4	2	3	1504.1970	29.9	1605.9269	II	(v_6=1, s)
			5	2	3	1448.5886	73.5			
			5	0	5	1481.7664	90.3			
			6	2	5	1408.3765	29.7			

IOP Conf. Series: Materials Science and Engineering **597** (2019) 012040 doi:10.1088/1757-899X/597/1/012040

6	2	5	5	2	4	1512.7898	9.3	1658.4992	II	(v_6=1, *s*)
			6	2	4	1443.3029	56.0			
			6	0	6	1487.0378	81.8			
			7	2	6	1401.4713	9.6			
7	2	6	6	2	5	1520.9864	31.1	1718.5367	II	(v_6=1, *s*)
			7	2	5	1437.8203	78.6			
			7	0	7	1492.3315	93.7			
			8	2	7	1394.5185	34.4			
3	3	1	3	3	0	1459.2608	8.9	1540.1487	II	(v_6=1, *a*)
			3	1	2	1478.3809	94.6			
			4	3	2	1422.2266	47.3			
4	3	2	3	3	1	1496.4982	54.6	1577.1957	II	(v_6=1, *a*)
			4	3	1	1458.2028	34.2			
			4	1	3	1477.7535	94.7			
			5	3	3	1412.9429	37.6			
5	3	3	4	3	2	1505.4160	18.7	1623.3382	II	(v_6=1, *a*)
			5	3	2	1455.6316	36.4			
			5	1	4	1478.5716	84.1			
			6	3	4	1403.9847	18.9			
6	3	4	5	3	3	1514.1296	40.6	1678.3826	II	(v_6=1, *a*)
			6	3	3	1451.3903	76.5			
			6	1	5	1481.0923	87.9			
			7	3	5	1395.5711	47.4			
7	3	5	6	3	4	1522.7468	20.7	1742.1002	II	(v_6=1, *a*)
			7	3	4	1445.9450	73.1			
			7	1	6	1485.0609	80.0			
			8	3	6	1387.8546	29.3			

5. Conclusion

We performed the first analysis of the high resolution IR spectra of the $^{15}NHD_2$ molecule in the region of 1000 – 1800 cm^{-1}, where the fundamental v_6 band is located. More than 550 transitions belonging to this band were assigned in the experimental spectra with the maximum values of quantum numbers $J^{max.}/K_a^{max.}$ equal to 11/6. Information obtained from the experimental data will be used as initial data in the weighted fit procedure for determination of spectroscopic parameters of (v_6=1, *s*) and (v_6=1, *a*) states.

Acknowledgments

The research was funded by the Russian Science Foundation, project 18-72-00032.

References

[1] Taban I M, Schutte W A, Pontoppidan K M, Van Dishoeck E F 2003 *Astron. Astrophys.* **399** 169–175A reference

[2] Hargreaves R J, Li G, Bernath P F 2011 *Astrophys. J.* **735** 111–21

[3] Busquet G, Palau A, Estalella R, Girart J M, Sanchez-Monge A, Viti S, Ho P T P, Zhang Q 2010 *Astron. Astrophys.* **517** L1–4

[4] Tine S, Roue E, Falgarone E, Gerin M, Pineau des Forets G 2000 *Astron. Astrophys.* **356** 1039–49

[5] Fehrensen B, Hippler H, Quack M 1998 *Chem. Phys.* **298** 320–8

[6] Cane E, Di Lonardo G, Fusina L, Nivellini G, Tamassia F, Villa M 2013 *J. Quant. Spectrosc. Radiat. Transfer* **119** 1–11

[7] Irwin P G J, Calcutt S B, Sihra K, Taylor F W, Weir A L, Ballard J, Johnston W B 1999 *J. Quant. Spectrosc. Radiat. Transfer* **62** 193–204

IOP Conf. Series: Materials Science and Engineering **597** (2019) 012040 doi:10.1088/1757-899X/597/1/012040

[8] Yurchenko S N, Barber R J, Tennyson J 2011 *Mon. Not. R. Astron. Soc.* **413** 1828–34
[9] Ulenikov O N, Bekhtereva E S, Gromova O V, Fomchenko A L, Kwabia Tchana F 2017 *J. Quant. Spectrosc. Radiat. Transfer* **202** 210–9
[10] Fomchenko A L, Kuznetsov A V, Bekhtereva E S 2017 *Rus. Phys. J.* 60 279–83
[11] Ballentine L E 2000 *Quantum mechanics: A modern development* (Singapore: World Scientific Publishing Co. Pte. Ltd.)

IOP Conf. Series: Materials Science and Engineering **597** (2019) 012041 doi:10.1088/1757-899X/597/1/012041

Effect of gold and nickel co-additives on gas-sensitive characteristics of SnO_2 thin-film on exposure to hydrogen and nitrogen dioxide

A V Almaev[1,2], N N Yakovlev[1] and E V Chernikov[3]

[1]Department of Semiconductor Electronics, Tomsk State University, Tomsk634050, Russia
[2]Laboratory of Functional Electronics, Tomsk State University, Tomsk634050, Russia
[3]Laboratory of Physics of Semiconductor Devices, Kuznetsov Siberian Physical Technical Institute, Tomsk State University, Tomsk634050, Russia

E-mail: nik_mr_x@mail.ru

Abstract. The results of investigation of the gas-sensitive properties of sensors based on the tin dioxide thin films with combined additives of gold and nickel obtained by the DC - magnetron sputtering are presented. The investigated sensors are characterized by a high response to low concentrations of NO_2 of 0.45 – 10.23 ppm at temperatures of 50 – 150 °C with response time of 10 s. The sensitivity of sensors to hydrogen appears at the temperature of 250 °C. The hydrogen sensors are characterized by high reproducibility of the measurement results. The obtained results are explained by the synergistic effect of gold and nickel additives, as well as the ability of the Ni to prevent the interaction of hydrogen with lattice oxygen atoms in the subsurface part of tin dioxide.

1. Introduction

The development of air monitoring systems and methods for monitoring the human exhaled mixture require the creation of sensors of low concentrations of gases characterized by high sensitivity, selectivity, stability of parameters and the low energy consumption. The most promising direction in this area is the development of sensors based on thin films of metal oxide semiconductors. This work is devoted to thin-film gas sensors with a sensitive layer based on polycrystalline tin dioxide with gold and nickel additives. Hydrogen and nitrogen dioxide were chosen as detected gases.

Hydrogen sensors based on SnO_2 thin films can replace H_2 leakage monitoring systems in nuclear power plant premises and become the basis of safety for hydrogen energy. For these purposes, it is sufficient to develop high-speed devices that are sensitive to the effects of a gas concentration equal to 0.1–0.2 of the lower explosive limit (LEL). According to the state standard for hydrogen LEL is 4 vol. %, so conversion into part per million equals $4 \cdot 10^4$ ppm. The more difficult task is the development of measuring instruments for low hydrogen concentration below 10^3 ppm. Such devices are of interest for laboratory and medical purposes. In particular testing of a mixture exhaled by a person for the presence of hydrogen at concentrations from 5 to 120 ppm is the "gold standard" for the diagnosis of lactose, fructose, sorbitol and xylitol intolerance [1,2]. The increased concentration of hydrogen in the exhaled mixture also indicates the presence of the syndrome of excessive growth of bacteria in the small intestine, cirrhosis of the liver, problems with the pancreas [2].

IOP Conf. Series: Materials Science and Engineering **597** (2019) 012041 doi:10.1088/1757-899X/597/1/012041

Interest in the development of sensors of low concentrations of nitrogen dioxide is due to the need for environmental monitoring of urban air. According to sanitary norms, the maximum permissible concentration of the working area of this gas is 1 ppm. In addition, the detection of nitrogen oxides at a level of 10 ppb permits to detect the traces of explosives in the air. Such sensitivity is achieved by an expensive and cumbersome system based on the lidar principle using the laser fragmentation/laser-induced fluorescence method [3]. Cheap and mobile nitrogen oxides sensors are needed for the broad introduction of such systems for monitoring and security.

The principle of operation of gas sensors is based on the phenomenon of reversible chemisorption of gases on the surface of SnO_2. Sensors based on "pure" SnO_2 are characterized by low gas responses and long response times [4, 5]. The sensor response to hydrogen is the ratio G_H/G_0, where G_H is the conductance of the sensor in the atmosphere containing hydrogen; G_0 is the conductance of the sensor in the atmosphere of clean air. Hydrogen is a reducing gas and its chemisorption on the surface of SnO_2 electrons receive to the conduction band of the semiconductor and the conductivity of the sensor increases. As a response of the sensor to the effect of nitrogen dioxide we take the ratio G_0/G_{ND}, where G_{ND} is the conductance of the sensor in an atmosphere containing nitrogen dioxide. Nitrogen dioxide is an oxidizing gas and when NO_2 is chemisorbed to the SnO_2 surface, electrons are captured from the conduction band of the semiconductor and the conductivity of the sensor decreases. The response time t_{res} is the time interval when the conductance of the sensor reaches $0.9G_{gst}$ on exposure to H_2 and $1.1G_{gst}$ on exposure to NO_2, where G_{gst} – stationary conductance on exposure to a certain gas concentration. A promising way to improve the characteristics of the sensors is the modification of the bulk and surface of the sensitive layers with gold and nickel additives.

There is not general consensus on the question of the effect of an addition of nickel on the properties of gas sensors. The authors [6-10] came to the conclusion that modification of films and other nanoscale structures of tin dioxide with nickel contributes to increasing of sensor resistance in clean air and response to a reducing gases, the displacement of the operating temperature in the low-temperature range. The high resistance of the films is caused by the formation of deeper impurity levels in the forbidden gap of SnO_2, as well as an increase in the surface density of chemisorbed oxygen. The addition of nickel is presented in the form of oxide and on the border with SnO_2 microcrystals can form a heterojunction that prevents the percolation of electrons. As a result of that the sensors have high resistance. The increase in nickel concentration contributes to the formation of fine-grained developed structure of tin dioxide films. The addition of nickel not only can be integrated into the crystal lattice of SnO_2, but is able to form clusters with its own crystal structure.

Gold due to its properties is not able to form compounds with oxygen and other gases, so the influence of Au on the characteristics of sensors is explained only by its catalytic properties [11,12]. It is noted that metal clusters of gold activate special forms of chemisorbed oxygen O^{2-}in the three-phase boundary Au-SnO_2-gas, that at operating temperatures are not the main on the surface, but are extremely reactive with atoms of reducing gases. It should be noted that the sensitivity to NO_2 tin dioxide shows only with the introduction of gold in its structure [11,12].

According to [13-18], sensors based on metal oxide films and nanostructures can be used for NO_2 detection. The most common sensing elements of NO_2 sensors are thin WO_3 films that demonstrate sensitivity to gas at the room temperatures. However, to increase the response to the gas and reduce the response time, the semiconductor must be heated to high temperatures. To improve the performance of the sensors, the WO_3 sensing element is modified with gold. This allows to increase the response and reduce the response time. Other materials (ZnO, SnO_2:In) demonstrate a short response time and a high response when exposed to a concentration of NO_2 in the range of 10 – 100 ppm, that is not suitable for solving the problems described above. In [16] presents research of thick films SnO_2 modified by joint additives Au and NiO. Such films were obtained by precipitation from the solution. It is shown that the combined introduction of Au and Ni additives significantly increases the response of sensors to NO_2 and reduces their operating temperature *T*. Mechanism of this synergistic effect of additives is unclear and requires further research. Study the system Au–Ni/SiO_2 by the method of x-ray photoelectron spectroscopy has shown [16] that joint presence of Au^0 and Ni^{2+} leads to a partial flow of the charge

IOP Conf. Series: Materials Science and Engineering **597** (2019) 012041 doi:10.1088/1757-899X/597/1/012041

from the oxidized surface of nickel atoms and the appearance of additional charge on the Au nanoparticles, that is manifested in the shift of position of the electronic Au 4*f* level towards higher energies. Selective adsorption on charged centers can provide a significant effect in changing the electrical conductivity of materials. Most of the results are devoted to sensors made by thick-film technology that has a number of inconveniences, for example, does not allow combining sensitive elements with microelectronic technology.

Thus, the aim of this work is to research the gas-sensitive properties of SnO_2 thin films modified with joint additives Au and Ni under the influence of low concentrations of hydrogen and nitrogen dioxide.

2. Experiment

SnO_2 films were obtained by direct current magnetron sputtering of a target consisting of alloy Sn and Sb (Sb – 0.49 at.%). To introduce additives into the bulk of films pieces of the corresponding metal were placed on the surface of the target. On the basis of special studies, the optimal ratio S_m/S_{Sn} was estimated, where S_m is the area of metal pieces, S_{sn} is the area of the sprayed part of the target. For the studied samples S_{Au}/S_{Sn}=3·10^{-3} and S_{Ni}/S_{Sn}=3·10^{-3}. On the surface of the films by the same method of magnetron sputtering dispersed Au layers were sprayed for 30 s. The finished samples were annealed in air at the temperature of 723 K for 24 hours.

To research the electrophysical and gas-sensitive parameters of the samples the installation described in [19] was used. The conductivity of the sensors was measured using an automated stand with the temperature controller of the devices of the devices and the humidity of the gas-air mixture in the measuring chamber. The measuring stand provides measurement of conductivity with an error of 1 %, measurement of the level of relative humidity with an error of 3 %, the temperature of the sensors with an accuracy of 0.5 °C. In this work, all studies were carried out at a relative humidity of 30 – 35 %.

3. Results and discussions

The introduction of Ni additives into the bulk of SnO_2 films leads to a significant increase in the sensor resistance in the atmosphere of pure air R_0 compared to films only with gold additives in the bulk and on the surface. So for thin-films Au/SnO_2:Sb,AuR_0 at room temperature is 0.25 MΩ , and for Au/SnO_2:Sb,Au,Ni – 2.5 MΩ. The increase in film resistance is due to the formation of deeper levels localized in the band gap of the semiconductor. The $R_0(T)$=const·exp[$\Delta Ea/kT$] ratio is performed for partially compensated semiconductor in the sensor operating temperature range. Figure 1 shows the dependence oflnR_0 on the 1000/temperature for the samples. The dependence has a characteristic form of tin dioxide thin films. In the low temperature region there are two linear sections corresponding to donor levels in the forbidden zone of SnO_2. The activation energies of donor levels were calculated using linear approximation of these sections. For the studied sensors ΔEa_1 = 0.23±0.01 eV,ΔEa_2 = 0.50±0.01 eV. R_0 rises due to an increase in the surface density of chemisorbed oxygen in the form of O^- [10] with increasing the temperature from 210 °C. The growth of R_0 continues up to T=500 °C. The presence of such a large area of growth of R_0 due to an increase in the negative charge on the surface does not contradict the data of [20]. However, it can be concluded that the addition of nickel leads to growth of the surface density of oxygen ions due to the formation of high-temperature adsorption centers. According to many works [6-8], not only non-stoichiometric tin atoms, but also non-stoichiometric nickel atoms can act as adsorption centers.

Temperature dependences of sensor response at exposure to 100 ppm of H_2 and 10.23 ppm of $NO_2$are shown in figure 2. Sensors demonstrate high sensitivity to NO_2 at low operating temperatures range from 50 to 150 °C. The sensitivity of the sensors to hydrogen appears at higher temperatures. The response to hydrogen rises with increasing T and reaches a maximum at 500 °C. Based on the analysis of temperature dependences of response and resistance of sensors it can be concluded that the interaction of SnO_2 films with H_2 occurs with the participation of previously chemisorbed O^- ions that surface density increases with the introduction of nickel additives. As a result of the reaction between H_2 and chemisorbed O^- ions the negative charge on the SnO_2 surface decreases and the conductivity of the sensors increases. The mechanism of interaction between NO_2 and the surface of SnO_2 films modified with gold and nickel includes two processes: 1 – the formation of negatively charged nitrite ions by adsorption of NO_2 to the center and the capture of an electron from the conduction band of SnO_2; 2 – the interaction of NO_2 with chemisorbed molecular oxygen in the form of O^{2-}. In the first and in the second case NO_2^-appears on the surface, but as a result of the first process, a negative charge on the surface of the semiconductor increases and the conductivity of the sensors decreases.

It should be noted that the conductivity of the sensors after pumping of hydrogen at all operating temperatures is recovered to initial value. The heating of the sensor at T = 400 – 500 °C for 10 to 30 s is required to full recovering of sensor conductivity after exposure to NO_2. Strong links with the semiconductor surface are formed at chemisorption of the nitrogen dioxide molecules. So to clean the surface heating of the sensors is required. Water molecules and hydroxyl groups are formed on the surface at the hydrogen interacts with SnO_2, for desorption of these products of reactions temperatures above 200 °C are sufficient [21].

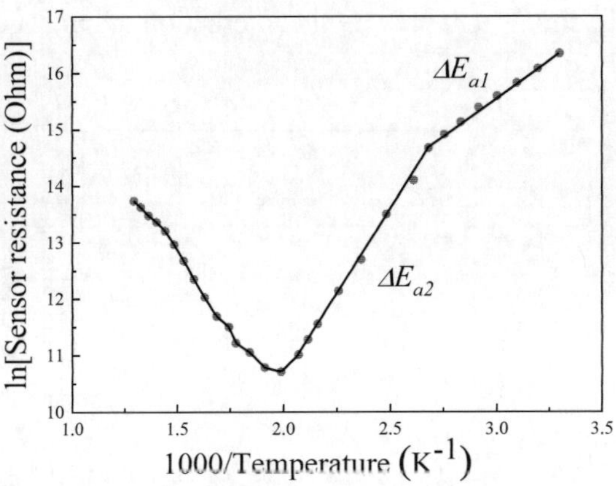

Figure1. Temperature dependence of sensor resistance in clean air.

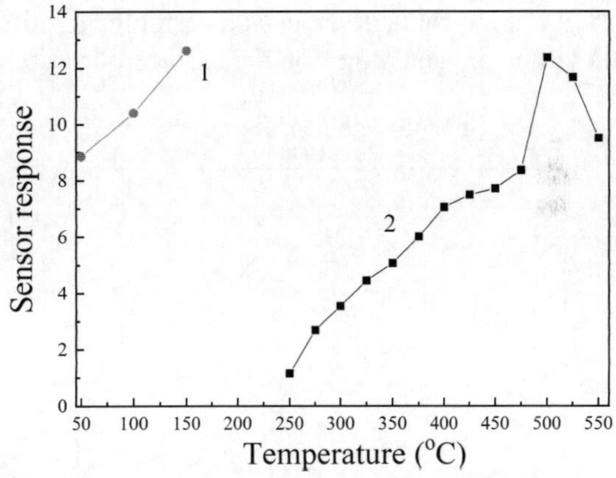

Figure 2. Temperature dependence of sensor response: 1 – at exposed to 10.23 ppm NO_2; cu. 2 – at exposed to 100 ppm H_2.

The dependences of the sensor response on the concentration of NO_2 at T=50 °C and the hydrogen concentration at T=425 °C are shown in figure 3. Analysis of the experimental data showed that G_0/G_{ND} with increasing NO_2 concentration rises exponentially with $G_0/G_{ND}=\exp(b_{ND}*n_{ND}+A_{ND})$, where A_{ND}andb_{ND} are constants; n_{ND} is the concentration of nitrogen dioxide. At T=50 °C A_{ND} = 0.15±0.03, b_{ND} = (0.199 ± 0.005) ppm^{-1}. The response of the sensor G_H/G_0 with increasing hydrogen concentration rises according to the power law $G_H/G_0=A_H*n_H^{bH}$, where A_H and b_H are constants, n_H is the concentration of hydrogen. At T=425 °C A_H = (0.29 ± 0.01) ppm^{-1}, b_H = 0.70 ± 0.01. The power dependence of the sensor response on the hydrogen concentration is observed in many works. For the same films, the exponential dependence of response on $n_{ND}$is observed for the first time. By increasing the operating temperature of the sensors to 100 – 150 °C, the dependence of G_0/G_{ND} on n_{ND} becomes linearthatis observed in many works [13-15] in this area of T.

Analysis of the concentration dependence of the response under the influence of hydrogen concentrations up to 1.4 vol. % showed that in the atmosphere of pure air the ratio of the width of the space charge region due to the chemisorption of oxygen and the width of the conducting channel SnO_2 is 0.95 that is a known condition for a high response to reducing gases [11].

The estimates showed that at T = 50 °C and n_{ND} = 0.45 ppm the response time of the sensors is 10 s. it is worth noting that in [13] the response times at exposure to 0.9 ppm NO_2 on samples based on tungsten trioxide and T = 230 °C were 115 – 400 s. For sensors based on thin films of SnO_2:In at T = 230 °C and n_{ND} = 50 ppm t_{res} = 115 s. [14], for sensors based on nanorods of ZnO at T = 230 °C and n_{ND} = 7.5 ppm t_{res} = 120 s. [15]. Also in the works response times – 55 and 26 s are specified at temperatures

140 and 160 °C for thin films SnO_2/C [17] and α-MnO_2 [18], respectively, but do not specify the gas concentration. It is known [21] that the response times are determined by two processes, the rapid adsorption of NO_2 molecules to the semiconductor surface and their long diffusion into the semiconductor bulk. At T = 50 °C diffusion is negligible. Given the large difference in t_{res}, it can be concluded that the introduction of Au and Ni additives compared to other modifiers is an effective way to improve the performance of sensors.

At exposure to 100 ppm hydrogen and T = 425 °C t_{res} of sensors studied was 23.8 s. This value is not outstanding in comparison with the literature data.

The results of the study of short-term reproducibility of sensor characteristics under the influence of 100 ppm H_2 and T = 425 ° C are shown in figure 4. The portion of the hydrogen was supplied to chamber after 20 minutes of the start of operation of the sensor. After establishing a stationary value of G_H clean air was pumped through the measuring chamber. G_0 and G_H values were measured every 20 minutes. From figure 4 it is seen that G_0 and G_H and the response of the sensors change slightly. As in [22], the stability of the characteristics is determined by nickel additives that prevent the interaction of hydrogen with lattice oxygen atoms on the surface of SnO_2.

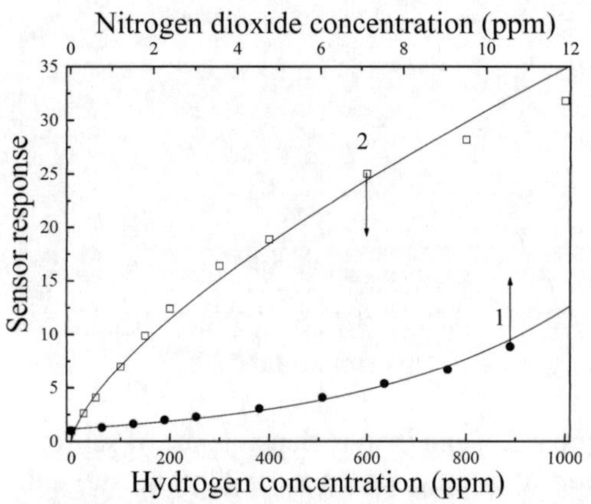

Figure 3. The concentration dependence of the sensor response: 1 – at T = 50 °C and exposure to NO_2; 2 – at T = 425 °C and exposure to H_2.

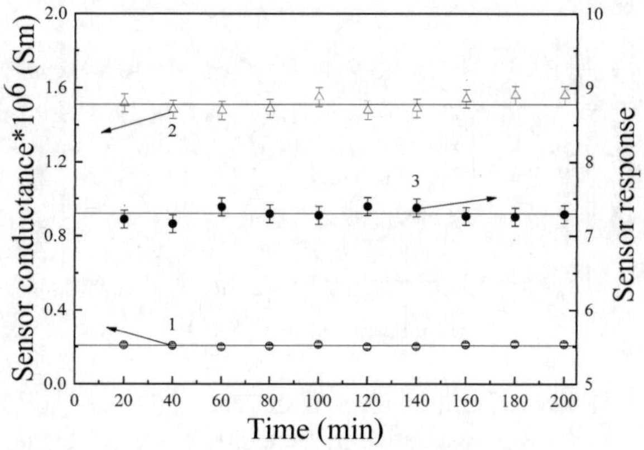

Figure 4. The reproducibility of the sensor characteristics at periodic exposure to 100 ppm H_2: 1 –G_0; 2 –G_H; 3 –G_H/G_0.

Studies of the effect of long-term tests on the conductivity of the investigated sensors in clean air, at n_H=100 ppm and at n_H=1000 ppm have shown that G_0 and G_H do not have any tendency to change with increasing of service life.

4. Conclusion

The mechanism of interaction between NO_2 and the surface of SnO_2 films modified with gold and nickel includes two processes: 1 – the formation of negatively charged nitrite ions by adsorption of NO_2 to the center and the capture of an electron from the conduction band SnO_2; 2 – the interaction of NO_2 with chemisorbed molecular oxygen in the form of O_2^-.

The joint modification of SnO_2 thin films with gold and nickel additives results in the sensitivity of sensors to NO_2 at low temperatures of 50 – 150 °C with response times at the level of 10 s. This behavior of sensors is due to the synergistic effect of additives. Gold, because of their catalytic properties increases the density of the adsorption centers on the surface of SnO_2 for NO_2, on the other hand super-stoichiometric nickel atoms are also the centers of adsorption for oxygen and NO_2.

The sensitivity of sensors to hydrogen appears at a temperature of 250 °C, the maximum response is observed at 500 °C. Hydrogen sensors based on SnO_2 thin films with gold and nickel additives demonstrate high reproducibility of characteristics due to the ability of Ni additives to prevent the interaction of hydrogen and lattice oxygen atoms in the near-surface area of tin dioxide.

References

[1] Katwal G, Rao B M and Varghese O K 2016 *Proc. IEEE Sensors (Orlando, FL, USA)*(IEEE) 16597370
[2] Eisenmann A, Amann A, Said M, Datta B and Ledochowski M 2008 *J. Breath Res.* **2** 046002
[3] Bobrovnikov S M, Vorozhtsov A B, Gorlov E V, Zharkov V I, Maksimov E M, Panchenko Yu N and Sakovich G V 2015 *Russ. Phys. J.* **58** 14
[4] BarbosaM S 2017 *Sens. Actuators* **239** 253
[5] Seal S 2002 *JOM* **54** 35
[6] Bagal L K, Patil J Y, Mulla I S and Suryavanshi S S 2012 *Ceram. Int.* **38** 6171
[7] Nguyen Van Hieu, PhungThi Hong Van and Le Tien Nhan 2012 *Appl. Phys. Lett.* **101** 253106-1
[8] Vlachos D S, Papadopoulos C A and Avaritsiotis J N 1996 *Appl. Phys. Lett* **69** 650
[9] Soleimanpour A M, Khare Sanjay V, Jayatissa and Ahalapitiya H 2012 *Mater. Today.* **4** 4651
[10] McAleer J F, Moseley P T, Norris J O W and Williams D E 1987 *J. Chem. Soc.* **83** 1323
[11] Sevastyanov E Y, Maksimova N K, Chernikov E V, Novikov V A, Rudov F V and Sergeychenko N V 2012 *Semiconductors.* **46** 801
[12] Korotcenkov G, Brinzari V and Cho B K 2016 *Microchim Acta.***183** 1033
[13] Anisimov O V, Maksimova N K, SevastYanov E Yu, Chernikov E V, Naiden Yu P, Novikov V A and Rudov F V 2010 *J. Phys. Chem.***84** 1220
[14] Kaur J, Roy S C and Bhatnagar M C 2007 *Sens. Actuators* **126** 478
[15] Wang X, Sun F, Duan Y, Yin Z, Luo W, Huang Y A and Chen J 2015 *RSC***3** 11397
[16] Badalyan S M, Alikhanian A S,Rumyantseva M N, Nikolaev S A, Marikutsa A V, Smirnov V V and Gaskov A M 2010 *Inorg. Mater.* **46** 232
[17] Zhang R, Liu X, Zhou T, Wang L and Zhang T 2018 *J. Colloid Interface Sci.* **524** 76
[18] Pritamkumar V 2018 *Appl. Surf. Sci.* **442** 178
[19] Gaman V I, Maksimova, N K, Almaev A V and Sergeychenko N V 2016 *Key Eng. Mater.* **683** 353
[20] Korotcenkov G, Brinzari V and Golovanov V 2004 *Sens. Actuators* **98** 41
[21] Gaman V I 2008 *Russ. Phys. J.* **51** 425
[22] Sevast'yanov E Yu, Maksimova N K, Potekaev A I, Sergeichenko N V, Chernikov E V, Almaev A V and Kushnarev B O 2017 *Russ. Phys. J.* **60** 1094

Microstructure features of Ni-based and Ti-based alloys formed by method of wire-feed electron beam additive technology

D A Gurianov, K N Kalashnikov, V Utyaganova, E S Khoroshko and A V Chumaevskii

Institute of Strength Physics and Materials Science of the Siberian Branch of the Russian Academy of Sciences, pr. Akademicheskii, 2/4, Tomsk, 634021, Russia

Corresponding author: desa-93@mail.ru

Abstract. In the present study, a microstructural investigation of wire-feed EBAM-manufactured nickel-based and titanium-based alloys were conducted by producing single wall sample with 16 and 19 vertical layers, respectively. It was shown that in obtained material microstructural and elemental gradient presents. The results of the research show that dendrites and grains grow epitaxial in the direction of temperature gradient. Non-directional dissipation of heat on the edge of the sample leads to formation of equiaxed structure. Chosen parameters allow to produce low-defective samples by EBAM technology.

1. Introduction

Additive manufacturing is a group of processes used to obtain products from their three-dimensional models. These processes implement a layer-by-layer deposition and until forming a nearly net shape product. Additive technologies are classified [1] by type of the feed material (powder or wire) and heat source (laser, electron beam, arc). Metals, polymers, ceramics and composites can be used as source materials. The additive manufacturing is characterized by the local layer-by-layer deposition of the material until forming a compact product. The homogeneity of melting and solidification of the material is the most important aspect for obtaining qualitative and durable products. The layer-by-layer growing is a complicate system of heat and mass transfer processes. As a rule, such a complicate process may result in appearance of defects such as hot cracks, interlayer separation, pore formation due to incomplete melting and deformation in cooling. Deformation is caused by difference in temperature expansion coefficients between wire and substrate materials [2, 8]. The final microstructure determines the mechanical properties of the product and in its turn depends on the technological parameters of the product formation such as energy density, growing speed and wire-feed rate. Selection of optimal parameter values is necessary for obtaining products with mechanical properties comparable or superior to those of products obtained by traditional methods of casting and metalworking. Due to the fact that the microstructure of the material determines its mechanical properties, control and optimization of the microstructure is an important and urgent task [3, 5, 14].

In accordance with the terminology adopted in American Society for Testing and Materials the wire-feed electron-beam additive manufacturing of metallic components, relates to a class known as Directed Energy Deposition (DED-technologies). The DED-systems are characterized by a high productivity, effective energy input, supplying raw materials directly to the molten pool, possibility of applying material to existing parts/substrates and low cost consumables [6]. DED-technologies allow obtaining large nearly net shape components, which require minor mechanical treatment. The DED technologies are most widely used in the aerospace, defense, and petrochemical industries.

The wire-feed additive manufacturing allows making products during layer-by-layer process of metal wire deposition. The deposition is carried out sequentially using some suitable deposition strategy to form a three-dimensional shape. This technology can be used to obtain products from a wide range of materials, including titanium alloys, nickel and copper-nickel alloys, aluminum alloys, alloyed steels, tungsten, etc. [8]. Titanium-based alloys are also used both in aircraft and in biomedical industries. Whereas the nickel-based superalloys are used for production of complex and critical parts of the hot section in a modern gas-turbine engine-building industry for aviation and energy. Nowadays expensive nickel-based superalloys with directional solidification are used for production blades of gas-turbine engines. These superalloys contain up to 10wt.% of rare-earth metals such as rhenium, ruthenium, cerium, yttrium, lanthanum. However, the introduction of rhenium contributes to the selection of undesirable topologically close-packed phases that adversely affect the properties of the material. The modern technology of casting products with directional solidification is also expansive and the yield of products does not exceed 80%. It occurs due to the fact that new crystallization seed can form on the walls of injection molds, due to the high growth rate. Additive manufacturing doesn't have such problem. In case of damage to the product from such superalloy, it is more profitable to apply repairing rather than replace the entire part [15]. Most modern papers [3, 4, 13] dedicated to the organization of directional solidification in the additive process use powdered feed material and a laser as a heat source. This approach is characterized by specific defects - residual pores [4], to eliminate which it is necessary to carry out hot isostatic pressing, which increases the cost of production. The use of electron-beam additive manufacturing makes it possible to organize directional solidification in a formed product and the use of the considered economically doped nickel superalloy allows reducing production costs. Despite the advantages, such a combination of material and technology is completely not considered in modern literature. In addition, the structure of the product obtained by this method is close to the cast state, which is important for nickel-based superalloys and titanium-based alloys.

The purpose of this study is to identify the characteristics of the formation of the macro and microstructure of the material of metal products obtained using wire-feed electron beam additive manufacturing.

2. Material and methods

Nickel-based superalloy ZhS6U (brand chemical composition presented in table 1) and titanium-based alloy Ti-6Al–4V wire filaments were used as raw materials. Formation of vertical walls was carried out in a vacuum chamber with chosen parameters of accelerating voltage (25-35 kV), electron beam current (30-50 mA) and beam scan frequency (1000 Hz). Products obtained on the installation were described in the previous work [7]. A Ti-6Al–4V alloy wall consisted of 20 layers and had a height 23 mm (Fig. 1). A wall made of ZhS6U alloy consisted of 16 layers and had a height of 19 mm. Transverse and longitudinal polished and etched metallographic cross-sections were prepared for investigation of macro- and microstructure of material of walls. Cross-sections of products from an alloy of ZhS6U were etched in the Marble's reagent ($CuSO_4$ – 8 g, HCl – 40 ml, H_2O – 40 ml) within 10 seconds [9]. Cross-sections of products from an alloy of Ti-6Al-4V were etched in Kroll's reagent (HF – 2 ml, HNO_3 – 6 ml, H_2O – 92 ml) within 15 seconds [10]. Metallographic investigations were carried out using Olympus LEXT OLS 4100 laser confocal scanning microscope. Microstructure was studied on scanning electron microscope LEO EVO 50.

Table 1. Chemical composition (% wt.) of alloy ZhS6U [11]

W	Ni	Al	Co	Cr	Mo	Nb	Ti
9.5-11.0	54.3-62.7	5.1-6	9-10.5	8-9.5	1.2-2.4	0.8-1.2	2-2.9

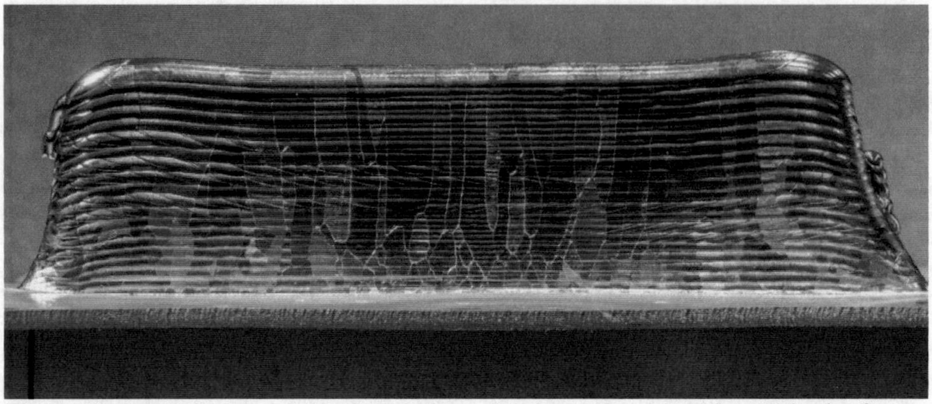

Figure 1. General view of the product from titanium alloy Ti-6Al – 4V in the form of a vertical wall.

3. Results and discussion

The macrostructure of the material of the titanium alloy wall in longitudinal section is presented in Figure 2. As can be seen, the epitaxial growth of columnar grains occurs along almost the full wall's height. The concentration of quasi-equiaxial grains increases as approaching to the edges of the sample. This phenomenon can be explained by the fact that three dimensional heat dissipation occurs at the boundary of each of the deposited layers. It is caused by heat removal into the vacuum chamber space in the form of radiation, as well as through the previously formed layers of the product into the substrate and a cooled worktable. The radiation component of cooling makes a significantly less contribution into heat removal for metal volumes of material far from the surface. In this case, the prevailing direction of heat removal flux is directed along the temperature gradient to the substrate and the cooled worktable that is why the columnar grains are formed.

It is known [4] that there is a dependence of the grain morphology and the microstructure on the ratio temperature gradient (G)/solidification rate (R) (Figure 3). Thus, as the value of G decreases and the value of R increases, the morphology of the elements of the microstructure changes in the following order: planar, cellular, cellular dendrites, columnar dendrites, equiaxial (for titanium-based alloys: columnar, mixed, equiaxed grains). On this basis, it can be assumed that in the newly applied layers the ratio $G/R<1$, which leads to the formation of grains with a morphology close to equiaxial. A similar ratio holds for the edges of the grown wall. Directed grain growth occurs when $G/R>1$. The columnar grains have an average length of 16.07 ± 4.63 mm and a width of 2.39 ± 0.77 mm. Grains with a morphology close to equiaxed have an average diameter of 0.86 ± 0.29 mm. In addition, pores of up to 0.6 mm in size are present in the investigated sample.

IOP Conf. Series: Materials Science and Engineering **597** (2019) 012042 doi:10.1088/1757-899X/597/1/012042

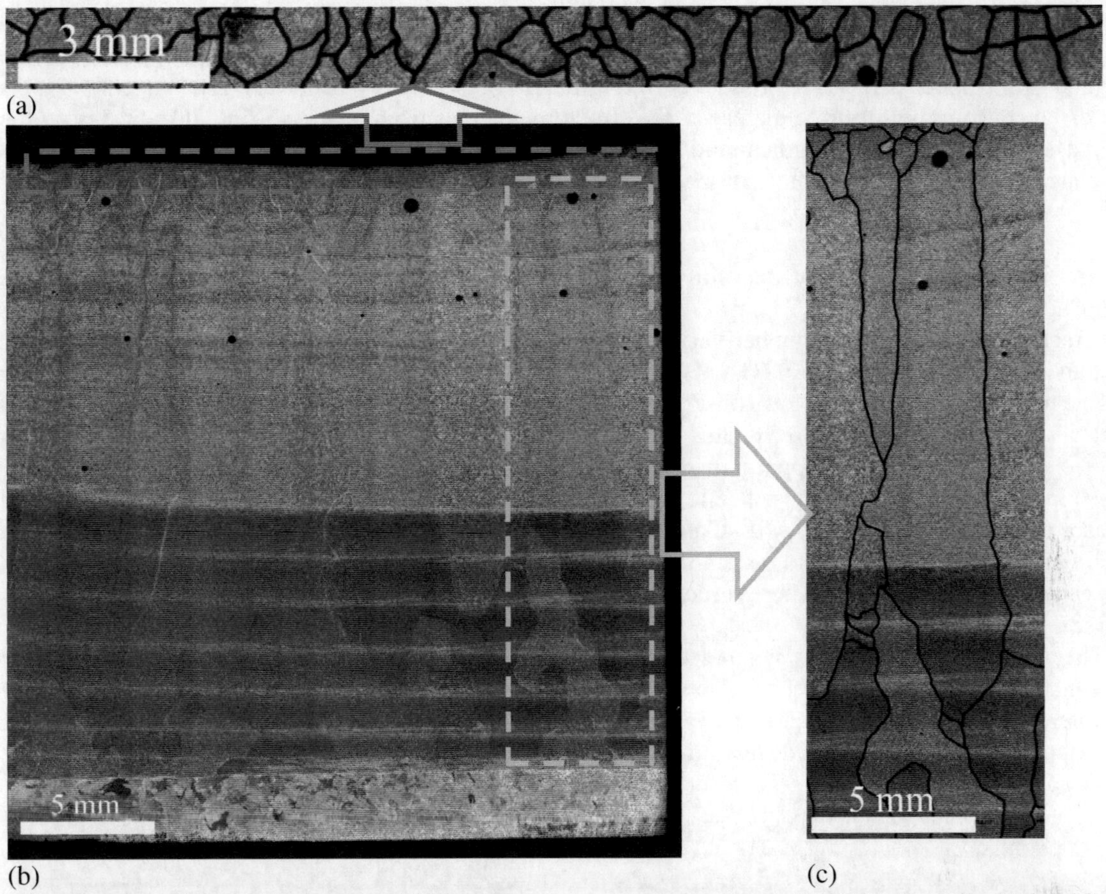

Figure 2. Metallographic images of macrostructure of titanium alloy Ti-6Al–4V wall obtained by wire-feed electron-beam additive technology. (a) – equiaxed grains in top of sample, (b) – panoramic images, (c) – columnar grains in bulk of sample.

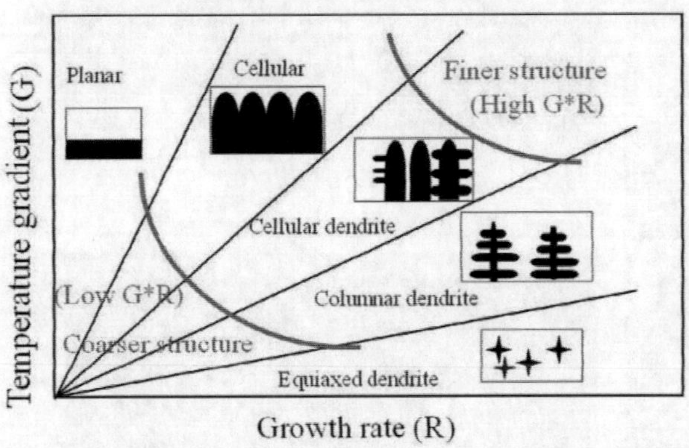

Figure 3. The effect of temperature gradient (G) and growth rate (R) on the solidification morphology and size [4].

Images of the microstructure of the heat-resistant nickel-based alloy ZhS6U are presented in Figure 4. Analysis of the entire cross-sectional area of the wall shows that a thin layer of irregular dendrites is on the border with the substrate, then dendrites form with a morphology close to equiaxial, and then dendrite growth begins mainly in the direction of additive growth (shown in Figure 4 by arrows). Dendrites grow in one direction in groups of 5-30 pieces; the difference in the direction of growth of groups can reach 90°. The morphology of the elements of the structure changes from directed to equiaxed, as in the case of titanium alloy, as approaching to the edge. The same transition can be observed at the interlayer boundary. The SEM images of the ZhS6U alloy show a skeletal structure formed in the upper part of the product by the primary and secondary dendrites' arm system [12]. As the distance from the substrate increases, the structure changes from coarse to fine. It indicates on the increase in the temperature gradient and crystallization rate, according to [4]. It is known that (G × R) is related to the primary dendrite arm spacing according to equation as follows:

$$\lambda = a(G \times R)^{-n} \quad (1)$$

where λ (μm) is the primary dendrite arm spacing, a (°C) is the parameter proportional to the solidification interval ΔT, G (°C / cm) is the temperature gradient, R (m/s) is the solidification ratio, n is the fractional dimensionless number varying from 1/4 to 1/2. Due to these two facts, it can be argued that an increase in the product of (G × R) values is indeed taking place (G×R).

The microstructure image allows observing gray (dendrite arms) and dark gray (interdendritic spaces) areas. The secondary phase precipitates represented by white particles along the boundaries of arms have a needle shape (in the upper part of the sample) and a complex morphology near the substrate. These phases belong to the γ/γ'-eutectic, carbides (MC, M_6C, $M_{23}C_6$) and topologically close-packed phases (Laves - Ni_3Nb, σ - CrCo, μ - $Co_7(Mo,W)_6$). It is also noticeable that these phases are thinned at the boundaries of the layers and become gray. Elemental analysis suggests a lower Al and Ti and high Co contents in the dendrites whereas increased contents of Ti, W and Nb is detected in the secondary phases.

The presence of pores was not detected on the investigated cross-section views. However, the presence of cracks was established. The most probable mechanism of cracks formation is the blocking of interdendrite spaces with carbides in the solidification process, which leads to the formation of micropores from which cracks develop during the shrinkage [13].

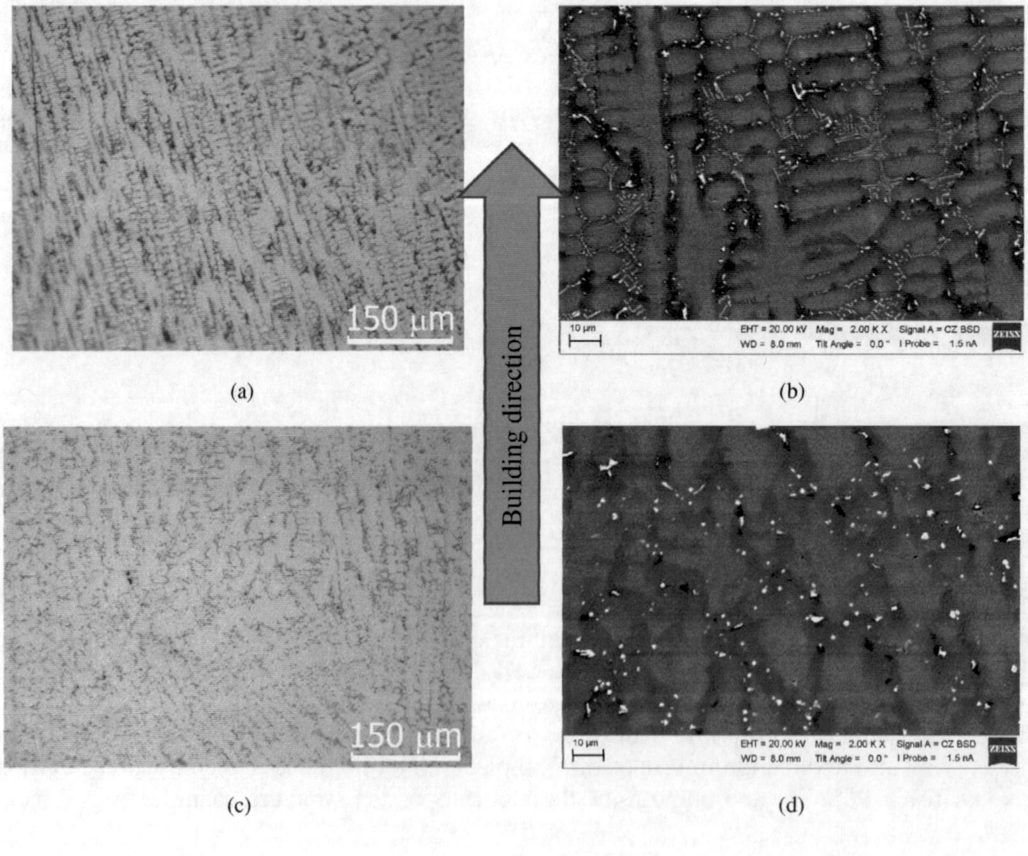

(a) (b)

(c) (d)

IOP Conf. Series: Materials Science and Engineering **597** (2019) 012042 doi:10.1088/1757-899X/597/1/012042

Figure 4. Metallographic (a, c) and SEM (b, d) images of macro- and microstructure in transverse section of vertical wall from alloy ZhS6U. Top area – (a, b), bottom area – (c, d).

4. Conclusions

Both macro- and microstructure of products made of Ti-6Al-4V and ZhS6U alloys using the additive wire-feed electron-beam technology were investigated. The results of the study show that titanium grains and dendrites in a nickel-based superalloy grow epitaxially along the direction of the temperature gradient through the previously formed layers of the product into the substrate and the cooled worktable. In addition, it was shown that directional solidification occurs only in the bulk of the samples. The reason for such an effect is the omnidirectional three-dimensional heat dissipation at the boundary of the sample, including one by radiation cooling. It is established that nickel alloy specimens are less defective than titanium ones.

Acknowledgments

This work was performed within the frame of the Fundamental Research Program of the State Academies of Sciences for 2013-2020, line of research III.23.

References

[1] Vayre B., Vignat F., Villeneuve F. 2012 *Mechanics and Industry.* **13(2)** 89
[2] Vastola G., Zhang G., Pei Q.X., Zhang Y.W. 2015 *Additive Manufacturing.* **7** 57
[3] Sahoo S., Chou K. 2016 *Additive Manufacturing.* **9** 14
[4] Karimi P., Sadeghi E., Åkerfeldt P., Ålgårdh J., Andersson J. 2018 Materials and Design 160 427
[5] Tarasov S., Filippov A., Savchenko N., Fortuna S., Rubtsov V., Kolubaev E., Psakhie S. 2018 *International Journal of Advanced Manufacturing Technology.* **99** 2353
[6] Kalashnikov K.N., Khoroshko E.S., Kalashnikova T.A., Chumaevskii A.V., Filippov A. 2018 *Journal of Physics Conference Series.* **1115** 042049
[7] Filippov A., Fortuna S.V., Gurianov D.A., Kalashnikov K.N. 2018 *Journal of Physics Conference Series.* **1115** 042044
[8] Kolubaev A.V., Tarasov S., Filippov A., Denisova Yu.A., Kolubaev E.A., Potekaev A.I. 2018 *Russian Physics Journal.* **61(5)** 1491
[9] Gontcharov A., Yuan Tian, et. al. 2019 *Journal of Engineering for Gas Turbines and Power.* **141** / 041031-1
[10] Yuwei Zhai, Haize Galarraga, Diana A. Lados 2016 *Engineering Failure Analysis.* **69** 3
[11] Kablov E., Petrushin N., Parfenovich P. 2018 *Metal Science and Heat Treatment.* 60 106
[12] J. A. Brooks, A. Thompson, and J. Williams 1984 *Weld. J.* **63** 71
[13] Zhipeng Zhou, Lan Huang, et. al. 2018 *Materials and Design.* **160** 1238
[14] Nikonov A.Yu., Zharmukhambetova A.M., Ponomareva A.V., Dmitriev A.I. 2018 *Phys. Mesomech.* **21(1)** 43
[15] Lei Z., Lu N., Yu X. 2019 *Journal of Manufacturing Processes.* **42** 11

IOP Conf. Series: Materials Science and Engineering **597** (2019) 012043 doi:10.1088/1757-899X/597/1/012043

Obtaining the bimetallic composition by the electron beam freeform fabrication

D A Gurianov, K N Kalashnikov, K S Osipovich and A V Chumaevskii

Institute of Strength Physics and Materials Science, Tomsk, Russian Federation

E-mail: desa-93@mail.ru

Abstract. This paper is devoted to obtaining the multimaterials using the electron-beam wire-feed additive manufacturing technology. The sample of titanium alloy Ti-6Al-4V and stainless steel 321 composition was obtained. It was found that steel and Ti-alloy form a transitional intermediate layer between themselves. The Ti-alloy layer does not undergo major changes, while the layer of stainless steel undergoes significant cracking.

1. Introduction

Electron beam additive manufacturing (or the electron beam freeform fabrication) is a promising technology for producing various products from metal wire or rod [1]. At the moment, the features of the formation of products made from materials such as Ti-6Al-4V alloy, stainless steel 321 or 316, etc. are rather widely studied [2-5]. But formation of multimaterials in modern literature has been poorly studied.

In a number of works the formation of bimetals from titanium alloys and stainless steels was investigated. In [6], the formation of a bimetal of a Ti-6Al-4V alloy and stainless steel was studied using an intermediate layer of NiCr and laser additive technology. The authors showed that without such a transition layer a bimetallic compound cannot be obtained by laser additive method, since in this case numerous cracks due to residual stresses will form.

In the works [7-10], functional gradient materials of the Ti-Al system, which are used for manufacturing turbo-compressor blades, exhaust valves and other rocket-space products, were investigated. According to the results of these articles, gradient material with a composition from pure titanium to 100% titanium aluminide TiAl is obtained using the two-wire arc additive technology. Also, other polymetallic compounds are known [11-15]. These polymetallic products are of the greatest interest among polymetallic metamaterials. However, the manufacture of such materials by the method of wire electron beam additive manufacturing (EBAM) has not been studied. Among other things, the production of multimetals based on titanium alloys, rather than technically pure titanium, also remains virtually unexplored. In this regard, the study of the formation of polymetallic metamaterials based on titanium alloys is a very urgent and promising task.

2. Materials and Methods

Sample preparation was carried out using an experimental EBAM laboratory equipment at the Institute of Strength Physics and Materials Science SB RAS, Russia (Figure 1). Ti-6Al-4V alloy and AISI 321 stainless steel 1 mm diameter wires were used for EBAM deposition at process parameters shown in Table 1.

Table 1. Parameters of the manufacturing process

Parameter	Value
Voltage, kV	25
First layer beam current (Ti-6Al-4V), mA	35
Last layer beam current (Steel 321), mA	30
Feed rate (Ti-6Al-4V), mm/min	220
Feed rate (Steel 321), mm/min	250

Templates were cut off the obtained sample, and then subjected to mechanical grinding, polishing with diamond paste and chemical etching in the reagent 1 (30 ml HNO_3 + 0.5 ml HCl + 70 ml CH_3COOH) during 30 seconds and in the reagent 2 (100 ml HNO_3 + 1 ml HF) during 2 seconds. The templates were examined using a metallographic microscope Altami MET-1S and a scanning electron microscope SEMTRAC mini.

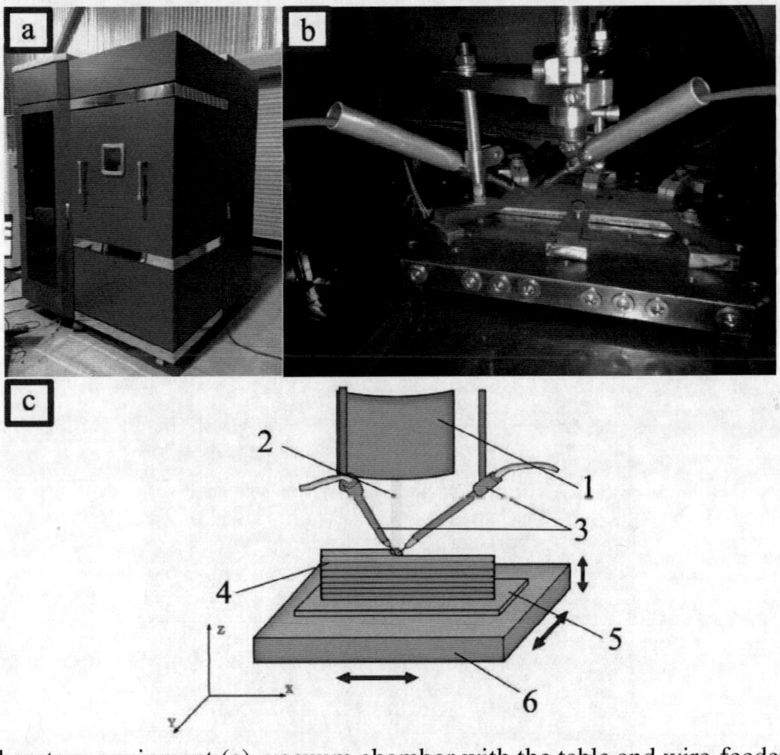

Figure 1. Laboratory equipment (a), vacuum chamber with the table and wire-feeders (b) and the scheme of the EBAM process (c). 1 – Electron beam gun, 2 – electron beam, 3 – wire-feeder, 4 – wall, 5 – substrate, 6 – table

3. Results and discussion

The sample of obtained bimetallic composition is shown in Fig. 2. As can be seen from Figure 2, primary Ti-6Al-4V wall was built successfully, despite its partial mixing with the substrate. This zone of primary wall (zone *I*) is represented by almost typical cast structure including primary β-grains and thin plates of α-phase, and practically free of macrodefects such as pores and cracks. The direction of grains growth depends on the direction of the heat removal, and in this case, it does not have a clear direction due to poor cooling conditions.

IOP Conf. Series: Materials Science and Engineering **597** (2019) 012043 doi:10.1088/1757-899X/597/1/012043

A layer of stainless steel (zone ***II***) deposited on the primary wall has not formed at all. As shown in Fig. 2, all deposited material stack on the right side of the wall. In addition, it is clear that the material has undergone significant cracking throughout the applied volume.

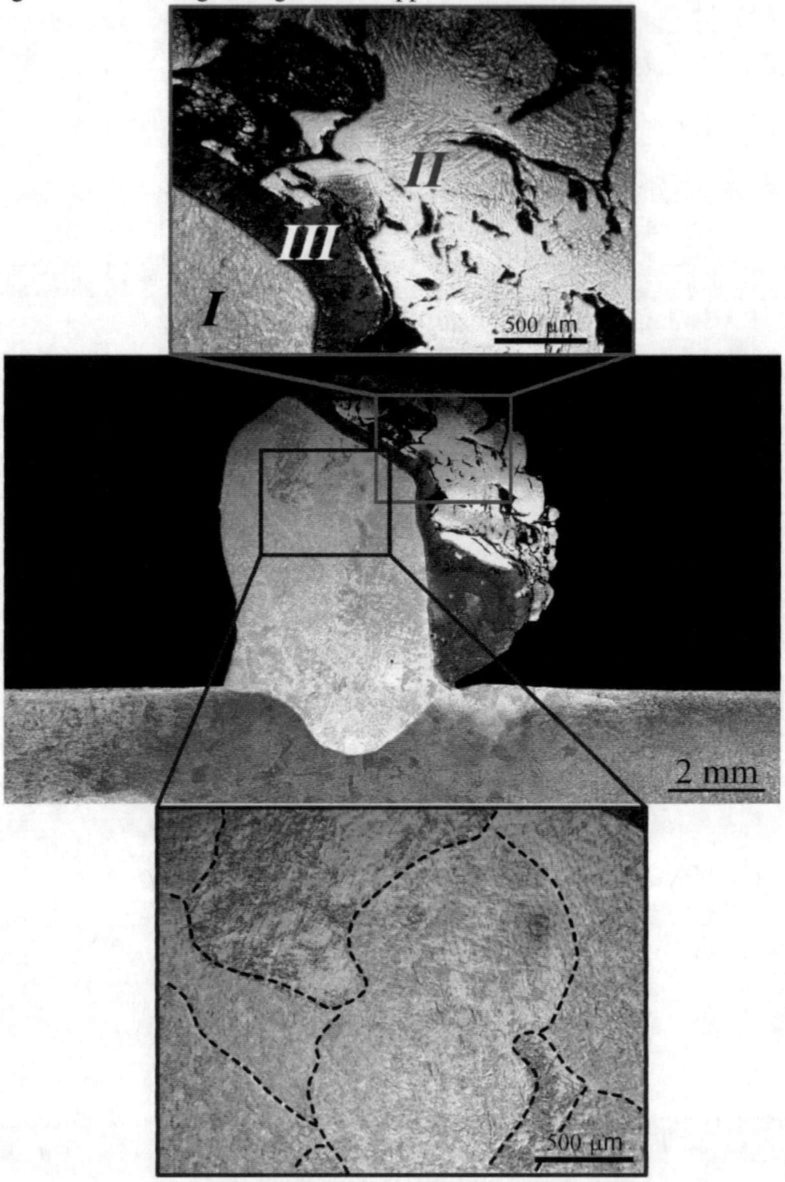

Figure 2. Metallographic image of the sample. I, II, III – zones of different materials, dashed lines – boundaries of primary β-grains in Ti-6Al-4V.

Most interesting is the intermediate layer formed between the two materials. As can be seen from Figure 2 this layer (zone ***III***) is etched differently than other zones and, therefore, has a structure different from these zones. As shown by energy dispersive analysis, the zone ***III*** is mainly represented by titanium (≈50–60%) and iron (≈15–20%). Apparently, this layer is an intermetallic compound. Its structure is also represented by dendrites growing from steel to titanium alloy. The results of energy dispersive analysis of the zone ***III*** are presented in table 2. Figure 3 shows the areas in which the elemental composition was studied. It is worth noting that in the first area Ni is present, and in area 2 it is absent, and they are also distinguished by the presence of aluminum. This suggests that the alloying elements of titanium alloy and steel form two different types of compounds. It can be assumed that the formation of such a layer and intermetallic compounds leads to the embrittlement of the stainless steel layer and its cracking during the crystallization process.

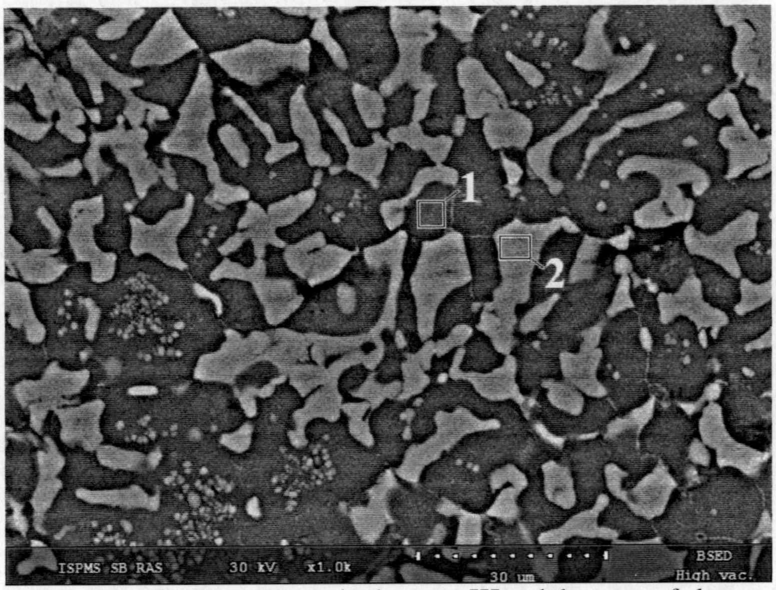

Figure 3. SEM image of microstructure in the zone ***III*** and the areas of elemental analysis

Table 2. The results of energy dispersive analysis, wt.%

Area	Ti	Al	Fe	V	Ni	Si
1	59.500	9.589	14.773	4.981	4.812	6.344
2	54.506	12.010	18.088	5.478	0	9.919

4. Conclusions

Studies of the steel-titanium bimetallic sample showed that the formation of such materials is hampered by the formation of brittle intermetallic compounds, which leads to cracking and deterioration of the stainless layer. In this case, the primary wall of titanium alloy does not undergo significant changes. It was revealed that a transition layer mainly represented by Ti and Fe and having a dendritic structure is formed between the materials joined by the EBAM method, this layer has practically no defects, however, it contributes to the embrittlement of the overlying layers due to the presence of intermetallic compounds.

Acknowledgments
This work was financially supported by the Russian Federation Ministry of Education and Science (agreement no 14.610.21.0013, project identifier RFMEFI61017X0013).

References

[1] DebRoy T, Wei H L, Zuback J S, Mukherjee T, Elmer J W, Milewski J O, Beese A M, Wilson-Heid A, De A, Zhang W 2018 *Prog. Mater. Sci.* **92** 112-224
[2] K N Kalashnikov et al 2018 *J. Phys.: Conf. Ser.* **1115** 042049
[3] Tarasov S Yu, Filippov A V, Savchenko N L, Fortuna S V, Rubtsov V E, Kolubaev E A, Psakhie S G 2018 *Int. J. Adv. Manuf. Tech.* **99** 2353-63
[4] Brandl E, Schoberth A, Leyens C 2012 *Mater. Sci. Eng.* A **532** 295-307

IOP Conf. Series: Materials Science and Engineering **597** (2019) 012043 doi:10.1088/1757-899X/597/1/012043

[5] Nikonov A Yu, Zharmukhambetova A M, Ponomareva A V, Dmitriev A I 2018 *Phys. Mesomech.* **21(1)** 43-50
[6] Sahasrabudhe H, Harrison R, Carpenter C, Bandyopadhyay A 2015 *Addit. Manuf.* **5** 1-8
[7] Ma Y, Cuiuri D, Shen C, Li H, Pan Z 2015 *Addit. Manuf.* **8** 71-77
[8] Lin X, Yue T M, Yang H O, Huang W D 2006 *Acta Mater.* **54(7)** 1901-15
[9] Wang J, Pan Z, Ma Y, Lu Y, Shen C, Cuiuri D, Li H 2018 *Mater. Sci. Eng.* A **734** 110-119
[10] Ma Y, Cuiuri D, Hoye N, Li H and Pan Z 2014 *J. Mater. Res.* **29(17)** 2066-71
[11] Besisa D H A, Ewais E M M 2016 *Advances in functionally graded ceramics – processing, sintering properties and applications Advances in Functionally Graded Materials and Structures* ed F Ebrahimi (InTech, Rijeka) 1–31
[12] Kan W H, Albino C, Dias-da-Costa D, Dolman K, Lucey T, Tang X, Chang L, Proust G, Cairney J 2018 *Mater. Charact.* **136** pp 196–205
[13] Forth S C, Favrow L H, Keat W D, Newman J A 2003 *Eng. Fract. Mech.* **70(15)** 2175–85
[14] Madec C, Le Gallet S, Salesse B, Geoffroy N, Bernard F 2018 *J. Mater. Process. Technol.* **254** pp 277–282
[15] Wilson J M, Shin Y C 2012 *Surf. Coat. Technol.* **207** 517–522

IOP Conf. Series: Materials Science and Engineering **597** (2019) 012044 doi:10.1088/1757-899X/597/1/012044

Modification of physical-mechanical properties of NiTi alloy by electron beam in surface melting mode

F A D'yachenko[1], S N Meisner[1], E V Yakovlev[2], A A Atovullaeva[3] and L L Meisner[1, 3]

[1]Institute of Strength Physics and Materials Science SB RAS, Tomsk, Russia
[2]Institute of High-Current Electronics SB RAS, Tomsk, Russia
[3]National Research Tomsk State University, Tomsk, Russia

E-mail: frozennonetroll@mail.ru

Abstract. This paper presents a research data on investigation of physical-mechanical properties of the surface of NiTi shape memory alloy after low-energy high-current pulsed electron beam (LEHCPEB) treatment in surface melting mode. Investigation of physical-mechanical and functional properties (Young's modulus E_{IT}, dynamic hardness H_{IT}, plasticity characteristic δ_H, and parameter of shape recovery ratio η) was held by method of instrumented nanoindentation test. Our data suggest that after LEHCPEB treatment physical-mechanical properties of the surface layers of ~500 nm thickness change in depth from thc surface without critical influence to the pseudoelastic properties of NiTi alloy.

1. Introduction

Methods of electron beam treatment modify the structure and, accordingly, change the physical-mechanical properties of the surface layers of metallic materials [1]. It was found [2], that the morphology of surface and structure of the surface layers of nickel-titanium alloy (NiTi) were changed by the low-energy high-current pulsed electron beam treatment (LEHCPEB). However, as was shown in study [3], the influence of LEHCPEB with a large number of pulses n leads to a change of physical-mechanical properties of the surface layers of more than ~2 μm thickness.

Currently, it was established [4, 5] that the modified surface layers of NiTi alloy are residual elastic stress concentrators. It is possible to reduce the level of residual stresses in modified layers of the material by decreasing the number of pulses n of electron beam and/or decreasing the energy density E_s up to a limiting value. This mode of LEHCPEB treatment is called a surface melting mode.

It can be assumed that the use of LEHCPEB treatment in surface melting mode will allow achieving optimal physical-mechanical characteristics. This, in turn, will not have a critical impact on the important functional (*pseudoelastic*) properties of the NiTi alloy (shape memory effect and superelasticity, SME-SE).

Here, using the LEHCPEB in surface melting mode, we investigate formed modified surface and its physical-mechanical properties.

2. Experimental

2.1. Material

The material investigated here is commercial NiTi alloy (MATEK-SMA, Russia) produced as rolled sheets by vacuum induction melting (VIM). The chemical composition of the alloy was: Ti (balance); 55.08 Ni; 0.051 C; 0.03 O; 0.002 N (wt %).

Samples were cut to dimensions of 10x10x1 mm by laser, mechanically grinded and polished (abrasive grit: P120, P320, P600, P1200), electrolytically polished (acid solution: 3 part CH_3COOH + 1 part $HClO_4$), and washed in an ultrasonic bath with a distilled water (duration t = 15 min).

Experimental studies of surface of initial NiTi alloy and NiTi alloy after electron beam treatment in surface melting mode were studied on 8 samples (4 – initial NiTi; 4 – after LEHCPEB).

2.2. LEHCPEB-treatment

Electron beam modification in surface melting mode of the surface of NiTi alloy was carried out on the modified automatic setup RITM-SP (Microsplav, Russia) [6]. Working characteristics of the setup are: the current of electron beam is up to 25 kA; the accelerating voltage is up to 19 kV; the beam diameter is up to 10 cm; the working gas (argon) pressure is 0.02–0.2 Pa. Parameters of surface melting mode are: the number of pulses of electron beam n = 5, the pulse duration τ = 2–2.5 μs, the beam energy density E_s = 1.1 J/cm^2.

2.3. Methods of investigation

The study of the morphology of surface of NiTi samples before and after electron beam treatment was carried out by the method of optical metallography on the setup Axiovert 200MAT (Zeiss, Germany) using the optical light and dark-field and function of the differential interference contrast (DIC).

Investigation of physical-mechanical parameters of modified surface layers was examined by method of instrumented nanoindentation test on a Nano Hardness Tester (CSM, Switzerland). As a result of the experiment, according to the standarts ISO 14577 [7], a series *P-h* diagrams were obtained (depenences of the depth of the indener h on the maxium load P). These diagrams represent information of local deformation under the indenter. Using the Oliver-Pharr method [8] and standart ISO 14577 [7], the Young's modulus E_{IT}, dynamic hardness H_{IT}, and static Vickers hardness HV were determined.

The plasticity characteristic δ_H of surface layer shows the propensity of the material to deform under the action of the load. This parameter δ_H was estimated [9]:

$$\delta_{\mathrm{H}} = 1 - 14{,}3\,(1 - \nu - 2\,\nu^2)\frac{HV}{E_{IT}} \tag{1}$$

where ν – Poisson's ratio; HV – Vickers hardness; E_{IT} – Young's modulus determined by instrumented nanoindentation.

Pseudoelastic properties of surface layers were estimated using the parameter η. This parameter characterizes the degree of contribution of pseudoelastic recovery of area print of indenter [10]:

$$\eta = \frac{h_{max} - h_{plastic}}{h_{max}} \times 100 \tag{2}$$

where h_{max} – maximum indentation depth under P_{max}; $h_{plastic}$ – residual depth after unloading.

3. Experimental results and discussion

3.1. Morphology of surface

Figure 1 shows optical metallographic images of the surface of NiTi before and after LEHCPEB treatment in surface melting mode (each image from one tested sample of NiTi alloy before and after LEHCPEB). Before electron beam irradiation of initial NiTi sample (figure 1*a*, *b*, *c*, *d*) the morphology of surface is determined by the rolling texture. In figures 1*c*, *d* obtained in the DIC and dark-field regimes, it is clearly seen that inclusions are located on the surface. It was established that the size of the inclusions varies from ~0.5 up to 5 μm.

In studies [11, 12], it was found that inclusions on the surface of the NiTi alloy are the phases of: Ti(C, O) – titanium carbides and oxides; $Ti_4Ni_2O_x$ – oxides based on the Ti_2Ni intermetallic phase. The presence of inclusions of these phases on the surface of the NiTi alloy is associated with the stages of its technological production (rolled sheets by vacuum induction melting).

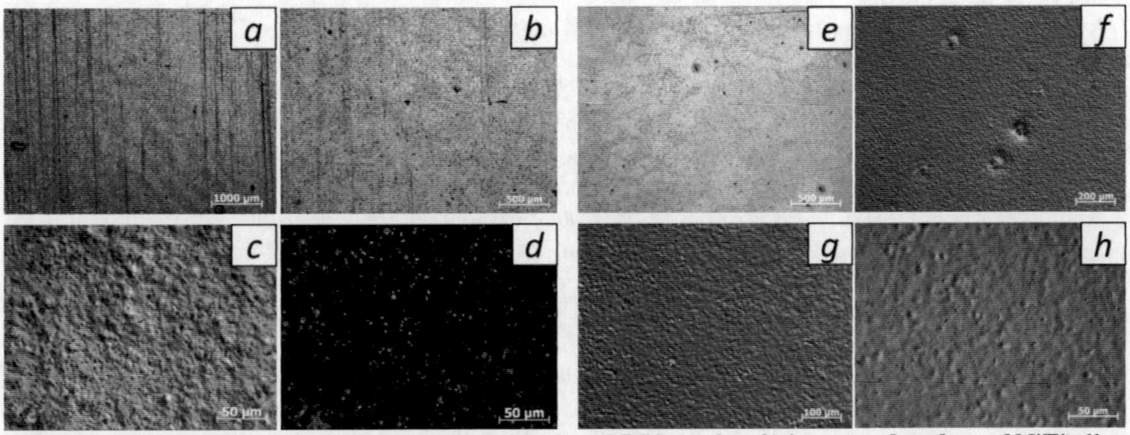

Figure 1. Optical light-field (*a*, *b*, *e*), dark-field (*d*) and DIC (*c*, *f*, *g*, *h*) images of surface of NiTi alloy: *a*, *b*, *c*, *d* – initial NiTi (before LEHCPEB treatment); *e*, *f*, *g*, *h* – NiTi after treatment

After LEHCPEB treatment of the surface of NiTi alloy (figure 1*e*, *f*, *g*, *h*) a crater-like relief is seen on the surface, as well as a small number of inclusions that were not melted by the electron beam. As a result of LEHCPEB treatment, according to [12], the occurrence of crater-like relief is due to the presence if inclusions on the surface of initial NiTi alloy. These inclusions were not completely melted under the impact of the electron beam in surface melting mode.

3.2. Physical-mechanical properties

The physical-mechanical properties of surface layers of NiTi samples before and after LEHCPEB treatment can be addressed from figure 2 showing Young's modulus E_{IT} (*1*), dynamic hardness H_{IT} (*2*), and plasticity characteristic δ_H (*3*) as a function of maximum indentation depth h_{max} under the increasing load *P* from 5 up to 300 mN.

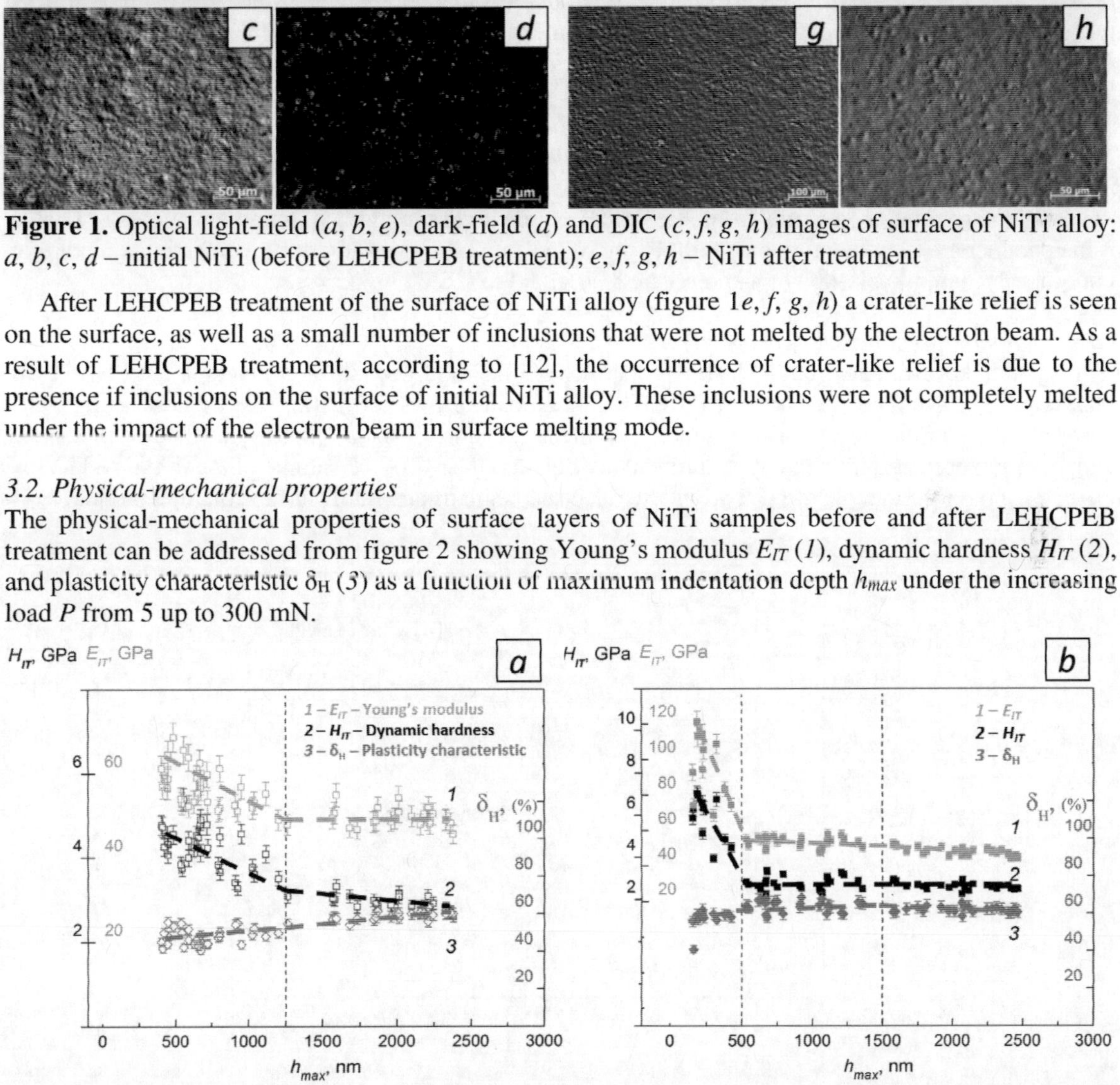

Figure 2. Dependences of E_{IT} (*1*), H_{IT} (*2*) and δ_H (*3*) on the maximum indentation depth h_{max} for: *a* – initial NiTi (before LEHCPEB treatment); *b* – after LEHCPEB treatment in surface melting mode

Near the surface the parameters of Young's modulus E_{IT} and dynamic hardness H_{IT} of NiTi alloy before LEHCPEB treatment take a values: E_{IT} = 60 GPa, H_{IT} = 4 GPa (figure 2*a*, *curves 1, 2*). These parameters change linearly to the values of: E_{IT} = 50 GPa, H_{IT} = 2.5 GPa at a depth *h* of more than ~1.25 μm thickness. High values of physical-mechanical parameters at a depth *h* up to ~1.25 μm are associated with mechanical hardening of surface. This is due to the manufacturing of NiTi alloy that was fabricated by VIM and subsequent mechanical grinding of the surface of the NiTi samples before electron beam treatment.

Analysis of the physical-mechanical parameters of the NiTi samples after LEHCPEB treatment showed that modification of surface in melting mode leads to a change of strength characteristics in a depth from the surface. In figure 2*b* (*curves 1, 2*), it can be seen that electron beam treatment led to the hardening of the surface to a depth *h* of ~500 nm. However, at a depth *h* of more than ~500 nm, the values of the physical-mechanical parameters correspond to the parameters that represent the properties in bulk of initial NiTi sample. It is important to note the LEHCPEB treatment in surface melting mode did not lead to a highly change of parameters to a depth greater than ~ 500 nm. A further linear horizontal view of the curves of the physical-mechanical parameters E_{IT} (h_{max}), H_{IT} (h_{max}), and δ_H (h_{max}) at a depth more than ~500 nm (figure 2*b*, *curves 1, 2, 3*) means that the mechanical hardening of surface due to the manufacturing of NiTi alloy and subsequent mechanical grinding of the surface of samples was relaxed after melting of surface layers by LEHCPEB treament.

The change in the values of the plasticity characteristic δ_H before and after LEHCPEB treatment is presented in figure 2*a*, *b* (*curves 3*). It was found that in modified layer of ~500 nm thickness figure 2*b* (*curve 1*), the parameter δ_H varies from 40 to 59% relative to the initial NiTi sample, which has a value of the parameter δ_H = 48% at given depth. In the NiTi sample before and after treatment at a depth *h* of more than 1.5 μm, the plasticity characteristic δ_H is equal to 0.6%.

3.3. Functional (pseudoelastic) properties

In this study to estimate pseudoelastic properties of the modified surface layers after LEHCPEB treatment, we calculated parameter η [10]. This parameter allows estimating the pseudoelastic shape recovery of the print, which formed under the indenter. Figure 3 shows the dependence of the shape recovery ratio of the print on the maximum indentation depth h_{max} for the sample of the initial NiTi alloy (*curve 1*) and for the sample of NiTi alloy after electron beam treatment in surface melting mode (*curve 2*).

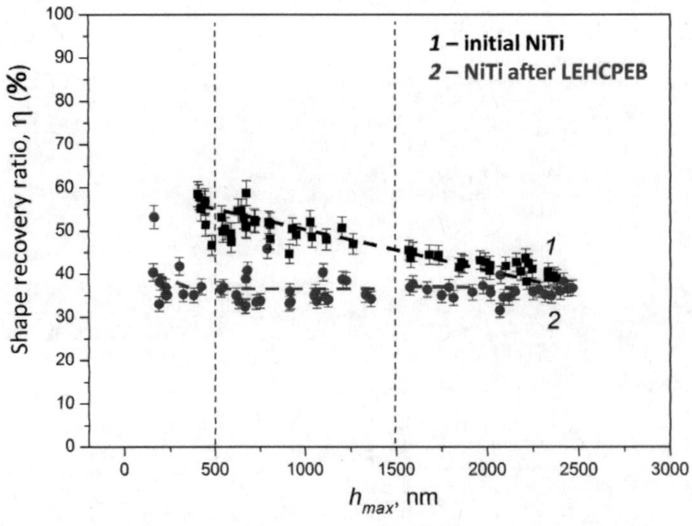

Figure 3. Dependences of shape recovery ratio η on the maximum indentation depth h_{max} for: *1* – initial NiTi (before LEHCPEB treatment); *2* – after LEHCPEB treatment in surface melting mode

It is seen, that after treatment the parameter of shape recovery ratio η varies from 50 to 40% inside a layer of no more than ~ 500 nm thicknesses. At a depth from ~500 nm and more than 2 μm, the parameter η does not change and retains its value and equal to the value of the parameter η in the volume of the initial NiTi sample before electron beam treatment. This result means that LEHCPEB treatment in surface melting mode did not have a critical influence on the pseudoelastic property of the NiTi shape memory alloy.

4. Conclusion
The main conclusions obtained in the work are as follows:

– the mechanical hardening of surface due to the manufacturing of NiTi alloy and subsequent mechanical grinding of the surface of samples was relaxed after melting of surface layers by LEHCPEB treament;

– the physical-mechanical properties of the surface layers of ~500 nm thickness change in depth from the surface without critical influence to the pseudoelastic properties of NiTi alloy.

Acknowledgments
Electron beam treatment of the surface of NiTi alloy was supported by grant of the Russian Science Foundation No. 18-19-00198 (26.04.2018); analysis of morphology and physical-mechanical properties of surface of NiTi alloy before and after treatment were performed with the frame of the Fundamental Research Program of the State Academies of Science for 2013–2020, line of research III.23.

References
[1] Poate J M, Foti G, and Jacobson D C 1983 *Surface Modification and Alloying: by Laser, Ion, and Electron Beams* (New York: Plenum Press)

[2] Zhang K M, Zou J X, Grosdidier T, Gey. N, Yang D Z, Hao S Z and Dong S 2007 *J. Alloys Compd.* **434** 682–685

[3] Meisner S N, Yakovlev E V, Semin V O, Meisner L L, Rotshtein V P, Neiman A A and D'yachenko F 2018 *Appl. Surf. Sci.* **437** 217–226

[4] Meisner L L, Semin V O, Mironov Yu P, Meisner S N and D'yachenko F A 2018 *Mater. Today Commun.* **17** 169–179

[5] Dong C *et al.* 2003 *Surf. Coat. Technol.* **163** 621–624

[6] Markov A B, Mikov A V, Ozur G E and Padei A G 2011 *Instrum. Exp. Tech.* **54(6)** 862–866

[7] International Standard Organization, ISO 14577-1:2002, Metallic Materials – *Instrumented indentation test for hardness and materials parameters – Part 1: Test method*

[8] Oliver W C and Pharr G M 2004 *J. Mater. Res.* **19(1)** 3–20

[9] Milman Yu V 2008 *J. Phys.* D: *Appl. Phys.* **41** 074013

[10] Crone W C, Shaw G A, Stone D S, Johnson A D and Ellis A B 2003 *SEM Annual Conference Proceedings* 1–6

[11] Coda A, Zilio S, Norwich D and Sczerzenie F 2012 *J. Mater. Eng. Perform.* **21(12)** 2572–77

[12] Meisner L L, Rotshtein V P, Markov A B, Meisner S N, Yakovlev E V and D'yachenko F 2016 *Procedia Struct. Integrity* **2** 1465–72

IOP Conf. Series: Materials Science and Engineering **597** (2019) 012045 doi:10.1088/1757-899X/597/1/012045

Quasi-viscous deformation mode in nonequilibrium nanocrystalline structures

I I Sukhanov[1,2], D S Zaytsev[2], I A Ditenberg[1,2,3] and A N Tyumentsev[1,2,3]

[1]Institute of Strength Physics and Materials Science SB RAS, Tomsk, 634055, Russia
[2]National Research Tomsk State University, Tomsk, 634050, Russia
[3]Siberian Physical-Technical Institute, Tomsk, 634050, Russia

E-mail: suhanii@mail.ru

Abstract. Based on the analysis of the elastic-stressed state of partial disclinations of nanodipoles the rate of the quasi-viscous deformation mode in nanocrystalline structures is estimated. It's shown that velocity of movement of point defects (V) and nanodipole propagation ($\dot{L}$) within the quasi-viscous deformation mechanism, depending on the activation energy of the migration of point defects (vacancies, interstitial atoms) and deformation temperature, can reach high values close to the speed of sound in metals.

1. Introduction

At present, the study of non-equilibrium states is obtained considerable interest, in particular, at the description of processes of nanostructured materials formation, the features of their plastic deformation and destruction. These states largely determine the unique complex of physicomechanical properties of such materials (low-temperature and high-speed superplasticity, the phenomenon of anomalous mass transfer, etc.). In [1, 2], after intense plastic deformation on Bridgman anvils on a wide class of metallic materials (pure Ni, V-based alloys, Mo-Re solid solution, etc.), nanobands of deformation localization were discovered with the crystal lattice reoriented with dimensions (width) from several to several tens of nanometers. In this case the carriers of the rotational mode of deformation are various compensated disclination configurations, including dipoles of partial disclinations. In contrast to the traditional collective dislocation models, the motion of these defects can be described within the quasi-viscous mechanism of plastic deformation by fluxes of nonequilibrium point defects.

In this work, based on the analysis of the elastic-stressed state of partial disclination nanodipoles, the rates of the quasi-viscous deformation mode depending on the type (vacancies and interstitial atoms) of these defects are estimated.

2. Experimental materials and procedures

The calculation of the stress fields and nanodipole energies was conducted in the software Maple 17 using the partial wedge disclination stress tensor with the direction of the defect lines along the z axis, which has the form [3, 4]:

$$\sigma_{ij}^{(e)z} = D\omega g_{ij}^{(e)z} , \qquad (1)$$

IOP Conf. Series: Materials Science and Engineering **597** (2019) 012045 doi:10.1088/1757-899X/597/1/012045

$$g_{ij}^{(e)z} = \begin{pmatrix} \ln(r) + \frac{y^2}{r^2} + \frac{\nu}{1-2\nu} & -\frac{xy}{r^2} & 0 \\ \frac{xy}{r^2} & \ln(r) + \frac{x^2}{r^2} + \frac{\nu}{1-2\nu} & 0 \\ 0 & 0 & 2\nu\ln(r) + \frac{\nu}{1-2\nu} \end{pmatrix}. \quad (2)$$

where $D = G/[2\pi(1-\nu)]$, $r^2 = x^2 + y^2$, G – shear modulus, ν – Poisson's ratio, ω – Frank's vector (power of disclinations).

Based on the experimental data, the moduli of pure nickel were used as elastic moduli of the continuum in which the analysis of elastically stressed states was performed. To simplify the calculation, only wedge disclinations were considered. The visualization of stress fields and gradients of these fields was made using the graphical tools of the Maple 17 package.

3. Results and discussion

The above mentioned nanobands of reorientation with sizes from several to several tens of nanometers are characterized by a high (hundreds of degrees per micrometer) elastic curvature of the crystal lattice. Plastic accommodation of such structural states is associated with significant mass transfer effects. The effects of anomalous mass transfer can be realized within the quasi-viscous deformation mechanism (Figure 1) [1, 2]. The reorientation of the crystal lattice in the nanoband is realized by quasi-viscous mass transfer from the OPQ compression regions in the ORS extension region by non-equilibrium fluxes (generated during plastic deformation) point defects in fields of high local pressure gradients. For small values of the crystal lattice reorientation ($\varphi \approx 1°$) the expressions for the average velocity of point defects (V) in the x direction and the average velocity of nanodipole propagation ($\dot{L}$) within this model have the form [1, 2]:

$$D \approx \nu r^2 \exp\left(-\frac{E^m}{kT}\right), \quad (3)$$

$$V \approx \frac{D}{kT} \Omega \frac{\partial P}{\partial x}, \quad (4)$$

$$\dot{L} \approx \frac{2cD}{kT} \frac{\Omega}{\varphi} \frac{\partial P}{\partial x} \quad (5)$$

Here $\nu \approx 10^{13}$ s^{-1} – Debye frequency and $r \approx 3\times10^{-10}$ m – length of an elementary jump of a point defect, c – concentration of nonequilibrium point defects; D – their diffusion coefficient; k – Boltzmann constant; T – deformation temperature; Ω – atom volume (vacancies); φ – angle of reorientation of the nanobands; $\partial P/\partial x$ – pressure gradient in the direction of mass transfer.
Integral averaging of pressure gradients for a nanodipole of partial disclinations with size l has the form:

$$\overline{\frac{\partial P}{\partial x}} = \frac{1}{l} \int_{\delta-\frac{l}{2}}^{\frac{l}{2}-\delta} \left(\frac{\partial P}{\partial x}\right) dx \quad (6)$$

Here, the parameter $\delta \approx 0.3$ nm characterizes the size of the singular regions of the stress fields near the lines of disclination type defects. The maximum values of the pressure gradients are reached in the plane of the nanodipole of partial disclinations and were $\overline{\partial P/\partial x} \approx 0.015\ E$ nm^{-1} (E – Young's modulus). Due to the high localization of stress fields, the above-mentioned values decrease to $\overline{\partial P/\partial x} \approx 0.003\ E$ nm^{-1} as the distance from this plane increases.

IOP Conf. Series: Materials Science and Engineering **597** (2019) 012045 doi:10.1088/1757-899X/597/1/012045

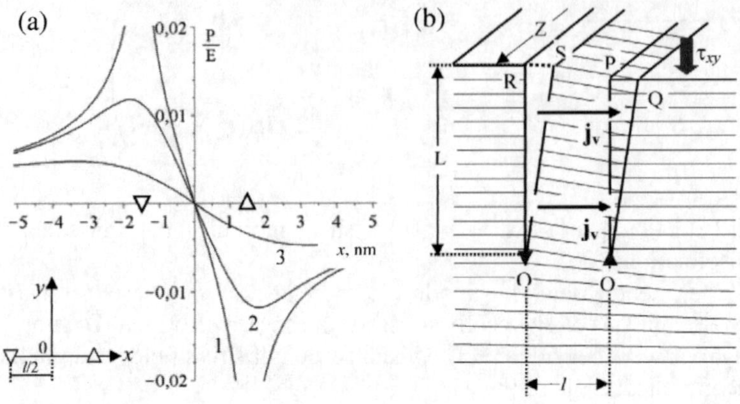

Figure 1. (a) – pressure change $P = (\sigma_{11}+\sigma_{22}+\sigma_{33})/3$ in the direction of its maximum gradient in the dipole plane (curve 1) and at distances of 1 nm (curve 2) and 3 nm (curve 3) from this plane. (b) – scheme of reorientation of the crystal lattice by diffusion streams of point defects in the process of movement of the partial disclinations dipole.

As can be seen from relations (3), (4) and (5) the rate of plastic deformation in addition to the elastic-stressed state of the continuum medium is largely determined by the diffusion properties (diffusion activation energy E^m) and the concentration of point defects. In the case of mass transfer by fluxes of non-equilibrium vacancies (E_v^m = 0.6 eV) at their concentrations $c_v \approx 10^{-4}$ for nanodipoles and local pressure gradients presented in Figure 1 (a) the velocity of movement of point defects V and movement of the nanodipole of partial disclinations $\dot{L}$ in depending on the temperature of deformation was estimated (Table 1). The characteristic values of the velocity of a quasi-viscous deformation mode for this type of point defects lie in the interval $V = 0.3\times10^{-6}..3.2$ m/s.

Table 1. The speeds of movement of point defects (V) and the propagation of a nanodipole of partial disclinations ($\dot{L}$), depending on the activation energy of point defects and deformation temperature.

			Vacancies $E_v^m = 0.6$ eV		
T (K)	293	473	673	873	1073
V (nm/s)	371.5	1.92×10^6	1.07×10^8	8.81×10^8	3.17×10^9
$\dot{L}$ (nm/s)	4.26	2.2×10^4	1.22×10^6	1.01×10^7	3.63×10^7
			Interstitial atoms $E^m = 0.1$ eV		
V (nm/s)	1.45×10^{11}	4.057×10^{11}	5.9×10^{11}	6.75×10^{11}	7.03×10^{11}
$\dot{L}$ (nm/s)	1.67×10^9	4.65×10^9	6.76×10^9	7.73×10^9	8.06×10^9

$c_v \approx 10^{-4}$; $\overline{\partial P}/\partial x \approx 0.015E$ nm^{-1}.

Another type of flux carrier of point defects involved in plastic deformation is interstitial atoms, the activation energy of diffusion of which is significantly lower than the similar energy for vacancies (E^m = 0.1 eV). It was demonstrated that in the low-temperature region in this case the values of V and $\dot{L}$ are tens of times higher than those for vacancies (Figure 2). Despite the fact that the interstitial atoms correspond to a weaker temperature dependence of the velocity of point defects motion, the magnitude of the rate of plastic deformation is about hundreds of meters per second. In [5, 6], the method of molecular dynamics was used to estimate the generation and accumulation of point defects (vacancies, bi-vacancies, and interstitial atoms) in single crystals with fcc structure taking into account the interactions between ensembles of point defects and a dislocation ensemble during plastic deformation. It was shown that at the stages of developed plastic deformation, the concentration of generated point defects in the absence of annihilation mechanisms on dislocations can reach very large values of $c = 10^{-}$

^{3}–10^{-2}. With such values of concentration, the magnitude of the velocity of the quasi-viscous mode reaches the value of the speed of sound in metals already at temperatures of local heating of the shear zones of metallic samples that are deformed on Bridgman anvils $T \geq 473$ K. The question of the limitation of the propagation speed of nanobands within the quasi-viscous deformation mechanism still remains open and requires further analysis.

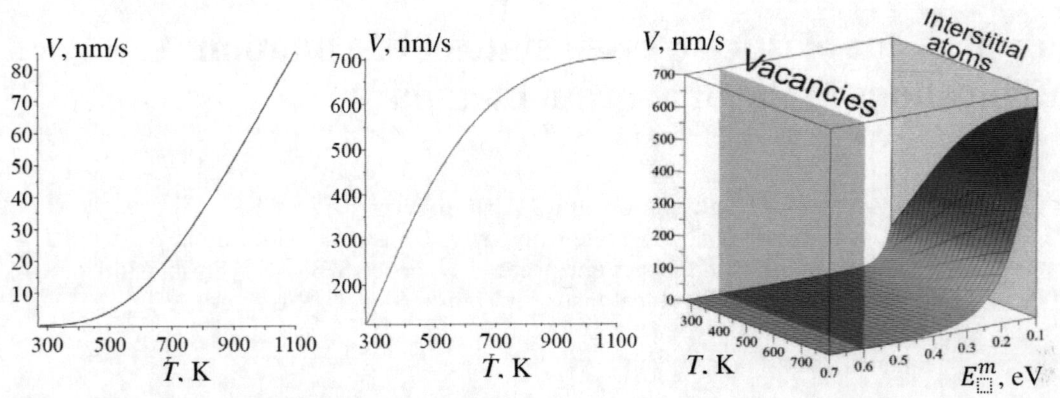

Figure 2. The rate of the quasi-viscous deformation mode, depending on the activation energy of vacancy migration and the strain temperature.

The presented rates of the quasi-viscous deformation mode are approximate and do not take into account a number of many factors and conditions of deformation. Among them: an increase in the concentration of nonequilibrium point defects, in particular, when they are generated by cores of moving disclinations; change in the relative contribution of pair vacancies; non-equilibrium boundaries of nanograins and partial disclinations as potential sources of point defects, etc.

4. Conclusion

It is shown that a quasi-viscous mechanism of deformation by fluxes of non-equilibrium point defects can be realized on the example of pure Ni. Based on the analysis of the elastic-stressed state of a partial disclination nanodipole, theoretical estimates were made of the velocity of the movement of point defects V and the motion of the nanodipole of partial disclinations $\dot{L}$. It has been demonstrated that, within the quasi-viscous deformation mechanism, depending on the activation energy of migration of point defects (vacancies, interstitial atoms) and deformation temperature, the velocity of the quasi-viscosity deformation mode can reach high values close to the speed of sound in metals.

Acknowledgments
Theoretical analysis of stress fields of disclination configurations was performed within the frame of grant provided by the Russian Science Foundation (Project No. 17-19-01374); study of quasi-viscous deformation mechanism was performed within the frame of the Fundamental Research Program of the State Academies of Sciences for 2013-2020, line of research III.23.

References

[1] Tyumentsev A N, Ditenberg I A, Korotaev A D and Denisov K I 2013 *Phys. Mesomech.* **16 4** 319
[2] Tyumentsev A N, Ditenberg I A and Sukhanov I I 2018 *Russ. Phys. J.* **60 10** 1832
[3] Romanov A E, Vladimirov V I 1992 *Dislocations in Solids*, ed Nabarro F R N 1992 **9** 191
[4] Romanov A E and Kolesnikova A L 2009 *Prog. Mater. Sci.* **54 6** 740
[5] Starenchenko V A, Cherepanov D N, Solov'eva Y V and Popov L E 2009 *Russ. Phys. J.* **52 4** 398
[6] Starenchenko V A, Cherepanov D N and Selivanikova O V 2015 *Russ. Phys. J.* **58 4** 446

IOP Conf. Series: Materials Science and Engineering **597** (2019) 012046 doi:10.1088/1757-899X/597/1/012046

Analysis of the elastic-stressed state of disclination configurations in the nanograin boundary

D S Zaytsev[1], I I Sukhanov[1,2] and I A Ditenberg[1,2,3]
[1] National Research Tomsk State University, Tomsk, 634050, Russia
[2] Institute of Strength Physics and Materials Science SB RAS, Tomsk, 634055, Russia
[3] Siberian Physical-Technical Institute, Tomsk, 634050, Russia

E-mail: deserza@gmail.com

Abstract. Disclination models of grain boundaries in crystals are considered. The features of the spatial distribution of the configuration stress fields depending on number of disclinations and magnitude of the common Frank's vector are described. The estimate of the energy of the configurations based on the calculations in a wide range of values of the bending-torsion tensor 0.2…1.8 degrees per nanometer was obtained. It was shown that the energy of the equidistant model is higher than the energy of the other two with irregular distribution of disclinations.

1. Introduction

It is known that the main elements of the defective structure of metallic materials are grain boundaries. Depending on the nature of long-range stress fields and excess free energy, a special class of grain boundaries can be distinguished – non-equilibrium boundaries [1-3]. Such boundaries are of great interest for the description and development of ideas about the nature of nanocrystalline materials, which are characterized by the presence of a significant fraction of the grain boundary phase.

Unfortunately, at present, there are only a few works with a quantitative assessment of the parameters of structural states with a high density of defects at various scale levels, and there is a need for conducting complex experimental studies with a theoretical interpretation of the structural elements of the boundary. The generalization of experimental results presented in [4] demonstrates that disclination configurations play a decisive role in the formation of nanocrystalline states in metals and alloys of different classes. Thus, the theoretical description of the grain boundary phase of nanocrystalline materials within the framework of disclination model representations is an important task of modern materials science.

In the present work, a simulation and analytical study of the features of stress fields and the energies of the boundaries of disclination-type grains in nanocrystalline materials is conducted.

2. Experimental materials and procedures

The computing experiment was carried out using the program environment Maple 17. Elastic properties of the modelled media were characterized by modules of elasticity of pure nickel (E = 204 GPa G = 79 GPa, where E – Young's modulus, G – shear modulus). Disclination configurations were considered within the framework of the continual method. The charge carriers in these configurations were partial wedge disclinations of the same sign. During modelling the size of the common Frank's vector did not

IOP Conf. Series: Materials Science and Engineering **597** (2019) 012046 doi:10.1088/1757-899X/597/1/012046

change and was equal to $\Omega = 10°$. To calculate the stress fields and elastic energy, we used the explicit form of the stress tensor components. [5]. In case of wedge disclinations with Frank's vector ω_z directed along axis z the stresses field have the form:

$$\chi = \frac{G\omega_z}{4\pi(1-\nu)} r^2 \ln(r), \tag{1}$$

$$\sigma_{xx} = \chi_{,yy},\ \sigma_{yy} = \chi_{,xx},\ \sigma_{xx} = \chi_{,yy},\ \sigma_{xy} = -\chi_{,xy},\ \sigma_{zz} = \nu(\sigma_{xx} + \sigma_{yy}). \tag{2}$$

Here χ – function of stress of Erie, ν – Poisson ratio, G – shear modulus, $r^2 = x^2 + y^2$.

The analysis of the energies of disclinational configurations was carried out using the Gauss numerical integration in a 30-point adaptive variation. The integration values were estimated over a set of 61 estimation points using the Kronrod formula.

3. Results and discussion

In the present work, three schemes of disclination distribution of a charge, which can describe structure of borders of nanograins, were considered. The arrangement of linear defects in the first disclination configuration had equidistant character, at which the distance between two disclinations is constant (Figure 1 a). In the second configuration, the distance between disclinations had the irregular exponential character (Figure 1 b). In the third configuration the distance between disclinations charges increased in an arithmetic progression (increase with a constant step), and also had irregular character (Figure 1 c). Power of disclinations depended only on their quantity N like $\omega_i = \Omega/N$. The sizes of the flat disclination concentration varied from several to several tens nanometers, but did not exceed the nanograin size R = 100 nm. The specified configurations were chosen based on the analytical form of the stress tensor (2).

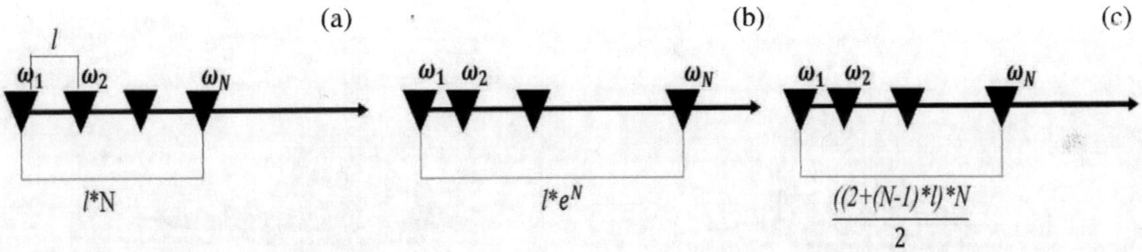

Figure 1. Schemes of disclination configurations: equidistant (a), exponential (b), arithmetic progression (c).

On the basis of the analysis of elastic stress it was found that fields of stress are mainly localized in the plane of the disclinations (Figure 2). The areas through which the defect lines pass correspond to local peaks, which are characterized by divergence of the stress tensor elements. In the case of a uniform charge distribution, due to the superposition of the stress fields, the maximum pressure values are reached in the center of the disclination configuration and at N = 10 are $P \approx E/55$ (Figure 2 a). In the exponential disclination scheme, a significant part of the stresses is observed in the region of the maximum density of disclinations and is equal to $P \approx E/26$ (Figure 2 b). For a configuration with an arithmetic progression of the stress field, similarly, as in the case of the exponential configuration, localized in places with maximum density of disclinations (Figure 2 c). These configurations are uncompensated, since they consist of single sign disclinations. For this reason, they do not observe relaxation of stress fields over the entire grain size. The average value of stress gradients for all configurations presented above is $P \approx E/35$. An increase in the number of disclinations N while maintaining the common Frank's vector Ω leads to a "blurring" of stress fields. Thus, at N = 10 the maximum pressure values no longer exceed $P \approx E/26$.

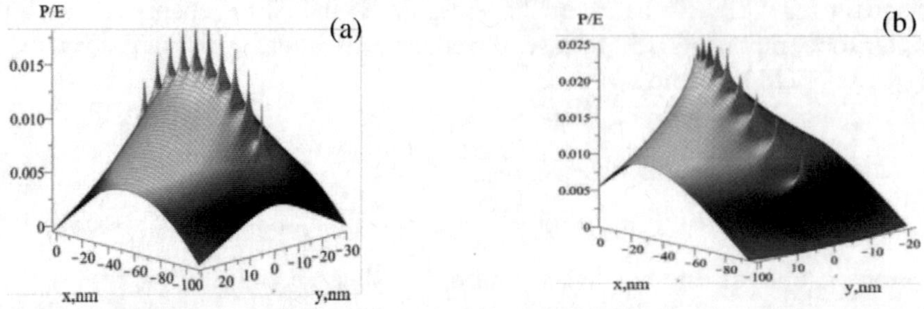

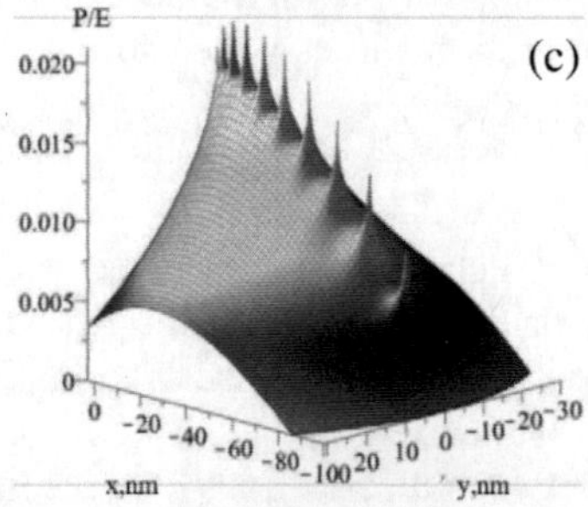

Figure 2. Spatial distribution of fields of stress for equidistant (a) exponential configuration (b) and configurations with an arithmetic progression (c). Number of disclinations $N = 10$, common Frank's vector $\Omega \approx 10°$.

The study of the spatial distribution of elastic energy showed that most of the energy in all configurations is concentrated near the lines of disclination type defects (Figure 3). In contrast to the equidistant case, in configurations with an irregular pattern of charge distribution, the features of the superposition of stress fields provide a high density of elastic energy in a region of high density of disclinations (Figure 3 b, c).

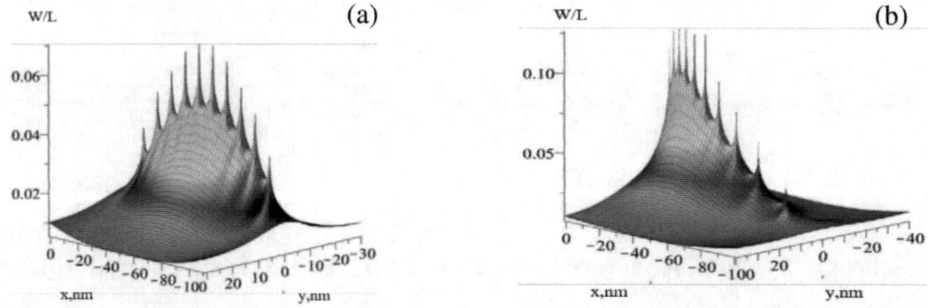

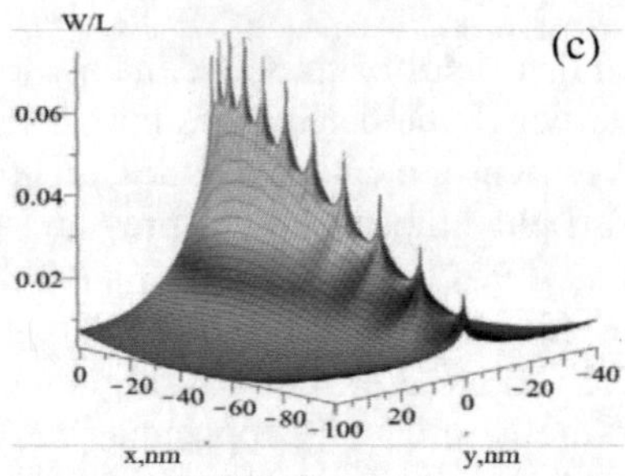

Figure 3. Spatial distribution of elastic energy for equidistant (a), exponential configuration (b) and configuration with arithmetic progression (c). Number of disclinations $N = 10$, common Frank's vector $\Omega \approx 10°$.

Figure 4 shows the dependences of the energies of all the disclination configurations on the value describing the curvature of the crystal lattice – the bending-torsion tensor. It is shown that regardless of the method of constructing configurations these dependencies have a similar character. The energy of the equidistant configuration is slightly lower than the energy of the other two in a wide interval of values of the bending-torsion tensor. This indicates that the equidistant configuration is an energetically more stable system. Therefore, based on this configuration, it is possible to construct a relatively stable model of the boundary of the nanograin. To build a more non-equilibrium boundary one should use irregular disclination configurations, for example, a configuration with an exponential distribution of disclinations.

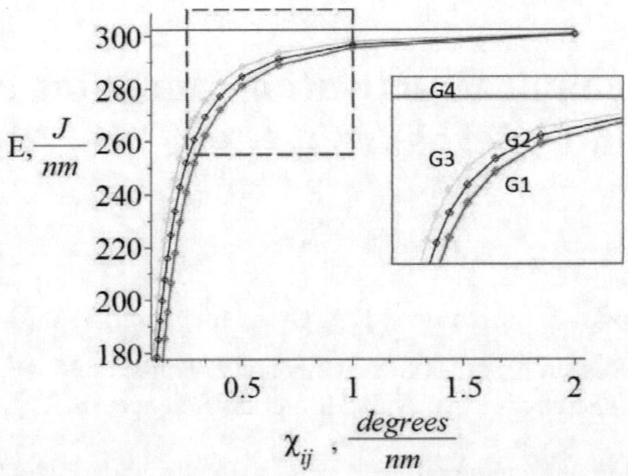

Figure 4. Comparison of the energies of disclination configurations depending on the magnitude of the component of the bending-torsion tensor. G1 – curve of energy for equidistant configuration, G2 – curve of energy for exponential configuration, G3 – curve of energy for configuration with the arithmetic progression, G4 – energy wedge disclination with Frank's vector $\Omega \approx 10°$

In the case of increasing the distance between disclinations ($\chi \to 0$), the share of exchange interaction (superposition) decreases, which leads to the degeneration of these configurations into a system of individual noninteracting disclinations (Figure 4). In other case, as the density of the disclination charge increases ($\chi \to +\infty$), there is an asymptotic tendency of the total energy of the configurations to the disclination energy with the with Frank's vector $\omega \approx 10°$.

4. Conclusion

The paper discusses disclination configurations that simulate the boundaries in nanocrystalline materials. In the study of the elastic-stress state of these configurations, features of the stress fields were found depending on the number of disclinations and the magnitude common Frank's vector. Based on the performed calculations, the estimate of the configuration energies was obtained in a wide range of values of the bending-torsion tensor, which is equal to 0.2…0.10 degrees per nanometer. It was found that the energy of the equidistant model is lower than the energy of the other two with irregular distribution of disclinations. This indicates that the equidistant configuration is an energetically more stable system.

Acknowledgments
This work was performed within the frame of the Fundamental Research Program of the State Academies of Sciences for 2013-2020, line of research III.23.

References

[1] Pumphrey P H and Gleiter H 1975 *Phil. Mag.* **32 4** 881
[2] Valiev R Z, Murashkin M Y, Ganeev A V and Enikeev N A 2012 *Phys. Met. Metallogr.* **113 13** 1193
[3] Nazarov A A and Murzaev R T 2018 Comput. Mater. Sci. **151** 204
[4] Tyumentsev A N, Ditenberg I A, Korotaev A D and Denisov K I 2013 *Phys. Mesomech.* **16 4** 319
[5] Romanov A E and Kolesnikova A L 2009 *Prog. Mater. Sci.* **54 6** 740

IOP Conf. Series: Materials Science and Engineering **597** (2019) 012047 doi:10.1088/1757-899X/597/1/012047

Influence of mechanical activation duration on morphology, microstructure and microhardness of 3Ni – Al-system powder compound

D A Osipov[1], I V Smirnov[1, 2], K V Grinyaev[1, 2] and V V Melnikov[1]

[1] National Research Tomsk State University, Tomsk, 634050, Russia
[2] Institute of Strength Physics and Materials Science SB RAS, Tomsk, 634055, Russia

E-mail: osipov_ff_tsu@mail.ru

Abstract. Investigation of influence of mechanical activation duration in high-energy planetary ball mill on morphology, microstructure parameters and microhardness of 3Ni – Al powder system was conducted. Main stages of transformation of microstructure and mechanical properties are revealed.

1. Introduction

Currently, mechanical activation (MA) is widely used in powder technologies both for milling and for mixing the components of metal powder mixtures, and for the direct synthesis of intermetallic compounds [1 – 4]. Investigation of the influence of MA conditions and duration on morphology, microstructure and strength properties level of precursors and resulting materials is still a topical task.

In the present work, using the example of 3Ni-Al powders mixture, we studied the influence of the duration of MA in energy-stressed planetary ball mills on morphology, microstructure parameters, and microhardness of the resulting precursors.

2. Experimental materials and procedures

The study was conducted using mixture of 3Ni powders (99.85 %) – Al (98 %) after MA in the AGO-2 planetary ball mill in argon. The volume of each of the two steel drums of the mill is 160 cm^3. Balls diameter is 8 mm, balls mass in each drum is 200 g, sample mass is 10 g. The centrifugal acceleration of the balls is 400 m/s^2 (40 g). Water cooling is used to prevent heating in the MA process. The duration of mechanical activation was 1, 1.5, 2.5, 3.5, 4.5, 5.5 and 10.5 minutes.

Morphology of 3Ni – Al system precursors was studied using a Tescan Vega 3 SBH scanning electron microscope (20 kV). Energy dispersive analysis was performed using FEI Quanta 200 3D (30 kV). The microstructure parameters were determined by X-ray diffraction (XRD) analysis on a Shimadzu XRD 6000 diffractometer. X-ray analysis of the phase composition, sizes of coherent scattering regions (CSR) and microdistortions ($\Delta d/d$) was performed using the PCPDFWIN database and the full-profile analysis program POWDER CELL 2.4.

Microhardness (H_μ) was determined by the Vickers method on a Neophot 21 instrument with a load of 0.5 N and exposure time of at least 10 s. After mechanical activation, measurements were performed on individual powders and their conglomerates, for which compacts were prepared based on a mixture of precursors with epoxy glue.

3. Results and discussion

The initial (before mechanical activation) state of the powders used was studied in [5]. It is shown that nickel powders have a lamellar form with a developed dendritic structure and are characterized by sizes from 5 to 15 μm. While aluminum powders have a globular shape with a smoothed surface relief, and their sizes are in the range from 30 to 100 μm.

After mixing in a mortar, 3Ni-Al system consists of small and large particles of nickel and aluminum with the initial morphology (Figure 1 a).

Figure 1. Influence of MA duration on morphology of powder compound 3Ni-Al. a) – mortar mixing; MA: b) – 1 min; c) – 1.5 min; d) – 2.5 min; e) – 3.5 min; f) – 4.5 min; g) – 5.5 min; h) – 10.5 min.

Powder after MA lasting 1 minute is characterized by the combination of fine powder into larger lamellar conglomerates (Figure 1 b), which consist of both the original powder and fragments of their milling. The size of such conglomerates reach a length of 200 μm, and the width is about 50 μm.

With an increase in processing time of up to 1.5 minutes (Figure 1 c), both the processes of milling the powders into smaller powders and their combination into large (about 100 μm) conglomerates of a highly distorted form still occur. Based on the data of scanning electron microscopy, including the results of energy dispersive analysis, it was established that fine nickel powders are hammered into soft aluminum particles and their conglomerates. Thus, already at the initial stages of processing, the processes of mechanical mixing of the initial mixture components are activated.

After 2.5 minutes of MA, an increase in the size of conglomerates is observed, which, when compacted, are transformed into a plate-like form (Figure 1 d). The sizes of such conglomerates are 100-150 μm, and their thickness does not exceed several tens μm.

MA lasting 3.5 minutes is characterized by the formation of large (up to 200 μm) compacted conglomerates, which acquire a globular shape (Figure 1 e). In this case, there is a sharp decrease in the number of individual small particles, ranging in size from several μm to several tens of μm. These features indicate the active consolidation processes.

Subsequent processing (4.5 minutes) is characterized by the beginning of intensive dispersion of large conglomerates into spherical particles with sizes ranging from several tens to 100 μm (Figure 1 f). Increasing the processing time to 5.5 minutes contributes to the further grinding of conglomerates whose dimensions do not exceed 50 μm (Figure 1 g). In addition, the appearance of particles of lamellar form, ranging in size from several μm to several tens of μm, is observed.

After 10.5 minutes of MA, large (up to 150 μm) and small (up to several tens of μm) conglomerates are observed, which consist of particles with a size of about several μm (Figure 1 h).

Figure 2 shows the X-ray diffraction patterns obtained from precursors after MA of different duration.

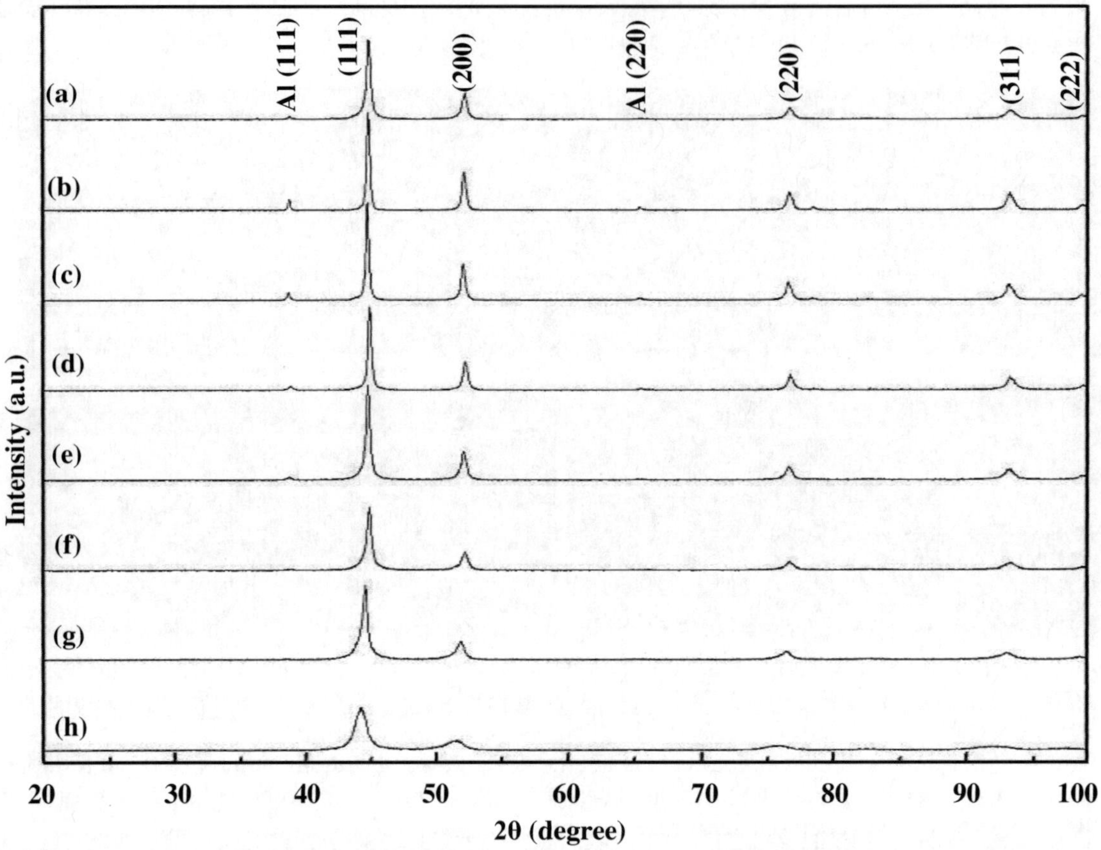

Figure 2. X-ray diffraction patterns after MA of different duration. a) – mortar mixing; MA: b) – 1 min; c) – 1.5 min; d) – 2.5 min; e) – 3.5 min; f) – 4.5 min; g) – 5.5 min; h) – 10.5 min.

As can be seen (Figure 2), an increase in the activation time of precursors is accompanied by a significant broadening of diffraction peaks, which is associated with changes in the size of coherent scattering regions (CSR) and the level of microdistortion (Δd/d). Analysis of the data obtained (Table 1) shows that with MA time increasing from 1 minute to 10.5 minutes, CSR size of the nickel component decreases from 66 nm to 24 nm, and the level of microdistortion increases from 5.2 to 14.3. At the same time, no changes in CSR and microdistortion were detected in the aluminum component in the entire processing time interval.

Table 1. CSR sizes and microdistortions level (Δd/d) of 3Ni – Al powder mixture precursors subject to MA duration (t).

t (min)		1	1.5	2.5	3.5	4.5	5.5	10.5
CSR (nm)	Ni	66	69	53	67	38	33	24
	Al				> 150			
$\Delta d/d \cdot 10^{-3}$	Ni	5.2	5.2	6.2	5.5	8.4	10.9	14.3
	Al				> 1			

Traces of Ni_3Al intermetallic phase are observed in precursors after 5.5 minutes of activation, which indicates the beginning of solid-phase synthesis directly in the MA process.

The changes above in the precursors morphology and structure are accompanied by changes in their microhardness (Table 2). As can be seen from the table, in the time interval from 1 to 5.5 minutes, the average values of microhardness vary almost linearly (from 2 to ~ 7.4 GPa). The highest value of H_μ = 8.34 GPa is reached after MA lasting 10.5 minutes. Comparison of microhardness values after 5.5 and 10.5 minutes of MA shows (Table 2) that a twofold increase in processing time contributes to an increase in H_μ only by 1 GPa. This character of hardening indicates the achievement of values of microhardness close to their maximum.

Table 2. Microhardness of 3Ni-Al precursors subject to MA duration [6].

t (min)	1	1.5	2.5	3.5	4.5	5.5	10.5
H_μ (GPa)	1.99	2.54	3.84	5.6	6.19	7.37	8.34

4. Summary

An increase in mechanical activation duration is characterized both by a change in the morphology and microstructure parameters. The mixing of aluminum and nickel components is accompanied by the transformation of conglomerates consisting of small particles. After 5.5 minutes MA, the synthesis of Ni_3Al intermetallic phase is activated in the powder mixture. The transformation of the microstructure and morphology of 3Ni-Al system precursors promotes an increase in the microhardness values with an increase in the duration of activation.

Acknowledgements

The authors are grateful to Dr. Korchagin M.A. and Dr. Ditenberg I.A. for facilitating experimental research and discussing the results. The investigations were conducted using the equipment of Tomsk Material Engineering Center of Collective Use of Tomsk State University.

This work has been supported by the Ministry of Education and Science of the Russian Federation, Project No. 3.9586.2017/8.9.

References

[1] Albiter A *et al* 2003 *Materials Science and Engineering: A* **347** 154-164
[2] Mason B A *et al* 2015 *Intermetallics* **66** 88-95
[3] Suryanarayana C 2001 *Progress in Materials Science* **46** 1-184
[4] Cherif A *et al* 2016 *Int. J. Adv. Manuf. Technol.* **86** 2937-43
[5] Ditenberg I A *et al* 2019 *Russian Physics Journal* **61(9)** 1674-80
[6] Ditenberg I A *et al* 2019 *Russian Physics Journal* **61(10)** 1947-49

Influence of oxidation duration on features of surface scale of V–Cr–Ta–Zr alloy

I V Smirnov[1, 2], **K V Grinyaev**[1, 2], **A S Tsverova**[2] **and I A Ditenberg**[1, 2]

[1] Institute of Strength Physics and Materials Science SB RAS, Tomsk, 634055, Russia
[2] National Research Tomsk State University, Tomsk, 634050, Russia

E-mail: smirnov_iv@bk.ru

Abstract. Investigation results of surface oxide scale features during the oxidation of V–Cr–Ta–Zr alloy in air are presented. Intensive oxidation processes in time interval under study lead to an increase in the scale thickness. Oxide scale structure is characterized by open porosity formed as a result of sticking together of small flakes of plate-shaped oxides (V_2O_5).

1. Introduction

In papers [1 – 3], high efficiency of chemical heat treatment (CHT) of vanadium alloys of different systems was demonstrated. Such processing contributes to a significant increase in high-temperature strength values and a significant (by several hundred degrees) increase in thermal stability of microstructure and properties. An integral stage of CHT is the formation of surface scale [4, 5], which can be formed during oxidation in air or in an oxygen-containing medium. The scale thus obtained is a source of oxygen diffusing deep into the metal during high-temperature vacuum annealing, the duration of which usually ranges from several hours to several hundred hours [4, 5]. At the same time, insufficient attention has been paid to the study of the features of structure formation and thickness of surface scale, depending on oxidation duration. In [6], the influence of structural state of V–Cr–Ta–Zr alloy samples on the oxidation rate during surface scale formation in CHT process was studied.

In this paper, we studied the effect of the oxidation duration on structure and thickness of surface scale on V–Cr–Ta–Zr alloy samples.

2. Experimental materials and procedures

We used V–Cr–Ta–Zr vanadium alloy produced by SC "VNIINM", whose chemical composition in weight (wt. %) and atomic (at. %) percent is given in Table 1.

Table 1. Chemical composition of V–Cr–Ta–Zr alloy.

Element	V	Cr	Ta	Zr	O	C	N
wt. %	base	6.80	6.10	0.79	0.052	0.031	0.009
at. %	base	6.99	1.80	0.45	0.174	0.138	0.034

Samples of this alloy were subjected to thermomechanical treatment (TMT) by traditional mode [7, 8], which assumes alternating deformation by rolling (up to $\varepsilon \approx 50$ %) at room temperature with one-hour anneals at 1000 °C in vacuum. After this treatment, one-hour annealing in vacuum at 1100 °C and deformation by rolling at room temperature ($\varepsilon \approx 50$ %) were carried out. The subsequent heat

IOP Conf. Series: Materials Science and Engineering **597** (2019) 012048 doi:10.1088/1757-899X/597/1/012048

treatment of the samples in air, leading to surface scale formation, was carried out at a temperature of 620 °C, the oxidation duration was 10, 30, 60, and 90 minutes. At this temperature an intensive formation of a metastable oxide (V_2O_5) on the sample surface occurs. Oxide scale composition was identified by TEM electron diffraction analysis. Mass and concentration of introduced oxygen were determined by weighting of samples before and after oxidation on an electronic laboratory balance A&DCOLTD GH-200 with an accuracy of no worse than 10^{-4} g.

Structural studies were performed using Philips CM30 TWIN transmission electron microscope (300 kV), scanning electron-ion FEI Quanta 200 3D (30 kV) and electron Tescan Vega 3 SBH (20 kV) microscopes. Microhardness of surface scale was determined by the prints of a Vickers diamond pyramid on a Neophot 21 instrument with a load of 0.5 N and exposure time of 15 s. At least 10 prints were made at each oxide scale thickness.

3. Results and discussion

The results of a detailed study of V–Cr–Ta–Zr alloy grain and defect structure after TMT preceding oxidation are presented in [6]. As can be seen (Figure 1 a), the microstructure is characterized by anisotropic structural state. The grains are elongated in the rolling direction (RD), forming strips of length from several tens of microns to hundreds of microns, and their width is 10 – 15 microns. Inside some grains a strongly gradient color is observed, indicating the presence of structural states with a continuous change of crystal lattice orientation. In addition, the low angle misorientation boundaries break up the grains into fragments of almost equiaxed shape and microbands, like those shown in Figure 1 b. The characteristic sizes of microbands range from a few tenths of a micron to several microns. The dislocation density reaches $2.5 \cdot 10^{10} – 3.0 \cdot 10^{10}$ cm^{-2} (Figure 1 c).

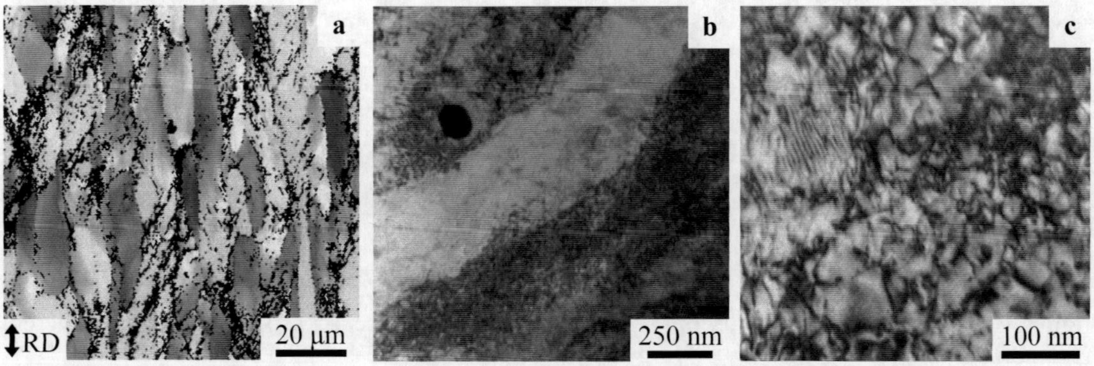

Figure 1. V–Cr–Ta–Zr alloy microstructure before oxidation [6].

Table 2 shows data on the change in mass (Δm) and concentration (C_O) of oxygen depending on the oxidation duration.

Table 2. Mass (Δm) and concentration (C_O) of introduced oxygen depending on oxidation duration (t).

t (min)	10	30	60	90
$\Delta m \cdot 10^{-6}$ (kg)	0.40	1.15	2.20	3.40
C_O (at. %)	0.28	0.85	1.58	2.26

Figure 2 shows an example of the change in oxide scale thickness on V–Cr–Ta–Zr alloy samples, depending on the duration of oxidation in air. End sections were obtained by brittle fracture of samples with oxide scale at liquid nitrogen temperature.

It is established that the thickness of the oxide scale formed is significantly inhomogeneous. For example, after 60 minutes of oxidation, local areas are observed (Figure 2c), in which the scale thickness varies from 1.8 μm to 2.8 μm. At the same time, for this treatment the scale thickness is

IOP Conf. Series: Materials Science and Engineering **597** (2019) 012048 doi:10.1088/1757-899X/597/1/012048

mainly in the range from 2 μm to 2.2 μm. Moreover, on the frontal SEM images it can be clearly seen that the scale formed consists of flakes of plate-shaped V_2O_5 oxide, the dimensions of which are in length from 3 to 7 μm and from 1 to 3 μm in width, with thickness from 0.05 to 0.3 μm (Figure 3).

At the same time, the increase in the oxidation duration contributes to the sticking of the flakes among themselves, as a result of which a false impression can be created about their size increase.

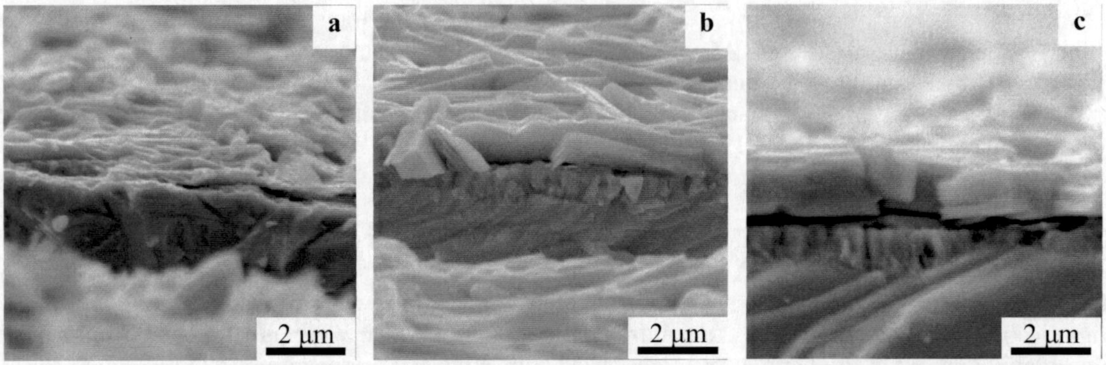

Figure 2. Electron-microscopic images of surface oxide scale end sections after different oxidation duration. a – 10 min; b – 30 min; c – 60 min.

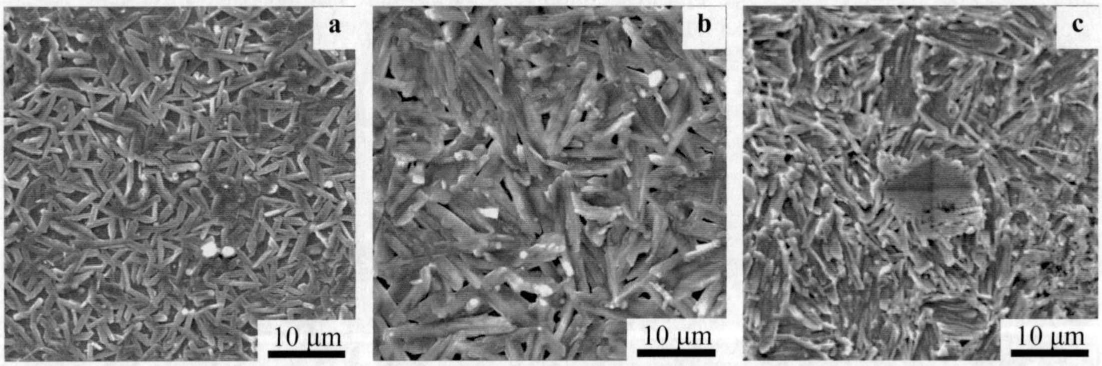

Figure 3. Frontal electron-microscopic images of surface oxide scale structure after different oxidation duration. a – 10 min; b – 30 min; c – 90 min.

Chaotic sticking of oxide flakes causes a non-uniform oxide layer to form. Note that the average thickness values are not informative at the above deviations. Thus, it is advisable to consider the characteristic intervals of the scale thickness depending on oxidation duration (Table 3).

Table 3. Thickness (h) of surface oxide scale depending on oxidation duration (t).

t (min)	10	30	60	90
h (μm)	0.4 – 1.0	1.0 – 2.0	1.5 – 3.0	2.5 – 4.0

In the process of end sections obtaining (Figure 2 and 4) by fracture at liquid nitrogen temperature, it was found that surface scale often cracks into thinner layers (Figure 4 a, b) or local chipping of some sites from the oxidized sample surface occurs (Figure 4 c). This fact testifies to its high brittleness and low adhesive strength with the surface. The averaged microhardness values of the scale after 90 minutes of oxidation reach 10 GPa with an indenter penetration depth of no more than 1.6 μm, which is 1.5 – 2.5 times lower than the scale thickness (Table 2). Note that the microhardness of the vanadium alloy prior to the oxidation does not exceed 2 GPa.

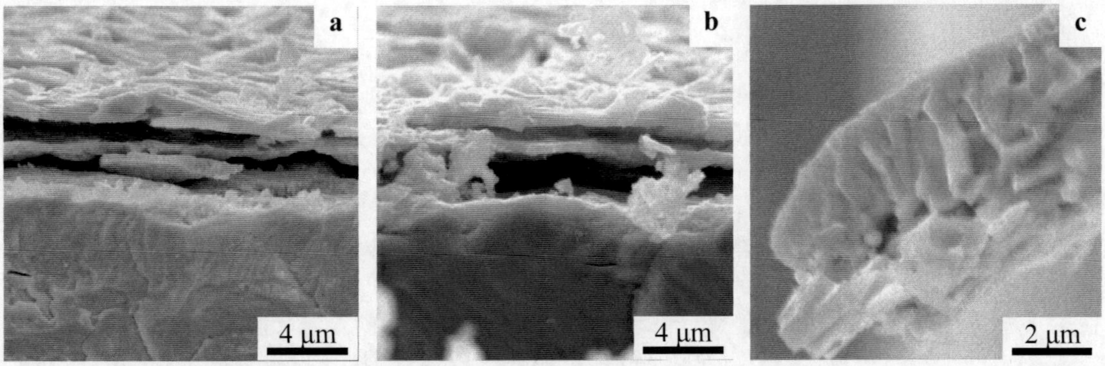

Figure 4. Electron-microscopic images of surface oxide scale end sections after different oxidation duration. a – 30 min; b – 60 min; c – 90 min.

One of the important features of surface scale formed during oxidation in air is the presence of open porosity, which is the result of the chaotic sticking of flakes together. This feature is clearly visible on the front (Figure 3) and end (Figure 4 c) SEM images. The pores on the faces in the depth of the Vickers pyramid imprint after measuring the microhardness (Figure 3 c) indicate the presence of porosity at different depths of surface scale. Moreover, the end electron microscopic image obtained from the breakaway fragment of scale shows pores throughout the entire thickness of the scale (Figure 4 c). The presence of open porosity ensures the supply of oxygen to the sample surface, on which the processes of nucleation and growth of flakes take place.

4. Summary

In the time interval studied, surface scale thickness on samples of V–Cr–Ta–Zr alloy significantly depends on the duration of oxidation in air. It is established that the surface scale consists of flakes of plate-shaped oxide (V_2O_5) and is characterized by the presence of open porosity, which provides oxygen supply to the sample surface

Acknowledgements

The authors are grateful to Professor A.N. Tyumentsev, Professor V.M. Chernov and M.M. Potapenko for facilitating experimental research and discussing the results.

Thermomechanical treatments and mechanical tests have been performed using the financial support from RFBR (Grant No. 18-08-00213_A). The microstructure studies were carried out within the framework of the Fundamental research program of the state academies of sciences for 2013 – 2020, research line III.23. The investigations were conducted using the equipment of Tomsk Material Engineering Center of Collective Use of Tomsk State University.

References

[1] Tyumentsev A N *et al* 2007 *Journal of Nuclear Materials* **367-370** 853
[2] Potapenko M M *et al* 2015 *Physics of Atomic Nuclei* **78** 1087
[3] Chernov V M *et al* 2015 *Nuclear Materials and Energy* **3-4** 17
[4] Horz G 1972 *Metallurgical Transactions* **3** 3069
[5] Birks N, Meier G H and Pettit F S 2006 *Introduction to High Temperature Oxidation of Metals* (New York: Cambridge University Press)
[6] Ditenberg I A *et al* 2018 *Russian Physics Journal* **61** 1506
[7] Tyumentsev A N *et al* 2004 *Journal of Nuclear Materials* **329-333** 429
[8] Ditenberg I A *et al* 2018 *Russian Physics Journal* **61** 936

IOP Conf. Series: Materials Science and Engineering **597** (2019) 012049 doi:10.1088/1757-899X/597/1/012049

Influence of V–Cr–W–Zr alloy microstructure on features of oxidation during chemical-heat treatment

I V Smirnov[1, 2], **K V Grinyaev**[1, 2] **and I A Ditenberg**[1, 2]

[1] Institute of Strength Physics and Materials Science SB RAS, Tomsk, 634055, Russia
[2] National Research Tomsk State University, Tomsk, 634050, Russia

E-mail: smirnov_iv@bk.ru

Abstract. Investigation results of microstructure influence on oxidation coefficient during chemical-heat treatment of V–Cr–W–Zr alloy are presented. It is shown that the formation of a high defect structural state in the alloy leads to a twofold increase in the oxidation coefficient and a significant increase in the oxygen concentration. The maximum effects of the microstructure on the oxidation processes are detected at short treatment times.

1. Introduction

The use of oxidation in the process of chemical heat treatment (CHT) is one of effective ways of modifying the structural-phase state of vanadium alloys, which significantly improve their physical-mechanical and chemical characteristics [1 – 6]. It was shown in [6] that, as a result of CHT, high density of fine-disperse high-stable particles of the second phases provide an increase in thermal stability of microstructure and mechanical properties of V–Cr–W–Zr system alloy. As is known [7], the formation of high defect structural states in metallic materials contributes to a significant acceleration of diffusion processes and the associated mass transfer effects. Thus, for the controlled production of the required volume fraction of particles of the desired dispersion, the questions of structural state influence on the processes of oxidation and redistribution of oxygen remain topical.

In the present work, we studied the influence of V–Cr–W–Zr alloy structural state on the oxidation coefficient during CHT.

2. Experimental materials and procedures

V–Cr–W–Zr vanadium alloy produced by SC "VNIINM" was used, the chemical composition of which in weight (wt. %) and atomic (at. %) percent is given in Table 1.

Table 1. Chemical composition of V–Cr–W–Zr alloy.

Element	V	Cr	W	Zr	O	C	N
wt. %	Base	4.23	7.56	1.69	0.02	0.02	0.01
at. %	Base	4.41	2.23	1.00	0.07	0.09	0.04

Samples of this alloy were subjected to thermomechanical treatment by mode II (TMT II) [8] without annealing (No. 1) and with one-hour stabilization annealing at temperatures ($T_{s.a.}$) 1200 °C (No. 2) and 1400 °C (No. 3). The samples had the same shape and surface area ($S \approx 175$ mm^2). Chemical heat treatment by internal oxidation method [6] was carried out on samples after various

IOP Conf. Series: Materials Science and Engineering **597** (2019) 012049 doi:10.1088/1757-899X/597/1/012049

combinations of TMT and stabilizing annealing (No. 1 – 3). The duration of oxidation (formation of surface scale) in air at 600 – 700 °C was 3 and 6 hours. The redistribution of oxygen inside the material was carried out by vacuum stepwise (from 600 to 1000 °C) annealing for 9 hours.

Electron backscatter diffraction (EBSD [9]) analysis was performed with scanning electron-ion microscope FEI Quanta 200 3D (30 kV). The Kikuchi patterns formed by backscatter electrons were automatically indexed by the program "TSL OIM data collection". The collected data was processed with the use of "TSL OIM analysis" software. Microstructural investigations were conducted using a transmission electron microscope Philips CM30 TWIN (300 kV).

Total oxygen concentration (C_O, at. %), achieved during CHT, was determined by the ratio $C_O \approx 100\Delta m(0.295m_1 + \Delta m)^{-1} + C_O^I$, where m_1 – sample mass before CHT, Δm – mass change during CHT, 0.295 – normalization coefficient for conversion to atomic percent, taking into account the alloy chemical composition before oxidation, $C_O^I \approx 0.07$ at.% – initial concentration of oxygen in the studied samples before CHT (Table 1).

3. Results and discussion

It was established that after treatment No. 1, the material is characterized by grains elongated in the rolling direction (RD) of more than 100 μm in length ($d_{\|}$) and 5 – 20 μm in width ($d_{\perp}$) (Figure 1 a). Gradient color is observed inside these grains, indicating the presence of small-angle misorientations. The dislocation density ($\rho_{d.}$) reaches $4{\cdot}10^{10}$ cm^{-2} (Figure 1 d).

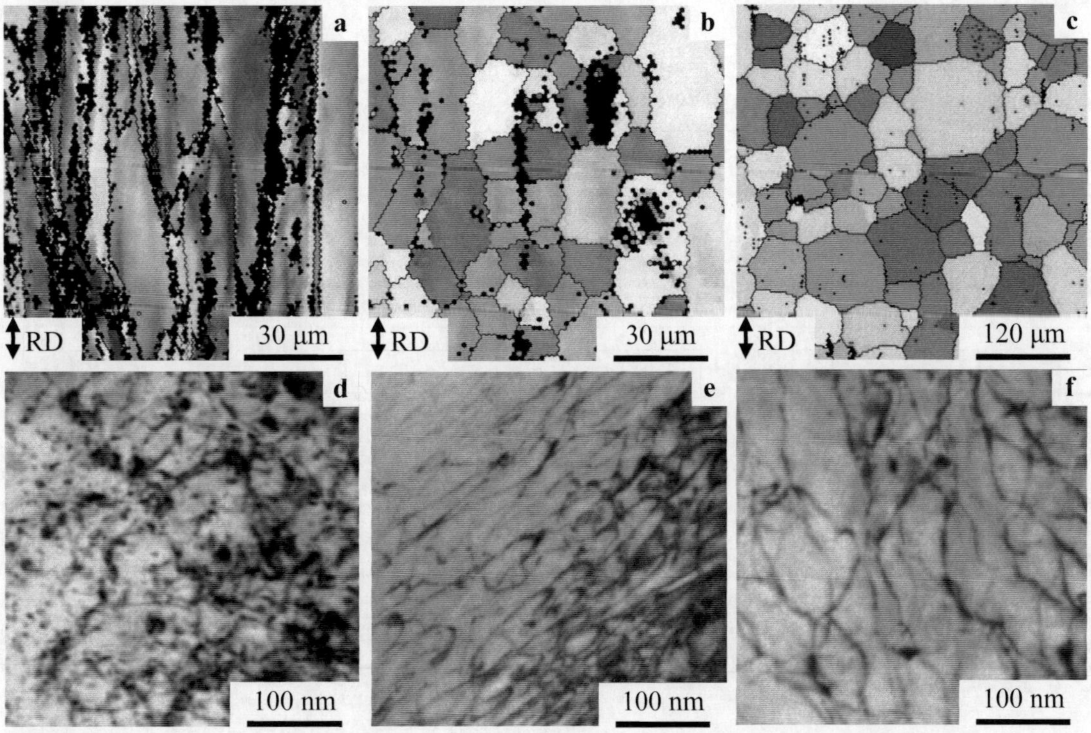

Figure 1. Orientation mapping of V–Cr–W–Zr alloy subject to TMT conditions: No. 1 (a), No. 2 (b), No. 3 (c) and corresponding defect substructure No. 1 (d), No. 2 (e), No. 3 (f).

Structural state No. 2 is characterized by the formation of fine (d ≈ 5 – 20 μm) almost equiaxed grains on the background of interlayers of coarse ($d_{\perp} \approx 10 - 20$ μm; $d_{\|} \approx 20 - 50$ μm) grains elongated in RD (Figure 1 b). In large and small grains, the alignment of the gradient color is observed. The dislocation density decreases to $2{\cdot}10^{10}$ cm^{-2} (Figure 1 e).

Processing No. 3 leads to a significant increase in grain size (d ≈ 20 – 130 μm) (Figure 1 c) and defect substructure relaxation (Figure 1 f). In all grains there is no gradient color; the dislocation density (Figure 1 f) does not exceed 10^{10} cm^{-2}.

The parameters of grain and defect structure depending on TMT conditions are summarized in Table 2.

Table 2. Parameters of grain (d, $d_\perp$, $d_\parallel$) and defect ($\rho_{d.}$) structure depending on TMT conditions.

Treatment	Grain sizes (μm)	$\rho_{d.}$ (cm^{-2})
No. 1 – TMT II (without annealing)	$d_\perp \approx 5 - 20$ $d_\parallel \approx 30 - 300$	$4 \cdot 10^{10}$
No. 2 – TMT II ($T_{s.a.}$ = 1200 °C)	$d_\perp \approx 10 - 20$ $d_\parallel \approx 20 - 50$ $d \approx 5 - 20$	$2 \cdot 10^{10}$
No. 3 – TMT II ($T_{s.a.}$ = 1400 °C)	$d \approx 20 - 130$	10^{10}

d – equiaxed grain diameter;
$d_\parallel$ and $d_\perp$ – elongated grains length along and perpendicular to RD.

Table 3 shows the values of C_O, depending on TMT conditions (structural state) and oxidation duration in air.

Table 3. Influence of TMT conditions and CHT duration (t_O) on mass change (Δm), oxygen concentration (C_O) and oxidation coefficient (k).

Treatment before CHT	t_O (h)	$\Delta m \cdot 10^{-6}$ (kg)	C_O (at. %)	$k \cdot 10^{-10}$ ($kg^2 \cdot m^{-4} \cdot s^{-1}$)
No. 1 – TMT II (without annealing)	3 6	1.75 2.39	1.41 1.84	1.4
No. 2 – TMT II ($T_{s.a.}$ = 1200 °C)	3 6	1.42 2.20	1.12 1.60	1.0
No. 3 – TMT II ($T_{s.a.}$ = 1400 °C)	3 6	1.07 1.72	0.89 1.38	0.7

It was established that during 3 and 6 hours of oxidation, the maximum mass increase (Δm) occurs in samples No. 1 (Table 3) without stabilization annealing, which are characterized by the most defect structural state (Table 2). In samples No. 2 (Table 3), stabilized at 1200 °C, Δm after 3 hours of oxidation is lower by 20 %, and after 6 hours – by 15 %. The minimum values of the mass increase are observed during the oxidation of samples No. 3 (Table 3), annealed at 1400 °C. Compared with state No. 1, Δm after 3 hours of oxidation is lower by 40 %, and after 6 hours – by 30 %.

According to [10], the consequence of oxygen concentration increase in the oxidation process is an increase in the sample mass normalized to a unit surface (ΔW = Δm/S), which is described by formula

$$\Delta W = \sqrt{k\, t_O}\,,$$

where: t_O – duration of heat treatment in air, k – oxidation coefficient.

Figure 2 shows the strait and quadratic dependences of the normalized mass change of samples after different treatments on the duration of oxidation in air. Inclination angles tangents of the approximated straight lines (Figure 2 b) represent the oxidation coefficients (k) of samples with different structural states.

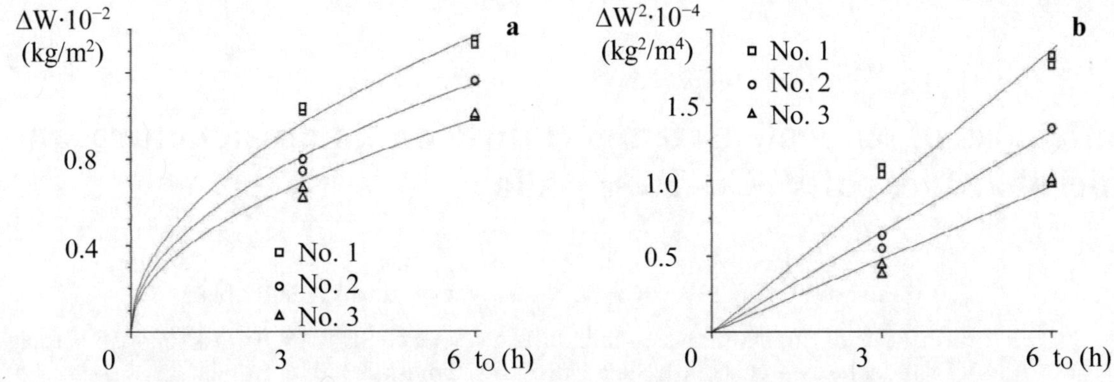

Figure 2. Strait (a) and quadratic (b) dependences of the normalized mass change (ΔW) on the duration of oxidation in air (t_O).

The presented results (Figure 2, Table 3) indicate that the oxidation coefficient (k) is sensitive to microstructure parameters (Figure 1, Table 2). The most intensive processes of internal oxidation occur in the high defect state No. 1, averaged (in the time interval from 0 to 6 hours) oxidation coefficient k of which is 1.4 times higher than after annealing at 1200 °C (No. 2) and 2 times higher than in state after annealing at 1400 °C (No. 3). Comparison of the data presented in Table 3 suggests that the maximum effect of the microstructure on the oxidation processes is detected at short treatment times (t_O = 3 h).

4. Summary

The formation of a high defect structural state in V–Cr–W–Zr alloy at the TMT stage contributes to a significant increase in the oxidation coefficient during CHT in air. At the same time, the oxidation coefficient is 1.4 times higher than in the structural state, stabilized at 1200 °C, and 2 times higher as compared to the state after recrystallization at 1400 °C.

Acknowledgements

The authors are grateful to Professor A.N. Tyumentsev, Professor V.M. Chernov and M.M. Potapenko for facilitating experimental research and discussing the results.

The work was carried out within the framework of the Fundamental research program of the state academies of sciences for 2013 – 2020, research line III.23. The investigations were conducted using the equipment of Tomsk Material Engineering Center of Collective Use of Tomsk State University.

References

[1] Tokiwai M and Morozumi S 1980 *Trans. Jpn. Inst. Met.* **21** 72
[2] Yeliseyeva O, Muroga T, Yao Z and Tsisar V 2009 *Journal of Nuclear Materials* **386-388** 696
[3] Chen J *et al* 2002 *Journal of Nuclear Materials* **302** 135
[4] Tyumentsev A N *et al* 2007 *Journal of Nuclear Materials* **367-370** 853
[5] Pint B A and DiStefano J R 2005 *Oxidation of Metals* **63** 33
[6] Chernov V M *et al* 2015 *Nuclear Materials and Energy* **3-4** 17
[7] Kolobov Yu R *et al* 2007 *Grain boundary diffusion and properties of nanostructured materials* (Cambridge: Cambridge International Science Publishing)
[8] Tyumentsev A N *et al* 2015 *Physics of Atomic Nuclei* **78(10)** 1092
[9] Adams B L *et al* 2009 *Electron Backscatter Diffraction in Materials Science* ed A J Schwartz, M Kumar, B L Adams and D P Field (Boston: Springer)
[10] Fromm E and Gebhardt E 1976 *Gases and Carbon in Metals* (Berlin/Heidelberg: Springer-Verlag)

Influence of annealing temperature on microstructure and microhardness of V–Cr–Ta–Zr alloy

I V Smirnov[1,2], A S Tsverova[2], K V Grinyaev[1,2] and I A Ditenberg[1,2]

[1] Institute of Strength Physics and Materials Science SB RAS, Tomsk, 634055, Russia
[2] National Research Tomsk State University, Tomsk, 634050, Russia

E-mail: smirnov_iv@bk.ru

Abstract. The effect of stabilizing annealing temperature on the features of microstructure and microhardness of V–Cr–Ta–Zr alloy after thermomechanical treatment by regime II was studied. Main stages of relaxation processes are revealed. Characteristic microstructure parameters and microhardness values are determined for each relaxation stage.

1. Introduction

One of the ways to increase the mechanical properties of vanadium alloys is the use of thermomechanical treatments (TMT), which provide a modification of the structural-phase state [1 – 7]. In work [7], it was shown that TMT by mode II (TMT II) of V–Cr–Ta–Zr-system alloy increases its strength values at room and elevated temperatures. A necessary step in optimizing TMT modes is to study the effect of stabilizing annealing temperature on the structural-phase state.

In this work, we studied the effect of annealing temperature on microstructure parameters and microhardness of V–Cr–Ta–Zr alloy after TMT II.

2. Experimental materials and procedures

We used V–Cr–Ta–Zr vanadium alloy produced by SC "VNIINM", whose chemical composition in weight (wt. %) and atomic (at. %) percent is given in Table 1.

Table 1. Chemical composition of V–Cr–Ta–Zr alloy.

Element	V	Cr	Ta	Zr	O	C	N
wt. %	base	6.80	6.10	0.79	0.052	0.031	0.009
at. %	base	6.99	1.80	0.45	0.174	0.138	0.034

Samples of this alloy were subjected to TMT II, which includes homogenizing one-hour annealing at 1400 °C and several rolling cycles ($\varepsilon \approx 20 - 30$ %) at room temperature with intermediate one-hour annealing at 600 – 700 °C [7]. To study the thermal stability after final deformation cycle of TMT II, the samples were subjected to annealing in the temperature range from 700 to 1500 °C. Microhardness (H_μ) was determined by the Vickers method on a Neophot 21 instrument with a load of 0.5 N and exposure time of 15 seconds. At least 10 prints were made for each structure state.

Electron backscatter diffraction (EBSD [8]) analysis was performed on scanning electron-ion microscope FEI Quanta 200 3D (30 kV) with the analytical system Pegasus in the hexagonal point staging mode. The program "TSL OIM data collection" has automatically indexed the Kikuchi

IOP Conf. Series: Materials Science and Engineering **597** (2019) 012050 doi:10.1088/1757-899X/597/1/012050

patterns formed by backscatter electrons. Collected data was processed using the "TSL OIM analysis" software. Structural studies were carried out on Philips CM 30 TWIN transmission electron microscope (300 kV).

3. Results and discussion

Grain structure of the alloy after TMT II before the stabilizing annealing (Figure 1) was studied in [7]. According to EBSD-analysis (Figure 1 a), it is characterized by grains, elongated in the rolling direction (RD), with a width of 3 to 35 μm and a length of 50 μm to several hundred μm. Gradient color is observed inside such grains, which indicates the presence of low-angle misorientation boundaries. According to the results of transmission electron microscopy, the microstructure is characterized by high heterogeneity. There are areas almost free of dislocations and areas where the scalar dislocation density ($\rho_{d.}$) reaches 10^{11} cm^{-2}. Heterophase structure of the alloy is characterized by the presence of three main fractions of second phase particles (Figure 1 b, c): large (1 – 2 μm), medium (50 – 400 nm) and nanoscale (2 – 10 nm). It is established that these particles of different fractions are zirconium-based carbides and oxycarbonitrides.

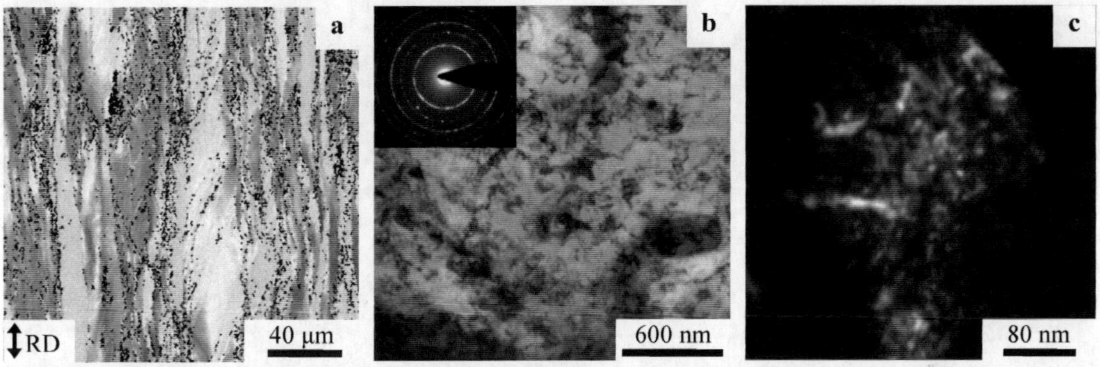

Figure 1. Microstructure of V–Cr–Ta–Zr alloy after TMT II before stabilizing annealing [7].

After annealing at 700 °C (Figure 2 a), structural state is similar to that which is formed immediately after TMT II (Figure 1 a). But at the same time, the dislocation density decreases to (3 – 7)·10^{10} cm^{-2}. Minor changes in the microstructure after annealing at 800 °C (Figure 2 b) are revealed by the appearance of areas with less pronounced gradient color or without it. In our opinion, this is a result of recovery processes, which are accompanied by a decrease in the dislocation density to (2 – 4)·10^{10} cm^{-2}. After annealing at 900 °C (Figure 2 c), near-equiaxed grains with sizes from 3 to 20 μm with a gradient-free color are formed against the background of the initial band structure. The dislocation density in such grains does not exceed 10^{10} cm^{-2}. Thus, at 900 °C primary recrystallization processes are activated.

Annealing at 1000 °C is accompanied by a sharp intensification of primary recrystallization processes in the entire volume of the samples (Figure 2 d). This leads to the formation of almost equiaxed grains with gradientless color, the dimensions of which are in range from 3 to 30 μm. The dislocation density in these grains does not exceed 10^{10} cm^{-2}. At the same time, on the background of these grains, individual large grains elongated in RD (width 20 – 30 μm, length 50 – 300 μm) are preserved in the material, inside which gradient color is still observed. Such grains are a legacy of the initial structural state.

Processes of collective recrystallization begin to occur at 1100 °C (Figure 2 e). In addition, at this temperature the relaxation of elongated initial grains into large fragments occurs, the dimensions (width 20 – 30 μm, length 40 – 70 μm) of which are several times lager than grain sizes (5 – 20 μm) after primary and collective recrystallization. There is no gradient color inside large and small grains, which indicates the relaxation of the defect substructure.

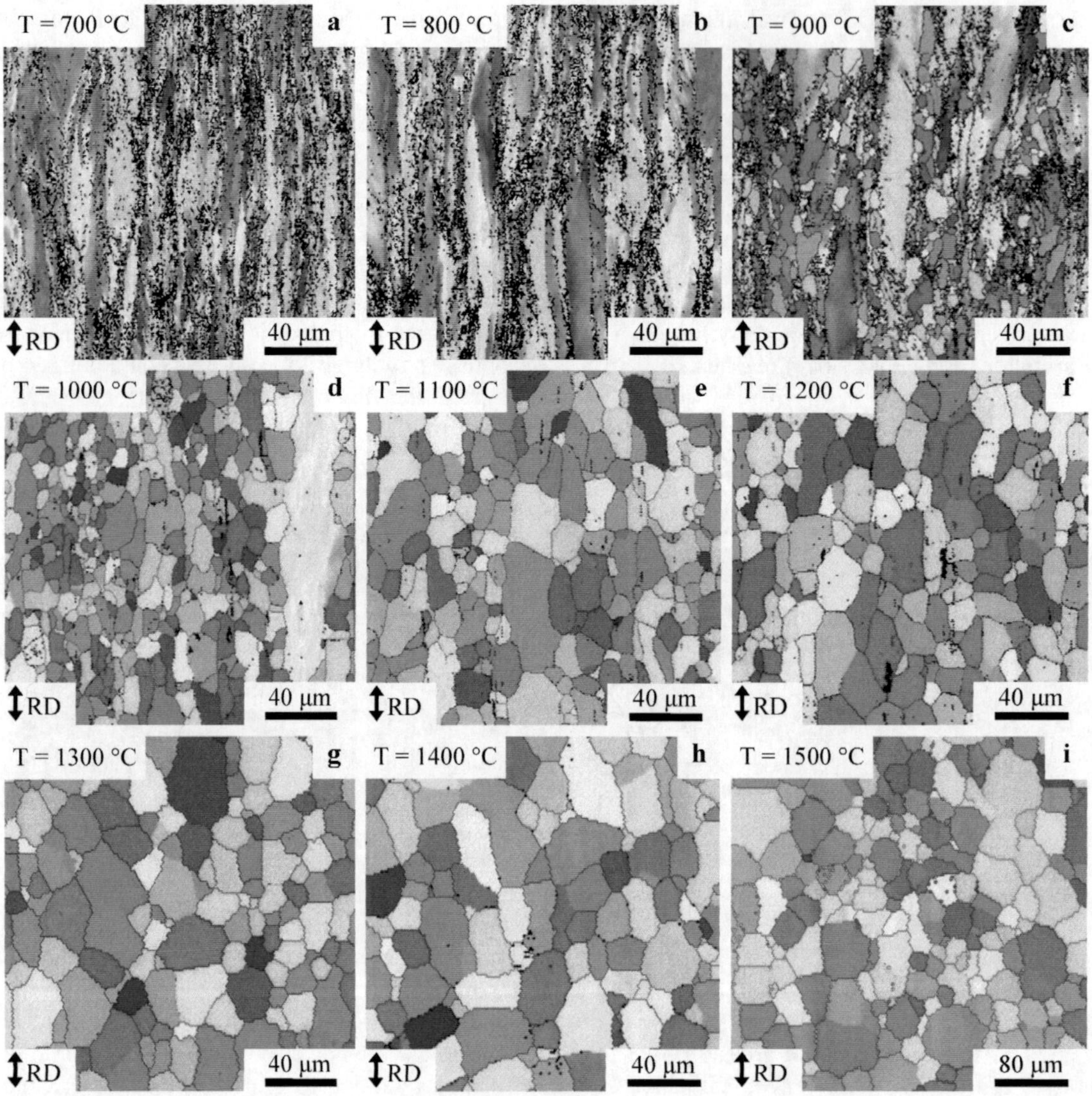

Figure 2. Grain structure of V–Cr–Ta–Zr alloy after TMT II and annealing. EBSD.

A further gradual increase in annealing temperature up to 1400 °C inclusive is accompanied by processes of collective recrystallization (Figure 2 e – h), at which the grain sizes increase in a monotonic manner. At the same time, the division into two main fractions of grains is preserved: small (10 – 25 μm) and large (30 – 50 μm).

After annealing at 1500 °C, coarse grains up to 100 μm in size appear on the background of relatively small grains (10 – 20 μm). The scatter between small and large grains increases significantly, which manifests itself during statistical processing, the results of which indicate the presence of two main fractions, characterized by intervals of 15 – 40 μm and 60 – 100 μm. Thus, in the process of annealing at 1500 °C, secondary recrystallization is activated, which leads to substantial grain growth.

Table 2 presents the results of microhardness measurement before and after stabilizing annealing in the range from 700 °C to 1300 °C.

Table 2. Microhardness of V–Cr–Ta–Zr alloy subject to annealing temperature.

Annealing temperature (°C)	–	700	800	900	1000	1100	1200	1300
H_μ (GPa)	3.64	3.00	2.64	2.10	1.52	1.57	1.53	1.56
$\pm \Delta H_\mu$ (GPa)	0.10	0.09	0.08	0.09	0.09	0.10	0.04	0.07

Accompanied by recovery processes ($\rho_{d.}$ reduced to $(3 - 7)\cdot 10^{10}$ cm^{-2} and $(2 - 4)\cdot 10^{10}$ cm^{-2}) anneals at 700 and 800 °C lead to a decrease in the microhardness values by ≈ 20 % and 30 %, respectively. A substantial (more than 40 %) decrease in H_μ after annealing at 900 °C is associated with the onset of active relaxation of defect substructure until the appearance of small defect-free grains, the dislocation density of which is 10^{10} cm^{-2}. The completion of the primary recrystallization processes at 1000 °C is accompanied by a decrease in the microhardness to 1.5 GPa, which is ≈ 60 % of the initial values (Table 2). After heat treatments in the range from 1100 °C to 1300 °C, the values of H_μ are at the same level (Table 2) within the measurement error (≈ 10 %). Dislocation density after such annealing does not exceed 10^{10} cm^{-2}. It was established that the main processes of microstructure transformation are associated with collective recrystallization, which results in the redistribution of the volume fractions of large and small grains.

The study of the effect of annealing temperature on the stability of heterophase structure is one of the topical areas for further research.

4. Summary

As a result of investigation of annealing temperature effect on microstructure and microhardness values of V–Cr–Ta–Zr alloy, the main stages of relaxation processes were revealed. Annealings at temperatures of 700 – 800 °C are accompanied by recovery processes. Heat treatment at 900 °C is characterized by the onset of primary recrystallization, the active phase of which proceeds up to 1000 °C inclusive. An increase in the annealing temperature in the range of 1100 – 1400 °C is accompanied with collective recrystallization processes. The temperature of secondary recrystallization onset is 1500 °C.

Acknowledgements

The authors are grateful to Professor A.N. Tyumentsev, Professor V.M. Chernov and M.M. Potapenko for facilitating experimental research and discussing the results.

Thermomechanical treatments and mechanical tests have been performed using the financial support from RFBR (Grant No. 18-08-00213_A). The microstructure studies were carried out within the framework of the Fundamental research program of the state academies of sciences for 2013 – 2020, research line III.23. The investigations were conducted using the equipment of Tomsk Material Engineering Center of Collective Use of Tomsk State University.

References

[1] Tyumentsev A N *et al* 2004 *Journal of Nuclear Materials* **329-333** 429
[2] Chen J M *et al* 2004 *Journal of Nuclear Materials* **334** 159
[3] Chen J M *et al* 2008 *Journal of Nuclear Materials* **374** 298
[4] Fu H Y *et al* 2013 *Journal of Nuclear Materials* **442** S336
[5] Muroga T *et al* 2014 *Journal of Nuclear Materials* **455** 263
[6] Tyumentsev A N *et al* 2015 *Physics of Atomic Nuclei* **78(10)** 1092
[7] Ditenberg I A *et al* 2018 *Russian Physics Journal* **61(5)** 936
[8] Adams B L *et al* 2009 *Electron Backscatter Diffraction in Materials Science* ed A J Schwartz, M Kumar, B L Adams and D P Field (Boston: Springer)

IOP Conf. Series: Materials Science and Engineering **597** (2019) 012051 doi:10.1088/1757-899X/597/1/012051

Study into luminescence and photoconductivity of nano-structured ZnO obtained through sol-gel method

L V Mikhnev, E A Bondarenko, O M Chapura*, A A Skomorokhov, A I Nikonov, M S Malorodov, R V Pigulev and D L Gazdinsky

North-Caucasus Federal University, Pushkin Street 1, Stavropol 355009, Russian Federation

*Email: ChapurOL-7@mail.ru

Abstract. Sol-gel method from zinc acetate in water was employed to synthesize nano-sized zinc oxide. The structural features of the ZnO obtained samples, their spectral features of luminescence and photoconductivity were studied. The influence of the treatment temperature on the luminescence and photoconductivity of nano-sized zinc oxide was studied. The type and energy position of the intrinsic defects involved in the luminescence and photoconductivity processes were determined.

1. Introduction

In recent years, a wide range of possible applications of nanomaterials and nanoparticles of zinc oxide in particular has resulted in an interest in their physical and chemical properties. Nanoparticles of zinc oxide themselves and in composites offer potential for the creation of antibacterial agents [1, 2], active materials for toxic waste processing [3], as well as for the detection of fire-hazardous and poisonous gases [4]. Zinc oxide can be used in the ultraviolet range optoelectronic devices while due to the high binding energy of the exciton (60 meV), radiation can be generated at high temperatures reaching 550 K [5]. ZnO nanoparticles can be used in the development of ultraviolet radiation sensors and high-efficiency solar cells [6, 7].

Currently, there are many methods for developing zinc oxide nanostructures, namely – sonochemical [8], sol-gel [9], hydrothermal [10], etc. These methods in combination with the possibilities of varying the synthesis technological parameters make it possible to obtain structures based on ZnO of different morphology. At the same time, the fields of their application are not determined only by the geometric parameters of the nanoparticles, yet also by a set of physical and chemical properties of the resulting group of nanoparticles, while the latter is known to be significantly affected by the defective situation in the synthesized structures.

Most often, the composition of point defects involved in radiative recombination, is successfully evaluated through studying photoluminescence spectra [11-16]. At the same time, the study of photoconductivity spectra allows investigating electronic transitions in a material, including those involving defects that lead to the appearance of free nonequilibrium charge carriers. Therefore, analysis of the luminescence spectra and photoconductivity in a combination stand as a more detailed method for estimating the defect situation.

2. Materials and methods

IOP Conf. Series: Materials Science and Engineering **597** (2019) 012051 doi:10.1088/1757-899X/597/1/012051

Zinc acetate dihydrate ($Zn(CH_3COO)_2 \cdot 2H_2O$), 25% ammonia solution ($NH_3 \cdot H_2O$) and distilled water were used as the initial materials for synthesis ZnO nanoparticles. First, a 0.5 M aqueous zinc acetate solution was prepared. To the resulting solution, at intensive stirring, the ammonia solution was added dropwise, with continuous monitoring of the pH level. At pH ≈ 8, the process was stopped, since it is at this acidity value that zinc oxide is formed from its hydroxide. Next, the received gel was centrifuged at 2500 rpm, and the resulting precipitate was triple-washed with distilled water. This procedure is needed to remove various impurity ions remaining after the reaction taking place between zinc acetate and ammonia solution. Further on, the washed gel was dried at a temperature of 110 °C for 12 hours. For the synthesis of experimental samples 1, 2, 3, the powders obtained through drying were subjected to additional thermal treatment at temperatures of 175, 270 and 400 °C, respectively.

The structural features of the obtained particles were studied with X-ray diffraction and scanning electron microscopy, using the PANalytical Empyrean and TESCAN MIRA3 devices, respectively. The measurements of optical, luminescent and photoelectric properties were done on an automated installation based on a monochromator MDR-41 (Lomo JSC). When studying the photoconductivity spectra, the powder was placed in a sandwich-type cell consisting of two quartz glasses with transparent antimony doped tin dioxide electrodes applied. The thickness of the active layer was controlled by fluoroplastic liners and was around 100 μm. Prior to the measurements, the cell with the sample had been kept in the dark for 12 hours.

3. Results and discussion

Figure 1 shows the outcomes of studying the structural features of zinc oxide based nanoparticles obtained via the sol-gel method.

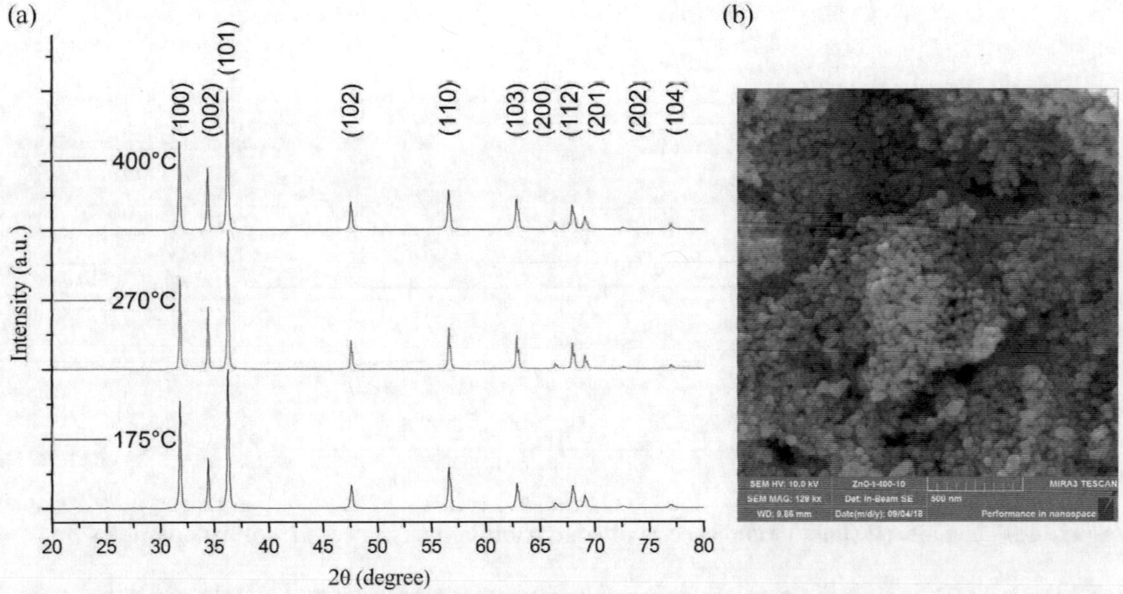

Figure 1. XRD patterns of ZnO samples annealed at different temperatures (a) and an electron microscopic image of a ZnO sample subjected to thermal treatment at a temperature of 400 °C (b).

As can be seen from figure 1, the diffractograms of ZnO samples subjected to thermal treatment at different temperatures, display peaks typical of zinc oxide in wurtzite-type structure (JCPDS card no. 36-1451). Besides, as the calcination temperature increases, the intensity and narrowing of the peaks in the diffractograms do up, which is associated with an increase in ZnO nanoparticles dimensions due to particles aggregation during sintering and improved crystallinity of the resulting samples. The estimates of the crystallite size, performed using the Debye-Scherrer formula for the (101) plane, yielded the values of 12, 30 and 38 nm for samples 1, 2 and 3, respectively.

As the results of the analysis of electron-microscopic images showed, the obtained zinc oxide nanoparticles have a shape that is close to spherical, figure 1, (b).

The optical band gap (ΔE_g) of nano-sized ZnO was evaluated through studying the diffuse reflection spectra employing the Kubelka-Munk function following the method described in [17], figure 2. The results of the evaluation numerical values can be seen in Table 1.

Table 1. Major features of nano-sized ZnO obtained through sol-gel method.

# of sample	T_{samp}, C	ΔE_g, eV	The average particle size, nm
Sample 1	175	3.45	12
Sample 2	270	3.29	30
Sample 3	400	3.27	38

The displacement of the absorption edge into the region of lower energies (observed from diffuse reflection spectra), and hence the width of the band gap, along with increasing processing temperature is associated with an increase in the size of the ZnO nanoparticles [18,19].

The luminescence spectra were measured at excitation into the fundamental absorption region with 3.96 eV quanta (the band excreted by a monochromator from the spectrum of a high-pressure mercury lamp was used as the source of excitation).

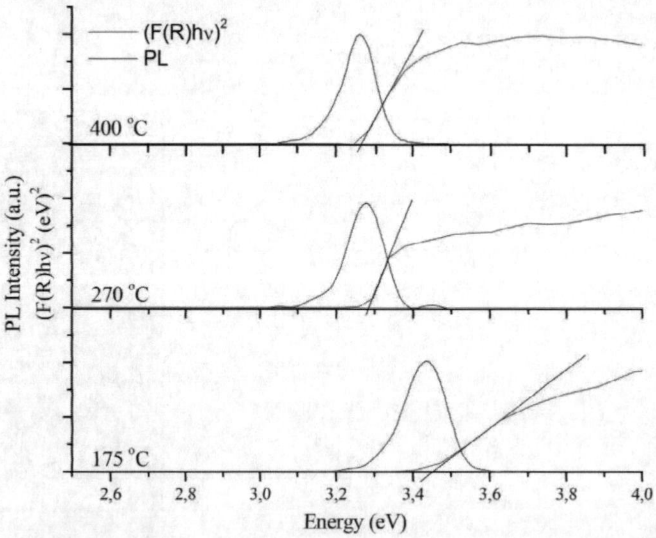

Figure 2. Tauc plots ($(F(R)h\nu)^2$) and spectra of edge photoluminescence (PL) of nanoparticles ZnO.

All the samples had luminescence in two non-overlapping ultraviolet and visible spectral regions. In the ultraviolet region, this was the edge luminescence (figure 2, curves b), where the presence of ZnO is explained by radiative recombination of free excitons [5,11]. Note to be made that the position of the luminescence maximum peak basically coincides with the values of ΔE_g obtained when analyzing the diffuse reflection spectra.

The maximum of the edge luminescence spectrum shifts to the long-wave region along with an increase in the processing temperature, which also indicates a decrease in the width of the band gap. The increase in the intensity of this luminescence at increasing temperature can be attributed to improved crystallinity of the samples.

The visible luminescence spectra in figure 3have a non-elementary form and can be decomposed into separate Gaussian components. For all the samples, the luminescence spectra consist of two elementary bands (in figure 3 they are shown as 1 and 2, respectively), whereas there is a change in the contribution magnitude and the maximums position of these bands along with an increase in the processing temperature.

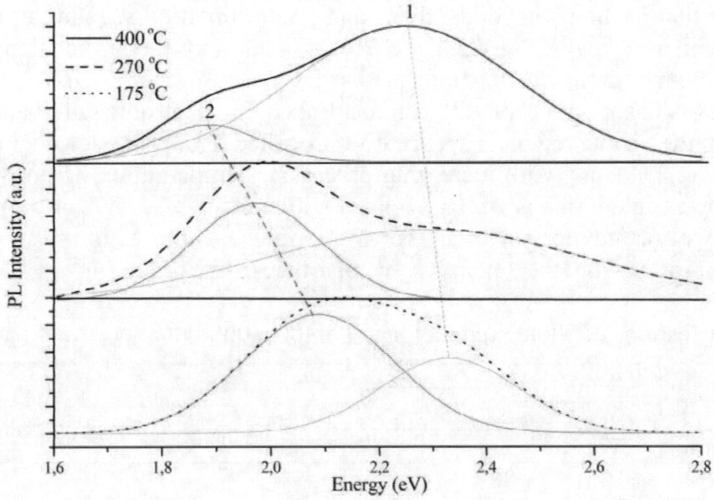

Figure 3. Photoluminescence spectra of nanoparticles ZnO (λ_{ex}=365 nm).

The maximum of luminescent band 1 is located at a wavelength of 530 nm, while its contribution goes up along with an increase in the processing temperature. For band 2, when the sample processing temperature is increased, the maximum is shifted to the long-wave region (for sample 1 the value obtained was 587 nm; for sample 2 – 635 nm; for sample 3 – 642 nm), whereas its contribution to the visible light intensity decreases.

To determine the luminescence and photoconductivity centers, measurements of the photoconductivity spectral features were carried out (figure 4, curves a) and photoexcitation (figure 4, curves b) of nano-sized ZnO. The photoexcitation spectra were measured by recording the radiation at the luminescence maximum.

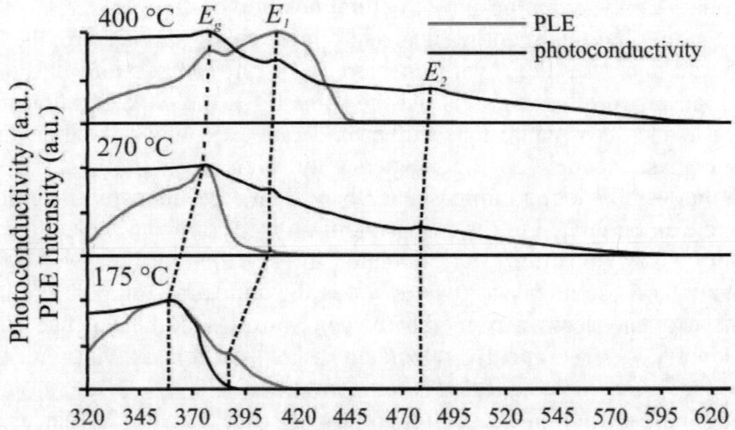

Figure 4. Spectra of photoconductivity and photoexcitation (λ_{em}=530 nm) of nanoparicles ZnO.

At the present time, two reasons have been described for the disturbed current transmission during photoexcitation of powder structures based on zinc oxide. The first reason is due to deep capture levels the presence in ZnO [20], while the second one is related to the effect that adsorbed oxygen has on the conductivity mechanisms [21]. At the same time, we believe that the nature of the current transmission does not have a significant effect on the position of the local maxima in the photoconductivity spectrum.

The figure show that in the photoconductivity and photoexcitation spectra, all the samples reveal typical peaks indicated as E_g and E_1; besides, the photoconductivity spectra for samples 2 and 3 reveal a peak in the 470-490 nm region, which is indicated as E_2.

The E_g peak in its position (Table 2) can be linked to electronic interband transitions. The photoconductivity in the visible region is apparently associated with point defects in zinc oxide, which is also mentioned in [22]. Along with increasing processing temperature, the position of these peaks shifts to the long-wave region; this is well associated with a change in the band gap size. An increase in photoconductivity observed along with an increase in the annealing temperature may be due to the hydroxyl groups desorption from the zinc oxide nanoparticles surface.

Table 2. Major features of photoexcitation and photoconductivity spectra of nano-sized ZnO.

# of sample	Peak						Relative photoconductivity value
	E_g		E_1		E_2		
	nm	eV	nm	eV	nm	eV	
Sample 1	361	3.43	387	3.2	–	–	0.17
Sample 2	375	3.3	407	3.04	479	2.58	0.19
Sample 3	377	3.29	410	3.02	484	2.56	1

When assessing a defective situation, it can be assumed that defects with a low formation energy should prevail. Such defects are vacancies of zinc V_{Zn} and oxygen V_O [19, 24], while at processing temperatures above 200 °C other defects [23, 24] should be expected, such as interstitial oxygen atoms O_i, Zinc in the Zn_i interstitium, as well as atoms in anti-structural positions of O_{Zn}, Zn_O. Of all the defects listed above, under abundance of oxygen, Zn_O reveals a very high formation energy [24], so it is unlikely to have a significant effect on zinc oxide photoluminescence and photoconductivity. Zinc in the interstitial Zn_i, as is described in [23], has a very low activation energy for migration, and hence its concentration should be minimal. Therefore, the following defects should be responsible for luminescence and photoconductivity in the obtained samples – zinc V_{Zn} oxygen V_O vacancies and interstitial oxygen atoms O_i, oxygen in the anti-structural position of O_{Zn}.

Since in our case the thermal treatment was carried out in the air atmosphere, the concentration of oxygen vacancies should decrease along with an increase in the temperature due to vacant oxygen sites being filled with adsorbed oxygen atoms. At the same time, the concentration of zinc vacancies should go up due to the release of metal ions to the oxide surface, which in some cases leads to the formation of a metallic phase that delivers the powders gray color.

Given the considerations above, the luminescence band 1, whose intensity increases along with an increasing temperature, can be related to the transition of an electron from the conduction band to the zinc vacancy. A similar energy position was calculated in [25] for the free levels of V_{Zn}^0 and V_{Zn}^{-1}. The luminescence band 2, whose intensity goes down as the annealing temperature increases, can be related to the transition of an electron from the oxygen vacancy to the valence band. The energy position of the free level V_O^{+1}, theoretically calculated in [25], correlates well with the experimental data that we obtained for ZnO sample annealed at temperature of 175 °C.

The shift in the second band luminescence maximum is probably associated with a change in the radiative recombination path. As the processing temperature rises, the concentration of oxygen vacancies decreases, while the concentration of defects of another type increases simultaneously, which may be O_i^{-1}, whose recombination, involving the free level, can give red luminescence [26].

IOP Conf. Series: Materials Science and Engineering **597** (2019) 012051 doi:10.1088/1757-899X/597/1/012051

An increase in the thermal treatment temperature leads to a significant increase in the contribution that the E_1 peak has to the photoconductivity, and apparently, it can be related to zinc vacancies. In its energy position [25], it coincides with the transition of electrons from the occupied level V_{Zn}^{-1} to the conduction band.

The photoconductivity peak E_2, which can be seen in samples annealed at above 270 °C, can be associated with defects of O_i, O_{Zn}. The concentration of interstitial oxygen and zinc atoms in an anti-structural position should go up along with increasing temperature under conditions rich in oxygen, since they perform a compensating role, similar to zinc vacancies [24]. The O_i, and O_{Zn} defects are deep acceptors and the photoconductivity peak E_2 is associated with electron transition from the O_i, O_{Zn} level to the conduction band. Mention to be made that the position of the occupied O_i^{-2} level calculated in [27] fits the data obtained in the data we obtained through our study.

4. Conclusion

The study was aimed at investigating the spectral features of the luminescence and photoconductivity of nano-sized ZnO obtained by the sol-gel method. The outcomes show along with an increase in the processing temperature, the optical width of the band gap decreases, which is probably due to an increase in the crystallite size of zinc oxide nanostructures. The obtained zinc oxide samples proved to have luminescence in the near ultraviolet and visible regions. The luminescence in the ultraviolet region is explained with the appearance of free excitons radiative recombination. The luminescence in the visible region is associated with the recombination of photoexcited charge carriers involving the levels of intrinsic point defects of zinc oxide. Zinc oxide nanostructures were revealed to possess photoconductivity where three characteristic peaks can be distinguished, of which the first one is associated with the interband mechanism of nonequilibrium charge carriers emergence, whereas the second and third ones – with intrinsic defects involved in increasing conductivity upon visible radiation absorption.

Given the results of the impact that the thermal treatment temperature has on the photoluminescence and photoconductivity spectra, possible transitions of charge carriers involving defects levels and allowed bands, were analyzed. Based on this, visible luminescence is associated with electrons transition from the conduction band to the V_{Zn}^{0} and V_{Zn}^{-1} (band 1) vacancies, and from the V_O^{+1} oxygen vacancies to the valence band (band 2). At the same time, an increase in the thermal treatment temperature results in a shift of the 2 band maximum to the long-wave region, associated with the occurrence of recombination involving O_i^{-1} defects. Photoconductivity in the visible region is associated with the occurrence of nonequilibrium charge carriers due to electron transition from the occupied levels of V_{Zn}^{-1} (band E_1) and O_i^{-2}, O_{Zn} (band E_2) to the conduction band.

References

[1] Raghupathi K R, Koodali R T and Manna A C 2011 *Langmuir* **27** 4020–28
[2] Shu Z, Zhang Y, Yang Q and Yang H 2017 *Nanoscale Res. Lett.* 12:135
[3] Hoffmann M R, Martin S T, Choi W and Bahnemannt D W 1995 *Chem. Rev.* **95(1)** 69–96
[4] Hsua C-L, Chenb K-C, Tsaic T-Y and Hsueh T-J 2013 *Sens. and Actuators B:Chemical* **182** 190-196
[5] Özgür Ü, Alivov Y I, Liu C, Teke A, Reshchikov M A, Doğan S, Avrutin V, Cho S-J and Morkoç H 2005 *J. of Appl. Phys.* **98** 041301
[6] Lim J W, Hwang D K, Lim K Y, Kang M, Shin S-C, Kim H-S, Choi W K and Shim J W 2017 *Solar Energy Materials and Solar Cells* **169** 28–32
[7] Fang J, Fan H, Tian H, Dong G 2015 *Mater. Charact.* **108** 51–7
[8] Jung S-H, Oh E, Lee K-H, Yang Y, Park C G, Park W and Jeong S-H 2008 *Cryst. Growth Des.* **8** 26–9
[9] Hayat K, Gondal M A, Khaled M M, Ahmed S and Shemsi A M 2011 *Appl. Catal.* A: General **393** 122–9
[10] Baruwati B, Kumar D K and Manorama S V 2006 *Sensors and Actuators* B: Chemical **119** 676–

IOP Conf. Series: Materials Science and Engineering **597** (2019) 012051 doi:10.1088/1757-899X/597/1/012051

82
[11] Djurisic A B, Ng A M C and Chen X Y 2010 *Progress in Quantum Electronics* **34** 191–259
[12] Zhang L, Yin L, Wang C, Lun N, Qi Y and Xiang D 2010 *J. Phys. Chem.* C **114** 9651–8
[13] Li M, Xing G, Xing G, Wu B, Wu T, Zhang X and Sum T C 2013 *Phys. Rev.* B **87** 115309
[14] Kawabata K, Nanai Y,· Kimura S and Okuno T 2012 *Appl Phys.* A **107** 213–20
[15] Camarda P, Messina F, Vaccaro L, Agnello S, Buscarino G, Schneider R, Gerthsen D, Lorenzi R, Gelardi F M and Cannas M 2016 *Phys. Chem. Chem. Phys*. **18** 16237–44
[16] Han N, Wu X, Chai L, Liu H, Chen Y 2010 *Sens. and Actuators* B **150** 230–8
[17] Stone F S 1983 UV-visible diffuse reflectance spectroscopy applied to bulk and surface properties of oxides and related solids *Surface Properties and Catalysis by Non-Metals* ed by J P Bonnelle, B Delmon, E G Derouane (Dordrecht: Springer), chapter 8 pp 237-272
[18] Chithira P R and John T T, 2017 *J. of Luminescence* **185** 212–8
[19] Koao L F, Dejene F B and Swart H C 2014 *Materials Sci. in Semiconductors Process*. **27** 33–40
[20] Nayak J, Kasuya J, Watanabe A and Nozaki S 2008 *J. Phys.: Condens. Matter.* **20** 195222
[21] Prades J D, Hernandez-Ramirez F, Jimenez-Diaz R, Manzanares M, Andreu T, Cirera A, Romano-Rodriguez A and Morante J R 2008 *Nanotechnology* **19** 465501
[22] Mishra S K, Srivastava R K and Prakash S G 2012 *J. Alloys and Comp.* **539** 1–6
[23] Janotti A and Van de Walle C G 2009 *Rep. Prog. Phys.* **72** 126501
[24] Vidya R, Ravindran P, Fjellvag H, Svensson B G, Monakhov E, Ganchenkova M and Nieminen R M 2011 *Phys. Rev.* B **83** 045206
[25] Patterson C H 2006 *Phys. Rev.* B **74** 144432
[26] Lee C T 2010 *Materials* **3** 2218–59
[27] Hu J and Pana B C 2008 *J. Chem. Phys.* **129** 154706

Study intra-4f transitions of Sm^{3+} ions under direct photoexcitation in TiO_2

O M Chapura*, L V Mikhnev, E A Bondarenko, A A Skomorokhov, R A Goncharov and A A Tsagikyan

North-Caucasus Federal University, Pushkin Street 1, Stavropol 355009, Russian Federation

*Email: ChapurOL-7@mail.ru

Abstract. A systematic study of photoluminescence transitions of samarium ion (Sm^{3+}) incorporated in titanium dioxide (TiO_2) powder with anatase structure has been performed. Perceptible changes of luminescence under direct excitation as compared with indirect excitation of Sm^{3+} in titanium dioxide (TiO_2:Sm^{3+}) have been detected. Measurements of photoluminescence excitation spectra for all detected radiative transition of Sm^{3+} were accomplished. The values of asymmetry ratio were calculated for different excitation wavelengths. The dependence of luminescent intensity on Sm^{3+} doping concentration in TiO_2 host was studied under direct and indirect excitation. Thus, evidences of the existence of the several Sm^{3+} sites in anatase TiO_2 have been presented in this work.

1. Introduction

Titanium oxide (TiO_2) is a promising material for solar energy [1], optics [2] and photocatalysis [3] because of its relatively high chemical stability, high refractive index and unique photocatalytic properties. Due to the fact that titanium oxide is a semiconductor with a large band gap, especially in the form of anatase (the band gap is 3.2 eV), it is an also attractive host for activation by rare earth ions. The introduction of rare-earth ions in titanium oxide leads to a significant modification of its properties, which expands of its potential application. As shown in [4, 5], the introduction of rare-earth ions leads to an enhancement of the photocatalytic activity of titanium oxide. In addition, the potential of using rare-earth doped titanium oxide as material for anode in dye-sensitized solar cell [6] and gas sensors [7, 8] was previously demonstrated. The introduction of rare-earth ions into the crystal lattice of titanium oxide makes it possible to obtain efficient orange phosphors, which can be used in light-emitting diodes [9].

Due to the large energy gap between the metastable $^4G_{5/2}$ level and the nearest lower level $^6F_{11/2}$, the samarium ion (Sm^{3+}) exhibits strong photoluminescence in the visible region. Recently, special attention has been paid to the research of the radiation efficiency under direct excitation of Sm^{3+} in various host due to the development of highly efficient GaN laser diodes, which have the emission band overlapped with one of the absorption bands of the Sm^{3+} [10]. The luminescent properties of Sm^{3+} in titanium oxide (TiO_2:Sm^{3+}) have been previously reported in the works [11–16]. Special mention should be made that the works presented earlier are dedicated to the study of host-sensitized luminescence. Thus, to the best of our knowledge, there is a lack of the papers dealing with research of Sm^{3+} luminescence under direct activator excitation.

In this way, the purpose of this work is to study the luminescence parameters of $TiO_2:Sm^{3+}$ under direct photoexcitation of samarium ion.

2. Materials and methods

2.1. Synthesis of nanoparticles $TiO_2:Sm^{3+}$

$TiO_2:Sm^{3+}$ samples were obtained by sol-gel method based on hydrolysis and peptization of titanium tetraisopropylate in anhydrous alcohol [17]. The initial reagents for synthesis were titanium tetraisopropylate ($Ti(OC_3H_7)_4$), isopropyl alcohol (C_3H_7OH), concentrated nitric acid (HNO_3), distilled water (H_2O), samarium oxide (Sm_2O_3). Reagent grade chemicals and distilled water with 18.2 MΩ·cm resistivity were used.

Initially, Sm_2O_3 was dissolved in hot nitric acid to form nitrate solution, which then evaporated until drying-out. Then the dry samarium nitrate was used for preparation of 0.05 M alcoholic solution. At the next stage of the synthesis $TiO_2:Sm^{3+}$, an alcoholic solution of titanium tetraisopropylate were prepared under constant magnetic stirring. Next, an alcoholic solution of samarium nitrate and distilled water was successively added dropwise to the titanium tetraisopropylate solution. After turbidity of the solution, nitric acid was added to it. The amount of titanium tetraisopropylate was selected so that its final concentration was 0.25 M. To study the dynamics of concentration quenching, samarium was taken in the range from 0.125 to 1.25 mol% compared to the concentration of titanium tetraisopropylate. Nitric acid and water were taken in relation to titanium tetraisopropylate as 2:1 and 4:1, respectively. The resulting sol was kept at room conditions for 48 hours until complete gelation.

The gel was dried at 100 °C for 24 hours. At the last stage, the sample was annealed for 2 h at a temperature of 700 ° C in oxidation atmosphere. The annealing was carried out using a laboratory electric furnace SNOL 7.2 / 1100.

2.2. Characterization methods

The structural properties of the obtained particles were studied by X-ray diffraction using the PANalytical Empyrean diffractometer. The diffraction patterns were collected using Cu Kα irradiation source (λ_{ex}=0.15406 nm). The method of scanning electron microscopy (SEM) was used to characterize the morphology of the synthesized samples. SEM images were obtained on a TESCAN MIRA3 microscope.

The photoluminescence, photoluminescence excitation and diffuse reflection spectra were measured on an automated installation assembled on the basis of a MDR-41 and MDR-23U monochromators (Lomo JSC). All spectral measurements were performed at room temperature. A high-pressure Xenon lamp was used as a source of excitation radiation in photoluminescence spectra measurement. The spectra of diffuse reflection and photoluminescence excitation were measured using a tungsten halogen lamp. The photoluminescence spectrum and photoexcitation of luminescence were corrected to instrumental response. MgO powder was used as a reference sample in all spectral measurements.

3. Results and discussion

3.1. Structural investigation of $TiO_2:Sm^{3+}$

Figure 1 shows of typical XRD pattern and SEM image of titanium dioxide particles doped Sm^{3+} and obtained via the sol-gel method. Figure 1(b) shows, that $TiO_2:Sm^{3+}$ powder is a mixture of agglomerated nanoparticles with the size below 100 nm.

IOP Conf. Series: Materials Science and Engineering **597** (2019) 012052 doi:10.1088/1757-899X/597/1/012052

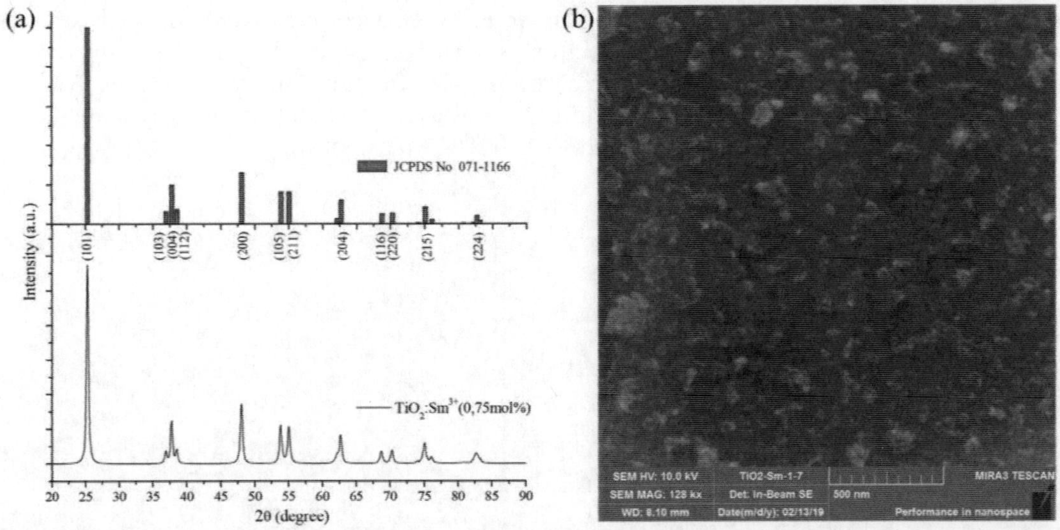

Figure 1. XRD pattern $TiO_2:Sm^{3+}$ 0.75 mol% nanocrystalline powder and the standard card (JCPDS No 071-1166) for anatase TiO_2 (a) and an electron microscopic image of $TiO_2:Sm^{3+}$ 0.75 mol% (b).

An analysis of the XRD patterns of the experimental samples showed the presence of the anatase phase of titanium dioxide (JCPDS No 071-1166). No trace of any others impurity phases such as samarium oxide were detected. The crystallite sizes calculated using the Debye-Scherrer formula decrease from 41.5 nm to 18.6 nm for samples contained 0.125 mol% and 1.25 mol% Sm^{3+} respectively. The results of measuring crystallite sizes for all experimental samples are presented in Table 1. This dependence of crystallite sizes is typical for titanium oxide particles doped with rare-earth ions [18].

Table 1. Crystallite size in dependence with Sm^{3+} concentration

Sm^{3+} concentration, mol%	Crystallite sizes, nm
0.125	41.5
0.25	36.4
0.5	27.6
0.125	23.5
0.25	21.2
0.5	18.6

3.2. Optical properties of $TiO_2:Sm^{3+}$

The estimation of the band gap for $TiO_2:Sm^{3+}$ was carried out on the basis of the Tauc plot for direct semiconductors presented in Figure 2(a). In the present measurements, the Kubelka–Munk function has been used as a measure of absorption of $TiO_2:Sm^{3+}$ samples.

The optical band gap of all $TiO_2:Sm^{3+}$ samples were found to be around 3.2 eV [19]. In context with the crystallite sizes of experimental samples, that is much bigger than exciton Bohr radius of TiO_2 [20], we can make conclusion that the effect of quantum confinement on the band gap of our examples can be neglected.

IOP Conf. Series: Materials Science and Engineering **597** (2019) 012052 doi:10.1088/1757-899X/597/1/012052

Along with the fundamental absorption region, an activator absorption band with a maximum of 2.53 eV (488 nm) was also detected on the spectrum, shown in the inset of Figure 1, a. This relatively wide band can be correlated with the transition of the Sm^{3+} ion from the $^6H_{5/2}$ state to $^4I_{11/2}$+$^4I_{13/2}$ states. The presence of activator bands in the region of 411 nm was also shown previously in the work [21]. We associated this absorption band with the $^6H_{5/2}\rightarrow{}^6P_{3/2}+{}^6P_{5/2}$ transition.

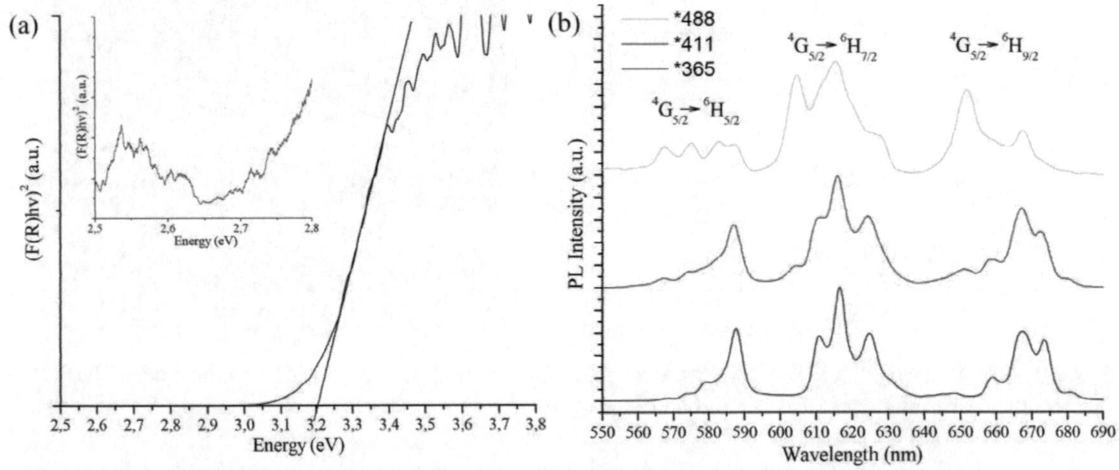

Figure 2. Tauc plot for nanoparticles of TiO_2:Sm^{3+}(0.75mol%) sample (a) and emission spectra of TiO_2:Sm^{3+}(0.75mol%) under direct (λ_{ex}=411 and 488 nm) and indirect (365 nm) excitation of Sm^{3+}.

Figure 2(b) shows the emission spectra obtained by excitation into activator absorption bands (411 and 488 nm) and fundamental absorption band (365 nm). As can be seen from Figure 2(b) emission spectrum obtained by indirect photoexcitation into consists of three groups of sharp spectral lines with the main maxima at 589, 616 and 667 nm. The intraconfigurational 4f-4f transitions of the Sm^{3+} from excited level $^4G_{5/2}$to stark splitting levels $^6H_{5/2}$, $^6H_{7/2}$ and $^6H_{9/2}$ well correspond to these maxima. However, in the case of direct excitation, one additional maximum appear to each of the multiplets from the shortwave side.

To reveal the nature of these additional peaks, the photoluminescence excitation spectra were measured for each of the observed spectral lines. The analysis of the resulting excitation spectra showed the presence of two types of photoexcitation spectral distributions, the shape of which is shown in Figure 3.

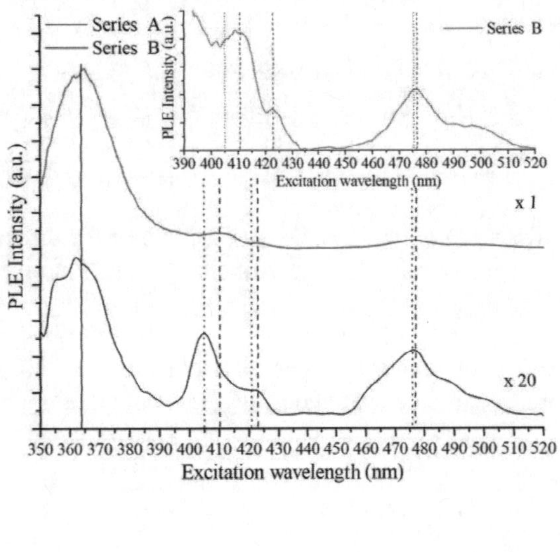

Figure 3. Photoluminescence excitation (PLE) spectrum for samarium series A and B in TiO_2:Sm^{3+}(0.75mol%) sample.

Therefore, it is possible to divide the emission lines into two series A and B with difference in the shape of the photoexcitation spectra. Series A consist of additional spectral lines (567, 604 and 648 nm), appearing under direct excitation. All spectral lines observed under indirect photoexcitation are included in series B.

Special mention should be made of the difference by the efficiency of activation of luminescence centers under excitation into the fundamental absorption band. Because, the lines of series A aren't observed in the photoluminescence spectra (Figure 2(b)) under excitation in fundamental absorption band, it can be concluded that the excitation efficiency of luminescence centers responsible for these transitions by indirect mechanism is small compared to the luminescence centers responsible for the spectral series B.

As can be seen from Figure 3, the excitation spectra of these two series have a different position of the bands in visible region. It should mention that the shape of the photoexcitation bands has a non-elementary character. Due to the fact that integrated intensity of band is produced by interplay of several transitions, the shift of the local maxima of the photoexcitation spectrum can be caused by the redistribution of the contributions of transitions Sm^{3+}. It should mention that probability of electric-dipole and magnetic-dipole transitions in different way depends from symmetry of luminescent center [22].

These differences indicate that, there are two sites of `Sm^{3+} in anatase TiO_2 with different photoluminescent characteristics. These luminescent centers were respectively named site A and site B.

To reveal the nature of the coordination environment of the luminescence center and its symmetry, the asymmetry ratio was determined upon different excitation wavelengths. Calculations of the asymmetry ratio have been performed to the procedure described in the work [23].However in the case of Sm^{3+}, asymmetry ratio is defined as integral intensity ratio of induced electric dipole transition $^4G_{5/2}\rightarrow{}^6H_{9/2}$ and magnetic-dipole transition $^4G_{5/2}\rightarrow{}^6H_{5/2}$ [10]. The analysis of the asymmetry ratio dependence, presented in Figure 4a, shows that the luminescent center responsible for additional spectral lines (site A) has value of the asymmetry ratio higher than the other luminescent center (site B). Thus, it can be concluded that Sm^{3+} placed in site A has lower symmetry than in site B.

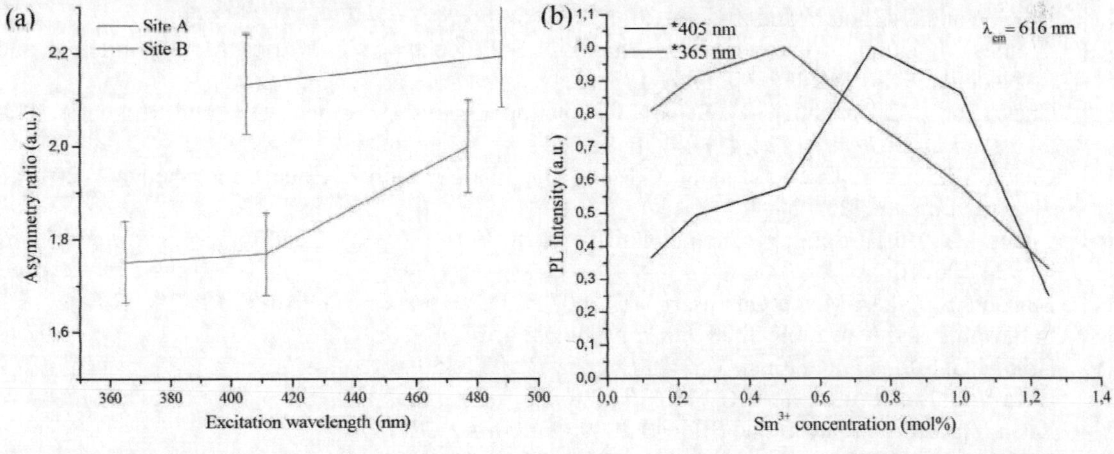

Figure 4. Asymmetrical ratio of TiO_2:Sm^{3+}(0.75mol%) as function of excitation wavelength (a) and normalized concentration dependences of $^4G_{5/2}\rightarrow{}^6H_{7/2}$ (λ_{em}=616 nm) intensity upon excitation wavelengths 365 and 405 nm (b).

Figure 4(b) shows normalized concentration dependences of the luminescence of the most intensive transition $^4G_{5/2}\rightarrow{}^6H_{7/2}$ (616 nm) under different excitation wavelengths 365 and 405 nm. As seen from figure 4(b) optimal concentration for Sm^{3+} shift depending on excitation wavelengths. So, optimum Sm^{3+} concentration for indirect excitation was determined 0.5 mol% whereas for direct Sm^{3+}

excitation (405 nm) is about 0.75 mol%. This difference can be explained by the existence of an additional energy transfer pathway, that inverse to the indirect mechanism of photoexcitation. Thus, there is probability transfer energy from Sm^{3+} to the TiO_2 host that can lead to energy dissipation.

4. Conclusion

The nanocrystalline TiO_2:Sm^{3+} with anatase structure was prepared by sol-gel methods. TiO_2:Sm^{3+}. The luminescence spectra of Sm^{3+} in TiO_2 host were obtained under direct and indirect photoexcitation. Additional spectral lines for each transition Sm^{3+} in visible region were observed upon direct excitation. The spectral distributions of luminescence photoexcitation for all detected spectral lines were measured. The symmetry of luminescent centres local surrounding was estimated using the asymmetry ratio. Photoluminescence intensity as a function of doping concentration has been studying under direct and indirect excitation. Numerous experimental evidences of two Sm^{3+} sites in the anatase crystal lattice were obtained.

References

[1] Roose B, Pathak S and Steiner U 2015 *Chem. Soc.Rev.* **44** 8326–49
[2] Huang H, Xie Y, Zhang Z, Zhang F, Xu Q and Wu Z 2014 *Appl. Surf. Sci.* 2014 **293** 248–54
[3] Liu Y, Li Z, Green M, Just M, Li Y Y and Chen X 2017 *J. Phys. D: Appl. Phys.* **50(19)** 193003
[4] Reszczyńska J, Grzyb T, Wei Z, Klein M, Kowalsksa E, Ohtani B and Zaleska-Medynska A 2016 *Appl. Catal. B-Environ.* **181** 825–37
[5] Stengl V, Bakardjieva S and Murafa N 2009 *Mater. Chem. Phys.* **114(1)** 217–26
[6] Liu M, Hou Y and Qu X 2017 *J. of Mater. Res.* **32** 3469–76
[7] Eltermann M, Utt K, Lange S and Jaaniso R 2016 *Opt. Mater.* **51** 24–30
[8] Eltermann M, Kiisk V, Berholts A, Dolgov L, Lange S, Utt K and Jaaniso R 2018 *Sens. Actuators, B* 265 556–64
[9] Zhu C, Lv C, Gao Z, Wang C, Li D, Ma X and Yang D 2015 *Appl. Phys. Lett.* **107** 131103
[10] Lupei A, Tiseanu C, Gheorghe C, Voicu F 2012 *Appl. Phys. B* **108** 909–18
[11] Hu L, Song H, Pan G, Yan B, Qin R, Dai Q, Fan L and Li S and Bai X 2007 *J. Lumin.* **127** 371–6
[12] An G, Yang C, Zhou Y and Zhao X 2012 *Phys. Status Solidi A* **209(12)** 2583–8
[13] Vranjes M, Kuljanin-Jakovljević J, Ahrenkiel S P, Zekovića I, Mitrića M, Šaponjića Z and Nedeljkovića J M 2013 *J. Lumin.* **143** 453–8
[14] Chakraborty A, Debnath G H, Saha N R, Chattopadhyay D, Waldeck D H and Mukherjee P J. 2016 *Phys. Chem. C* **120(41)** 23870–82
[15] Chang M, Song Y, Chen J, Zhang X, Meng D, Zhua H , Shib Z, ZouaH and Shenga Y 2017 *J. Alloy Compd.* **725** 724–38
[16] Chang M, Zou H, Song Y, Chen J, Cui L, Sheng Y and Zheng K 2017 *J. Phys. Chem. Solids* **124** 100–10
[17] Mahshid S, Askari M and Ghamsari M S 2007 *J. Mater. Process. Technol.* **189** 296–300
[18] Setiawati E and KawanoK 2008 *J. Alloy Compd.* **451** 293–296
[19] Brik M G, Sildos I and Kiisk V 2010 *Physica B* **405** 2450–6
[20] Luca V 2009 *J. Phys. Chem. C* **113** 6367–80
[21] Kiisk V, Reedo V, Sild O and Sildos I 2009 *Opt. Mater.* **31** 1376–9
[22] K Binnemans 2015 *Coord. Chem. Rev.* **295** 1–45
[23] Kolesnikov I E, Povolotskiy A V, Mamonova D V, Kolesnikov E Y, Kurochkin A V, Lähderanta E and Mikhailov M D 2018 *J. Rare Earth* **36** 474–481

IOP Conf. Series: Materials Science and Engineering **597** (2019) 012053 doi:10.1088/1757-899X/597/1/012053

Method for extraction of gold and silver from technogenic pyrite raw materials

V I Sachkov, R A Nefedov and R O Medvedev
Innovation and Technology Center, National Research Tomsk State University, 36 Lenin Street, Tomsk 634050, Russian Federation

E-mail: rodionmedvedev7@gmail.com

Abstract. The paper contains the results of hydrometallurgical processing of technogenic pyrite raw materials. The method is based on selective extraction of gold and silver from ore by solution, which contains sodium sulfite and sodium sulfate. The most promising results were obtained utilizing solution with sodium sulfite and sodium sulfate concentration of 25 g/L and leaching time – 1 hour. Under these conditions, the extraction of gold was 67%. The use of sodium sulphate and sodium sulfite allows to extract gold and partially silver from refractory sulfide ores.

1. Introduction

In connection with the depletion of reserves of rich and easily enriched gold-bearing ores, it is necessary to involve in the production technogenic pyrite raw materials, which are promising and easily available [1–3]. At the same time, complex processing of technogenic pyrite raw materials solves the problem of environmental hazards during storage in landfills [4–6]. The problem of waste disposal is acute for all of Russia's regions [7]. According to the data, the amount of nonrecycled waste in Russia is estimated at approximately 82 billion tons [8]. The gold and silver reaches are respectively 2.5 and 20 g/t [9]. The content of components is enough for their industrial extraction. Complex economical processing of technogenic deposits is possible with the creation of a rational environmentally processing technology [10]. The paper studies influence of the effect of time on the sulfate-sulfite leaching of gold and silver from technogenic pyrite raw materials.

2. Materials and methods

2.1. Ore

In the current work, we used a pyrite raw material from Sverdlovsk Region, Russia. Raw materials are tailings of flotation concentration of sulfide copper-zinc ores. Pyrite technogenic raw materials are light brown spherical odds agglomerates.

2.2. Leaching tests

The method was based on the leaching of raw materials using sodium sulfite and sodium sulfate. In this paper, the effect of time leaching on the target extraction components was studied. The concentration of sodium sulfate and sodium sulfite was 25 g/l. The time ranged from 1 to 122 hours. The masses of the studied samples were 100g. The mass ratio of solid: liquid = 1:2. The process was carried out in 1000 ml polyethylene agitators with constant stirring at room temperature. After leaching, filtration was

IOP Conf. Series: Materials Science and Engineering **597** (2019) 012053 doi:10.1088/1757-899X/597/1/012053

carried out with the suction filter. The solid residue was rinsed by the water on a suction filter. Then the precipitate was dried at 80 ° C in the oven to constant weight. Next, we analyzed the content of gold and silver in the solid phase.

2.3. Use of analytical techniques
The elemental composition of the samples was investigated by the method of energy-dispersed X-ray spectroscopy using a Hitachi S-3400N electron microscope. The content of microimpurities in the solid phase was determined by inductively coupled plasma mass spectrometry using an ICAP 6200 Duo spectrometer. Particle size analysis was investigated by screen test.

3. Results and discussions
The mass and atomic content of technogenic pyrite raw materials obtained by elemental analysis are presented in table 1.

Table 1. The elemental composition of pyrite raw materials.

Element	Mass content, wt.%	Atomic content, atm.%
Iron (Fe)	41.2	23.3
Sulfur (S)	26.6	26.1
Oxygen (O)	17.7	35.2
Silicon (Si)	6.5	7.2
Calcium (Ca)	3.6	2.8
Aluminum (Al)	2.5	3.0
Magnesium (Mg)	1.8	2.3

The bulk of the pyrite raw material represented by compounds of iron oxides and iron sulfides. The raw materials contain significant amounts of silicon (Si), calcium (Ca), aluminum (Al), magnesium (Mg).

The results obtained by ICP - MS analysis of technogenic pyrite raw materials are presented in table 2.

Table 2. The results of the ICP - MS analysis of pyrite raw materials.

Element	Content, g\t
Gold (Au)	1.15
Silver (Ag)	13.6
Copper (Cu)	1200
Zinc (Zn)	1500
Arsenic (As)	1400
Barium (Ba)	3900
Antimony (Sb)	57
Lead (Pb)	380
Indium (In)	2
Germanium (Ge)	1.5
Selenium (Se)	110
Manganese (Mn)	110
Cobalt (Co)	120

The gold and silver content are 1.15 and 13.6 g/t respectively. Copper, zinc, arsenic and barium are found in raw materials in significant concentrations. The interfering components may be arsenic and copper. Copper can be a catalyst for the oxidation of sodium sulfite. This affects to the consumption of sodium sulfite. Mercury can form complex compounds with the components of the leach solution, which will also lead to increase the consumption of sulfite and sodium sulfate.

The granulometric composition is an important technical indicator for raw materials. Large particles of material will include particles of the target components. Particle size distribution of technogenic pyrite raw materials are presented in table 3.

Table 3. The results of the particle size analysis of pyrite raw materials.

Particle size, mm	Mass content, wt.%
>1	1.30
0.9 – 1	53.43
0.8 – 0.9	2.90
0.7 – 0.8	3.30
0.6 – 0.7	3.03
0.5 – 0.6	2.77
0.4 – 0.5	1.72
0.3 – 0.4	1.85
0.2 – 0.3	4.62
0.1 – 0.2	5.28
0.1<	21.11

As the table 3 shows more than half of the particles had a particle size of 0.9-1 mm. Particles with a size of <0.1 mm in the raw material contained at the level of 20%.

The experiments results are shown in the graphs of concentration silver and gold in the solid phase from the interaction time when concentration of sodium sulfate and sodium sulfite is 25 g/l (figure 1,2).

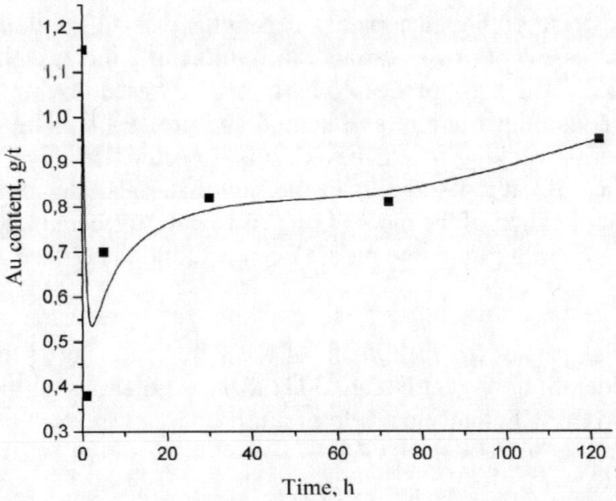

Figure 1. The dependence of gold content in the solid phase from the leaching time.

Analysis of the experimental data showed that with increasing process time, gold extraction decreases. The lowest gold content in the solid phase was achieved in the raw material samples in the first hour of processing. The degree of gold extraction in the first hour of leaching is 53%. Further, the degree of extraction is reduced to 20%. Then the gold content began to increase which may indicate the presence of natural sorbents in the raw materials. It may also be due to the decomposition of complex gold compounds.

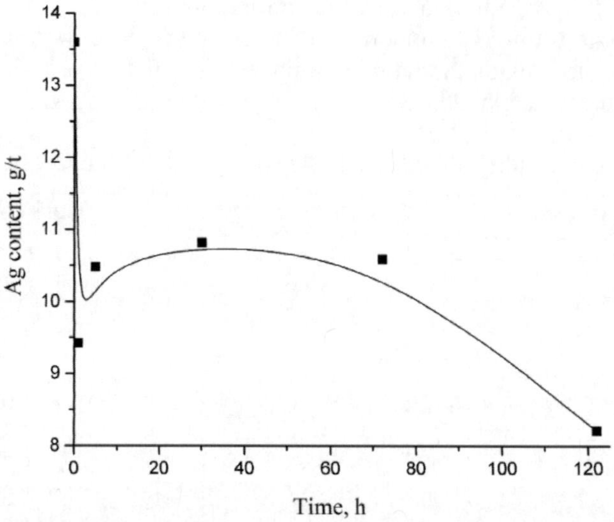

Figure 2. The dependence of silver content in the solid phase from the leaching time.

Figure 2 shows two pronounced extremums. According to the graph the silver content has a maximum in the point of 24 hours. In the first hour of the process there is a minimum on the graph (the silver content decreases rapidly). There is an upward trend of silver content from 5 to 24 hours. After 24 hours of processing, the silver content in the raw materials begins to decrease. The degree of silver extraction in the first hour of leaching is 26.5%, for 120 hours is 40%. Increasing the extraction time does not bring enough positive effect to the extraction of silver, to justify the increase in process time.

4. Conclusion

Technogenic pyrite raw materials with gold content – 1.15 g/t, silver – 13.6 g/t is a promising to produce target components. The processing of raw materials can significantly increase the extraction of precious metals. Raw materials can be efficiently processed by a method based on the interaction of pyrite raw materials with a solution containing sodium sulfate and sodium sulfite. The maximum extraction of components in the first hour of leaching was achieved. It is expected that this is due to the instability of the complex or the presence of natural sorbents in the raw materials. But the degree of extraction is insufficient for industrial application of the method. In this regard, further studies are aimed at studying the effect of the component concentration, the medium acidity, and the choice of oxidizing agents.

Acknowledgments

The results were obtained as part of the fulfillment of the state task of the Ministry of Education and Science of the Russian Federation, project No. 10.3031.2017 / 4.6 and with the support of the Tomsk State University Competitiveness Enhancement Program.

References

[1] Bocharov V A, Ignatkina V A, Lapshina G A and Khachatryan L S 2018 *Tsvetnye Met.* 27–34
[2] Chryssoulis S L and McMullen J 2005 *Dev. Miner. Process.* **15** 21–71
[3] Ivanik S A and Ilyukhin D A 2017 *J. Ind. Pollut. Control* **33** 891–7
[4] Kantemirov V D, Titov R S and Yakovlev A M 2019 *Metallurgist* **62** 1173–80
[5] Silva V P R, Passos F A C M, Domingos L M B, Faria R B, Castilhos Z C and Silva F A N G 2017 *Miner. Met. Mater. Ser.* **Part F7** 269–78
[6] Tugrul N, Moroydor Derun E and Piskin M 2007 *Powder Technol.* **176** 72–6
[7] Sizyakov V M, Bazhin V Y, Selishcheva T A and Vlasov A A 2014 *Metallurg* **1** 4–7

[8] Gilmutdinova R A, Michurin S V, Kovtunenko S V and Elizarieva E N 2017 *Adv. Curr. Nat. Sci.* **2** 68–73
[9] Aleksandrova T N, Romashev A O and Semenikhin D N 2015 *Metallurgist* **59** 330–8
[10] Tamara M, Valentin C, Tatiana I and Nadezhda G 2012 *26th International Mineral Processing Congress, IMPC 2012: Innovative Processing for Sustainable Growth - Conference Proceedings* pp 5391–400

IOP Conf. Series: Materials Science and Engineering **597** (2019) 012054 doi:10.1088/1757-899X/597/1/012054

Determination of a focusing system parameters in generation of THz radiation by bunch train

A Altmark[1], N Lesiv[2] and K Mukhamedgaliev[3]

[1]Department of Physics, St.-Petersburg State Electrotechnical University, St. Petersburg, Russian Federation
[2]Department of Electronic Instruments and Devices, St.-Petersburg State Electrotechnical University, St. Petersburg, Russian Federation
[3]Department of Photonics, St.-Petersburg State Electrotechnical University, St. Petersburg, Russian Federation

E-mail: kamilwork9@gmail.com

Abstract. It is present electron optics optimization for terahertz radiation generation by a sequence of short electron bunches passing through the dielectric loaded waveguide. In this method the superposition of individual bunch wakefields is used to selectively excite high frequency TM0n modes in a slow-wave structure. The main problem of this scheme is the transverse beam instability that limits the length of THz radiation wave packet. The main goal of this paper is a new parameter optimization algorithm for the FODO focusing system. The numerical modelling of the beam dynamics was performed in the original code "Dielectric waveguide" which utilizes analytic expression of the Green's function for the dielectric loaded waveguides with cylindrical and rectangular geometries.

1. Introduction

THz radiation has many applications in physics, chemistry and biology. There are a lot of methods to generate THz radiation using relativistic electron bunches [1]. The wakefield created by relativistic electrons in dielectric-loaded waveguides (DLW) has been studied for power THz radiation. When single bunch passes through DLW it generates wakefield radiation (Cherenkov radiation in a slow wave structure) [2]. This radiation consists of many TM modes and can be calculated analytically. A bunch train can be used to selectively excite specific modes [3, 4]. As it was pointed out in paper [3] this method has additional limitations due to beam break-up (BBU) instability. This paper contains calculations of focusing system parameters to increase the length of the wave packet.

The problem in hand has the following geometry. A train of electron bunches passes through a cylindrical dielectric tube metallized on the outside (figure 1). The radiation is generated by the sequence of relativistic electron bunches that pass along the waveguide axis. While single bunch generates a set of different TM0n-modes in the DLW, the bunch sequence allows selection of only one of the TM0n-modes via superposition. Tuning the bunch train spacing one can select different TM0n modes.

The main problem with this scheme is BBU-instability [6] due to a strong radial field. This field includes Hybrid Electro-Magnetic (HEM) modes that are excited when there is a radial deviation (offset) of the bunch from the waveguide axis. The instability of the bunch train is stronger than that of the single bunch, because bunches in the middle part of the sequence experience action not only from their own radial field but also from the fields of bunches moving ahead of them. This fact limits the length of the wave packet. One way to increase the efficiency of radiation generation is to control the transverse instability through a quadrupole focusing system. This system contains focusing and defocusing quadrupole cells and is known as the FODO system (Figure 1). This article represents a simple algorithm for FODO lattice optimization based on the numerical simulation of the beam dynamics.

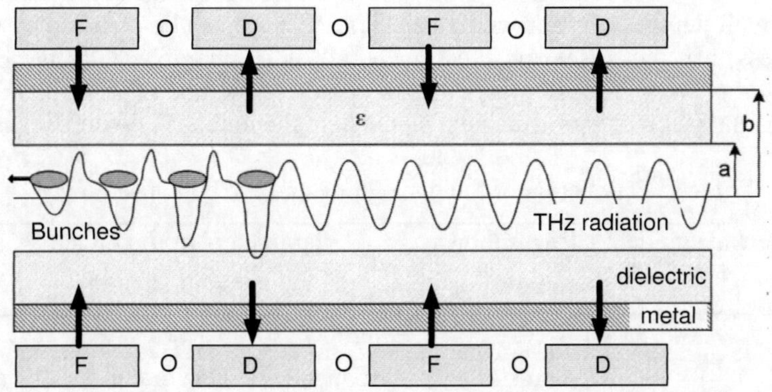

Figure 1. The sequence of bunches is passing through the dielectric waveguide with the FODO system.

2. "Dielectric waveguide" code

The "Dielectric waveguide" code is developed at the physics department of Saint-Petersburg Electrotechnical University "LETI". This code was developed for the purpose of calculating the wakefield created by high-current electron bunches in a dielectric waveguide [7]. This wakefield can be used for particle acceleration or as a source of THz radiation. The wakefield calculation is based on the explicit analytical formulas for the Green's function for DLWs with cylindrical and rectangular geometries. This method is also used in new module developed for the wakefield calculation from the sequence of bunches [6]. This module allows the user to control all parameters of the bunch train and prepare parameters for beam dynamics simulation.

The beam dynamics module works according to the macroparticle algorithm described in [8, 9]. It should be noted that the interaction between macroparticles includes not only Cherenkov radiation, but also the Coulomb field. The motion of the macroparticle is calculated using the Euler and Runge-Kutta methods. In addition, an external FODO quadrupole system can be included in the calculation of the beam dynamics. The program calculates particle motion using the equation of force and a predetermined time step.

3. Modeling of the FODO focusing system

The waveguide parameters for the THz radiation source were taken from our previous studies [7]. These parameters are presented in Table 1. The Green's function includes the TM and the HEM modes of the waveguide. We made a calculation for the simple sequence, which contains three bunches (Figure 2). We observed monochromatic radiation field Ez with the frequency of the TM02 mode (439 GHz) behind the bunch train, as well as the transverse field Fr. Frequencies of the HEM modes differ from the TM ones. So, the distances between bunches are selected to maximize generation of the TM mode. However, the distribution of the radial field depends on the spectrum of the HEM modes. This field can cause strong deflection within the bunch train. The maximal value of field Fr is at the tail of the third bunch (Figure 2, green line).

Design of the focusing system is based on maximization of propagation length. It is known that a quadrupole focusing system cannot reduce the phase space volume of the bunch. Incorrect selection of the FODO system can destroy the bunch train. The FODO focusing system is used to control the instability of a single bunch in the wakefield acceleration scheme [5]. The algorithm for calculating the FODO system for a series of Gaussian bunches (Table 1) is considered.

It is selected a simple focusing system, presented in figure 3, which consists of only two focusing cells and one defocusing cell. The system contains eight parameters for optimization (Figure 3): focusing gradients (G1, G2, G3), distances between cells (o1, o2), width of cells (z1, z2, z3).

The algorithm is based on variation of the focusing system parameters. The range of parameters is determined by the real physical constrains of the FODO system. At the first step, the algorithm makes calculations for a small number of macroparticles. This approach can be justified by small beam sizes: the radial size is less than the offset and the wavelength is comparable with the longitudinal bunch length. The optimization was implemented using the gradient method. The results achieved after the first stage give us initial parameters for the next step where the number of macroparticles is increased.

Table 1. Parameters of the waveguide and the bunch sequence.

Parameters of the waveguide	**Parameter value**	Parameters of the bunch	**Parameter value**
Inner radius “a”	600 um	Number of bunches	3
Outer radius “b”	850 um	Frequency of THz radiation	439 GHz
Relative permittivity	3.8	Distances between bunches	1320 um
TM01 mode frequency	149.5 GHz	Transverse beam size	100 um
TM02 mode frequency	439.5 GHz	Offset relative to the x	100 um
TM03 mode frequency	768.5 GHz	Longitudinal length	100 um
Length	10 cm	Energy	~75 MeV
Loss tangent	0.001	Charge	0.1 nC
Wall conductivity	5.7E+07 S/m	Number of macroparticles in bunch	200

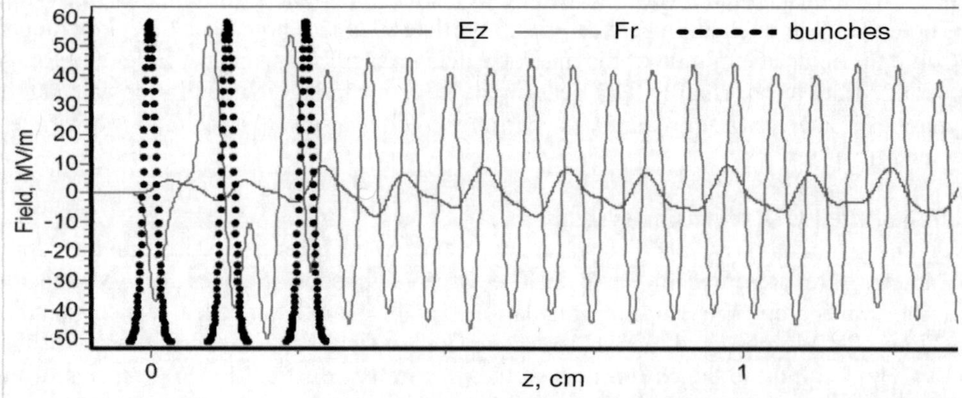

Figure 2. Longitudinal Ez and transverse Fr fields created by three bunches passing from rigth to left.

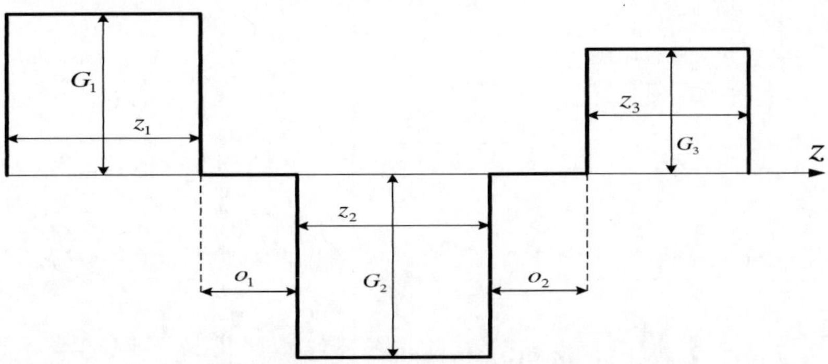

Figure 3. Longitudinal distribution of the FODO system with the input parameters of our algorithm.

4. Calculation results

First, we obtained the optimal parameters of the FODO system for the parameters (Table 2) of the dielectric waveguide listed in Table 1. These results show the dependence of the transverse bunch dimensions as a function of the longitudinal coordinate z. The focusing system allows the propagation distance to be increased to 18 cm. The results of the calculations iterated six times are shown in Figure 4. The longitudinal plane shows the up and down borders of the bunch train. These borders are defined according to Figure 5. The bunch path in the case that the FODO system is presented (or absent) as well (Figure 5, black trace, the propagation distance is 8 cm). These upper parameters are suitable for a certain offset range (0 < offset < 100um). Low offset value increases the propagation distance in the absence of FODO, but the presence of last one does not change the value of 18 cm. The offset increasing reduces the propagation length. For example: if the offset is 250 um, the propagation length without FODO is 5cm, with FODO it is 8 cm.

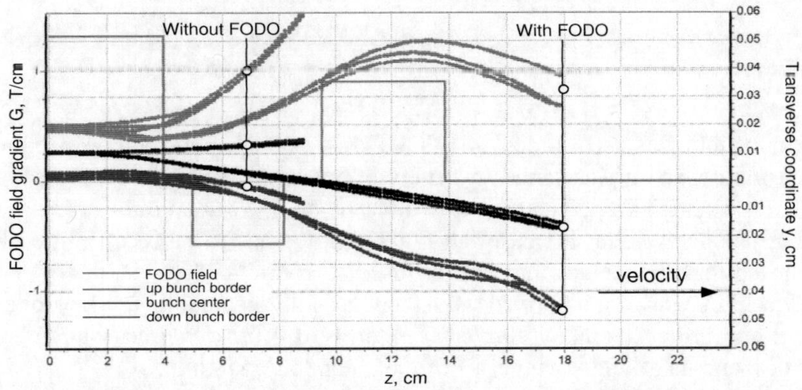

Figure 4. Calculation results for the bunch sequence passing from left to rigth.

The numerical simulation of beam dynamics at the end of propagation (17.78 cm) is presented in Figure 5. The calculation wais stopped when the first macroparticle in the sequence touched the dielectric channel (see transverse view in figure 5).

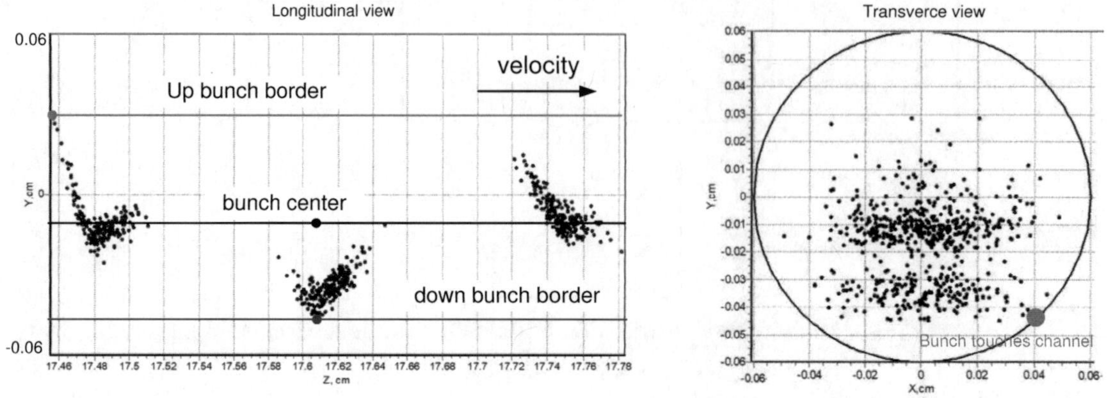

Figure 5. Longitudinal and transverse views of the bunch sequence at the end of the simulation

The HEM-mode amplitude grows as a function of the radial coordinate of the bunch. However, in our case the radial displacement of bunches does not change the spectrum of the Ez field (THz radiation). It remains monochromatic until electrons touch the dielectric tube. The longitudinal dynamics can cause the intervals between the bunches to change, but this is only possible for the low energy electrons, with energies below ~5MeV).

Table 2. Parameters of the optimal focusing system.

Parameter	Parameter value
G1, G2, G3	1.315, 0.5707, 0.8978 T/cm
o1, o2	1, 1.361 cm
z1, z2, z3	3.959, 3.21, 4.3 cm

5. Conclusion

An optimization algorithm was used to design a FODO lattice for bunch train propagation through the DLW. The bunch train spacing was tuned to generate monochromatic radiation at 439 GHz (TM02 mode). To selectively excite TM01, TM03 or any other mode, the bunch train spacing will have to be adjusted accordingly. However, in this case the FODO lattice has to be reconfigured for new bunch train spacing. The optimization algorithm can be used to redesign the system for new bunch train spacing. Our algorithm works for a fixed offset but is still acceptable for low ones. We plan to improve our algorithm further to make it universal in terms of beam dynamics calculation for different types of waveguides, and its efficiency in terms of the computational time.

Acknowledgement

This work is supported by the Ministry of Education and Science of the Russian Federation (project 3.6522.2017).

References

[1] Carr G, Michael C, Wayne R, George R and Williams G. 2002 High-power terahertz radiation from relativistic electrons *Nature* **420** p 153

[2] Galyamin S,1 Tyukhtin A, Antipov S and Baturin S 2014 Terahertz radiation from and ultra-relativistic charge exiting the openend of a waveguide with a dielectric layer *Optics Express* **22** 8902-8907

IOP Conf. Series: Materials Science and Engineering **597** (2019) 012054 doi:10.1088/1757-899X/597/1/012054

[3] Andonian G et a 2011 l Resonant excitation of coherent Cerenkov radiation in dielectric lined waveguides *Appl. Phys. Lett.* **98** 202901

[4] Antipov S et al 2012 Experimental Observation of Energy Modulation in Electron Beams Passing through Terahertz Dielectric Wakefield Structures *Phys Rev Lett.* **108** 144801

[5] Zholents A et al A preliminary design of the collinear dielectric wakefield accelerator 2016 *Nuclear Inst. and Methods in Physics Research* **829** p 190

[6] Schoessow P, et al 2010 Beam Breakup in Dielectric Wakefield Accelerating Structures: Modelling and Experiments, *Proceedings of IPAC'10* **4300**

[7] Altmark A and Kanareykin A 2016 The source of THz radiation based on dielectric waveguide excited by sequence of electron bunches *J. Phys.: Conf. Ser.* **732** 012037

[8] Sheinman I and Kanareykin A 2008 Transverse dynamics and bunch-to-bunch energy exchange in a dielectric-filled accelerating structure Technical Physics **53** 1350-1356

[9] Sheinman I, Kanareykin A and Sotnikov G 2012 Numerical and Analytical Methods of Modeling of Bunch Dynamics in Dielectric Filled Accelerating Structures, Proceedings Russian Particle Accelerator Conference *Proceedings of RUPAC2012* **266** ISBN 978-3-95450-125-0

IOP Conf. Series: Materials Science and Engineering **597** (2019) 012055 doi:10.1088/1757-899X/597/1/012055

Analysis of electron beam dynamics in a rectangular dielectric waveguide

A Altmark[1], N Lesiv[2] and K Mukhamedgaliev[3]

[1]Department of Physics, St.-Petersburg State electrotechnical University, St. Petersburg, Russian Federation
[2]Department of Electronic Instruments and Devices, St.-Petersburg State electrotechnical University, St. Petersburg, Russian Federation
[3]Department of Photonics, St.-Petersburg State electrotechnical University, St. Petersburg, Russian Federation

[2]E-mail: nikita.lesiv@gmail.com

Abstract. Wakefield acceleration of electron bunches is effective method developed for linear accelerator techniques. The main problem of this method is transverse beam dynamics which limits the propagation length of bunch. This work is dedicated to beam dynamics in dielectric waveguide with rectangular geometry. This type of waveguide has complex mode composition including LSM and LSE modes. The main purpose is a study of modes influence in rectangular dielectric waveguide on beam dynamics with different types of dismission of driver from waveguide's axis. It is shown that the LSM asymmetric modes make a major contribution to wakefield and transverse dynamics.

1. Introduction

The wakefield scheme is based on the acceleration of charged particles in the field of a traveling wave (Vavilov-Cherenkov radiation) created by a bunch (driver) with large charge and low energy. Dielectric waveguides with cylindrical geometry are widely used as an accelerator structure [1]. Rectangular dielectric waveguides have become widespread [1,2] due to the simplicity of manufacturing technology. The transverse and longitudinal sections of a rectangular waveguide are shown in figure 1. The driver 4 moves in the vacuum cavity 1 of the waveguide along the z axis parallel to the dielectric plates 2. A dielectric waveguide covered with a metal shell 3 has vertical walls along the x axis without dielectric layers. The Vavilov-Cherenkov radiation generated by the driver has strong component E_z using as accelerating field and transverse components of electromagnetic field which define deflecting force. In addition to the above, there is a possibility of changing the distance between dielectric plates in dechirpers [3] for tuning of wakefield spectrum. Also, rectangular waveguide is adapted to pass wide bunches with high charge through vacuum gap 1 [1].

The beam dynamics of bunches is calculated for determination of propagation distance of the driver before it touches the dielectric wall 2. This value allows to calculate the amount of energy that an accelerated bunch (witness) can receive. Numerical simulation of the beam dynamics is one of the problems for rectangular waveguide due to complex mode composition in compare with cylindrical one: symmetrical and asymmetrical LSM and LSE modes. In this article, we consider the contribution

IOP Conf. Series: Materials Science and Engineering **597** (2019) 012055 doi:10.1088/1757-899X/597/1/012055

of all above modes to the dynamics of the bunch and clarify cases when only part of the modes can be considered.

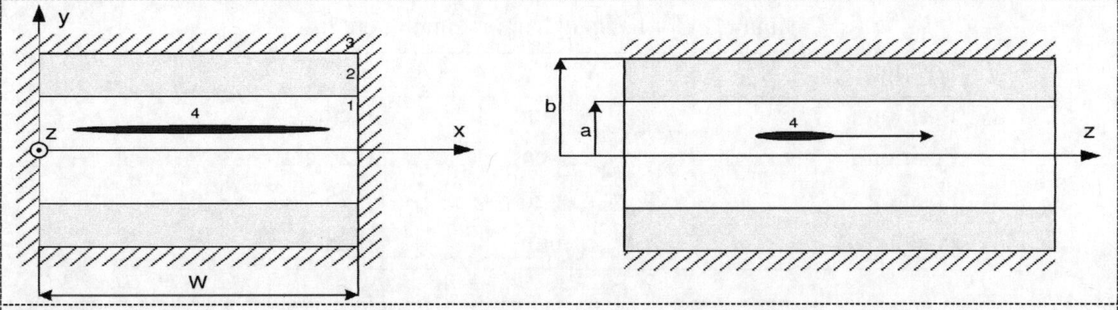

Figure 1. Longitudinal (left) and transverse (right) views of a rectangular waveguide.

2. Methods

The calculations presented in this paper are based on the previously obtained analytical expression of the Green function for a rectangular waveguide [2]. The wakefield is calculated using an integral convolution of charge distribution function with a Green function which includes LSM (Longitudinal Section Magnetic, H_y=0) and LSE (Longitudinal Section Electric, E_y=0) modes. The contribution of these modes depends on the transverse coordinate of the bunch. Each of the mentioned types of modes is also classified into symmetric and asymmetrical with respect to the plane relative to the coordinate $y = 0$.

The dynamics of a bunch of charged particles is calculated by the method of macroparticles [4]. This method is based on the representation of a bunch with a charge Q in the form of N particulates with a charge Q/N. Such a partition allows us to carry out numerical simulation of the dynamics of a real bunch with savings in computational resources. The number of macroparticles is defined by geometrical dimensions of the bunch, its charge and transverse coordinates. It can take values from 100 to 5000. For optimization of the calculating time the dynamics, the number of LSE and LSM modes is selected based on the longitudinal length of the bunch.

The longitudinal wake field E_z has a braking effect on the central part of the bunch, while the deflecting forces F_x and F_y act on the tail part of the bunch. The sign of the deflecting force depends on the direction and magnitude of the displacement of the center of the bunch relative to the axis of the waveguide.

3. Results and Discussion

Table 1 presents the parameters of the waveguide and the leading electron bunch corresponding to the accelerator facility of the Argonne National Laboratory [5].

Table 1. Parameters of the waveguide (see figure 1) and an electron bunch for numerical simulation.

Parameters of the waveguide	**Parameter value**	Parameters of the bunch	**Parameter value**
a	0.5 cm	Charge	100 nC
b	0.7 cm	Longitudinal length	0.1 cm
w	1 cm	Y length	0.035 cm
Permittivity of plates	3.75	X length	0.035 cm
LSM_{11} asymmetric mode frequency	6.93 GHz	Energy	50 MeV

Table 2. Variants of transverse positions of bunch in waveguide (figure 1) for beam dynamics simulation

Positions	X	Y
Position 1	0.25 cm	0 cm
Position 2	0.25 cm	0.25 cm
Position 3	0.5 cm	0.25 cm
Position 4	0.5 cm	0 cm

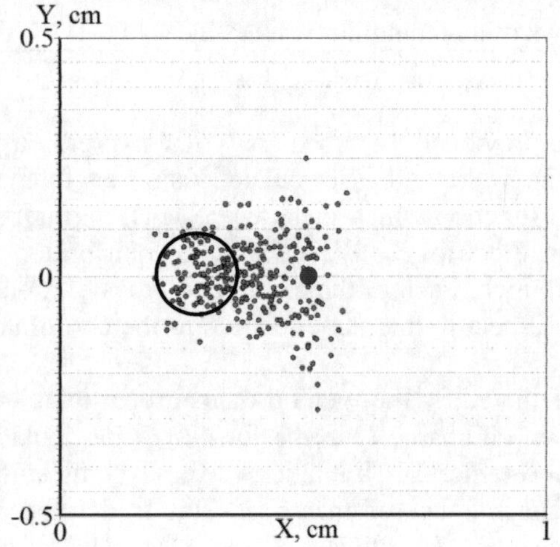

Figure 2. The cross section of the bunch after beam dynamics simulation for position 1. Only LSM asymmetrical modes are considering. Solid circle shows initial position of bunch, blue point is center of waveguide (position 4)

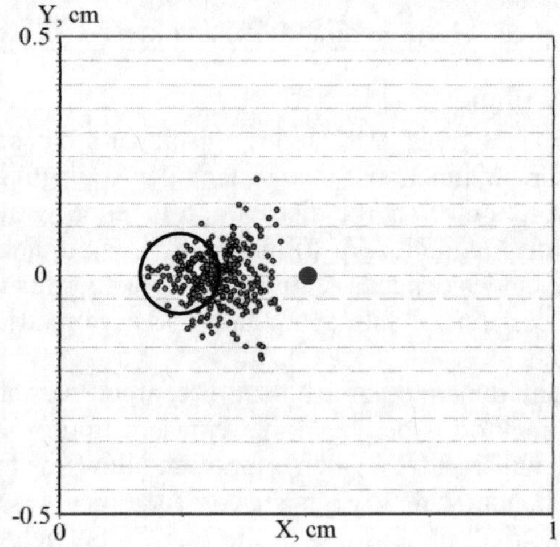

Figure 3. The cross section of the bunch after beam dynamics simulation for position 1. Only LSE symmetrical modes are considering. Solid circle shows initial position of bunch, blue point is center of waveguide (position 4)

The original Dielectric Waveguide program [6] was used to analyze the effect of individual types of modes on the transverse beam dynamics. This code allows to calculate Green function and beam dynamics. Table 2 present different initial transverse positions of bunch which can be dismissed in X and Y directions. Transverse force F_r is acting on bunch in case of transverse dismission (offset). So, the position 4 coincides with waveguide center and allows bunch to pass for long distances. Figures 2 and 3 present the results of modeling the dynamics of bunch for position 1 (offset only in X direction). Because of symmetry of bunch relative to Y coordinate only two types of modes can be excited: LSM asymmetrical modes and LSE symmetrical modes. Both modes act on bunch as focusing force namely to center of waveguide. From figures 2 and 3 it can be seen that influence LSM asymmetrical modes is higher that LSE ones.

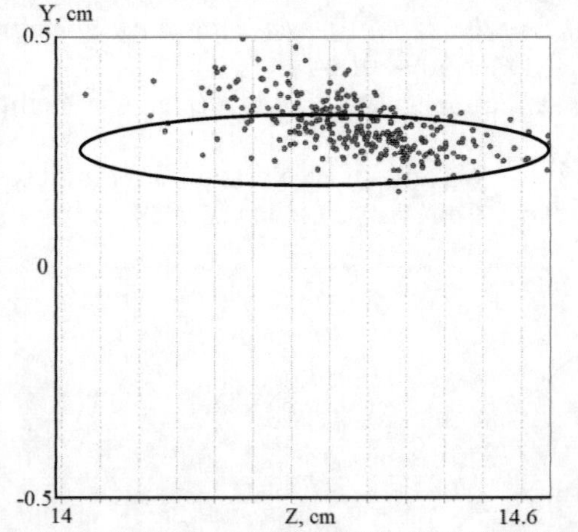

Figure 4. The longitudinal view (end of simulation) of the bunch passing from left to right for initial transverse position 3. Only LSE symmetrical modes are considering. Solid line shows initial position of bunch.

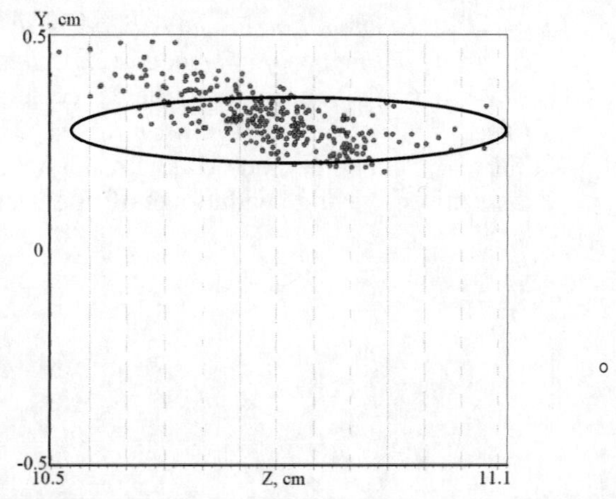

Figure 5. The longitudinal view (end of simulation) of the bunch passing from left to right for initial transverse position 3. Only LSM symmetrical modes are considering. Solid line shows initial position of bunch.

Positions 2 and 3 mean offsets in Y direction. Therefore, it is necessary to consider all types of LSE and LSM modes. The calculations show that LSM modes act on bunch stronger that LSE ones. In case of LSM mode macroparticles touch dielectric plate after 11 cm (figure 5) of passing in compare with 14 cm of LSE modes (figure 4). Also need to add that difference between position 2 and 3 is increase of contribution LSE modes in case of offset growth in coordinate X. In addition to the above that beam dynamics calculation with contribution of all modes shows that the tail of bunch touch dielectric on after passing of 6.3 cm.

4. Conclusion

In conclusion, it should be noted that the advantage of the rectangular wake structure consists not only in the simplicity of its manufacturing technology, but also in the ability to pass bunches of large charges by increasing their length along the X axis. The disadvantages of a rectangular structure compared to a cylindrical structure include the absence of symmetry about the central axis. This circumstance complicates the analytical expression and increases the time for calculating the dynamics due to the large number of modes that must be considered.

Acknowledgments

The work was supported by the Ministry of Education and Science of the Russian Federation (project 3.6522.2017).

References

[1] Andonian G, Stratakis D and Babzien M 2012 Dielectric wakefield acceleration of a relativistic electron beam in a slab-symmetric dielectric lined waveguide *PRL* **108** 244801

[2] Baturin S, Sheinman I, Altmark A and Kanareykin A 2013 Transverse operator method for wakefields in a rectangular dielectric loaded accelerating structure *Phys. Rev. ST Accel. Beams* **16** 051302

[3] Nie Y, Xia G and Pacey T 2018. Simulations of an energy dechirper based on dielectric lined waveguidesin a slab-symmetric dielectric lined waveguide *Nuclear Inst. and Methods in Physics Research* **A 909**

IOP Conf. Series: Materials Science and Engineering **597** (2019) 012055 doi:10.1088/1757-899X/597/1/012055

[4] Justin O 2016 *Simulation of electron beam dynamics in the 22 MeV accelerator for a coherent electron cooling proof of principle experiment.* BNL-113300-2016-TH 4

[5] Gai W 2016 Two beam wakefield acceleration at argonne wakefield accelerator facility *Proceedings of IPAC2016 Busan Korea*

[6] Altmark A, Lesiv N and Mukhamedgaliev K 2018 Wakefield acceleration in multilayer dielectric waveguide with usage of ring beam *J. Phys. Conf. Ser.* **1145 (1)** 012039

IOP Conf. Series: Materials Science and Engineering **597** (2019) 012056 doi:10.1088/1757-899X/597/1/012056

Determination of the properties and loading efficiency of encapsulated BSA-FITC and dexamethasone for drug delivery systems

E Chudinova[1], M Surmeneva[1], A Koptyug[2], V Sokolova[3], O Prymak[3], S Bouckercha[3], M Epple[3], R Surmenev[1]

[1]Physical Materials Science and Composite Materials Centre, *National Research Tomsk Polytechnic University*, 30 Lenin Avenue, Tomsk 634050, Russian Federation
[2]Department of Mechanical Engineering and Quality Technology, SportsTech Research Centre, *Mid Sweden University*, Akademigatan 1, SE 831 25, Östersund, Sweden
[3]Inorganic Chemistry and Center for Nanointegration Duisburg-Essen (CeNIDE), *University of Duisburg-Essen,* Essen 45117, Germany

[1]E-mail: rsurmenev@mail.ru

Abstract. In this work porous microparticles of calcium carbonate were synthesized with bovine serum albumin - fluorescein isothiocyanate conjugate (BSA-FITC) and dexamethasone, and then used for encapsulation in polymer microcapsules by means of layer-by-layer assembly (LbL). The properties of the obtained microcapsules were characterized by scanning electron microscopy, dynamic light scattering, infrared-, ultraviolet- and visible spectroscopy. According to the performed DLS measurements, an average hydrodynamic diameter ranged from 4 to 8 µm and zeta-potential for all types of capsules was determined as -18 and -21 mV. BSA-FITC was encapsulated using this approach yielded a loading efficiency of 49 % protein. This value calculated for dexamethasone was of 38%. The microcapsules filled with an encapsulated drug may find applications in the field of biotechnology, biochemistry, and medicine.

1. Introduction

Encapsulation and release of pharmaceutically active substances are increasingly important tasks in biomedical research. Microencapsulation provides the means to alter colloidal and surface properties, provide environmental protection and control over the release kinetics of the encapsulated drugs [1]. The primary reason for microencapsulation is found to be either for sustained or prolonged drug release [2].

A promising type of drug delivery systems is represented by hollow polyelectrolyte capsules [3]. Their fabrication presents the layer-by-layer (LbL) self-assembly of polyelectrolyte thin films and yields hollow capsules that have a diameter in the range of micrometers or even below [4, 5]. Briefly explained, the LbL technique is based on the alternating adsorption of charged species onto an oppositely charged substrate, using electrostatic interactions as the driving force. The main advantage of the LbL technique is the ease of manipulation and the unmet degree of multifunctionality [6], allowing one to tailor the surface with different kinds of functional groups [7], lipids [8], nanoparticles [9].

In case of this method, materials are entrapped within or adsorbed to sacrificial template particles prior to LbL coating. Here, inorganic calcium carbonate ($CaCO_3$) microparticles represent a valuable category of a template for loading, as they provide an excellent loading capacity for biological materials

IOP Conf. Series: Materials Science and Engineering **597** (2019) 012056 doi:10.1088/1757-899X/597/1/012056

together with rather "biofriendly" processing techniques [10, 11]. These inorganic components will provide lower-risk materials in various practical applications.

As the materials for loading can be used also proteins, drug microcrystals, enzymes and other biomolecules. Thus, for the synthesis of microcapsules and determination of the amount of a loaded component as model protein bovine serum albumin – fluorescein isothiocyanate conjugate (BSA-FITC) molecules can be chosen for encapsulation into the calcium carbonate microcapsules in the process of the capsule preparation [12-14]. Besides, dexamethasone can be used for loading in capsules as a model drug because of its potential use in specialized delivery systems in treating peritumoral edema associated with brain tumors and for treating and preventing retinal diseases. Dexamethasone is a glucocorticoid that is clinically used for its anti-inflammatory and immunosuppressive effects [15, 16].

Therefore, the main goal of this work was the synthesis of microcapsules with BSA-FITC and dexamethasone and determination of their composition, loading efficiency for the further possibility to determine the release of the encapsulated substance from the capsule.

2. Materials and methods

The $CaCO_3$ microparticles were synthesized to serve as a core of the capsules; BSA-FITC and dexamethasone were also encapsulated in the core of the capsules. Briefly, 0.615 mL of $CaCl_2$ solution (1 M) was added to a 2.5 mL of distilled water, after that 0.615 mL of Na_2CO_3 solution (1 M) was rapidly added under vigorous stirring (1100 rpm). This suspension was mixed during 25 minutes in order to form spherical $CaCO_3$ microparticles. The synthesis with BSA-FITC (Sigma, A9771, 4 mg mL^{-1}) and dexamethasone (KRKA, 4 mg mL^{-1}) was performed in the same conditions; they were added instead of the water. Then, the $CaCO_3$ microparticles were suspended in an aqueous solution of PAH (2 mg mL^{-1}) containing 0.15 M NaCl. Polymers poly(styrenesulfonate) (PSS) and poly(allylamine hydrochloride) (PAH) were used for formation of layers. The PAH was allowed to adsorb onto three types of microparticles for 10 min using a shaker with 180 rpm of rate. After that, the dispersion was centrifuged and washed three times with deionized water to remove the non-adsorbed polyelectrolyte. Next, the deposition of PSS was carried out on the PAH-coated three type of $CaCO_3$ particles using the same manipulation. Such a procedure of PAH/PSS deposition was repeated six times in order to form a core-shell structure. Afterwards, the particles were centrifuged and washed three times in deionized water.

The hydrodynamic diameter (HDD) of 3 different types of capsules was measured by Dynamic Light Scattering (DLS) using a Malvern Zetasizer Nano ZS and Nanoparticle Tracking Analysis (NTA) using a NanoSight LN 10. Zeta-potential of the obtained microparticles was also measured by DLS. The morphology of the microparticles was estimated using an ESEM Quanta 400 FEG instrument operating in a high vacuum with gold/palladium-sputtered samples. Ultraviolet and visible (UV-Vis) spectroscopy (Cary 300 Bio UV-Vis-spectrophotometer, Varian) was used to calculate the amount of the encapsulated microparticles in the obtained hybrid capsules. The chemical structure of the capsules was examined by Infrared (IR) spectroscopy (Vertex 70, Bruker). For that, the capsules were dried in an oven at 55 degrees for 12 hours, then 2 mg of capsules and 200 mg of KBr were taken, everything was thoroughly mixed in a mortar and the tablets were formed using a hydraulic system.

3. Results and discussion

SEM images of the synthesized capsules with BSA-FITC and dexamethasone are shown in Figure 1. It should be noted that morphology of the prepared capsules is close to spherical. According to the obtained micrographs, synthesized microparticles have a consistent size, with an average diameter for both type of capsules around 2 μm and pronounced roughness of their outer surface. These structural and morphological characteristics of $CaCO_3$ microparticles are directly determined by the mechanism of their formation from supersaturated solutions. It was found earlier [17, 18] that the growth of $CaCO_3$ microparticles is a kind of colloidal aggregation of primary $CaCO_3$ nanoparticles, instantaneously formed with mixing of calcium chloride and sodium carbonate solutions, into spherical micron-sized microparticles, which further slowly recrystallize in water into rhombohedral calcite crystals.

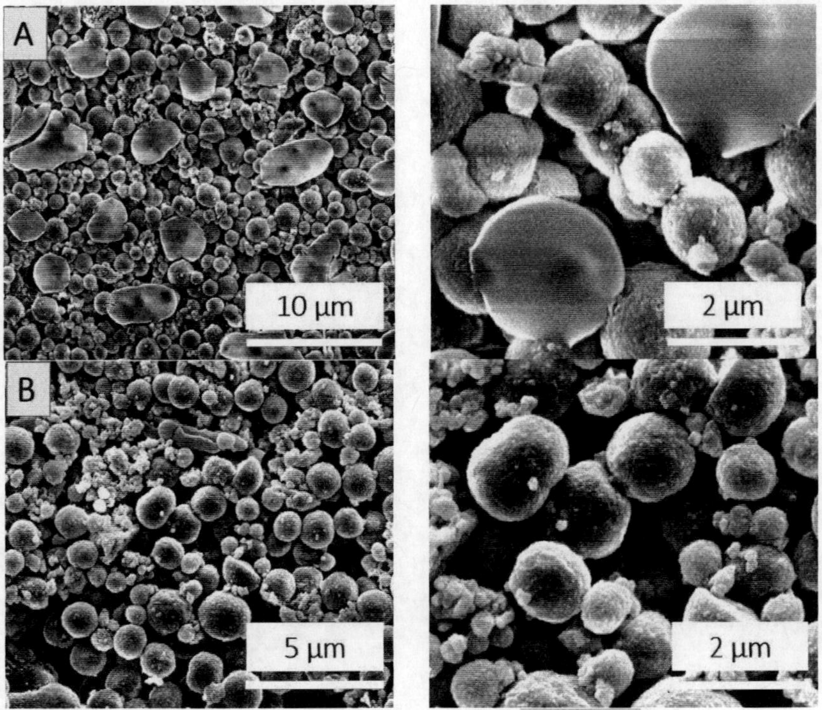

Figure 1. SEM image of microcapsules with BSA-FITC (row "A") and dexamethasone (row "B") prepared by LbL synthesis.

Results of DLS analysis presented in Table 1 showed that an average HDD of synthesized microcapsules was in the range from 4 to 8 μm (Table 1).

Since the outer layer of the capsules was negatively charged PSS, the zeta-potential of the control $CaCO_3$ microcapsules, microcapsules with BSA-FITC and dexamethasone was -18.1, -20.9 and -20.5 mV, respectively (Table 1).

Table 1. Data of DLS measurements

Type of microcapsules	HDD (μm)	Zeta-potential (mV)
Control $CaCO_3$ capsules	3.8±0.2	-18.1±1.0
Capsules with BSA-FITC	4.6±0.2	-20.9±0.9
Capsules with dexamethasone	8.4±0.3	-20.5±2.5

To determine the absorption maxima for PSS, BSA-FITC and dexamethasone, a series of standard solutions were prepared. Initially, the firstly-prepared solution containing capsules was taken. Then the series of 50% dilutions were prepared from it.

According to the obtained data, the absorbance maximum of PSS (control), BSA-FITC and dexamethasone was observed at 226 nm (Figure 2a), 280 nm and 486 nm (Figure 2b) and 241 nm (Figure 2c), respectively. These values of maximum correspond to the references [19-21].

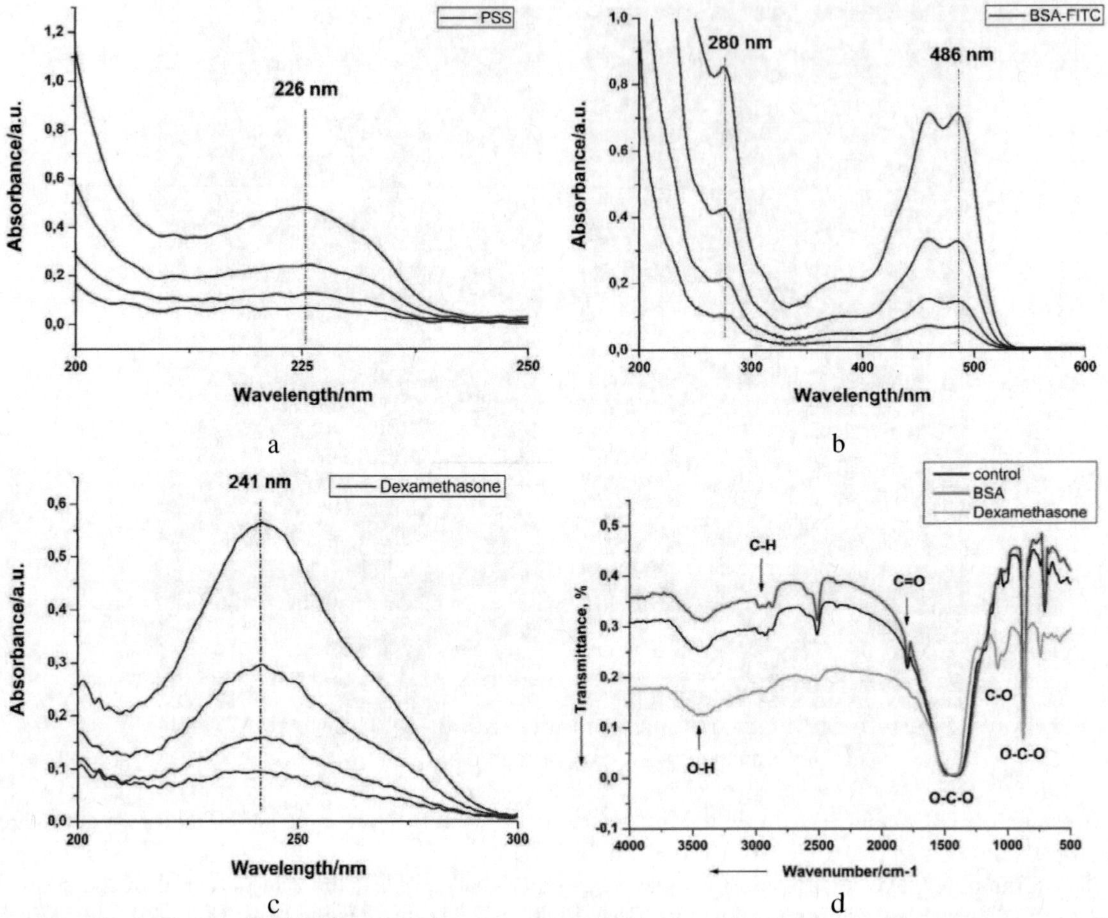

Figure 2. The absorbsion spectra of the control $CaCO_3$ microcapsules (a), BSA-FITC (b), dexamethasone (c), recorded as different concentrations, respectively, and their IR spectra (d).

After that, the concentrations of the components were determined using the standard calibration curves. According to the calibration curve calculations, the concentration of BSA-FITC in the initial solution was determined to be 4.79 mg/mL. While the concentration of this component in the supernatant after the synthesis of the core was 1.89 mg/mL. Thus, the concentration difference was 2.10 mg/mL, which was 48% of the effective load. This value means how much protein was loaded into the core.

An analysis of the selected literary sources over the past three years has revealed an active interest in using BSA-FITC as a model protein. The release of BSA-FITC from the polyelectrolyte microcapsules was also investigated elsewhere [22]. The protein yield was obtained within 38% at a temperature of 22ºC. The authors noted that that value depended on the thickness of the shell of the capsules, the method of encapsulation, as well as the composition of the salt, pH and temperature. Similar studies were also presented in the paper of Bolzinger M. A. et al., where the loading efficiency is correlated with the values obtained in this experiment. However, the authors provide ways to increase efficiency by reducing the volume of internal and external phases [23].

According to similar calculations for dexamethasone and PSS (control capsules) using a calibration curve, the loading efficiency was of 36 and 51 %, respectively.

It is known in literature that the efficiency of dexamethasone loading is about 51% [24], 77% [25], when similar method of encapsulation - layer-by-layer adsorption is used.

The IR spectra of the control $CaCO_3$ microcapsules, BSA-FITC and dexamethasone are shown in figure 2d. As a result of this method of analysis the following chemical bonds such as C – H, C – O, O – H, O – C – O, C = O are detected.

The IR spectra of the BSA-FITC- and dexamethasone-loaded $CaCO_3$ microcapsules reflect the characteristic absorption bands typical for BSA-FITC, dexamethasone and calcium carbonate without any new bands, which indicates physical adsorption of $CaCO_3$ microcapsules [26-29].

4. Conclusion

In this study, calcium carbonate spherical capsules with loaded BSA-FITC- and dexamethasone have been synthesized and studied. According to the DLS results, the diameter of the prepared microcapsules with BSA-FITC, dexamethasone and control $CaCO_3$ microcapsules was 4.6 ± 0.2, 8.4 ± 0.3 and 3.8 ± 0.2 μm, respectively. Zeta-potential of each type of capsules was determined, the determined values were in the range from -18 to -21 mV. The concentration of the encapsulated components was measured using UV spectroscopy, thus, the loading efficiency of BSA-FITC, dexamethasone and PSS (control capsules) was 48, 36 and 51 %, respectively, which corresponded well to the literature data for capsules prepared using the same method. The chemical bonds C – H, C – O, O – H, O – C – O, C = O were determined using infrared spectroscopy. The future experiments are aimed to deposit the microcapsules on the surface of titanium alloy scaffolds and to perform *in vivo* and antibacterial experiments.

Acknowledgments

The work was supported by Russian Science Foundation (No. 15-13-00043). Ekaterina Chudinova also acknowledges the personal support from the German-Russian Interdisciplinary Research Center (G-RISC). We acknowledge Professor G.B. Sukhorukov for fruitful discussions.

References

[1] Agnihotri N, Mishra R, Goda C and Arora M 2012 Microencapsulation–a novel approach in drug delivery: a review *Indo Global J. of Pharm. Sci.* **2/1** 1

[2] Bansode S S, Banarjee S K, Gaikwad D D, Jadhav S L and Thorat R M 2010 Microencapsulation: a review *Int. J. Pharm. Sci. Res* **1/2** 38

[3] Vilela C, Figueiredo A R, Silvestre A J and Freire C S 2017 Multilayered materials based on biopolymers as drug delivery systems *Expert opinion on drug delivery* **14/2** 189

[4] Guzman E, Mateos-Maroto A, Ruano M, Ortega F and Rubio R G 2017 Layer-by-Layer polyelectrolyte assemblies for encapsulation and release of active compounds *Adv. Colloid Interface Sci.* **249** 290

[5] Richardson J J, Cui J, Bjornmalm M, Braunger J A, Ejima H and Caruso F 2016 Innovation in layer-by-layer assembly *Chem. Rev.* **116/23** 14828

[6] Angelatos A S, Katagiri K and Caruso F 2006 Bioinspired colloidal systems via layer-by-layer assembly *Soft Matter* **2/1** 18

[7] De Geest B G, Jonas A M, Demeester J and De Smedt S C 2006 Glucose-responsive polyelectrolyte capsules *Langmuir* **22/11** 5070

[8] De Geest B G, Stubbe B G, Jonas A M, Van Thienen T, Hinrichs W L, Demeester J and De Smedt S C 2006 Self-exploding lipid-coated microgels *Biomacromolecules* **7/1** 373

[9] Shchukin D G, Ustinovich E A, Sukhorukov G B, Möhwald H and Sviridov D V 2005 Metallized polyelectrolyte microcapsules *Adv. Mater.* **17/4** 468

[10] Nayar S, Sinha M, Basu D and Sinha A 2006 Synthesis and sintering of biomimetic hydroxyapatite nanoparticles for biomedical applications *J. Mater. Sci. : Mater. Med.* **17/11** 1063

[11] Vallet-Regi M and González-Calbet J M 2004 Calcium phosphates as substitution of bone tissues *Prog. Solid State Chem.* **32/1-2** 1

[12] Appel E A, Tibbitt M W, Webber M J, Mattix B A, Veiseh O and Langer R 2015 Self-assembled hydrogels utilizing polymer–nanoparticle interactions *Nat. Commun.* **6** 6295

[13] Jalani G, Naccache R, Rosenzweig D H, Haglund L, Vetrone F and Cerruti M 2016 Photocleavable hydrogel-coated upconverting nanoparticles: a multifunctional theranostic platform for NIR imaging and on-demand macromolecular delivery *J. Am. Chem. Soc.* **138/3** 1078

[14] Kim B S, Oh J M, Kim K S, Seo K S, Cho J S, Khang G, Lee H B, Park K and Kim M S 2009 BSA-FITC-loaded microcapsules for in vivo delivery *Biomaterials* **30/5** 902

IOP Conf. Series: Materials Science and Engineering **597** (2019) 012056 doi:10.1088/1757-899X/597/1/012056

[15] Correia C R, Pirraco R P, Cerqueira M T, Marques A P, Reis R L and Mano J F 2016 Semipermeable capsules wrapping a multifunctional and self-regulated co-culture microenvironment for osteogenic differentiation *Sci. Rep.* **6** 21883
[16] Roila F, Ruggeri B, Ballatori E, Del Favero A and Tonato M 2013 Aprepitant versus dexamethasone for preventing chemotherapy-induced delayed emesis in patients with breast cancer: a randomized double-blind study *J. Clin. Oncol.* **32/2** 101
[17] Volodkin D V, Petrov A I, Prevot M and Sukhorukov G B 2004 Matrix polyelectrolyte microcapsules: new system for macromolecule encapsulation *Langmuir* **20/8** 3398
[18] Wang G, Siggers K, Zhang S, Jiang H, Xu Z, Zernicke R F, Matyas J and Uludağ H 2008 Preparation of BMP-2 containing bovine serum albumin (BSA) nanoparticles stabilized by polymer coating *Pharm. Res.* **25/12** 2896
[19] Friedrich R B, Ravanello A, Cichota L C, Rolim C M B and Beck R C R 2009 Validation of a simple and rapid UV spectrophotometric method for dexamethasone assay in tablets *Quim. Nova* **32/4** 1052
[20] Mundargi R C, Potroz M G, Park J H, Seo J, Tan E-L, Lee J H and Cho N-J 2016 Eco-friendly streamlined process for sporopollenin exine capsule extraction *Sci. Rep.* **6** 19960
[21] Wang W, Liu S, Li C, Wang Y and Yan C 2018 Dual-target recognition sandwich assay based on core-shell magnetic mesoporous silica nanoparticles for sensitive detection of breast cancer cells *Talanta* **182** 306
[22] Freitas S, Merkle H P and Gander B 2005 Microencapsulation by solvent extraction/evaporation: reviewing the state of the art of microsphere preparation process technology *J. Controlled Release* **102/2** 313
[23] Bolzinger M, Bordes C, Gauvrit J and Briançon S 2007 Improvement of a bovine serum albumin microencapsulation process by screening design *Int. J. Pharm.* **344/1-2** 16
[24] Joshi A B and Srivastava R 2009 Polyelectrolyte coated calcium carbonate microparticles as templates for enzyme encapsulation *Adv. Sci. Lett.* **2/3** 329
[25] Jayant R, McShane M and Srivastava R 2009 Polyelectrolyte-coated alginate microspheres as drug delivery carriers for dexamethasone release *Drug delivery* **16/6** 331
[26] Chen W, Qian C, Zhou K-G and Yu H-Q 2018 Molecular spectroscopic characterization of membrane fouling: a Critical Review *Chem* **4/7** 1492
[27] Facchinetto S E, Bortolotto T, Neumann G E, Vieira J C, Menezes B B d, Giacomelli C and Schmidt V 2017 Synthesis of Submicrometer Calcium Carbonate Particles from Inorganic Salts Using Linear Polymers as Crystallization Modifiers *J. Braz. Chem. Soc.* **28/4** 547
[28] Wang C, He C, Tong Z, Liu X, Ren B and Zeng F 2006 Combination of adsorption by porous CaCO3 microparticles and encapsulation by polyelectrolyte multilayer films for sustained drug delivery *Int. J. Pharm.* **308/1-2** 160
[29] Zhang J, Liu Y, Luo R, Chen S, Li X, Yuan S, Wang J and Huang N 2015 In vitro hemocompatibility and cytocompatibility of dexamethasone-eluting PLGA stent coatings *Appl. Surf. Sci.* **328** 154

IOP Conf. Series: Materials Science and Engineering **597** (2019) 012057 doi:10.1088/1757-899X/597/1/012057

Influence of phase composition on Portevin-Le Chatelier effect in Al-Mg alloys

Yu Borisova, D Yuzbekova and A Mogucheva
Belgorod State University, Belgorod 308015, Russia

E-mail: borisova_yu@bsu.edu.ru

Abstract. The temperature-strain rate domain of the Portevin-Le Chatelier (PLC) effect has been determined for three ultrafine grained aluminum alloys: Al-3Mg (A1), Al-4,57Mg-0,35Mn-0,2Sc-0,09Zr (A2) and Al-5,4Mg-0,52Mn-0,1Zr (A3) (wt.%). The apparent activation energy of the serrated yielding was estimated. It was found that the presence of dispersed particles leads to a decrease in the activation energy and the temperature–strain rate domain of the PLC effect becomes narrower.

1. Introduction

Non-heat treatable Al-Mg alloys are widely used in various industries, in particular, in aircraft, automotive and shipbuilding industries, and operated at cryogenic temperature. Excellent combination of corrosion resistance, weldability, high strength, and good ductility is the main reason for their wide use [1-2].

For a long time, Al-Mg alloys have been used as the common materials for studying the Portevin-Le Chatelier (PLC) effect [3-4]. These alloys are characterized by the instability of plastic deformation in a wide temperature/strain rate interval [4-6]. The PLC effect in Al-Mg alloys can be observed at the room temperature in a certain strain-strain rate interval [7-8] and it is characterized by strain localization in the form of narrow bands of intense shear.

The studying of the conditions for the appearance of the PLC effect is important to prevent its developing during the thermomechanical processing of Al-Mg alloys [9]. However, there is a lack of information on the effect of dispersed particles of the second phase on the PLC effect in aluminum alloys [3]. The purpose of the present study is to establish the influence of the phase composition on the region of existence of the PLC effect in Al-Mg alloys.

2. Materials and method

To reveal the influence of phase composition on PLC effect the following aluminum alloys were investigated: alloy A1: Al-3Mg (wt.%); alloy A2: Al-4.57Mg-0.35Mn-0.2Sc-0.09Zr (wt.%); alloy A3: Al-5.4Mg-0.52Mn-0.1Zr (wt.%). Alloy A1 was prepared by semi-continuous casting. The ingots were further homogenized at 500°C for 4 h, followed by slow cooling in the furnace. Then, billets with a cross-section of 20×20 mm^2 and a length of 110 mm were subjected to a two-stage ECAP at a temperature of 300°C to a strain of ~ 8 and at a temperature of 200°C to a strain of ~ 4. Alloys A2 and A3 were produced by continuous casting. The ingots of the alloy A2 were subjected to homogenization annealing at 360–380°C for 12 h. For the formation of a fine-grained microstructure, ECAP was carried out at 300°C to a strain of ~ 12. The obtained ingots of alloy A3 were subjected to homogenization annealing at 360°C for 6 h. The ECAP of the A3 alloy was carried out at 300°C to a strain of ~ 12.

The microstructural investigations and elemental analysis were performed by transmission electron microscopy (TEM) by using a JEOL JEM-2100 microscope equipped with an INCA energy dispersive x-ray spectrometer. The flat «dog-bone» shaped specimens with a 16 mm gauge length and 3×1.5 mm2 cross-section were cut parallel to the last extrusion axis of the pressed billets. The uniaxial tensile tests were carried out by using universal testing machine Instron 5882 in the air at temperatures from 398 K to 173 K and strain rates ranging from 1.0×10^{-5} s^{-1} to 2.1×10^{-5} s^{-1}.

3. Results and discussion

3.1. Microstructure after ECAP processing

The bimodal microstructure was observed in alloy A1 after ECAP. Mean size of coarse and fine grains in this condition is 20 and 1.4 μm, respectively. While, ECAP leading to the formation of the uniform fine-grained microstructure with a mean grain size of 1.1 and 1.5 μm in alloys A2 and A3 respectively.

TEM observation did not reveal any precipitates in alloy A1 (figure 1(a)). Whereas it was found that the particles of Al_6Mn and $Al_3(Sc,Zr)$ distributed homogeneously throughout the aluminum matrix in alloy A2 (figure 1 (b)). It is known that the fine particles of $Al_3(Sc,Zr)$ precipitated in aluminum alloys have a coherent interface. The mean size of these particles in alloy A2 is about 10 nm (figure 1b). The mean size of Al_6Mn particles (~80 nm) is rather larger than those of $Al_3(Sc,Zr)$. Only Al_6Mn particles with a mean size of 22 nm were found in alloy A3 (figure 1 (c)).

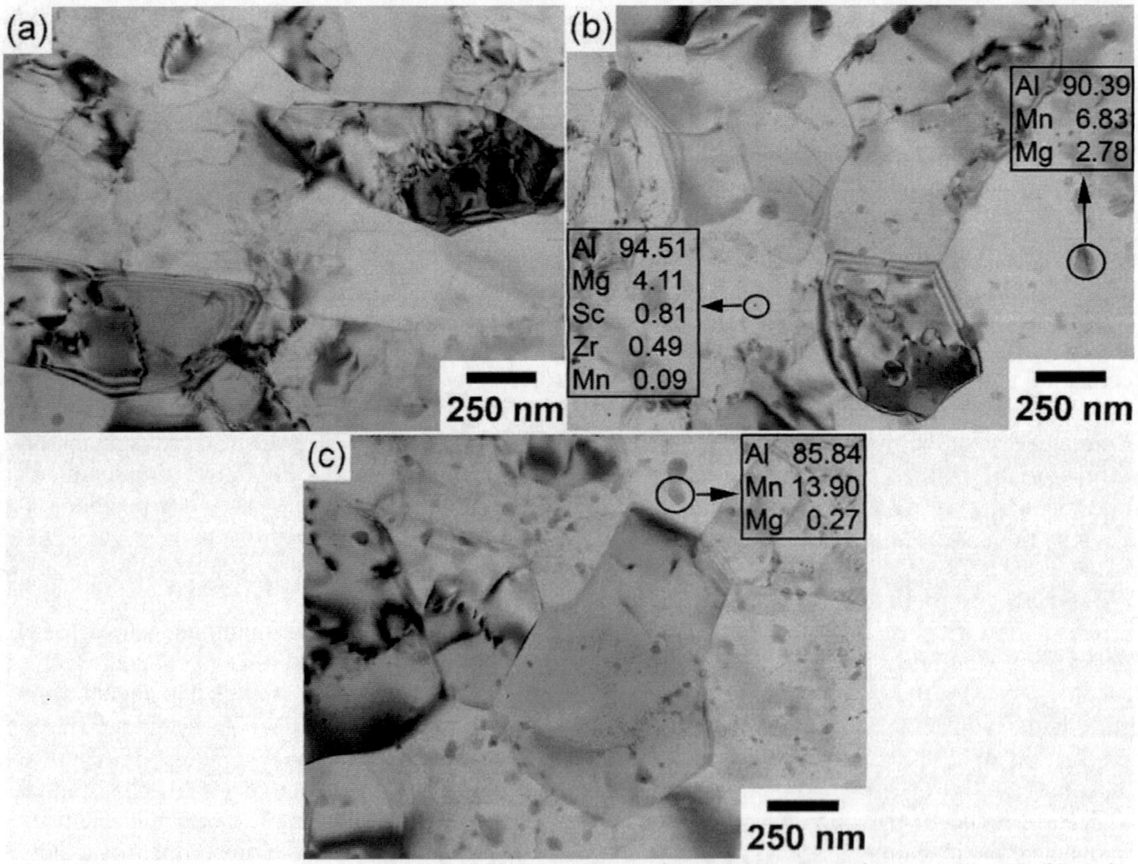

Figure 1. Microstructure of studied aluminum alloys after ECAP processing: A1 (a), A2 (b) and A3 (c). The chemical compositions are given in wt.%.

The presence of dispersed second phase particles precipitated during homogenization enhances the refinement of microstructure. Thus, the ECAP processing of alloy A1 leading to formation of relatively coarse grains. It is obvious that the dispersed particles of Al_6Mn determines the rate of grain refinement, mean size of grains and homogeneity of microstructures formed during ECAP processing.

3.2. Strain rate vs temperature domains of serrated flow in the studied Al-Mg alloys

The analysis of stress-strain curves was performed to determine the presence of PLC effect under different strain rate/temperature conditions in studied alloys after ECAP processing. Figure 2 shows the PLC domains, where solid and open symbols indicate the presence and absence of serrations respectively. It is obvious that the temperature interval of PLC domains of studied alloys tends to increase with decreasing the strain rate.

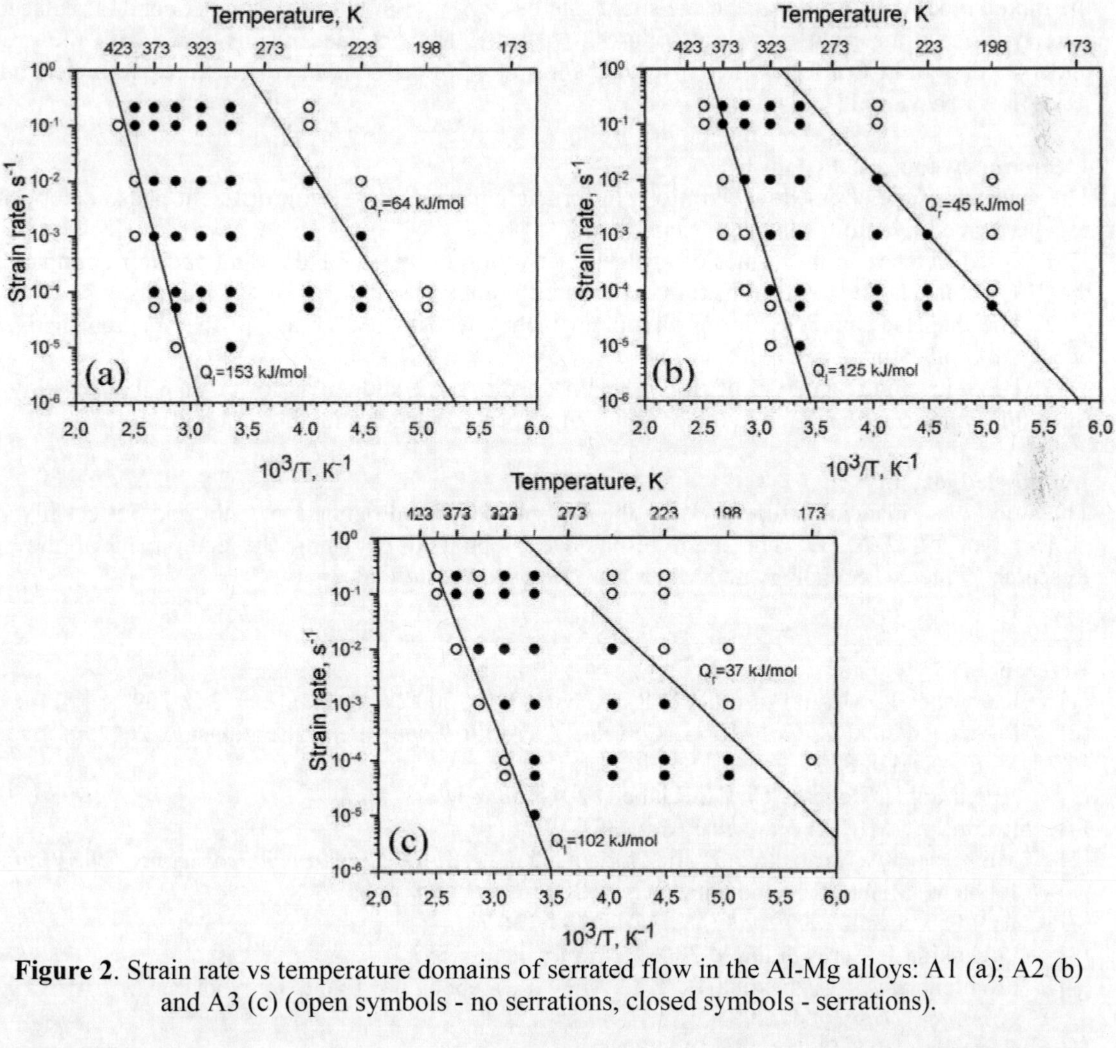

Figure 2. Strain rate vs temperature domains of serrated flow in the Al-Mg alloys: A1 (a); A2 (b) and A3 (c) (open symbols - no serrations, closed symbols - serrations).

In figure (1) it is clearly seen that for each $\dot{\varepsilon}$, the temperature interval of instability extends slightly, as shown by the delimiting boundaries between stable and unstable behavior. Such boundaries can be described by the following equation according to Cottrell theory [10]:

$$\dot{\varepsilon} = \frac{4b\rho C_V D_O}{l}\exp\left(\frac{-Q}{kT}\right), \quad (1)$$

where b, ρ, C_v, D_O, l, Q and k represent the Burgers vector, the dislocation density, the vacancy concentration, the diffusion frequency factor, the effective radius of the solute atmosphere, the activation energy for solute migration, and the Boltzmann constant, respectively.

The apparent activation energy of the PLC effect can be evaluated as a slope of the $\ln\dot{\varepsilon} - 1/T$ dependencies. The obtained activation energies of the stress serrations of alloy A1 is 64 and 153 kJ/mol for the right (Q_R) and left (Q_L) boundary respectively. It was found that activation energies of stress serrations of alloys A2 an A3 is somewhat lower than that of alloy A1. For alloy A2 the value of QR is 45 kJ/mol while QL is about 125 kJ/mol and for alloy A3 Q_R and Q_L is 37 and 102 kJ/mol respectively. The different values of Q_R and Q_L for the same alloy at high and low temperatures arise from the difference in mechanisms controlling the unstable flow. The transition from type C serrations at high temperatures to type A serrations at low temperatures was observed for alloy A1. In contrast, the mixed type A+B takes place at the strain rate of 1×10^{-3} s^{-1} in alloys A2 and A3 and is replaced by pure type A on the right boundary (figure 2 (b), (c)). The obtained results agree with the present understanding of PLC effect, which involve the influence of different microstructure parameters on the jerky flow behavior [11].

4. Summary and conclusions

The analysis of the stress flow serrations in three aluminum alloys with different phase composition was performed. The following results are found:

1. The difference in the values of apparent activation energies for the right and left boundaries of the PLC domain is due to the transition between different types of stress flow serrations.
2. The dispersed particles in an aluminum matrix lead to a decrease in the apparent activation energy for both boundaries.
3. The temperature interval of PLC domain of alloys A2 and A3 is lower than that of alloy A1. Such difference can be related to the presence of dispersed particles in alloy A1 and A2.

Acknowledgement

The study was financially supported by the Russian Science Foundation, Belgorod State University project No. 17-72-20239. The main results were obtained by using the equipment of the Joint Research Center, «Technology and Materials», Belgorod State University.

References

[1] Kaibyshev R, Musin F, Lesuer D R and Nieh T G 2003 *Mater. Sci. Eng.* **342** 169
[2] Perez-Bergquist S, Gray III G, Cerreta E, Trujillo C and Perez-Bergquist A 2011 *Mater. Sci. Eng. A* **528** 8733
[3] Aboulfadl H, Deges J, Choi ., Raabe D 2015 *Acta Mater.* **86** 34
[4] Robinson J M 1994 *Int. Mater. Rev.* **39** 217
[5] Mogucheva A, Yuzbekova D and Borisova Yu 2018 *AIP Conference Proceedings* **2051** 020201
[6] Jobba M, Mishra R K and Niewczas M 2015 *Int. J. Plast.* **65** 43
[7] Král R, Lukáč P 1997 *Mater. Sci. Eng.* **234-236** 786
[8] Lebedkina T, Lebyodkina M 2008 *Acta Mater.* **56** 5567
[9] Mogucheva A, Yuzbekova D, Kaibyshev R, Lebedkina T and Lebyodkina M 2016 *Metall. Mater. Trans. A* **47** 2093
[10] Cottrell A 1953 *Philos. Mag.* **44** 829
[11] Lebedkina T, Lebyodkin M, Lamark T, Janecek M and Estrin Y 2014 *Mater. Sci. Eng* **615** 7

IOP Conf. Series: Materials Science and Engineering **597** (2019) 012058 doi:10.1088/1757-899X/597/1/012058

Synthesis of Ti_3SiC_2-based composites by spark plasma sintering of preceramic papers

E B Kashkarov[1], M S Syrtanov[1], E P Sedanova[1], A S Ivashutenko[1], A M Lider[1] and N Travitzky[1,2]

[1] National Research Tomsk Polytechnic University, 634050 Tomsk, Russia
[2] Friedrich-Alexander-Universität Erlangen-Nürnberg, 91054 Erlangen, Germany

E-mail: egor_kashkarov@mail.ru

Abstract. This paper is devoted to fabrication of Ti_3SiC_2-based ceramic materials from preceramic paper using spark plasma sintering (SPS) method. The synthesis temperature and pressure were 1373-1473 K and 30 MPa, respectively. The phase composition, microstructure and elemental composition were analysed using X-ray diffraction, scanning electron microscopy and energy-dispersive X-ray spectroscopy, respectively. The effect of temperature on the sintering process as well as on the phase and microstructure of the sintered materials was investigated.

1. Introduction

In the last decades much attention has been paid to developing composite ceramic materials based on non-oxide carbon compounds. Among them, the synthesis of MAX-phase ceramics has a great interest due to specifics of its crystalline structure providing unique combination of properties such as high melting temperature, thermal shock resistance, high elastic modulus, oxidation and corrosion resistance, thermal conductivity and mechanical manufacturability. The MAX-phases are a relatively new class of refractory materials described by the formula $M_{n+1}AX_n$, where M– transition metal, A –IIIA-IVA periodic element, X – carbon or nitrogen. Different techniques are implemented to synthesize MAX-phase ceramic materials such as hot isostatic pressing, spark plasma sintering and self-propagating high-temperature synthesis [1, 2].

In the recent years spark plasma sintering is widely used for MAX-phase synthesis due to its great potential for manufacturing single-phase dense materials in relatively short time at considerably lower temperature. Such materials possess superior mechanical properties in comparison with the ones produced by traditional processes of hot isostatic pressing or other sintering techniques [3]. For the first time, Ti_3SiC_2 MAX-phase was synthesized by Zhang et al. using spark plasma sintering of Ti, Si, C powders [4]. This research group obtained the purity of the synthesized product of more than 98%. It has been also shown that the initial powder ratio (Ti, Si and C), powder purity, synthesis temperature and heating rate significantly influences on the forming ceramic composites based on Ti_3SiC_2 [5]. Different approaches were implemented to decrease the synthesis temperature of Ti_3SiC_2 such as addition of trace amount of aluminum [6, 7]. Different compositions and powder mixtures have been used and investigated during SPS. The vast majority of research is aimed at decreasing the synthesis temperature, manufacturing high purity and density composites as well as the properties of synthesized MAX-phase ceramics.

Novel approach based on application of preceramic papers as a feedstock for SPS is suggested in the present research. Preceramic papers are promising materials for manufacturing of complex shaped bodies as well as for the synthesis of graded structures [8–10]. This is due to the possibility to construct multilayer (laminar) structures from preceramic papers with the potential to modify their composition [11]. At the same time, the kinetic processes of MAX-phase synthesis from preceramic papers by SPS remain unstudied. The aim of this research is to fabricate Ti_3SiC_2-based MAX-phase ceramics by spark plasma sintering of Ti_3SiC_2 preceramic papers.

2. Materials and Methods

2.1. Synthesis of Ti_3SiC_2-based ceramics

The preceramic paper sheets with the ceramic filer (Ti_3SiC_2 powder) were fabricated by a laboratory dynamic hand-sheet former (Dynamic hand-sheet former D7, Sumet Systems GmbH, Germany). During paper processing the aqueous feedstock suspension was sprayed on a perforated rotating wire. The concentration of the fibers in the water-based solution was set to 1.2 wt.%. The concentration of the ceramic filler in preceramic papers was set to 90 wt.%. After dewatering paper sheets with a dimension of 94×25×0.05 cm³ were dried at 110 °C for 15 min, and then calendered.

The final preceramic paper was cut into the round sheets of 20 mm in diameter. The spark plasma sintering of the preceramic paper sheets was conducted using HPD25 (FCT Systeme GmbH, Germany). The stack of 12 layers of preceramic paper is placed between two graphite punches in a graphite die. A graphite foil is introduced at the punch/sample interface to ensure a good electrical contact. A constant pressure of 30 MPa is applied during the whole sintering cycle. The temperature is measured during sintering using a pyrometer. The composition of the preceramic paper sheets and SPS parameters are shown in Table 1.

Table 1. The composition of preceramic paper and SPS parameters.

Paper composition		Sintering parameters	
Cellulose fibers	13.35 %	Temperature	1373–1473 K
Ti_3SiC_2 powder	164.7 %	Heating rate	180 K/min
Cationic starch	1.8 %	Holding time	10 min
Anionic starch	1.8 %	Pressure	30 MPa
Retention aid	1.4 %	Atmosphere	Vacuum

2.2. Characterization

The phase composition was analyzed by X-ray diffraction (CuK_α radiation) using XRD 7000S diffractometer (Shimadzu, Japan) equipped with 1280-channel high-speed detector OneSight. The acceleration voltage and current were 40 kV and 30 mA, respectively. Diffraction patterns were recorded at following parameters: 2θ range – 5–80°; scan speed – 10.0°/min; sampling pitch – 0.0143°; exposure time – 21.49 s. The diffraction data was analyzed using Sleve+ program and ICDD PDF-4+ database. The following cards were used as the reference: Ti_3SiC_2 (#00-059-0189), TiC (#00-031-1400), TiSi (#04-008-9158), $TiSi_2$ (#03-065-2522). The microstructure and chemical composition were analyzed by scanning electron microscopy (SEM) using Vega 3 (Tescan, Brno, Czech Republic) equipped with energy-dispersive X-ray spectroscopy (EDX) attachment.

3. Results and discussion

3.1. Sintering process

The densification behavior was studied using the temperature measurement, current and differential effect for samples density during SPS. Figure 1 show densification curves of Ti_3SiC_2 sintered at 1373 K (a) and 1473 K (b) under 30 MPa load for 10 min. The current supplied to punches as a function of

IOP Conf. Series: Materials Science and Engineering **597** (2019) 012058 doi:10.1088/1757-899X/597/1/012058

sintering time is also shown in Figure 1. The density function was formed by value changing of sintered samples` calculated density depending on punch displacement values during sintering.

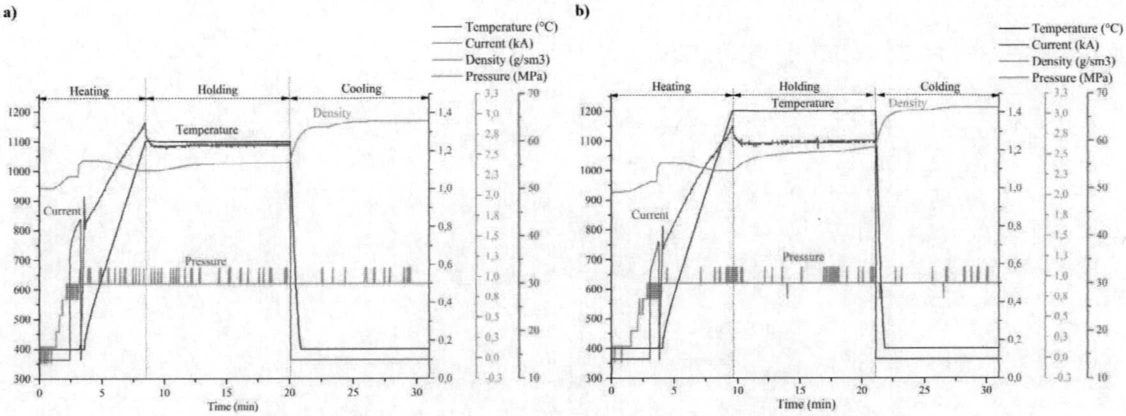

Figure 1. Densification curves of Ti_3SiC_2 sintered at 1373 K (a) and 1473 K (b) under 30 MPa for 10 min.

During spark plasma sintering, the temperature was controlled using a pyrometer. The working temperature range of the pyrometer is 673-3273 K. Therefore, the maximum current is passed through the entire assembly up to 673 K (400 °C) for 3 min heating, then the abrupt behavior of the current values occurs and the current flow is adjusted according to the temperature measurements. In a process of heating (7-10 min) under pressure the density of samples sintered at 1373 K and 1473 K changes in range of (2.1-2.4) g/cm^3. The sharp increase in the density occurs when the electrical current is passing through the moulds. The absorbed gases emitted from surface and volume of the samples and the material is thermally expanded under the heating process. When the temperature reaches 80% from the sintering temperature value, the shrinkage process of the sample is beginning. Nevertheless the density is reduced due to the process of thermal expansion prevails over the shrinkage process.

When the samples are heated to the sintering temperature (isothermal holding), the shrinkage process is the main and the density of the samples sintered at 1373 and 1473 K changes by 0.09 g/cm^3 and 0.28 g/cm^3, respectively.

After isothermal holding, the samples are cooled at a rate of approximately 500 K/min and then pressure is removed after 3-4 minutes. During the entire cooling process, the density of samples sintered at 1373 and 1473 K changes by 0.51 g/cm^3 and 0.49 g/cm^3, respectively. The increasing is caused by the ending of thermal expansion process. The final calculated density of Ti_3SiC_2-based composites sintered at 1373 K and 1473 K are 2.9 g/cm^3 and 3.1 g/cm^3 respectively.

3.2. Phase composition

Figure 2 shows the diffraction patterns of preceramic paper sheet and Ti_3SiC_2-based composites sintered at 1373 K and 1473 K. The results of X-ray analysis show that preceramic paper consist of Ti_3SiC_2, TiC, $TiSi_2$, TiSi crystalline phases and amorphous phase of organic binder. The content of Ti_3SiC_2 in the preceramic paper was more than 90 vol.%. The amorphous phase was not observed after sintering at 1373 K. However, the content of Ti_3SiC_2 phase (60.2 vol.%) significantly decreases along with the growth of TiC (17.3 vol.%), $TiSi_2$ (8.6 vol.%), TiSi (13.9 vol.%) phases. The decrease in the content of Ti_3SiC_2 phase indicates its partly decomposition. Increasing of TiC phase indicates the presence of high amount of carbon which reacts with titanium at 1373 K. Also the decomposition of Ti_3SiC_2 phase is accompanied by increasing of $TiSi_2$ and TiSi content. At the sintering temperature of 1473 K the content of TiC phase increases and the redistribution between $TiSi_2$ and TiSi phases occurs. The content of indicated phases was 22.2, 14.1 and 7.1 vol.%, respectively.

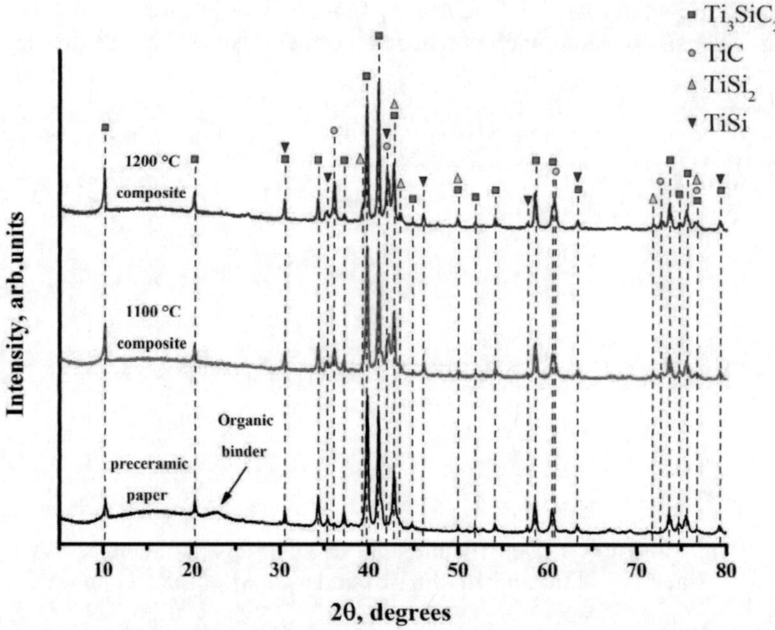

Figure 2. Diffraction patterns of preceramic paper sheet and Ti_3SiC_2-based composites sintered at 1373 K and 1473 K.

3.3. Microstructure and elemental composition

Figure 3 shows the SEM images demonstrating microstructures of the Ti_3SiC_2-based composites sintered at 1373 K (a) and 1473 K (b).

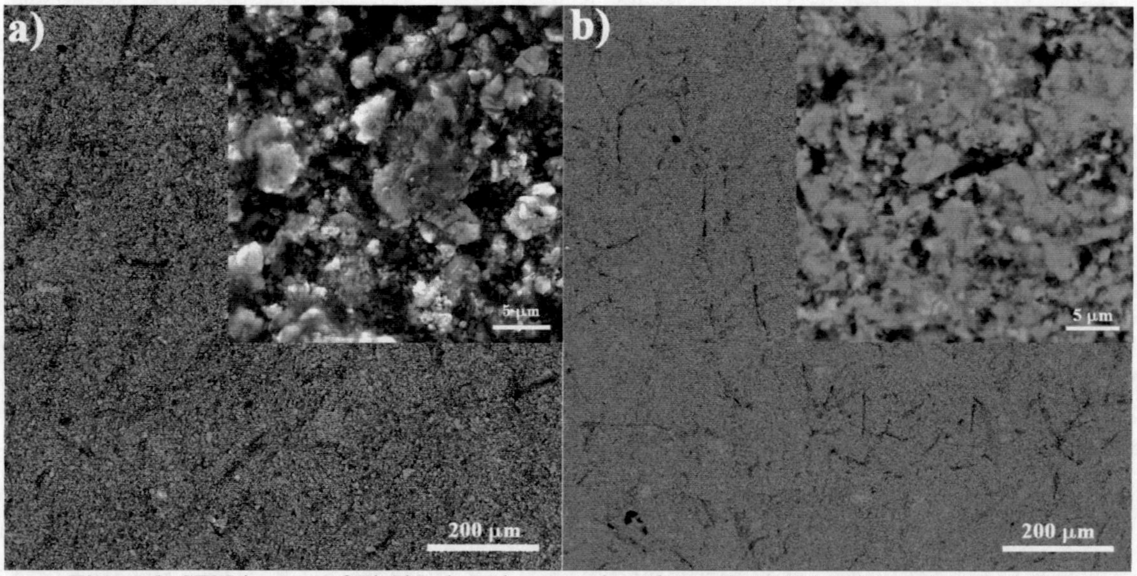

Figure 3. SEM images of Ti_3SiC_2-based composites sintered at 1373 K (a) and 1473 K (b). Insets: higher magnification images.

The analysis of the SEM images shows that microstructure of the sintered composites becomes denser with increasing sintering temperature. Also the traces of cellulose fibers were observed in the composite structure in both cases. Such traces form as the result of decomposition of cellulose fibers during sintering process. In this case, additional TiC phase can be formed during the reaction between the unbounded carbon from cellulose and the titanium from the Ti_3SiC_2 decomposition reaction. It should be noted that the pore size in the structure of Ti_3SiC_2-based composites sintered at 1473 K does not exceed 2-3 μm. According to EDX analysis the composition of the sintered materials was (44±2) at.% Ti, (23±3) at.% Si and (33±4) at.% C. The results of EDX well correlated with XRD measurements.

4. Conclusion

In this paper the Ti_3SiC_2-based MAX-phase composites were synthesized by spark plasma sintering of preceramic papers at 1373 and 1473 K. The sintered composites consist of multiphase Ti_3SiC_2, TiC, $TiSi_2$, TiSi crystalline structure. The content of Ti_3SiC_2 phase in the sintered at 1373 K composite was 60 vol.%. The partly decomposition of MAX-phase occurs during sintering, and this effect is more pronounced with temperature rise. According to SPS sintering process analysis and scanning electron microscopy, it was established that the microstructure of composite sintered at 1473 K becomes denser. Also the content of TiC and $TiSi_2$ phases increases as well as TiSi phase decreases. The elemental composition of sintered materials was presented by 44 at.% Ti, 23 at.% Si and 33 at.% C, which is well correlate with XRD data. The decomposition of cellulose fibers leads to the formation of specific traces in the structure of spark plasma sintered Ti_3SiC_2-based MAX-phase composites.

Acknowledgements

The work was supported by the Russian Science Foundation (grant No. 19-19-00192).
The authors are grateful to T. Yakich for the SEM analysis.

References

[1] Ying G, He X, Li M, Du S, Han W and He F 2011 *Mater. Sci. Eng.* **528** 2635
[2] Li S B, Yu W B, Zhai H X, Song G M, Sloof W G and Zwaag S 2011 *J. Eur. Ceram. Soc.* **31** 217
[3] Ghosh N C and Harimkar S P 2012 *Consolidation and synthesis of MAX phases by Spark Plasma Sintering (SPS): a review advances in Science and Technology of Mn+1AXn Phases* (Woodhead Publishing) pp 47–80
[4] Zhang Z F 2002 *J. Eur. Ceram. Soc.* **22** 2957
[5] Zhang Z F, Sun Z M and Hashimoto H 2004 *Mater. Sci. Technol.* **20** 1252
[6] Sun Z M 2006 *Scripta Materialia* **55** 1011
[7] Liang B Y, Jin S Z and Wang M Z 2008 *J. Alloys Compd.* **460** 440
[8] Schultheiß J, Dermeik J, Filbert-Demut I, Hock N, Yin N, Greil P and Travitzky N 2015 *Ceram. Intern.* **41** 12595
[9] Lorenz H, Thäter J, Matias Carrijo M M, Rambo C R, Greil P and Travitzky N 2017 *J. Mater. Res.* **32** 3409
[10] Carrijo M. M, Lorenz H, Rambo C. R, Greil P and Travitzky N 2018 *Ceram. Intern.* **44** 8116
[11] Krinitcyn M, Fu Z, Harris J, Kostikov K, Pribytkov G A, Greil P and Travitzky N 2017 Laminated Object Manufacturing of in-situ synthesized MAX-phase composites *Ceram. Intern.* **43** 9241

IOP Conf. Series: Materials Science and Engineering **597** (2019) 012059 doi:10.1088/1757-899X/597/1/012059

High-temperature oxidation resistance of Ti-implanted E110 alloy

E B Kashkarov, M S Syrtanov, A E Shevelev, A V Kurochkin and S K Pavlov

National Research Tomsk Polytechnic University, 634050 Tomsk, Russia

E-mail: egor_kashkarov@mail.ru

Abstract. This paper describes the effect of surface modification by high-intensity Ti-ion implantation on the high-temperature oxidation resistance of E110 zirconium alloy. The oxidation tests were performed in air at 873 K for 10 h and in water steam at 1373 K for 10 min. The microstructure, phase composition and depth distribution of elements were analysed using scanning electron microscopy, X-ray diffraction and glow-discharge optical emission spectroscopy, respectively.

1. Introduction

The accident at the Fukushima Daiichi power plant occurred in 2011 gave impetus to international research on the development of protective coatings for nuclear cladding tubes [1]. Besides zirconium claddings are resistant to corrosion and oxidation under the normal operating temperature, they cannot ensure the protection of light-water reactors under loss-of-coolant accident (LOCA) conditions. The accelerated corrosion reaction in water steam at 1373-1473 K leads to generation of explosive hydrogen and degradation of mechanical properties of nuclear fuel claddings. Currently, the deposition of protective coatings such as Cr, SiC, CrN, TiN/TiAlN and other are being developed to increase oxidation resistance of zirconium alloy [2–5]. However, the adhesion properties of protective coatings are the weaknesses of coating technology; therefore, surface modification by charged particle beams has also been developed. A high-intensity implantation of low-energy titanium ions can be implemented to form a deep modified layer providing enhanced mechanical and tribological properties of zirconium alloys [6]. However, the oxidation resistance of the Ti-doped surface modified zirconium alloy has not been studied. In the present work, the high-temperature corrosion behavior of E110 zirconium alloy with a surface modified layer formed by high-intensity ion implantation was considered.

2. Materials and Methods

2.1. Ion implantation

The samples with the size of 20×20×2 mm were cut from a cold rolled E110 (Zr–1Nb) zirconium alloy. The samples were grinded using a SiC abrasive paper and ultrasonically cleaned in acetone for 15 min. For surface modification, a high-intensity beam of low-energy Ti ions was used. To generate the direct current (DC) flow of metal plasma a vacuum arc evaporation system was used. The high-current density of metal ions was provided by the hemispherical ballistic focusing system. The ion beam formation, transportation and ballistic focusing system were described in [7]. The implantation parameters are shown in Table 1. According to energy-dispersive spectroscopy, the titanium has a gradient distribution up to 10 μm in depth with the maximum concentration of 55 at.% in the surface [8].

Table 1. Ion implantation parameters

Ion current density	Pulse duration	Duty factor	Temperature	Fluence
100 mA/cm^2	5 μs	0.5	973 K	5.6×10^{20} ion/cm^2

2.2. High-temperature oxidation and characterization
High temperature oxidation tests were performed in air atmosphere at 873 K for 10 h and in steam at 1373 K for 10 min. For the oxidation tests, the samples were placed in pre-heated chambers. For the steam oxidation test a furnace with evaporation equipment was used. The temperature was controlled by a thermocouple. After the oxidation tests, the weight gains were measured using analytical weighing-machine (Sartorius CP124 S) with an accuracy of 10^{-4} g. The samples after steam oxidation test were quenched in water from approximately 1073 K.

The phase composition was analyzed by X-ray diffraction (CuK_α radiation) using XRD 7000S diffractometer (Shimadzu, Japan) equipped with 1280-channel high-speed detector OneSight. The acceleration voltage and current were 40 kV and 30 mA, respectively. The diffraction data was analyzed using Sleve+ program and ICDD PDF-4+ database. The microstructure was analyzed by scanning electron microscopy (SEM) using S-2400N (Hitachi, Japan). For microstructural analysis, the samples were chemically etched in a solution of 45% HNO_3, 5% HF and 50 % H_2O.

3. Results and discussion

3.1. Weight gain
The samples were oxidized in atmosphere at 873 K to analyse the heat-resistance of the Ti-implanted alloy. The steam oxidation test was performed to evaluate the high-temperature resistance of the implanted alloy under the conditions simulating LOCA. Therefore, the weight gains of the surface modified samples were measured and compared with that of the untreated E110 alloy (Figure 1). It was found that the high-temperature oxidation resistance of the implanted sample depends on the oxidation conditions. After the oxidation in air, the weight gain of the surface-modified sample is higher by 20% in comparison with the E110 alloy. It indicates degradation of the oxidation resistance properties of zirconium alloy. On the contrary, the oxidation of the Ti-implanted alloy in steam is significantly lower than that of the untreated alloy. Such difference may be attributed to the specificity of the oxidation mechanisms as well as the difference in oxidation temperature. The presence of nitrogen during air oxidation of the modified layer can accelerate the oxidation process due to partial replacement of oxygen atoms in titanium and zirconium oxides. In turn, during the oxidation in stream, the main role is related to the interaction of oxygen and hydrogen with the material surface.

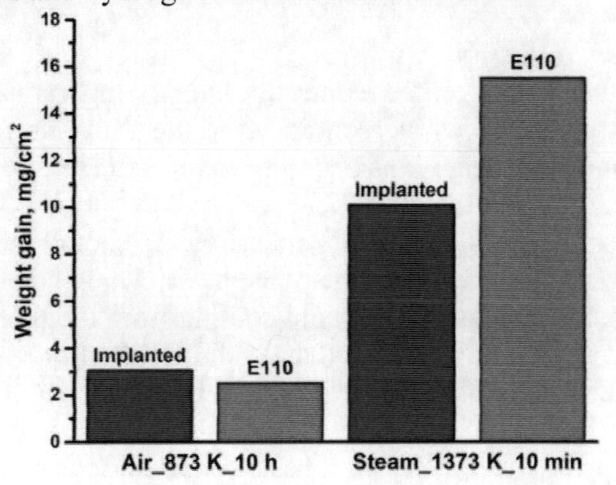

IOP Conf. Series: Materials Science and Engineering **597** (2019) 012059 doi:10.1088/1757-899X/597/1/012059

Figure 1. The weight gains of the samples after high-temperature oxidation in air and steam.

3.2. Phase composition

Figure 2 shows the diffraction patterns of Ti-implanted zirconium alloy after oxidation. According to X-ray analysis the as-implanted alloy consist of Zr-Ti solid solution ($Zr_{0.38}Ti_{0.62}$, $Zr_{0.5}Ti_{0.5}$), α-Zr phases with hcp structure and β-Zr phase with fcc structure. The phase composition of Ti-implanted zirconium alloy significantly changes after oxidation. The oxidation in air at 873 K for 10 h leads to the formation of monoclinic ZrO_2, hexagonal Zr_3O, hexagonal TiO_2 and orthorhombic $ZrTiO_4$ phases. The formation of orthorhombic $ZrTiO_4$ phase at slightly milder oxidation conditions was also observed in [9]. As a result of LOCA oxidation tests, in addition to monoclinic ZrO_2 and orthorhombic $ZrTiO_4$ phases, the tetragonal ZrO_2 phase was detected. It should be noted that the oxidation conditions directly affect the phase composition of Ti-implanted zirconium alloy. Moreover, the analysis of the diffraction patterns indicates the changes in oxides phase content depending on oxidation conditions. However, it is difficult to calculate the exact values of content due to superposition of its reflections.

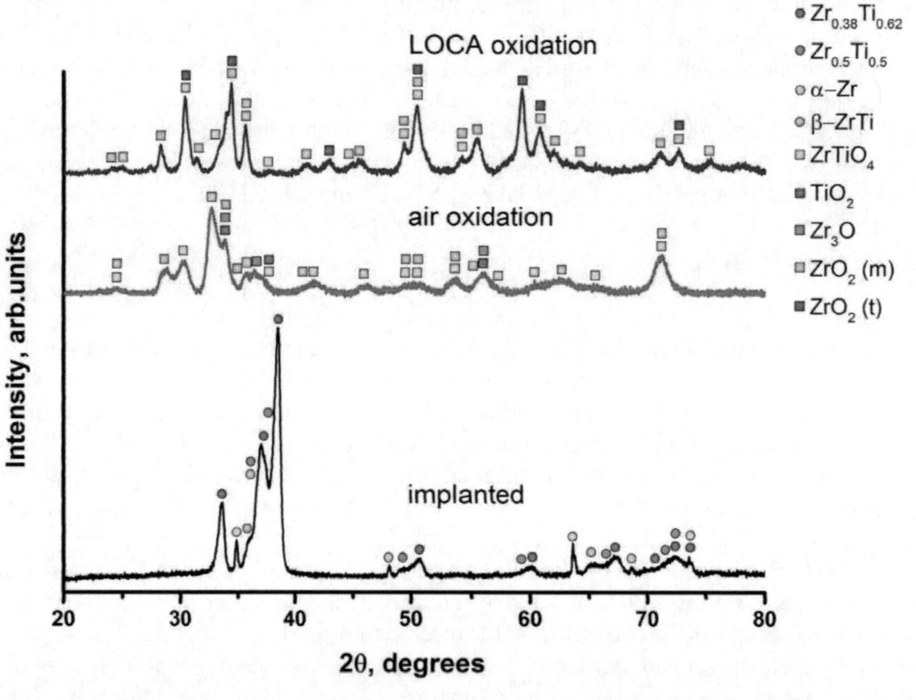

Figure 2. Diffraction patterns of Ti-implanted zirconium alloy after oxidation tests.

3.3. Microstructure and elemental distribution

Figure 3 shows the microstructure of the formed oxides after high-temperature oxidation test at 1373 K. The oxides have a columnar structure grown perpendicular to the alloy surface. The formation of high amount of cracks at the oxide/metal interface was observed in the oxidized E110 alloy. The propagation of these cracks apparently occurs across the grain boundaries of the primary β phase due to relaxation of tensile/compressive stresses at the oxide/metal interface. It should be noted that an oxygen-oversaturated solid solution α–Zr(O) forms under the oxide layer, which have a lamellar microstructure formed due to β→α phase transformations during rapid cooling from the β phase. The thickness of the oxide layer in Ti-implanted sample is lower than that of the E110 alloy. It well correlates with the measured weight gain data. Although small microcracks have been found at the metal/oxide interface in the implanted sample, this interface is fairly uniform and has the dense microstructure. It is also seen a non-uniform corrosion in the upper oxide layer, but it has a denser structure that apparently decrease the diffusion of oxygen and enhances the corrosion resistance.

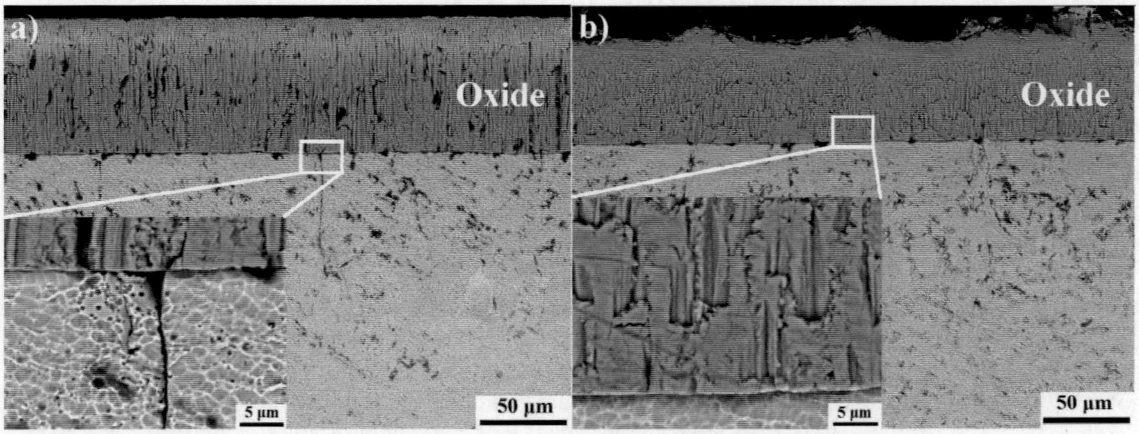

Figure 3. Cross-section SEM images of the steam oxidized E110 (a) and Ti-implanted E110 (b).

Figure 4 shows the depth distribution of elements in the oxidized samples. After oxidation in air, the thickness of the oxide layer was ~18 μm and ~22 μm for the untreated and Ti-implanted samples, which is in good agreement with the SEM data (not shown in the article). In both cases, the oxide compounds with porous microstructure are formed. It is also seen that the thickness of the formed oxide layer exceeds the thickness of the Ti-modified layer (~10 μm), which leads to the formation of ZrO_2 phases under $TiZrO_4$ oxide layer. The analysis of glow-discharge optical emission spectroscopy (GDOES) data indicates a partial diffusion of titanium into the alloy depth, especially during oxidation in stream. In the samples oxidized in steam, a thick oxide layer is formed with a fairly uniform distribution of oxygen over the depth. According to SEM results, the thickness of the oxide layer was 76-79 μm and 60-70 μm for the untreated and Ti-implanted alloys, respectively.

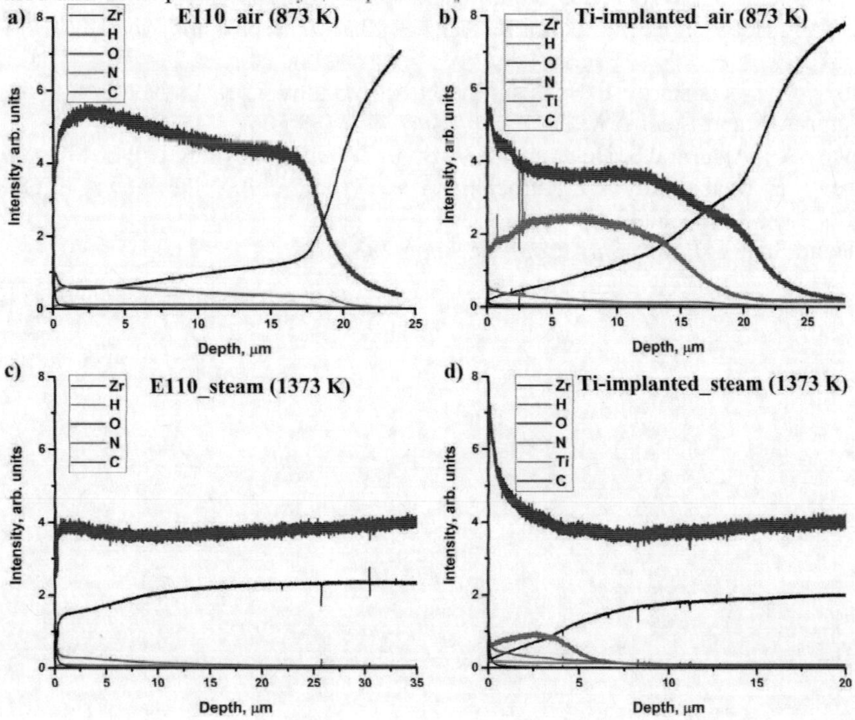

Figure 4. GDOES depth distribution profiles of elements in the oxidized samples.

4. Conclusion

The high-temperature oxidation resistance of E110 zirconium alloy surface modified by high-intensity implantation of low-energy Ti ions was investigated. The oxidation tests have shown the difference in oxidation kinetics of Ti-implanted alloy in air and steam. The corrosion resistance of the implanted alloy during oxidation in air atmosphere is significantly lower than that of the as-received E110 alloy. At the same time, an increase in the high-temperature oxidation resistance of the alloy was observed under LOCA oxidation test, where the weight gain was reduced to 10.1 mg/cm^2. The formation of orthorhombic $TiZrO_4$ oxide phase occurs in the implanted layer during oxidation in steam, which has a denser microstructure compared to ZrO_2 oxides. After rapid cooling of the samples after LOCA test, numerous cracks appeared at the oxide/metal interface in the untreated E110 alloy. The oxide layer in the implanted sample has a non-uniform structure on the surface, and there are practically no cracks at the oxide/metal interface. However, the relatively high oxidation rate of the surface Ti-doped alloy does not allow it to be considered as protective layer for fuel cladding.

Acknowledgements

The reported research was funded by the Russian Foundation for Basic Research and the government of the Tomsk region of the Russian Federation, grant No. 18-48-703034.

The authors are grateful to Y. Shanenkova for SEM measurements.

References

[1] Minin N 2018 *Energy Strategy Rev.* **21** 98
[2] Kashkarov E B, Syrtanov M S, Murashkina T L, Kurochkin A V, Shanenkova Y and Obrosov 2019 *Coatings* **9** (1) 31
[3] Sidelev D V, Kashkarov E B, Syrtanov M S and Krivobokov V P 2019 *Surf. Coat. Technol*. **369** 69
[4] Tang C, Stueber M, Seifert HJ and Steinbrueck M 2017 *Corros. Rev.* **35** 141
[5] Duan Z, Yang H, Satoh Y, Murakami K, Kano S, Zhao Z, Shen J and Abe H 2017 *Nucl. Energy Des.* **316** 131
[6] Ryabchikov A I, Kashkarov E B, Pushilina N S, Syrtanov M S, Shevelev A E, Korneva O S, Sutygina A N and Lider A M 2018 *Appl. Surf. Sci.* **439** 106
[7] Ryabchikov A I, Ananin P S, Dektyarev S V, Sivin D O and Shevelev A E 2017 *Vacuum* **143** 447
[8] Kashkarov E B, Ryabchikov A I, Kurochkin A V, Syrtanov M S, Shevelev A E, Obrosov A and Weiß S 2018 *Metals* **8** (12) 1081
[9] Cui W F and Shao C J 2015 *Surf. Coat. Technol.* **283** 101

IOP Conf. Series: Materials Science and Engineering **597** (2019) 012060 doi:10.1088/1757-899X/597/1/012060

Structure and mechanical characteristics of the hypereutectic silumin subjected to pulsed electron beam treatment

E A Petrikova[1], A D Teresov[1], M E Rygina[1,2] and Yu F Ivanov[1,2]

[1]Institute of High Current Electronics, Siberian Branch, Russian Academy of Sciences, 2/3, Akademicheskiy ave., Tomsk, 634055, Russia

[2]Tomsk Polytechnic University, 30, Lenina ave., Tomsk, 634050, Russia

E-mail: elizmarkova@yahoo.com

Abstract Silumin, aluminum with silicon alloy, is a promising material used for the manufacture of medium-loaded machine parts and mechanisms. High brittleness is one of the main drawbacks of hypereutectic silumin. Modification of a hypereutectic silumin (18–20 wt.% Si) was carried out by irradiating the samples with an intense pulsed electron beam. It was established that irradiation of cast hypoeutectic silumin by an electron beam leads to a significant reduction in the number of micropores, forming a high-speed cellular crystallization structure with a cell size of (0.4-0.6) μm. An irradiation mode allowing to increase the silumin surface layer hardness by more than 4 times, wear resistance - by 1.2 times, to increase ductility by 1.2 times in relation to the initial material was detected (35 J / cm^2; 200 μs, 20 imp. 0.3 s^{-1}).

1. Introduction

Silumin is aluminum based alloys with different silicon content. These alloys belong to the class of eutectic materials. Silumins are widely used in such industries as aircraft, enginery and shipbuilding for the manufacture of medium-loaded parts - pistons, bearings, housings, etc. An increased interest has been caused by the use of hypereutectic silumins in recent decades. Such materials are characterized by high hardness and wear resistance due to the high content of silicon. The main disadvantages of hypereutectic silumins include the high porosity of products, that origin is associated with a high content of hydrogen in ligatures, the presence in the structure of primary silicon crystallites with a transverse size of hundreds of micrometers [1-3]. These and some other features of the structural-phase state of the hypereutectic silumins lead to their increased brittleness, which significantly reduces the field of industrial application.

Modification of the material at the stage of formation or the use of various methods of casting, for example, with characteristics of centrifugal lithium is the main way to eliminate the mentioned disadvantages of hypereutectic silumins are used in industry. Both approaches have their drawbacks, so with centrifugal casting there is a limit on the weight (not more than 500 g) and the shape (ring) of the work piece [4].

Methods based on the use of concentrated energy flows (laser processing, plasma flow processing, powerful ionic and low-energy electron beams, the method of electric flashing in a magnetic field and some others) are used for the purpose of modifying the structure and properties of metals and alloys, including silumin in recent years [5-7]. Ultra-high speeds (up to 109 K / s) of heating to melting temperatures and subsequent cooling of a relatively thin surface layer of the material (10^{-7}-10^{-6} m), forming in it extreme temperature gradients (up to 107 ... 108 K / m) providing cooling the surface layer due to the heat sink into the main volume of the material at a speed of 104 ... 109 K / s, create conditions for the formation in the surface layer from one to hundreds of micrometers thickness of a nano- and submicrocrystalline structure. In total, this contributes to an increase in corrosion resistance, wear resistance and microhardness, fatigue life of the irradiated material, which is unattainable with traditional surface treatment methods [7–9].

The aim of this work is to analyze the results obtained in the study of the structure and properties of hypoeutectic silumin subjected to modification by an intense pulsed electron beam.

2. Materials and methods

The material under study was hypereutectic silumin with a silicon content of 18 to 20 wt.%. Modification of silumin was carried out in vacuum at a residual pressure of the working gas (argon) of 0.02 Pa on the SOLO setup [10]. Irradiation mode: the accelerated electron energy is 18 keV, the electron beam energy density is (15-35) J / cm^2, the pulse repetition rate is 0.3 s^{-1}, the duration of the electron beam exposure is 200 μs, the number of impact pulses is 20. The detected irradiation regimes, according to thermal calculations [11, 12], corresponded to the melting of the surface layer of silumin up to 100 μm thick. Part of the samples before irradiation was made in the form of blades for tensile testing in accordance with GOST 1497-84 [13]. The working area of the blades was irradiated from both sides by an electron beam. Tests of the initial and irradiated samples before destruction were carried out on an Instron3369 setup at a tensile rate of 0.2 mm / s. The structure of the samples after irradiation and the structure of the damage surface was investigated by scanning electron microscopy (SEM-515 Philips). Hardness tests were carried out on the device PMT-3.

3. Results

The structure of the samples of the eutectic silumin in the cast state (the state before electron beam irradiation) consists of eutectic grains, grains of primary silicon and intermetallic compounds [14]. Samples of silumin in the initial state contain a large number of pores that sizes vary from units to tens of micrometers. The initial sample hardness was 890 MPa, the wear parameter (inverse of wear resistance) was 5.86·10^{-4} mm^3 / N·m, and the friction coefficient was 0.44.

The irradiation of hypereutectic silumin samples with an intense pulsed electron beam leads to the formation of a high-speed cellular crystallization structure (Fig. 1a). The dimensions of solid solution based on aluminum cells vary within (0.4–0.6) μm (Figure 1, b).

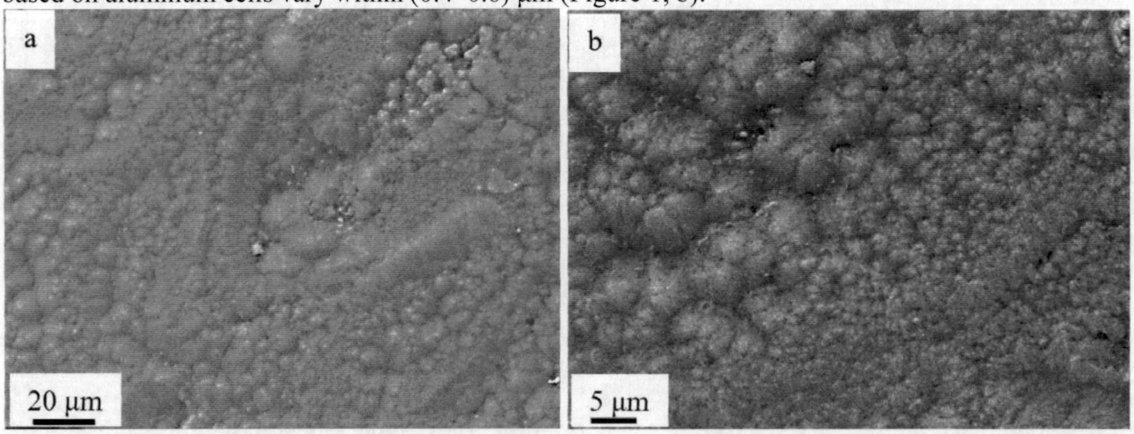

Figure 1. Structureofthesurfaceofthehypereutecticsiluminirradiatedwithanintensepulsedelectronbeam (35 J / cm^2; 200 μs, 20 pulses 0.3 s^{-1}).

The microhardness of the irradiated samples was 4580 MPa; wear parameter 4.96 ·10^{-4} mm^2 / N·m, friction coefficient 0.43. Thus, the completed tests showed that the irradiation of silumin with an electron beam led to a significant increase in the microhardness and wear resistance of the surface layer of the material.

The deformation curves of silumin subjected to an intense pulsed electron beam are shown in Figure 2. It is clearly seen that electron-beam treatment of silumin leads to an increase in the plasticity of the material with an increase in the energy density of the electron beam. Comparing the results obtained in tensile tests of silumin in the initial state and after irradiation, it is possible to note an increase in the plasticity of the irradiated samples by 1.2 times. The ultimate strength of the initial and electron beam irradiated (35 J / cm^2; 200 μs, 20 imp. 0.3 s^{-1}) samples are the same.

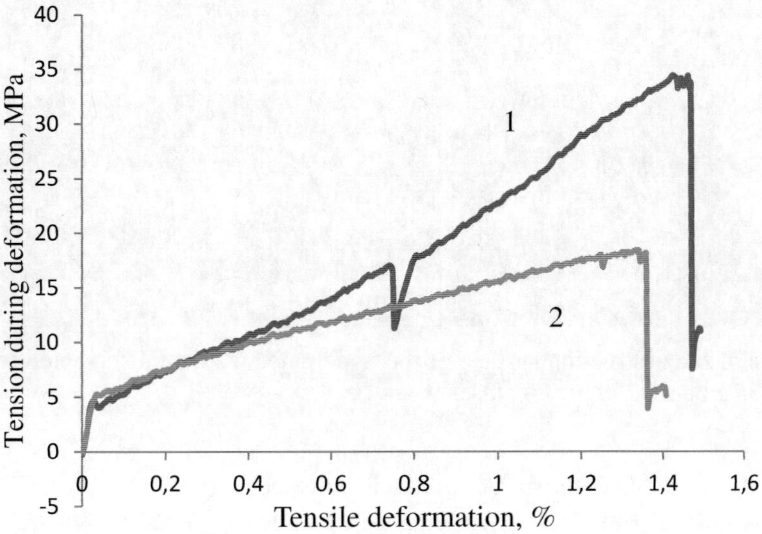

Figure 2. Deformation curves of the hypereutectic silumin subjected to electron beam irradiation (200 μs, 20 imp. 0.3 s^{-1}): curve 1 - 35 J / cm^2; curve 2 is 15 J / cm^2.

It should be assumed that the increase in plasticity of silumin, irradiated by an intense pulsed electron beam, is due to the transformation of its structure. Figures 3 and 4 show electron microscopic images of the destruction surface of the eutectic silumin in the initial state (Figure 3) and after irradiation with an intense pulsed electron beam (Figure 4).

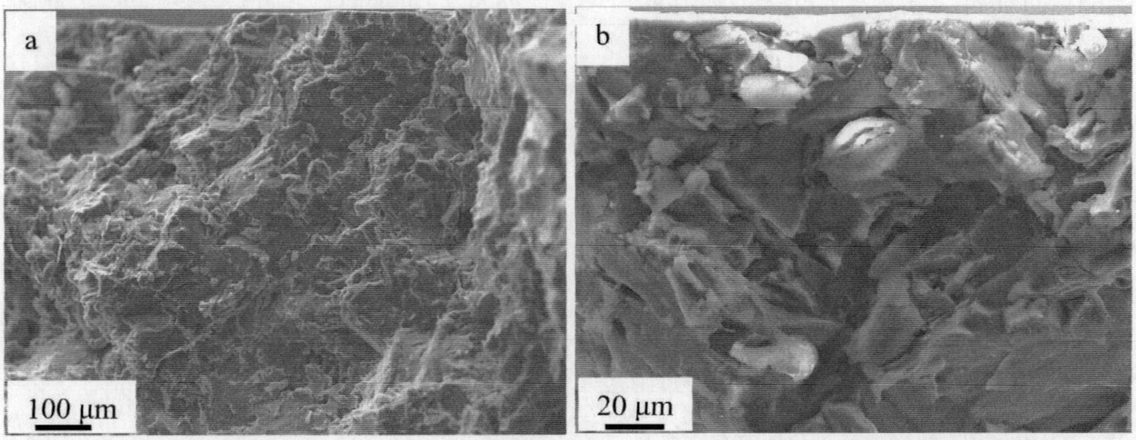

Figure 3. The structure of the surface destruction of the hypereutectic silumin in the initial state.

It is clearly seen that the fracture of silumin in the cast state is predominantly brittle (Figure 3, b). The irradiation of the silumin surface with an intense pulsed electron beam and the subsequent destruction as a result of the tensile test is accompanied by the formation of viscous fracture cells in the surface layer up to 100 μm thick (Figure 4, a). Cell sizes range is from 0.5 μm to 2 μm. It is obvious that the formation of such a structure is a consequence of the high-speed crystallization of the surface layer of silumin, which takes place during irradiation with an intense pulsed electron beam.

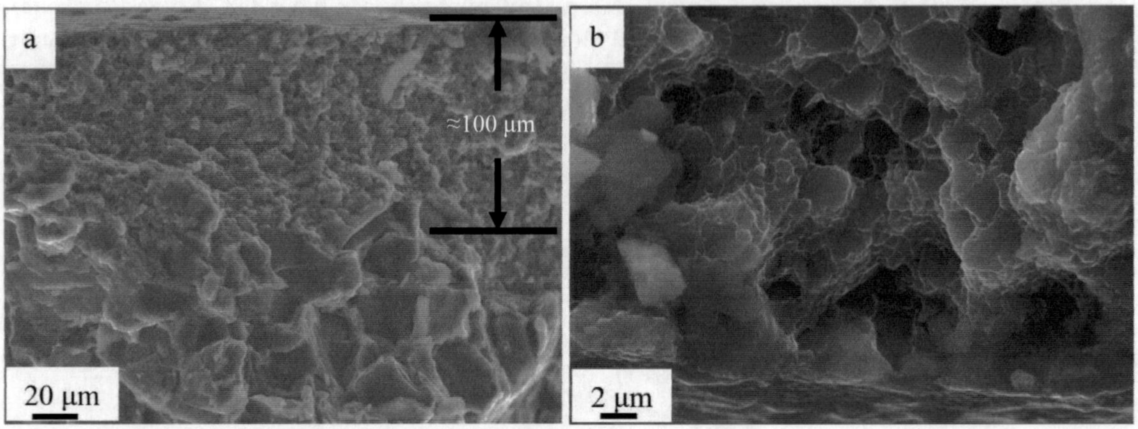

Figure 4. The structure of the destruction surface of the hypereutectic silumin subjected to an intense pulsed electron beam (35 J / cm^2; 200 µs, 20 pulses 0.3 s^{-1}).

4. Conclusion

It was established that the irradiation of the surface of cast hypereutectic silumin samples (silicon content (18–20) wt.%) by an intense pulsed electron beam is accompanied by a significant decrease in the number of micropores, as well as the formation of a high-speed cellular crystallization structure with a cell size (0.4–0.6) µm as a result of the conducted research. An irradiation mode of silumin with an intense pulsed electron beam (35 J / cm^2; 200 µs, 20 imp. 0.3 s^{-1}) was detected, which makes it possible to increase the hardness of the surface layer of the irradiated samples by more than 4 times, the wear resistance is 1.2 times relative to the initial material. The increase in these characteristics is associated with the grinding of the microstructure. The tensile tests of silumin in the initial state and state after irradiation with an electron beam (35 J / cm^2; 200 µs, 20 imp. 0.3 s^{-1}) revealed an increase in the plasticity of the irradiated samples by 1.2 times. The ultimate strength of the initial and electron-irradiated samples did not change.

Acknowledgments
This work was supported by the RFBR grant (project No. 19-52-04009 Bel_mol_a).

References

[1] Wladysiak R, Kozun A, Debowska, K andPacyniak T 2017 *Arch. foundry eng.*-**17 (2)** 137-44
[2] Yi LF, Yoshida N, Yamamoto T, Onda T and Chen Z.-Ch.2019 *J. mater. sci.* **54 (13)** 9933-44
[3] Nehrkorn S, Handel M, Dietrich D, Podlesak H, Nickel D, Wielage B and Lampke T2014 *Sur. Coat. Tech.* **242** 118–24
[4] Stetsenko V, Baranov K and Gutev A 2013 *Cast. and metals* **3 (72)** 116-7
[5] Mikheev R and Chernyshova T. 2009 *Metals* **(6)** 53-9
[6] Wong T and Liang G 1997 *J. Mater. Proc.Tech.* **63** 930-4
[7] Laskovnev A, Ivanov Y, Petrikova E, Koval N, Uglov V, Cherenda N, Bibik N and Astashincky V 2013 *Modification of the structure and properties of eutectic silumin electron-plasma processing* (Minsk: Belorussian navuka)
[8] Ivanov Y, Gromov V, Konovalov S and Aksenova K 2016 *Fatigue of silumin modified by electron beam processing* (Novokuznetsk:Poligraphist)
[9] Rotshtein V, Ivanov Yu and Markov A 2006 Surface treatment of materials with low-energy, high-current electron beams *Materials surface processing by directed energy techniques* ed Y

IOP Conf. Series: Materials Science and Engineering **597** (2019) 012060 doi:10.1088/1757-899X/597/1/012060

Pauleau: Elsevier chapter 6 pp 205-40
[10] Koval N and Ivanov Y 2008 *Rus. Phys. J.* **5** (51) 505-16
[11] Ivanov Yu, Petrikova E, Ivanova O, Ikonnikova I, Teresov A, Shugurov V and Krysina O 2015*Rus. Physics J.* **58** (3) 373-9.
[12] Ivanov Yu, Petrikova E, Ivanova O, Ikonnikova I and Tkachenko A 2015 *Rus. Phys. J.* **58** (4) 478-84
[13] GOST 1497-84 2005 Metals. Tensile test methods (Moskow:Standardinform)
[14] Rygina M, Ivanov Y, Laskovnev A, Teresov A, Cherenda N, Uglov V, Petrikova E and Astashinskaya M 2016 IOP Conf. Ser.: Mater. Sci. and Eng. **124** 1-5

IOP Conf. Series: Materials Science and Engineering **597** (2019) 012061 doi:10.1088/1757-899X/597/1/012061

Influence of Calcium-Phosphate Coating on Wettability of Hybrid Piezoelectric Scaffolds

A Zviagin[1], R Chernozem[1], M Surmeneva[1], K Loza[2], O. Prymak[2], M Ulbricht[3], M Epple[2] and R Surmenev[1,*]

[1] Physical Materials Science and Composite Materials Centre, National Research Tomsk Polytechnic University, 634050, Tomsk, Russia

[2] Inorganic Chemistry and Center for Nanointegration Duisburg-Essen (CeNIDE), University of Duisburg-Essen, 45141 Essen, Germany

[3] Technical Chemistry II, University of Duisburg-Essen, 45141 Essen, Germany

[*]E-mail: rsurmenev@mail.ru

Abstract. Herein, electrospun biodegradable scaffolds based on polycaprolactone (PCL), poly(3-hydroxybutyrate) (PHB) and polyaniline (PANi) polymers were fabricated. A calcium-phosphate (CaP) coating was deposited on the surface of the scaffolds via an improved soaking process. Influence of the deposition cycles and ethanol concentration in the solution on the relative increase of the scaffolds weight and water contact angle (WCA) are determined. The characterization of the molecular and crystal structure confirmed the formation of CaP phase. Importantly, WCA results showed that the pristine scaffolds have the hydrophobic surface, while the deposition of CaP coating onto scaffolds allows to significantly improve the surface wetting behavior, and infiltration of the water droplets into the CaP-coated scaffolds was observed. Thus, the fabricated hybrid biodegradable piezoelectric scaffolds can be utilized for regenerative medicine.

1. Introduction

The porous scaffold development is of significant importance in bone tissue engineering, which allows to provide the functional support of damaged tissues and ensure stimulation of bone regeneration. Numerous studies have reported successful application of biodegradable polymers for the scaffold fabrication [1]. In addition, relatively recently, the possibility of piezoelectric polymers for tissues recovery has been investigated [2]. Piezoelectric polymers are materials that can generate electrical charges in response to applied mechanical stress. Electrically charged surfaces influence on cell behavior, e.g. cell growth, adhesion and morphology of different cell types [3]. Poly(3-hydroxybutyrate) (PHB) is a promising biodegradable polymer for regenerative medicine that possesses piezoelectric properties, but the piezoelectric response of PHB is relatively low in comparison with other piezoelectric polymers as PVDF or piezoceramics [4]. In order to improve the piezoelectric response of PHB electrospun scaffolds, conductive polyaniline (PANi) can be added to PHB solution before the scaffolds fabrication process [5]. Nevertheless, such scaffolds have the hydrophobic surface, thereby limiting their successful application in bone tissue engineering [6]. To improve these properties, calcium phosphate (CaP) coatings are proposed, since CaP is the major component of the inorganic phase of bone tissue and its presence on the surface of the scaffolds can significantly increase the bioactivity of the material, thus, resulting in the repair of damaged bone tissue [7]. For CaP coating deposition, a soaking was used.

IOP Conf. Series: Materials Science and Engineering **597** (2019) 012061 doi:10.1088/1757-899X/597/1/012061

To prepare CaP coating onto fiber scaffold surface, a simple and low cost method, such as soaking can be successfully utilized [8]. Since the deposition is carried out in the media, this method allows to form a coating on the entire surface of the porous electrospun scaffold, unlike other deposition methods, such as, for example, physical vapor deposition. Therefore, the aim of this study is to deposit CaP coating on the porous piezoelectric electrospun scaffolds and to study the influence of the coating on the scaffolds wettability.

2. Materials and methods

2.1. Materials

Polycaprolactone (PCL, Mn = 80,000 g/mol), poly(3-hydroxybutyrate) (PHB, Mw = 300,000 g/mol) Polyaniline (PANi, emeraldine salt, Mw > 15,000 g/mol), calcium chloride ($ZnCl_2$), disodium hydrogen phosphate (Na_2PO_4) and ethanol 96 % (C_2H_5OH) were purchased from Sigma-Aldrich, Germany. Chloroform ($CHCl_3$) was purchased from EKOS-1, Russia.

2.2. Fabrication of electrospun scaffolds

PCL (9 wt.%) and PHB (6 wt.%) electrospun solutions were prepared by dissolving polymers in the $CHCl_3$. The solutions for composite PHB/PANi scaffolds were prepared by doping 1, 2 and 3 wt.% of PANi to PHB powder before being dissolved. Polymers were dissolved in an ultrasonic bath during 3 hours. A syringe pump (Aitecs 2016, Lithuania) was used to feed the solutions into the needle tip (inner diameter of 0.58 mm). Electrospinning was performed at room temperature, 6.5 kV of the applied voltage, 1.5 mL/h injection flowrate and 600 rpm of the cylindrical collector rotation speed.

2.3. CaP deposition process

The CaP formation on the scaffolds was carried out by an alternate soaking process and improved soaking process. The procedure in detail is reported elsewhere [8]. The calcium chloride and disodium hydrogen phosphate were dissolved in deionized water and in mixed solvent of ultrapure water and ethanol (25 v/v%), and with final concentrations of 0.5 M and 0.3 M, respectively. The ethanol was added to reduce the surface tension of the resulting solution. The samples were soaked into calcium solution and then rinsed in ultrapure water, further soaked into phosphate solution, and then rinsed, indicating one cycle for the soaking process. The samples were soaked for 10 minutes during the first cycle and 5 minutes during subsequent cycles. After the deposition process, the samples were dried in oven at the temperature of 50 °C for 2 hours.

2.4. Characterization of the fibrous scaffolds

The morphology of the scaffolds was characterized by scanning electron microscopy (SEM) (FEI Quanta 250 FEG ESEM, USA) operated at 8 kV. The average fiber diameter of the scaffolds was calculated by measuring 100 different fibers from each type of scaffolds using the software package, ImageJ. To study the phase composition and structure, X-ray diffraction was used (D8 Advance Bruker, Germany) with Cu Kα radiation ($\lambda = 0.154$ nm) in the 2θ range from 5° to 90° with a step size of 0.01° at 40 kV and 40 mA. Before XRD measurements, all samples were fixed on the tesa-film which then affected on the obtained patterns. The molecular bonds of the samples were studied using Fourier transform infrared (FTIR) spectroscopy (ALPHA FTIR Spectrometer, Germany) in the frequency range from 600 to 4000 cm^{-1} with a step size of 4 cm^{-1}. Water contact angle (WCA) in air was measured using the sessile drop method via a contact angle analyzer (OCA 15 Plus Data Physics Instruments GmbH, Germany) at room temperature. A minimum of 5 droplets (5 µL, 1 µL/s) of ultrapure Milli-Q water were examined for each sample.

3. Results and discussions

According to SEM results (Figure 1A-E), the morphology of electrospun scaffolds had a porous fibrous structure. The average fiber diameter of the scaffolds varied from 2.2±0.4 µm to 5.5±1.5 µm for

PHB/PANi2% and PCL scaffold, respectively (Figure 1F). The WCA measurements revealed the hydrophobic surface (WCA>90°) for all scaffold compositions (Figure 1A-E). The addition of PANi to PHB scaffolds (PHB/PANi2% and PHB/PANi3%) resulted in a slight increase in the WCA. This effect has been also observed by Zhou et al. [9].

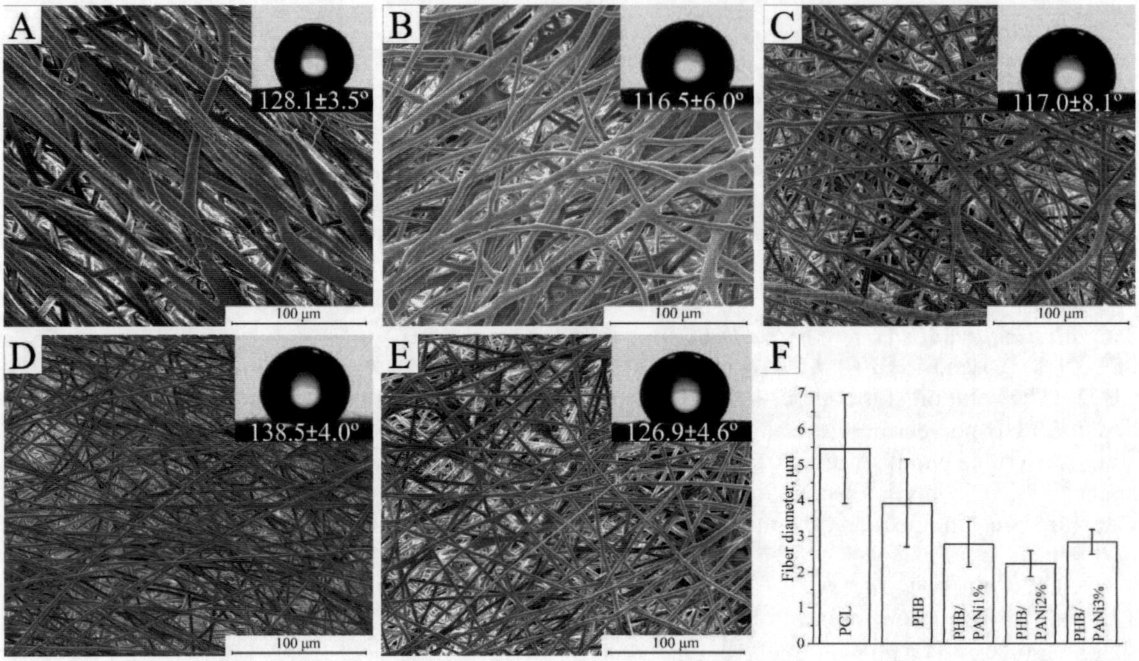

Figure 1. (A-E) SEM images of scaffold surface and insets of their WCA: A – PCL scaffold; B – PHB scaffold; C – PHB/PANi1% scaffold; D – PHB/PANi2% scaffold; E – PHB/PANi3% scaffold. (F) Average fiber diameters of the scaffolds

After investigations of the pristine scaffolds, an influence of CaP deposition parameters, such as the number of deposition cycles and ethanol concentration, on the relative increase in mass and WCA of PHB scaffolds were studied (Table 1).

Table 1. Changes in the relative increase in mass and WCA of PHB scaffolds due to varying the ethanol concentration and number of deposition cycles.

Number of deposition cycles	Ethanol concentration, v/v%	Changes of the relative sample mass, %	WCA, °
3	0	5.1	66.2±6.8
5	0	9.0	50.7±6.6
7	0	9.2	16.8±8.5
3	25	33.6	-
5	25	25.3	-
7	25	48.6	-

The increase of the number of deposition cycles led to increasing of the relative sample mass, probably, due to the increase of the CaP coating thickness/density or coated fibers surface area. CaP deposition from solutions containing 25 v/v% of ethanol also led to significantly increased relative sample mass. This effect can be due to the fact that the surface of the scaffolds with ethanol is better wetted compared to that of water. Besides, the results of WCA measurements revealed that water droplets immediately infiltrated into the scaffolds coated in ethanol, therefore, it was impossible to

IOP Conf. Series: Materials Science and Engineering **597** (2019) 012061 doi:10.1088/1757-899X/597/1/012061

correctly determine the initial WCA. However, this effect indicated that the wettability of the scaffolds coated with CaP in ethanol solution was improved in comparison with the scaffolds coated in water solution. Thus, the increase of the number of deposition cycles and the doping of ethanol lead to the decrease of WCA values of electrospun PHB scaffolds. Based on these results, 3 cycles of CaP deposition on the scaffolds from the solutions with 25 v/v% of ethanol was chosen for further studies.

The SEM results and energy-dispersive X-ray spectroscopy (EDS) of the scaffolds with deposited CaP coating presented in the Figure 2. The pores between the fibers were partially filled with the deposited CaP coating. In some places on the surface of the scaffolds, fibers with the finest diameter were completely coated with the CaP coating (Figure 2C-D). In case, when initial fibrous structure of the scaffolds should be preserved, the number of deposition cycles to prevent the formation of such a thick coating on the surface of the scaffolds should be reduced. EDS spectra showed the peaks corresponding to calcium and phosphorus as well as peaks of sodium and chlorine, which are byproducts of the reaction to prepare CaP coating. A high magnification image of the CaP coating (Figure 2F) showed that the coating has a rough surface which is similar with the surface of biphasic CaP bioceramics [10].

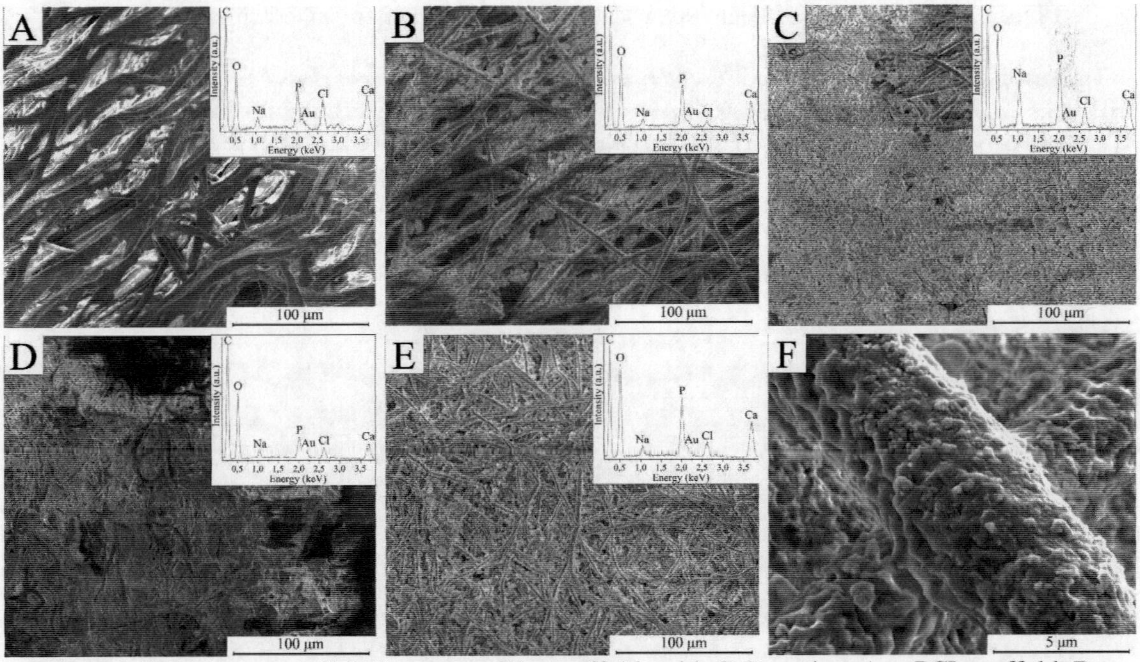

Figure 2. SEM images and EDS spectra of the scaffolds with CaP coating: A – PCL scaffold; B – PHB scaffold; C – PHB/PANi1% scaffold; D – PHB/PANi2% scaffold; E – PHB/PANi3% scaffold; F – high magnification of CaP coating on the PHB scaffold

To characterize the molecular and phase composition of the prepared coating, FTIR and XRD analyses were carried out. FTIR spectra exhibited all the characteristic peaks of PCL and PHB polymers for pristine scaffolds [11]. After the addition of PANi in PHB, new peak at 743 cm^{-1}, which attributes to the C-H out-of-plane bending on 1,2- aromatic rings in PANi structure, was observed [12]. In turn, after CaP deposition, FTIR spectra revealed the presence of typical phosphate peaks (PO_4^{3-}) at 559 (ν4), 959 (ν3) and 1019 (ν3) cm^{-1}, as well as the OH^- peak at 600 cm^{-1} from hydroxyapatite (HA) or amorphous CaP [8].

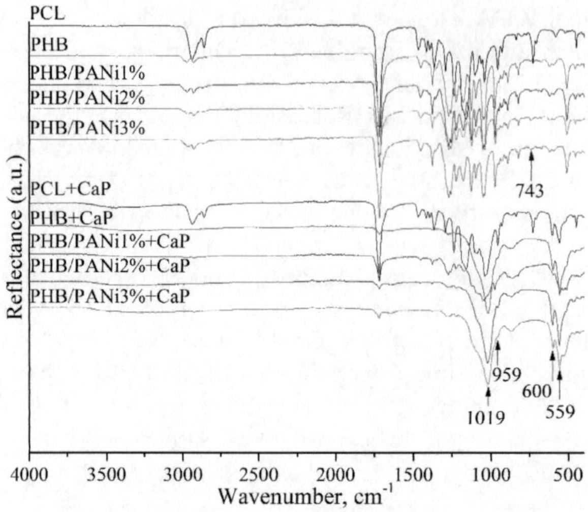

Figure 3. FTIR spectra of electrospun scaffolds before and after CaP coating deposition

The recorded XRD patterns are presented in Figure 4. The main peaks assigned to PCL [13] and PHB (#49-2212 PDF 4+) were observed. After the coating deposition, the peaks around 12°, 21° and 32° assigned to amorphous CaP and HA were observed [8]. Besides, XRD patterns revealed the presence of sodium chloride (NaCl) peak at 32° (JCPDS 5-0628) for the samples after CaP coating deposition. NaCl is a byproduct of the CaP production reaction, which can be removed via additional soaking.

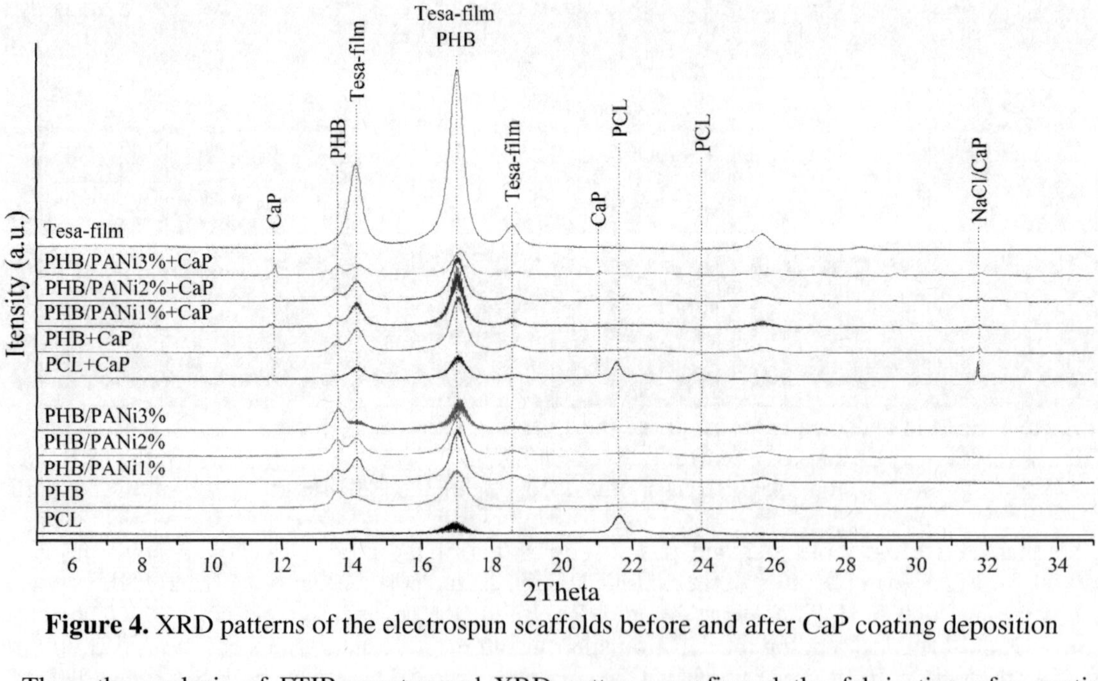

Figure 4. XRD patterns of the electrospun scaffolds before and after CaP coating deposition

Thus, the analysis of FTIR spectra and XRD patterns confirmed the fabrication of a coating corresponding to biphasic CaP bioceramics. Furthermore, this coating successfully improved the wettability of the biodegradable piezoelectric scaffolds.

4. Conclusion

IOP Conf. Series: Materials Science and Engineering **597** (2019) 012061 doi:10.1088/1757-899X/597/1/012061

Electrospun hybrid piezoelectric polymer scaffolds were coated with the CaP layer via an improved soaking process. It was shown that the addition of ethanol to the solutions before the deposition as well as increase of the number of deposition cycles from 3 to 7 allow to significantly increase the relative mass of the scaffold from 5 to 49 %, respectively. In addition, all the scaffolds before CaP coating deposition have the hydrophobic surface with WCA of 116.5-138.5°, while CaP coating allows to significantly improve the wettability of all studied scaffolds which allowed infiltration of water droplets into the scaffolds.

Acknowledgements
Andrei Zviagin acknowledges personal financial support from the Leonhard-Euler-Program of DAAD. The financial support from Russian Science Foundation 18-73-10050 is also acknowledged.

References
[1] Rogina A 2014 Electrospinning process: Versatile preparation method for biodegradable and natural polymers and biocomposite systems applied in tissue engineering and drug delivery *Appl. Surf. Sci.* **296** 221-30
[2] Vijaya M 2016 *Piezoelectric materials and devices: applications in engineering and medical sciences* (Boca Raton: CRC Press) p 186
[3] Tandon B, Blaker J J and Cartmell S H 2018 Piezoelectric materials as stimulatory biomedical materials and scaffolds for bone repair *Acta Biomater.* **73** 1-20
[4] Ramier J, Bouderlique T, Stoilova O, Manolova N, Rashkov I, Langlois V, Renard E, Albanese P and Grande D 2014 Biocomposite scaffolds based on electrospun poly (3-hydroxybutyrate) nanofibers and electrosprayed hydroxyapatite nanoparticles for bone tissue engineering applications *Mater. Sci. Eng. C* **38** 161-9
[5] Chernozem R V, Surmeneva M A and Surmenev R A 2018 Hybrid biodegradable scaffolds of piezoelectric polyhydroxybutyrate and conductive polyaniline: Piezocharge constants and electric potential study *Mater. Lett.* **220** 257-60
[6] Timin A S, Muslimov A R, Zyuzin M V, Peltek O O, Karpov T E, Sergeev I S, Dotsenko A I, Goncharenko A A, Yolshin N D and Sinelnik A 2018 Multifunctional scaffolds with improved antimicrobial properties and osteogenicity based on piezoelectric electrospun fibers decorated with bioactive composite microcapsules *ACS Appl. Mater. Interfaces* **10** 34849-68
[7] Bose S and Tarafder S 2012 Calcium phosphate ceramic systems in growth factor and drug delivery for bone tissue engineering: a review *Acta Biomater.* **8** 1401-21
[8] Watanabe J and Akashi M 2007 Formation of hydroxyapatite provides a tunable protein reservoir within porous polyester membranes by an improved soaking process *Biomacromolecules* **8** 2288-93
[9] Zhou X, Zhang Z, Men X, Yang J, Xu X, Zhu X and Xue Q 2011 Fabrication of superhydrophobic polyaniline films with rapidly switchable wettability *Appl. Surf. Sci.* **258** 285-9
[10] Lobo S E, Glickman R, Da Silva W N, Arinzeh T L and Kerkis I 2015 Response of stem cells from different origins to biphasic calcium phosphate bioceramics *Cell Tissue Res.* **361** 477-95
[11] Liau C P, Bin Ahmad M, Shameli K, Yunus W M Z W, Ibrahim N A, Zainuddin N and Then Y Y 2014 Preparation and characterization of polyhydroxybutyrate/polycaprolactone nanocomposites *Sci. World J.* **2014** 9
[12] Tang J, Jing X, Wang B and Wang F 1988 Infrared spectra of soluble polyaniline *Synth. Met.* **24** 231-8
[13] Balu R, Kumar T, Ramalingam M and Ramakrishna S 2011 Electrospun Polycaprolactone/Poly (1, 4-butylene adipate-co-polycaprolactam) blends: Potential biodegradable scaffold for bone tissue regeneration *J. Biomater. Tissue Eng.* **1** 30-9

Soil-concretes from Yakutia's clay materials

A E Mestnikov[1] and V I Fedorov[2]
North Eastern Federal university of M.K. Ammosov, 58, Belinskiy Str, Yakutsk, 677000 Russia

E-mail: valeriyif.ykt@gmail.rcom

Abstract: The article presents the results of experimental studies on the choice of the optimal technological parameters (the initial moisture of the soil-cement mixture, pressing force, amount of cement) for the preparation of samples of soil-concrete by semi-dry pressing. Research has shown that the predicted strength of the soil-concrete can be obtained not only by varying the amount of Portland cement injected, but also by selecting the pressing force and the initial moisture of the initial mixture. The obtained experimental data allow predicting the construction and operational properties of soil-concretes materials from clay raw materials of a specific field, to clarify the methods of manufacture and the area of their application (foundations of airfields, roads and engineering structures, building products and structures).
Keywords: concrete, soil, cement, particle, strength, water absorption, brick.

1. Introduction

Yakutia has huge reserves of natural mineral raw materials for the production of a wide range of building materials and products [1]. However, in most areas there are no fine (sand), coarse (crushed stone) and porous (expanded clay) aggregates for the production of effective concrete. In the conditions of the complete absence of basic building materials for remote areas of the North, materials using soil can be of practical interest.

As is known, available sandy-clay soils stabilized with cement or other binders and additives are used for the manufacture of soil-concrete, mainly to strengthen the foundations of airfields, roads and railways [2] and civil structures [3]. Concrete and clay materials have been previously used in rural construction, however, such materials are not widely used in modern building construction technology.

Analysis of publications devoted to the study and application of soil-concrete shows the increased interest of builders in the use of soil materials as bases for airfields, roads and engineering structures, as well as the complete absence of practical use of soil-concrete construction materials in modern residential construction [2-6]. The use of soil concrete in construction is one of the directions for soil stabilization. The appearance of unique additives, new specialized equipment allows you to use this method. This greatly simplifies and speeds up the process of performing work with a decrease in cost.

The problems of soil stabilization, improvement of the quality of soil concrete, improvement of the production technology of firing and unburned wall materials, selection of the optimal composition of raw mixes are described in numerous works [6-8].

At present, the types of soil-concrete are not fully investigated. Soil-concrete for the conditions of Yakutia is a new and promising material. The use of soil-concrete in the rural construction of Yakutia will make it possible to make extensive use of local building materials (soil, wood waste, etc.), reduce the cost and speed up the commissioning of buildings and structures for various purposes.

To this end, it is necessary to conduct comprehensive scientific and practical research on the development and implementation of the technology for the manufacture of soil-concrete based on local raw materials in relation to the extreme climatic conditions of Yakutia, taking into account the real possibilities and needs of various segments of the population.

The purpose of this work is an experimental study on the choice of optimal technological parameters (initial moisture content of the mixture, pressing force, amount of cement) for the manufacture of wall products from clay soils using the semi-dry pressing method.

2. Basic materials

The experimental studies used clay soils with the following characteristics:

- chemical composition, %: SiO_2 – 58.6; TiO_2 – 0.78; Al_2O_3 – 14.9; Fe_2O_3 – 3.4; MnO – 0.12; MgO – 3.63; CaO – 4.38; Na_2O – 2.18; K_2O – 2.68; P_2O_5 – 0.2; SO_3 – 0.14; loss on ignition – 6.5; H_2O – 1.08;
- granule size, %: particles less than 0.001 mm – 31.4; 0.05-0.002 mm – 51.9; more than 0.002 mm – 16.7;
- plasticity number - 7.3-13.1.

Cement grade CEM I 32.5B was used as a binder: residue on sieve No. 008 - 7%; normal viscosity - 26%; start setting - 1 hr 45 min; end of setting - 2 hr 50 min; compressive strength - 41 MPa; flexural strength - 7.3 MPa.

As the fine aggregate used sand according to GOST 8736-2014 with a particle size module of 1.2-1.4, organic aggregate - sawdust (pine).

Prevention of the composition of the soil-concrete is made in accordance with the requirements of GOST 27006-86. Tests of soil-concrete samples were carried out using standard methods, certified instruments and equipment. Evaluation of strength is made in accordance with the requirements of GOST 10180-2012. Determination of water absorption of samples was carried out according to the method GOST 12730.3-78.

The microstructure analysis of the soil was performed on a JEOL JSM-7100F scanning electron microscope. Statistical processing of experimental data was carried out using the programs MathCAD 2001i and ImageJ.

3. Result and discussion

The first stage of experimental research is the assessment of the particle size distribution of the used soil. In figure 1 shows micrographs of a sample of clay soil.

x100 x300 x500

Figure 1. The microstructure of clay soil

Analysis of micrographs shows (Figure 1) that soil particles have a high specific surface. And also there is a high content of dust and clay particles. Grains have an irregular shape and a rough surface.

Micrograph data were processed using the ImageJ program [9]. The results of processing are shown in the form of a histogram of the distribution of particle size of the soil (Figure 2).

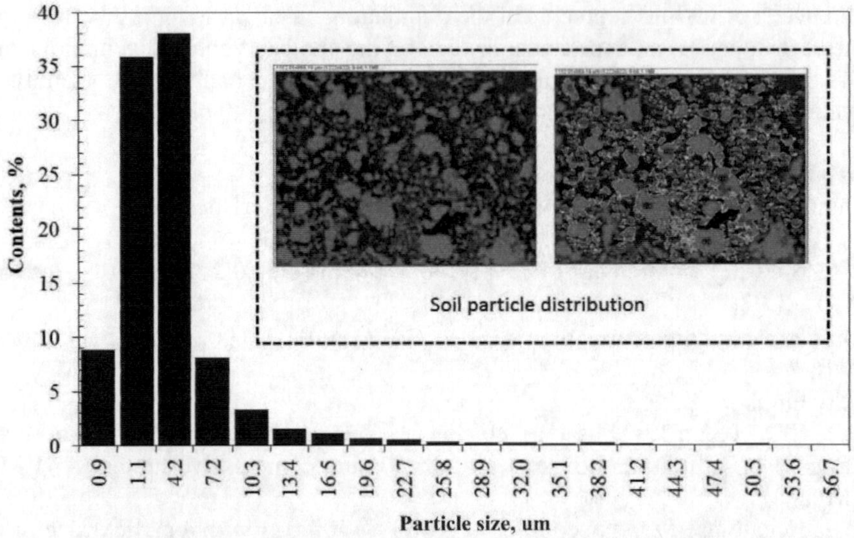

Figure 2. Histogram of the distribution of soil particles

An analysis of the histogram (Figure 2) shows that a distinct distribution peak is observed in the soil sample. This indicates a monodisperse distribution of particles, which is consistent with the image of the soil structure (Figure 1).

A visual analysis of the structure reveals the presence of numerous shapeless, relatively large soil particles (sand particles). First, these particles are distributed relatively evenly over the entire surface. The most optimal way to obtain building materials with desired performance properties is to apply a semi-dry pressing technology with simultaneous stabilization of the material by the introduction of cement.

Tests of samples of soil-concrete materials stabilized with Portland cement were carried out after 28 days of natural storage [10]. The composition of the mixture is the following, %: clay soil - 70, sand - 30, cement (3, 5 and 8% to the soil mixture). Experimental dependences of the strength and water absorption of soil-concrete samples on the pressing force 2.5–15.0 MPa were obtained at two initial moisture levels of the initial mixture of 10 and 15%, respectively (Table 1).

Table 1. Compressive strength and water absorption of soil-concrete samples

Cement content (by weight), %	Pressing force, MPa			
	2.5	5.0	10.0	15.0
At 10 % humidity				
3	3.2/21.7	4.1/17.7	6.9/19.5	9.0/14.8
5	4.8/15.6	5.5/16.8	7.8/13.8	12.4/15.6
8	5.0/14.8	7.0/13.2	8.7/10.5	13.2/12.9
At 15 % humidity				
3	6.3/14.8	6.9/15.6	7.6/13.2	8.9/13.6
5	6.5/13.1	7.7/11.3	8.6/13.3	12.6/14.2
8	7.7/12.4	10.1/11.3	12.5/10.8	14.1/12.0
To the line - compressive strength, behind the line - water absorption				

The maximum compressive strength of samples of 13.2-14.2 MPa is achieved with a pressing force of 15 MPa with an initial moisture content of a mixture of 10 and 15%, respectively. The maximum water absorption of 17.7-21.7% is observed for samples made by pressing with a minimum force of 2.5-5.0 MPa and a low cement content of 3% (Table 1).

Indicators of strength of samples made with a pressing force of 2.5-5.0 MPa are typical for soil-concrete obtained in removable formwork by tamping, and with a pressing force of 5.0-10.0 MPa – soil-concrete obtained by rolling with a roller up to 10 tons while strengthening ground road bases.

In the experimental indicators of water dehydration table. 1, there is some variation in the data, which does not prevent us from establishing that a decrease in the water absorption of the samples is facilitated by an increase in the amount of stabilizer added - Portland cement to a greater extent than an increase in the pressing force.

The increased values of the strength parameters of the soil-concrete are explained by the peculiarities of the process of compaction of samples at the stage of their manufacture. Compaction of samples is accompanied by physico-chemical processes in which the entire system is involved: solid phase (mineral particles), liquid (water) and gaseous (air). In the initial stage of pressing, the solid particles move in different directions, large pores that are formed from the grains at the moment of filling the form are destroyed, and the air is partially removed [11].

Increases the contact surface between the powder grains. As the pressure increases, the particles further condense. At the same time, the air that had not managed to retire is trapped between the grains of the powder and compressed. With further compaction of the powder, the movement of the grains occurs on their increased contact surfaces having aqueous films. At the last stage of pressing, the product is most compacted due to the further development of the contact surfaces [12].

4. Conclusions

1. Analysis of the granulometric composition of the soil studied showed its monodisperse composition.
2. Sand particles were found in the ground, while these particles are distributed evenly over the entire surface.
3. It is established that the introduction of cement into the structure of the soil-concrete allows improving the basic strength characteristics of the material.
4. In the manufacture of soil-concrete by the method of semi-dry pressing with a force of 10 MPa, the guaranteed brand of brick from ground-cement mixture (8-10% of cement) is M75 - M100, and the brand of brick the same clay raw material - M100-M150.
5. The experimental dependences of the compressive strength and water absorption of soil-concrete samples, made by the method of semi-dry pressing, on the values of the pressing force (2.5-15.0 MPa), the initial moisture content of the mixture and the amount of introduced cement will predict the construction and operational properties of soil-cement materials from clay materials.

References:

[1] Sevostyanova R F and Sitnikov V S 2018 *J. Journal of the Mining Institute* **234** 599
[2] Zege S O and Brid I I 2007 *J. Soil Mechanics and Foundation Engineering* **44** 143
[3] Murat M and Eyubhan A 2018 *J. Materials Journal* **115** 855
[4] Cong M and Bing C 2015 *J. Construction and Building Materials* **76** 61
[5] Jianguo F, Dongyuan W and Duo Q 2018 *J. Journal of Rock Mechanics and Geotechnical Engineering* **10** 791
[6] Mohammad R A and Ali T 2017 *J. Journal of Rock Mechanics and Geotechnical Engineering* **9** 623
[7] Estabragh A R, Ranjbari S and Javadi A A 2017 *J. Materials Journal* **114** 195
[8] Kharum M and Svintsov A P 2017 *J. Key Engineering Materials* **730** 358
[9] Udawatta R P and Anderson S H 2008 *J. Geoderma* **145** 381
[10] Estabragh A R, Khatibi M and Javadi A A 2016 *J. Materials Journal* **113** 709
[11] Venkatarama B V and Gupta A 2005 *J. Materials and Structures* **38** 639
[12] Cong M and Bing C 2015 *J. Construction and Building Materials* **76** 61

IOP Conf. Series: Materials Science and Engineering **597** (2019) 012063 doi:10.1088/1757-899X/597/1/012063

Coordinate dependence of the photo-response in metal – lithium niobate – metal sandwich structure

Y Perkov and V Ivanov

Department of Physics and Theoretical Mechanics, Far Eastern State Transport University, Khabarovsk, Russia

E-mail: tmeh@festu.khv.ru

Abstract. The light induced current in doped lithium niobate crystals with two electrodes of different metals (sandwich system) includes the photovoltaic and the thermoinduced effects. The anomalous coordinate dependences of the light response are revealed and investigated experimentally.

1. Introduction

The photo-induced effects are well known phenomena for the ferroelectric crystals [1]. A series of nonequilibrium phenomena in ferroelectrics, which do not have a definite theoretical description, has appeared recently. In paper [2], the thermovoltaic effect deals with the initiation of the electric voltage between the opposite faces of the semiconductive sample of samarium sulphide (*SmS*) at heating to 400–500 *K* is described. The effect has appeared in the presence of the donor impurity concentration gradient directed towards these faces [3].

In this paper the results obtained in experimental studies of the light induced currents in the thin-layer metal-lithium niobate-metal (MLM) structure are presented. The coordinate dependences of the light response were revealed and investigated experimentally. This effect is similar to photovoltaic effect, but it includes the thermal contribution [4]. The possible models of the investigated phenomena are discussed. The results obtained can be used for investigating the properties of the sandwiched ferroelectric structures [5-9], as well as for developing new types of the coordinate photo-detectors [10-15]. This type of detector does not require the modulation of radiation unlike to the classical pyroelectric coordinate detector [6].

2. Light induced currents in ferroelectrics

It is well known the photovoltaic effect (PVE) in lithium niobate with iron impurities. It has been established that the density of steady-state current J is proportional to the laser intensity $\Phi(t)$:

$$J_{\text{фr}} = \alpha K_G \Phi(t),$$

where K_G is the Glass constant dependent on the nature of absorbing centers and wavelength of radiation, α is the absorption factor [4].

In our work [5], the thermally-stimulated quasi-stationary current arising only in the highly-alloyed with iron crystal of lithium niobate with the evaporated electrodes from a pair of different metals was described. A sign of the thermo-electro-moving force (thermo-EMF) is determined by location of

IOP Conf. Series: Materials Science and Engineering **597** (2019) 012063 doi:10.1088/1757-899X/597/1/012063

electrodes applied by vacuum evaporation on the opposite faces of crystal and does not depend on orientation of the crystallographic axes of the sample in reference to electrodes.

In contrast with the classic pyroelectric effect, the value of observed thermal response is proportional to the crystal temperature increase and does not depend on the rate of its change. It was experimentally meased the coefficient characterizing the thermoinduced electro-mowing force (EMF) value in the mode of voltage U_n measurement:

$$P_{el} = (R_n S)^{-1} \Delta U_n / \Delta T, \text{ (A/K·sm}^2\text{)}, \quad (1)$$

where S is the area of crystal surface covered with sputtered (evaporated) electrode; $\Delta U_н$ is the change of voltage across load impedance R_n when the sample temperature changes by ΔT; ΔT is the difference between the initial and final temperatures of sample [5].

The study of the effect was performed using the slow modulation of the crystal temperature under the conditions of thermostat and the thermo-EMF was recorded under the stationary conditions (relaxation was not observed for observation time $\tau \sim 10^5$ s). The measurements showed that the thermo-voltaic effect has approximately equivalent values in the *Y*- and *Z*-cut crystals. Because the sufficiently high rates of heating [about 0.1 K/s] were used in the installation, the pyroelectric current for the polar *Z*-cut crystals reached the considerable value. Therefore, the principal measurements were performed for the *Y*-cut lithium niobate crystals where the thermal response could be observed in pure form.

The known contact phenomena in the sandwiched structures with ferroelectrics do not explain the observed quasi-stationary currents [5-8]. The presence of initial voltage U_n^0 across the sample allows us to suggest the non-equilibrium nature of the thermally stimulated current.

The electrets model of the MLM structure was proposed in our work [9]. In this model the thermoinduced changing of the conductivity of the crystal leads to change in the stationary current on the load impedance.

3. Research methods

The samples of doped lithium niobate crystals (iron concentration is equal to 0.3 wt%) with different electrodes (aluminum (*Al*) – chromium (*Cr*)) were studied. The metal electrodes were applied through the vacuum evaporation with a thickness varying from 0.1 to 1 μm.

The study of the light induced effect was performed by using the slow modulation of the light with the power *P* falling on the crystal (Figure 1).

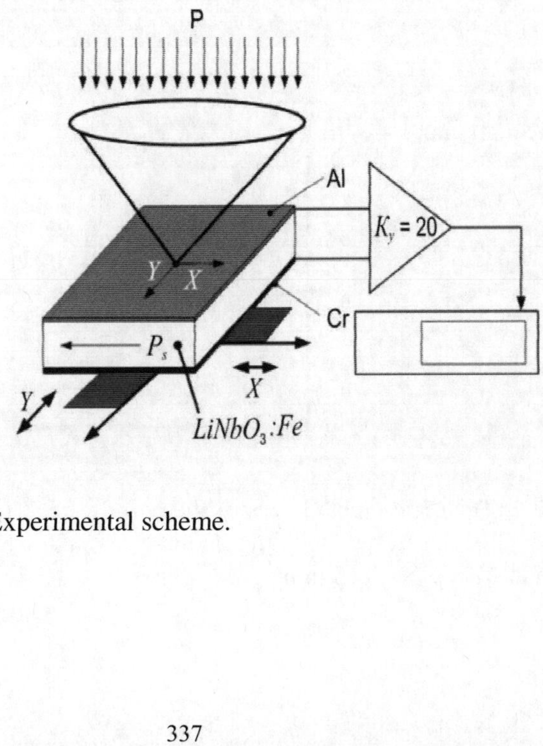

Figure 1. Experimental scheme.

IOP Conf. Series: Materials Science and Engineering **597** (2019) 012063 doi:10.1088/1757-899X/597/1/012063

As the sufficiently high rates of heating [about 0.1 *K*/s] were registered during the experiments, the pyroelectric current for the polar *Z*-cut crystals reached the considerable value. Therefore, the principal measurements were performed for the *Y*-cut lithium niobate crystals where the pyroelectric response could be observed in pure form.

The coefficient characterizing the light induced response was experimentally measured:

$$G = \Delta U_n / P, \text{ (V/W)}, \qquad (2)$$

where ΔU_n is the change of voltage across load impedance R_n when the sample is illuminated by the light, *P* is the power of the radiation.

The $LiNbO_3{:}Fe$ crystal ($0{,}3$ Wt.% Fe, $Al-Cr$, $2\times 2{,}5\times 0{,}13$ mm^3) was used. The preamplifier coefficient is *K*=20. The crystal was installed in the coordinate table, which allows you to seamlessly move crystal in two directions. The light spot diameter in the focus lens was 340 μm.

4. Experimental results

In some experiments He-Ne laser (0.63 μm) was used as a source of radiation. The dependence of light induced response from the coordinates of the laser spot on the crystal top surface was revealed.

The value of the stationary light response versus *X* (along polar axis) and *Y*-coordinates is shown in Figure2.

The coefficient G^* (*G* in arb. un.) reaches the greatest magnitude at the ends of the crystal and it takes different signs at the opposite ends of the sample (Figure 2a).

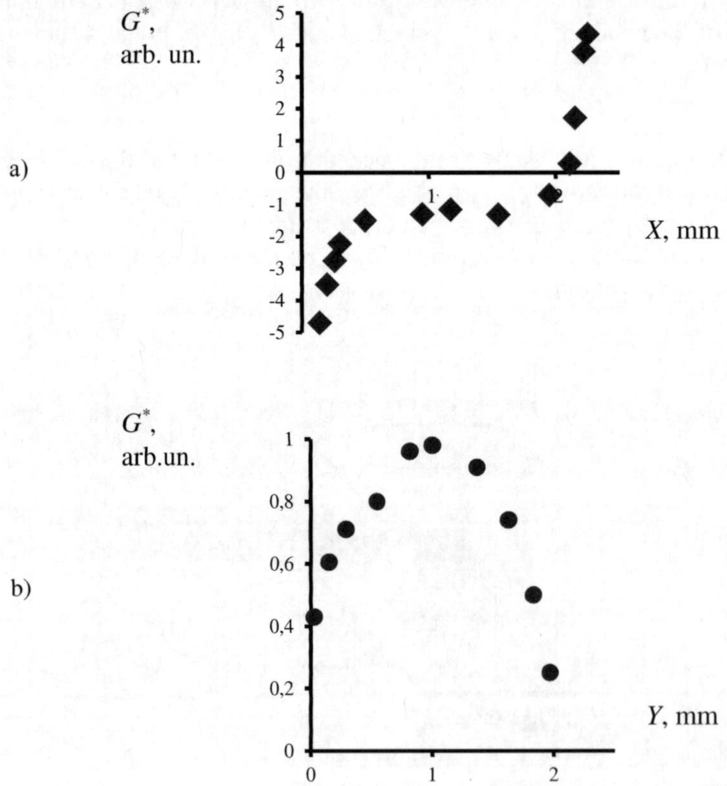

Figure 2. The coefficient G* versus the coordinate X (a) and the coordinate Y (b) of the laser spot (lithium niobate crystal; 0.3 wt% of Fe; Al–Cr electrodes; 2 × 2.5 × 0.13 mm^3).

IOP Conf. Series: Materials Science and Engineering **597** (2019) 012063 doi:10.1088/1757-899X/597/1/012063

The coefficient G^* versus *Y*-axis is changed monotonically (Figure 2b). This asymmetry can be associated with only one selected direction in crystal associated with the polar axis *X*.

Edge effects can be caused by the coordinate dependence of crystal temperature. These effects are confirmed experimentally by the thermographic registration (Figure 3). But these phenomena don't explain change of the response sign and the X-Y asymmetry (Figure 2).

a)

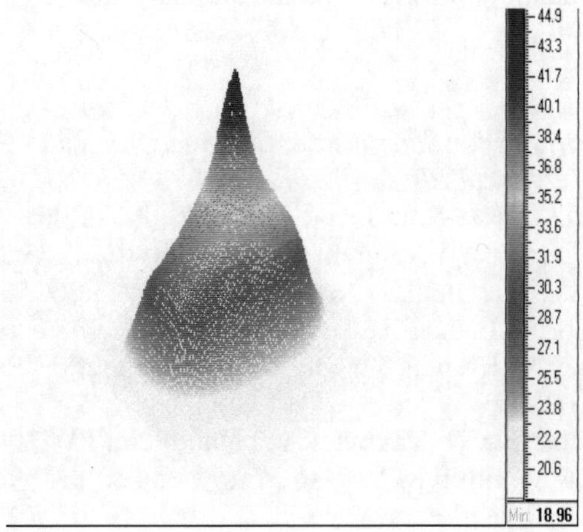

b)

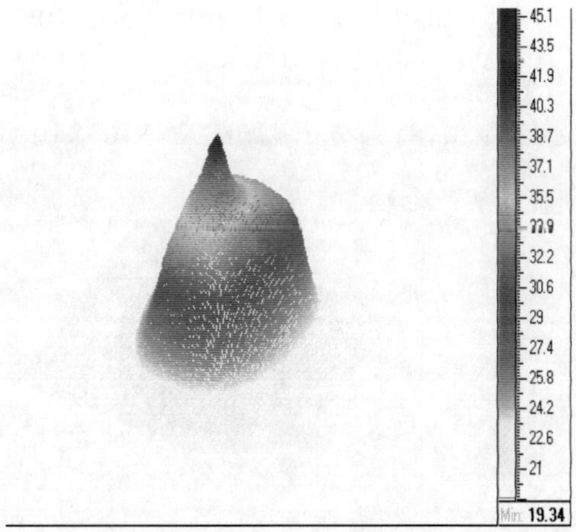

Figure 3. The temperature distribution on the surface of the lithium niobate crystal with He-Ne laser illumination: a) – the laser spot is on the sample edge, $T_{max} = 45.6\ ^0C$; b) - the laser spot is on the sample mid, $T_{max} = 42.1\ ^0C$.

The anomaly in the coordinate dependence can be explained by the existence of two response mechanisms (photovoltaic and thermovoltaic) with different signs. The photovoltaic response gives the contribution in the light induced current, because the electrodes are partially transparent. The coordinate dependence disappeared when the upper electrode was blackened. The detailed explanation of the describing experiments requires the further investigation.

5. Conclusions

The results obtained in experimental studies of the light induced currents in the thin-layer nonsymmetrical metal-lithium niobate-metal (MLM) structure are presented. In experiments with the local illumination of the $LiNbO_3:Fe$ crystal the anomalous dependence of light induced response from the coordinates of the laser spot on the crystal top surface was revealed. The temperature coordinate dependence confirmed experimentally, but thermal model cannot explain the change of response sign. However, the alleged explanation of the coordinate anomaly should include the photovoltaic effect and the thermovoltaic contribution.

References

[1] Serway R A 1990 *Physics for Scientists & Engineers* (3rd ed.) Saunders p 1150
[2] Kaminskii V 2013 *Technical Physics Letters* **39** 673
[3] Romero J J, Jaque D., Garcia-Sole J. and Kaminskii A A 2001 *Appl. Phys.Lett.* **78** 1961
[4] Glass A, Von der Linde D and Nergran T 1974 *Appl. Phys.* **25** 233
[5] Ivanov V, Karpets Yu, Marchenkov N and Zdorovtsev G 2005 *Proc. SPIE* **5851** 415
[6] Itskovskii M, Shchedrina L and Kladkevich M 1980 *Ferroelectrics* **29** 167
[7] Zdorovtsev G, Ivanov V, Karpets Yu and Klimentyev S 2007 *Proceedings of Tomsk Polytechnic University* **311** 102
[8] Pichugin V F, Bulycheva A A, Yakovlev V Yu and Kim I W 2005 *Physica status solidi (c)* **1** 208
[9] Barkan I B, Baskin E M, Entin M V 1980 *Phys. stat. sol. (a)* **59** 12
[10] Ivanov V I, Karpets Y M and Perkov Y O 2017 *Proc. SPIE* **10176** 1017625
[11] Kim I W, Park B C, Jin B M, et al. 1995 *Material Letters* **24** 157
[12] Perkov Y, Karpets Yu and Ivanov V 2019 *IOP Conf. Series: Materials Science and Engineering* **510** 012014
[13] Ivanov V I and Karpets Y M 2019 *Proc. SPIE* **11024** 110240D
[14] Pertsev N A, Petraru A, Kohlstedt H, Waser R,. Bdikin I K, Kiselev D and Kholkin A L 2008 *Nanotechnology* **19** 375703
[15] Perkov Y O and Ivanov V I 2018 *J. Phys. Conf. Ser.* **1038** 012114

IOP Conf. Series: Materials Science and Engineering **597** (2019) 012064 doi:10.1088/1757-899X/597/1/012064

Diagnostics of nanosuspension by the dynamic holography method

A Mjagotin, G Ivanova and V Ivanov

Department of Physics and Theoretical Mechanics, Far Eastern State Transport University, Khabarovsk, Russia

E-mail: tmeh@festu.khv.ru

Abstract. The theoretical analysis of the dynamic holograms efficiency in the nonideal nanoparticle gas is carried out. The electrostrictive mechanism of optical nonlinearity of the medium is due to the forces operating on the particles of the dispersed phase in gradient light field. The recording of the dynamic hologram is significantly non-linear process at high intensities of radiation. The amplitude of phase grating depends exponentially versus intensity but it saturates at the high filling factor due to the repulsion effect.

1. Introduction

Nonlinear optical diagnostics of materials involves methods based on different mechanisms of light induced modulation of optical constants of medium [1-3]. Multicomponent nanoparticles medium (liquid phase mixture, suspension, emulsion) is characterized by concentration perturbations caused by the redistribution component in a two-phase medium in a laser field [4-6]. The light induced change in the concentration of polymer nanoparticles provides various modifications of optical noncontact diagnostic methods in near real time. Such well known methods as the thermal lens and the dynamical holography are successfully used for the investigation of the mass transport phenomena in multicomponent liquids [7-9]. In a gradient light field the nanoparticles in the transparent medium are controlled by the electrostrictive forces, causing changes in their concentrations [10-12]. The medium is characterized by a cubic nonlinearity in this case that is correct only for small intensities of radiation [13-15]. For large radiation intensities the potential energy of particle is more than heat one and it requires consideration of non-linearity of the highest order.

The purpose of this paper is the theoretical analysis of the dynamic holograms efficiency in the nonideal nanoparticle gas in the case of the high intensity of radiation, when the concentration changing is much more than the initial value. This case is very interesting for the high intensity dynamic holography of dispersed liquid-phase media, as well as for optical diagnostics of nanomaterials [16-20].

2. Two-wave holographic scheme

The two-wave scheme is the simplest method of the holographic diagnostics. Dynamic hologram method (Forced Raleigh Scattering) based on the recording of the dynamic holograms in nonlinear material while reading it. Usually two references beams (with equal intensity) form the interference picture in the nonlinear material, thus $I = I_0(1+\sin Kx)$ is the intensity spatial distribution along the layer of the medium (x-axis), $K = 2\pi/\Lambda$ and Λ are the space wave vector and the period of the elementary hologram. The reading beam (with a different wavelength) is directed perpendicular to the wave vector of the hologram.

The next expression defines a diffraction efficiency of the hologram [1]:

$$\eta = I_{1p}/I_{0p}, \tag{1}$$

where I_{0p} is the intensity of the incident probe beam; I_{1p} is the intensity of the diffracted beam.

If medium is transparent and the phase modulation amplitude is small ($\varphi_1 << 1$) the diffraction efficiency of a thin phase hologram is [1]:

$$\eta = \{2\pi \Delta n L \lambda^{-1}\}^2, \tag{2}$$

where L is the thickness of a nanofluid layer, Δn is the amplitude of the space modulation of the refractive index.

In the case of non-resonant non-linearity (for weakly absorbing media) it is used the parameter - $n_2^{eff}\,[m^2/W]$, which characterizes the change of the refractive index of the medium under the influence of incident radiation (cubic nonlinear response):

$$n = n_0 + n_2^{eff} I, \tag{3}$$

where n_0 is the refraction index of the medium in the absence of radiation, I is the incident intensity, $n_2^{eff} = (\partial n / \partial I)$ is the coefficient of effective cubic nonlinearity.

In nanodispersive medium particle radius is much smaller than the radiation wavelength λ and the refractive index of the medium is proportional to the concentration of particles (for dilute systems):

$$n = n_1(1 + f\delta), \tag{4}$$

where $\delta = (n_2 - n_1)/n_1$; n_1 and n_2 are the refractive indices of the liquid and the dispersed phase, respectively; $f = (4/3)\pi a^3 C$ is the volume fraction of the dispersed phase, a is the radius of the nanoparticle, C is the concentration of nanoparticles [13].

It is sufficient to calculate the amplitude of the concentration grating C_1 for defining the amplitude of the space modulation of the refractive index in the material:

$$\Delta n = (\partial n/\partial C) C_1 . \tag{5}$$

3. Electrostrictive mechanism of optical nonlinearity

The electrostrictive mechanism of optical nonlinearity of the medium is due to the forces operating on the particles of the dispersed phase in gradient light field. Depending on the sign of the polarizability the nanoparticles may be drawn (if the refractive index of the dispersed phase material is greater than the dispersion medium) or pushed out (otherwise) from areas with greater intensity of the electromagnetic wave [1].

The system of balance equations for the concentration of particles for this case is written as follows [13]:

$$dC/dt = -\mathrm{div}(-D\nabla C + \gamma C(1 - \nu C)\nabla I). \tag{6}$$

IOP Conf. Series: Materials Science and Engineering **597** (2019) 012064 doi:10.1088/1757-899X/597/1/012064

where D is the diffusion coefficient of particle, $J = \gamma C \nabla I$ is electrostrictive flow, $\gamma = (2\beta b / cn_1)$, ($\beta$ and b - polarizability and mobility of nanoparticle, respectively), c is the light velocity, ν is the constant of the particle interaction.

The proposed phenomenological model involves the particles interacting (repulsion). This model gives the value $\nu = 1$ for the simple model of the hard sphere potential [11].

In case of substantial changes of concentrations of particles the original equation (6) can be solved in the stationary mode:

$$-D\nabla C + \gamma C(1 - \nu C)\nabla I = 0. \tag{7}$$

The solution of the equation (7) can be written as ($\nu = 1$):

$$C_R = (f_0 + \exp[-\alpha \mathrm{SinX}])^{-1}, \tag{8}$$

where $C_R = C / C_0$ is normalized concentration of nanoparticles, $X = x / \Lambda$ is dimensionless coordinate, $\alpha = I_0 / I_{th}$ is normalised intensity, $I_{th} = D\gamma^{-1}$ is some "thermal" intensity of radiation.

The case of the ideal gas of nanoparticles (without repulsion $\nu = 0$) gives the next expression:

$$C_R = \exp[\alpha \mathrm{SinX}], \tag{9}$$

The amplitude of the concentration grating depends exponentially versus intensity for low values of filling factor f_0. The expressions (8-9) give the same result for both cases (Figure 1).

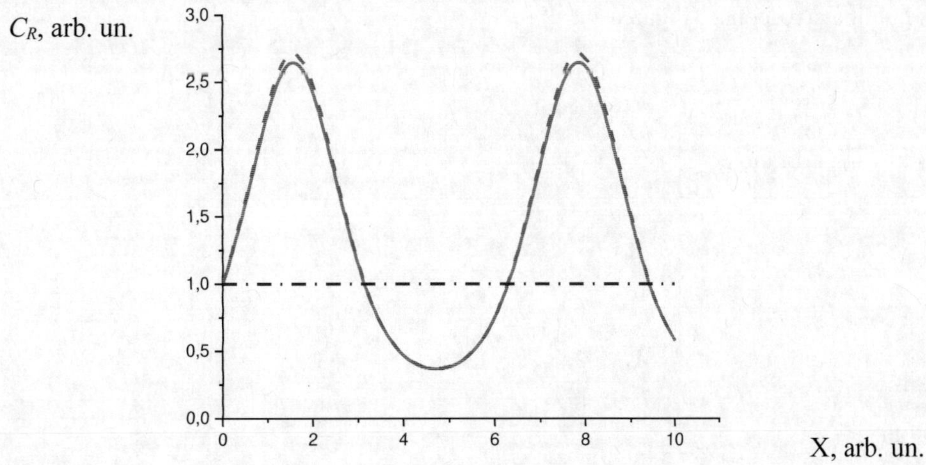

Figure 1. Concentration gratings of nanoparticles for intensity α=1 ($\alpha = I_0 / I_{th}$ is dimensionless intensity of radiation): in ideal gas (▬ ▬ ▬) according to (9); in the non-ideal gas (▬▬▬) according to (8). The initial concentration is constant for α=0 (▬ ▪ ▬); $f_0 = 0.01$ (low concentration).

IOP Conf. Series: Materials Science and Engineering **597** (2019) 012064 doi:10.1088/1757-899X/597/1/012064

High value of the initial filling factor f_0 leads to the essential difference for these two models. The amplitude of the concentration grating in the non-ideal gas is noticeably less than in the case of ideal gase (Figure 2).

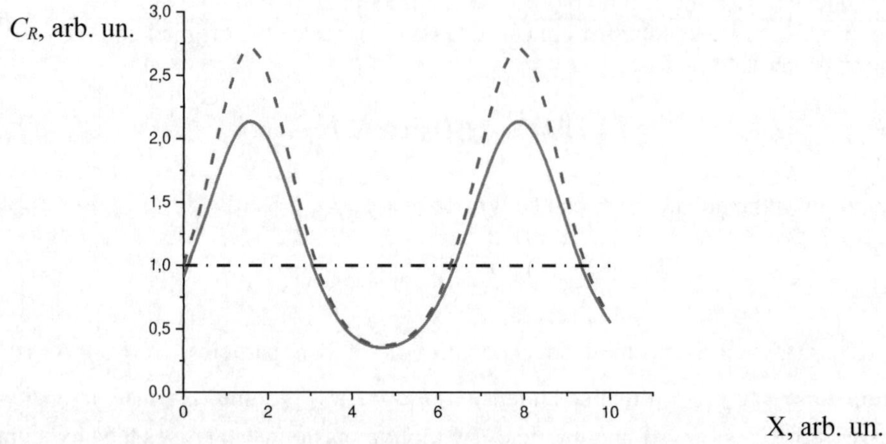

Figure 2. Concentration gratings of nanoparticles for intensity α=1: in ideal gas (▬ ▬ ▬) according to (9); in the non-ideal gas (───) according to (8). The initial concentration is constant for α=0 (▬ ▪ ▬); $f_0 = 0.1$ (large concentration).

The difference grows dramatically with the value of the intensity. The concentrations gratings are significantly nonsinusoidal for large initial concentration and high intensity. The ideal gas model (formula (8)) gives the high value of the concentration grating, but the hard sphere gas model gives much less nonlinear response (Figures 3).

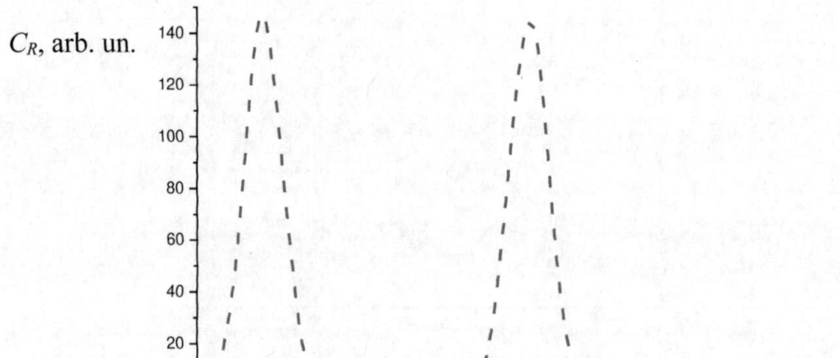

Figure 3. Concentration gratings of nanoparticles for intensity α=5: in the ideal gas (▬ ▬ ▬) according to (9); in the non-ideal gas (───) according to (8). The initial concentration is constant for α=0 (▬ ▪ ▬); $f_0 = 0.1$ (large concentration).

Thus the value of the nonlinearity (and the dynamic holograms efficiency) for the high intensity radiation is restricted by the repulsion effect in the case of the initial large particle concentrations.

4. Conclusions

The electrostrictive nonlinearity in the hard sphere gas model was theoretically investigated. The amplitude of the concentration grating depends exponentially versus intensity only for the low values of the intensity and initial filling factor and saturates for the high values of the both parameters.

References

[1] Shen Y R 1984 *The principles of nonlinear optics* (Wiley, New York) p 112-119
[2] Malacarne L C, Astrath N G C, Medina A N, Herculano L S, Baesso M L, Pedreira P R B, Shen J, Wen Q, Michaelian K H and Fairbridge C 2011 *Optics Express* **19** 4058
[3] Ivanov V I, Ivanova G D, Kirjushina S I and Mjagotin A V 2016 *Journal of Physics: Conference Series* **735** 012013
[4] Morozov K I 2009 *Phys. Rev. E* **79** 031204
[5] Leppla C and Wiegand S 2003 *Philosoph. Mag.* **83** 1989
[6] Ivanov V I, Ivanova G D and Khe V K 2017 *Proc. SPIE* **10176** 1017607
[7] Lee W, El-Ganainy R, Christodoulides D, Dholakia K and Wright E 2009 *Optics Express* **17** 10277
[8] Piazza R 2009 *Soft Matter* **4** 1740
[9] Vicary L 2002 *Philos. Mag.* B. **82** 447
[10] Ivanov V, Ivanova G, Okishev K and Khe V 2017 *IOP Conf. Ser.: Mater. Sci. Eng.* **168** 012045
[11] El-Ganainy R, Christodoulides D, Rotschild C and Segev M 2007 *Opt. Express* **15** 10207
[12] Chintamani P, Shalini M, Agnel P, Meera V, Tejas I H and Radha S 2014 *International Journal of Chemical and Physical Sciences* **3**(5) 44
[13] Ovseychook O O, Ivanov V I, Mjagotin A V and Ivanova G D 2018 *IOP Conf. Series: Journal of Physics: Conf. Series* **1038** 012091
[14] Philip J and Nisha M 2010 *Journal of Physics: Conference Series* **214** 012035
[15] Pizzoferrato R, Marinelly M, Zammit U, Scudiery F, Martelluchi S and Romagnoli M 1988 Optics Communications **68** 231
[16] Kamanina V, Kuzhakov P V, Serov S V, Kukharchik A A, Petlitsyn A A, Barinov O V, Borkovskii M F, Kozhevnikov N M and Kajzar F 2013 *Proc. SPIE* **8622** 86221B
[17] Ivanov V I , Ivanova G D, Okishev K N and Khe V K 2016 *Proc. SPIE* **10035** 100354Y
[18] Ishaaya A A, Vuong L T, Grow T D and Gaeta A L 2008 *Opt. Lett.* **33** 13
[19] Mjagotin A V, Ivanov V I and Ivanova G D 2017 *Proc. SPIE* **10176** 101761Z
[20] Polyakov P, Luettmer-Strathmann J and Wiegand S 2006 *Journal of Physical Chemistry B* **110** 26215

IOP Conf. Series: Materials Science and Engineering **597** (2019) 012065 doi:10.1088/1757-899X/597/1/012065

The formation of eyeball musculoskeletal stump using a Ni-Ti implant *in vivo*

E O Filippova[1,2], E A Gorbunova[2], O I Krivosheina[2] and N M Ivanova[1]
[1]Tomsk Polytechnic University, Lenin ave., 30, Tomsk, 634012, Russia
[2]Siberian State Medical University, Moskovskii tract, 2, Tomsk, 634012, Russia

E-mail: katerinabosix@mail.ru

Abstract. The purpose of this study is to investigate the possibility of forming an eyeball musculoskeletal stump using a Ni-Ti implant *in vivo*. Studies were performed on 18 Wistar rats. The animals were enucleated one of eyes. After the enucleation the musculoskeletal stump using Ni-Ti implant was formed. Ni-Ti implant was made from a thread of porous Ni-Ti (TN-10) with a thickness of 100 μm. Before implantation the Ni-Ti implant was sterilized by ethylene oxide. Results showed that the Ni-Ti implant surface was uniform and had not visible defects. The Ni-Ti implant surface was a hydrophobic with the mean value of $\theta = 93.5° \pm 1.4$. The Ni-Ti implant in the eye orbit was mobile. The Ni-Ti implantation into the eye orbit contributed the connective tissue with blood vessels formation and insignificant leucocytes infiltration (macrophages and lymphocytes infiltration). The study showed the possibility of forming an eyeball musculoskeletal stump using a Ni-Ti implant.

1. Introduction

Removal of the eyeball due to severe pathological conditions of the eye is a frequent ophthalmology operation. It is a serious reconstructive surgery due to which there is a pronounced shortage of tissue volume. Consequently it is necessary to place an implant from a biocompatible material into the orbit [1]. There are a number of materials offered as an orbital implant.

Synthetic and biological implants for the formation of the musculoskeletal stump are known [2]. Biological materials include a cartilage and subcutaneous fatty tissue of the human sole. Synthetic materials include hydroxyapatite, porous polyethylene, carbon felt, polytetrafluoroethylene, etc. [3]. However, biological implants tend to dissolve over time. It is necessary to create a bank of such materials which is associated with significant material and labor costs [4]. The disadvantages of the synthetic materials implantation are the exposure and rejection of implants (from 4 to 38%), their deformation, and in some case the high price [5].

Wang J.K. et al. revealed that late complications during the implantation of bioceramic materials also arise after the evisceration. They proposed to cover the implant with a Vicryl mesh for the prevention of exposure, adding anteriorly with a scleral patch graft following enucleation with primary or secondary implantation [6].

Thus, natural and synthetic hydroxyapatites currently occupy a leading position among orbital implants. However, the disadvantages of hydroxyapatite implants are the complex implantation technique and its high price (from 600 to 800 dollars). Therefore, ophthalmologists from different countries are actively engaged in research and development of biocompatible synthetic porous materials.

The purpose of the study is to investigate the possibility of forming an eyeball musculoskeletal stump using a Ni-Ti implant *in vivo*.

2. Material and methods

2.1. *Ni-Ti implant*

Ni-Ti implant was made from a thread of porous Ni-Ti (TN-10) with a thickness of 100 μm (TPU, Tomsk). The implant was round with a diameter of 4-5 mm (Figure 1).

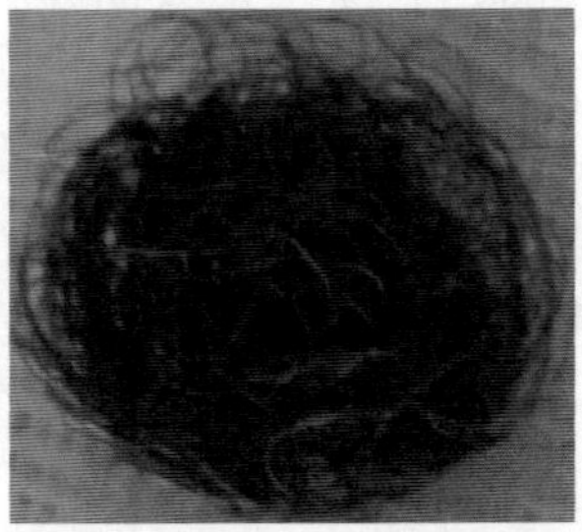

Figure 1. The Ni-Ti implant.

2.2. *Sterilization*

The Ni-Ti implant was sterilized by ethylene oxide using the gas sterilizer Steri-Vac 3M (USA). The ethylene oxide concentration was 750 mg/l. The temperature was 37°C. The sterilization time was 3 hours. The humidity was 70%.

2.3. *The surface topography of the Ni-Ti implant*

The surface topography was studied on the multipurpose correlator of optical, spectral and topographical surface objects properties "Centaur HR" (Russia). The surface roughness was estimated using the Gwyddion software. SEM of the Ni-Ti implant was obtained by using Hitachi S3400N Type II microscopc (Japan).

2.4. *The wetting angle of the Ni-Ti implant*

The wetting angle for purified water and glycerol were calculated using the sessile drop technique with the room temperature of 25±2°C, the "KRÜSS EasyDrop DSA 20" device (Germany) and the special software, with the measurement accuracy of ± 0.1°.

2.5. *Animals and treatments*

18 pubescent male Wistar rats (SSMU, Tomsk, Russia) weighing 250 g were used. All animals were healthy and free of ocular diseases. All procedures were approved by the Siberian Medical State University Life Science Ethical Review Committee (protocol № 3898 from November 24th, 2014).

The animals after an enucleation were implanted the Ni-Ti implant. All animals were instilled Tobramycin Drops (6 times per day), 0.1% Diclofenac Sodium Ophthalmic Solution (3 times per day) and 0.05% Vitabact (4 times per day) in the postoperative period.

The overall duration of the experiment comprised 21 days. Such methods as visual check and photographic registration were also used in course of the experiment. Sampling was performed on days 7, 14, and 21 after the start of the experiment for morphology studying.

2.6. *Histological processing specifics*

The enucleated animal eyeballs were put into the Carnoy's fixative solution for 2 hours, in course of the preparation for the light microscopy. Sections of tissues were stained with hematoxylin and eosin, according to the method of Van-Gizona. Light microscopy of the slides was conducted with 200x and 400x zoom using LOMO Biolam AU-12 (Russia) microscope. The microscope has an integrated digital camera ICC50 with the USB interface.

2.7. *Electron microscopy of tissues*

The study of the cytology structure was performed by transmission electron microscopy. The tissue was fixed in a 2.5% solution of glutaraldehyde for 2 hours at a temperature of 4°C. Then it was washed twice with cacodylate buffer, was fixed in 1% solution of four osmium oxides (in 0.1 M cacodylate buffer) for 2 hours, and washed with cacodylate buffer again. Ultrathin tissue sections of 60–100 nm thick were prepared on an "Ultrotome III" ultratome (LKB, Sweden). The resulting sections were examined in a JEM-100 CXII electron microscope (JEOL, Japan) with an aperture diaphragm of 25–30 μm at an accelerating voltage of 80 kV.

2.8. Morphometric analysis
The morphometric analysis of corneal specimens was carried out in histological sections stained with hematoxylin and eosin. Light microscopy with Avtandilov's eyepiece graticule analyzed 30 visual fields at 400× in each specimen to determine the volume percent (v/v %) of blood vessels and number of cells (macrophages, lymphocytes, plasma cells, and granulocytes).

2.9. Statistical processing of the research results
Statistical package IBM SPSS Statistics 20 was used for the statistical processing of the obtained results. A Student's t-test for unpaired values was used to analyze statistical significance between four groups. The Kruskal–Wallis test was used for the quantitative data. The Fisher's exact test was used for the nominal data. The dynamics analysis was conducted using the Wilcoxon signed rank test, which is used when comparing the mean value of two paired tests. Values of $p < 0.05$ were considered to be statistically significant.

3. Results

3.1. The surface topography of the Ni-Ti implant
Surface analysis of the Ni-Ti implant showed that the surface was uniform and had not visible defects (pores, cracks, craters, etc.) and a bimodal distribution (Figure 2).

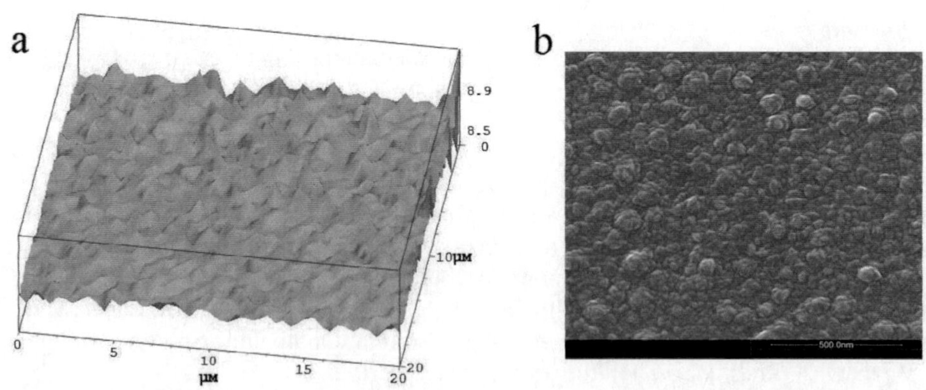

Figure 2. The surface topography of the Ni-Ti implant: a – AFM of implant surface; b - SEM of implant surface.

The surface roughness analysis showed that R_a of Ni-Ti implant surface was 35.0 ±5.4 nm, R_q was 42.1 ±6.2 nm, R_{sk} was 1.45. The ethylene oxide sterilization didn't influence on the surface roughness the Ni-Ti implant.

3.2. The wetting angle of the Ni-Ti implant
The measurement results of the wetting angle showed that the Ni-Ti implant surface was a hydrophobic with the mean value of $\theta = 93.5^o \pm 1.4$. The ethylene oxide sterilization didn't influence on the wetting angle of implant, the mean value of $\theta = 88.1^o \pm 6.0$.

IOP Conf. Series: Materials Science and Engineering **597** (2019) 012065 doi:10.1088/1757-899X/597/1/012065

3.3. Visual check

The conjunctival hyperemia of the eyelids and eye mucus discharge persisted on day 7 from the start of the experiment (Figure 3a). The Ni-Ti implant was round and mobile.

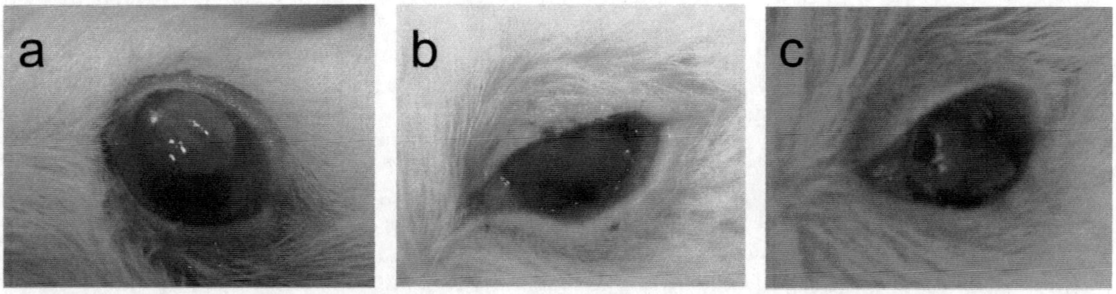

Figure 3. The visual check of eyes on day 7 (a), 14 (b) and 21 (c) from the start of the treatment.

The moderately pronounced conjunctival hyperemia persisted on day 14 from the start of the experiment. The Ni-Ti implant was round and mobile. The eye mucus discharge was absent (Figure 3b).

The moderately pronounced conjunctival hyperemia persisted on day 21 from the start of the experiment. The Ni-Ti implant was round and mobile. The eye mucus discharge was absent (Figure 3c).

3.4. Histology and electron microscopy results

The histology study showed the loose connective tissue with cells infiltration (macrophages, fibroblasts, lymphocytes, granulocytes) on day 7 from the start of the experiment (Figure 4a). Signs of lymphostasis were found in the lumen of the vessels.

The electron microscopy study showed that there were granulocytes (neutrophils) with di- and poly segmented nuclei and with many large lysosomes in the cytoplasm (Figure 4d). Macrophages with well-developed organelles (rough-surfaced endoplasmic reticulum with enlarged lumen cavities, mitochondria with normal structure) were encountered.

The eyeball cavity had the mature connective tissue with tightly arranged bundles of collagen fibers on day 14 from the start of the experiment (Figure 4b). Small areas of granulation tissue with a large number of newly formed vessels and diffuse cells infiltration (lymphocytes, monocytes, active fibroblasts, plasma cells) were observed between bundles of collagen fibers. The new formed blood vessels had thin walls.

Fibroblasts had the rough-surfaced endoplasmic reticulum, many ribosomes, mitochondrion with the destruction of the cristae and electron dense secretory granules. Euchromatin dominated in the nucleus of fibroblasts. The nucleolus was not visualized. The intercellular substance was determined around the fibroblasts where collagen fibrils were poorly ordered (Figure 4f).

The loose connective tissue was determined in the eyeball cavity on day 21 from the start of the experiment (Figure 4c). Collagen fibers were thin and randomly directed between which there were full-blooded arterioles, venules and cells (fibroblasts, macrophages, plasma cells and lymphocytes).

Cytoplasm fibroblasts had microvesicles, rough-surfaced endoplasmic reticulum with enlarged lumen cavities, mitochondrion. The fibroblast nucleus had also elongated shape. The euchromatin prevailed.

Endotheliocytes of the newly formed blood vessels had well-developed organelles (Figure 4g). An intercellular substance with a small number of thin collagen fibrils oriented randomly in different directions was found around blood vessels. Red blood cells were in the lumen of blood vessels.

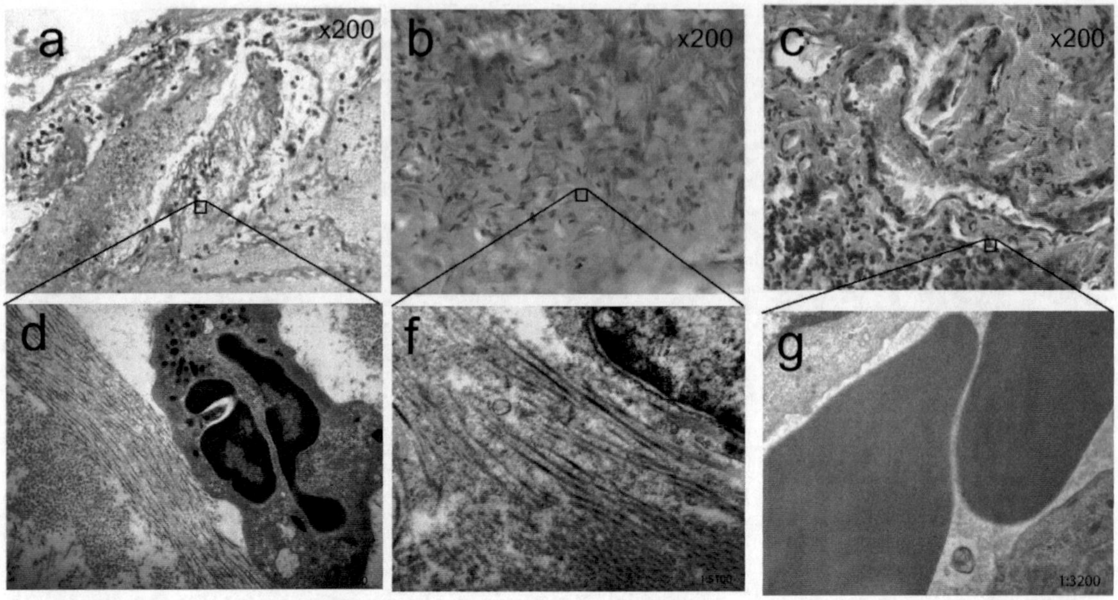

Figure 4. Histology tissue cross-sections (a, b, c) and electron microscopy pictures (d. f. g) on day 7 (a, d), 14 (b, f) and 21 (c, g) from the start of the treatment. Histology tissue cross-sections (a, b, c) are 200 magnification and hematoxylin and eosin stain.

3.5. Morphometric analysis

According the morphometric analysis the increasing of the cells infiltration of macrophages, lymphocytes and plasma cells were found on day 14 from the start of the experiment (Table 1). The number of granulocytes decreased by 2.5 times on day 14. It decreased by 7.6 times on day 21 from the initial value.

The statistically significant increasing of blood vessels number was noted on day 14 and day 21 day from the initial value (Table 1).

Table 1. Morphometric indicators of the eyeball musculoskeletal stump after Ni-Ti implantation.

	The day from the start of the experiment		
	7 day	14 day	21 day
Macrophages, cells	1078.8±191.9	2731.9±574.7*	985.7±264.9
Lymphocytes, cells	837.8±303.4	2012.1±353.3*	1034.9±297.9
Plasma cells, cells	147.9±75.3	295.7±131.4*	69.4±39.2*
Granulocytes, cells	342.3±207.5	135.9±95.9*	45.3±10.3*
Blood vessels (%)	0.02±0.014	2.4±1.2*	5.3±1.9*

Note: * - statistically significant differences according to the U test Mann-Whitney compared to the day 7.

The Ni-Ti implantation into the eye orbit contributes the connective tissue with blood vessels formation and leucocytes infiltration (macrophages and lymphocytes infiltration). The Ni-Ti implant into the eye orbit is mobile.

4. Conclusion

The surface of Ni-Ti implant is uniform and has not visible defects. The surface of Ni-Ti implant is a hydrophobic with the mean value of $\theta = 93.5^{o} \pm 1.4$. The Ni-Ti implant into the eye orbit is mobile. The implantation of Ni-Ti implant in the eye orbit contributes the connective tissue with blood vessels formation and insignificant leucocytes infiltration (macrophages and lymphocytes infiltration). The study showed the possibility of forming an eyeball musculoskeletal stump using a Ni-Ti implant.

Acknowledgments
This work was supported by the Ministry of Education and Science of the Russian Federation, the Federal Target Program (Agreement No. 14.575.21.0140, unique identifier RFMEFI57517X0140).

References

[1] Baino F, Potestio I 2016 *Mater. Sci. Eng. C. Mater. Biol. Appl.* **69** 1410
[2] Schellini S, Jorge E, Sousa R, Burroughs J, El-Dib R. 2016 *Orbit.* **35** (2) 78
[3] Perry A Adv. 1990 *Ophthalmic Plast. Reconstr. Surg.* **8** 75
[4] Feng K, Ying W 2017 *MATEC Web Conf.* **100** 04020
[5] Custer P L 2007 *Ophthalm. Plast. and Reconst. Surg.* **23** (1) 1
[6] Wang J-K, Lai P-C, Liao S L 2009 AJO. **147** (1) 162

IOP Conf. Series: Materials Science and Engineering **597** (2019) 012066 doi:10.1088/1757-899X/597/1/012066

Polylactic acid films implantation into the anterior chamber of eye *in vivo*

E O Filippova[1,2], N M Ivanova[1], V F Pichugin[1] and Z Sun[1]
[1]Tomsk Polytechnic University, Lenin ave., 30, Tomsk, 634012, Russia
[2]Siberian State Medical University, Moskovskii tract, 2, Tomsk, 634012, Russia

E-mail: katerinabosix@mail.ru

Abstract. The purpose of this research is the study of the influence of thin polylactic acid films implantation on the corneal morphology *in vivo* experiment. Studies were performed on 8 pubescent female Sylvilagus bachmani rabbits. Polylactic acid films were implanted into the anterior chamber of one animal eye. The feedstock for the films was obtained by dissolving polylactic acid in the chloroform. Before the implantation PLA films were sterilized. The overall duration of the *in vivo* experiment comprised 21 days. Such methods as visual check, photographic registration, non-contact ocular tonometry and optical coherent tomography of cornea were also used in course of the experiment. Sampling was performed on day 21 after the start of the experiment for morphology studying. Results showed that the implantation of the PLA films into the anterior chamber of eye does not induce an inflammatory reaction and does not increase an intraocular pressure. The study showed the possibility of PLA films using as a corneal implant.

1. Introduction

Polylactic acid (PLA) is a biodegradable aliphatic non-toxic and eco-friendly polyester and is widely used as medical implants in the form of anchors, screws, plates, pins, rods, and as a mesh [1 - 4]. A comprehensive literature search reveals the applications of PLA and its polymeric composites in medical fields such as: orthopedics, drug carriers, facial fracture repair, tissue engineering, antimicrobial agents, antitumor, ureteral stents, biomaterials, miscellaneous applications [1 - 4]. Of particular interest is the use of PLA as a corneal implant for the bullous keratopathy treatment. The bullous keratopathy is the severe cornea disease and occupies one of the leading positions among the causes of corneal weak vision. Primary and secondary processes of the degenerative and infectious diseases of the cornea play a significant role in the development of bullous keratopathy which lead to the death of endotheliocytes and dysfunction of the corneal endothelial layer [5]. This contributes to the development of corneal edema, reduced vision, the occurrence of recurrent erosions and pronounced pain symptom.

One of the directions in the treatment of this disease is the use of stem cells [5]. However, cultivation of stem cells and their landing on the inner surface of the cornea in order to replace the endothelial defect without using a substrate is a big surgical problem. There is a high probability of cell loss during surgical procedures. Using of scaffolds based on biodegradable thin films with controlled solubility as a temporary material for stem cells may be an alternative to existing insoluble polymers.

The purpose of this research is the study of the influence of thin polylactic acid films implantation on the corneal morphology *in vivo* experiment.

2. Material and methods

2.1. *Polylactic acid films*
The feedstock for the films was obtained by dissolving polylactic acid (PURASORB® PL 10, Netherlands) in the chloroform ($CHCl_3$). 1% solution was poured into Petri dishes (10 g) which were placed in a fume hood until complete evaporation of $CHCl_3$ and the formation of polylactic acid films.

2.2. *Sterilization*
The steam sterilization was carried out in a GPA-10 PZ (Russia) autoclave, operating at the temperature of 120°C - 121°C and pressure of 0.11 MPa for 20 min.

2.3. *The surface topography of polylactic acid films*
The surface topography was studied on the multipurpose correlator of optical, spectral and topographical surface objects properties "Centaur HR" (Russia). The surface roughness was estimated using the Gwyddion software. SEM of polylactic acid films was obtained by using Hitachi S3400N Type II microscope (Japan).

2.4. *The wetting angle of polylactic acid films*
The wetting angle for purified water and glycerol were calculated using the sessile drop technique with the room temperature of 25±2°C, the "KRÜSS EasyDrop DSA 20" device (Germany) and the special software, with the measurement accuracy of ± 0.1°. The surface energy was calculated using Owens, Wendt, Rabel and Kaelble method.

2.5. *Animals and treatments*
8 pubescent female Sylvilagus bachmani rabbits (SSMU, Tomsk, Russia) weighing 2.0-2.5 kg were used. All animals were healthy and free of ocular diseases. All procedures were approved by the Siberian Medical State University Life Science Ethical Review Committee (protocol № 3898 from November 24th, 2014).

Polylactic acid films were implanted into the anterior chamber of one animal eye. All animals were instilled Tobramycin Drops (6 times per day), 0.1% Diclofenac Sodium Ophthalmic Solution (3 times per day) and 0.05% Vitabact (4 times per day) in the postoperative period.

The overall duration of the experiment comprised 21 days. Such methods as visual check, photographic registration, non-contact ocular tonometry and optical coherent tomography (OCT) of cornea were also used in course of the experiment. Sampling was performed on day 21 after the start of the experiment for morphology studying.

2.6. *Optical Coherence Tomography (OCT) of cornea*
Optical coherent tomography of cornea was carried out on days 7, 14, and 21 after the start of the experiment on Cirrus HD-OCT 5000 (Germany).

2.7. *Non-Contact Ocular tonometry*
Non-Contact Ocular tonometry was done using Non-Contact Tonometer Huvitz HNT-7000 (South Korea).

2.8. *Histological processing specifics*
The enucleated animal eyeballs were put into the Carnoy's fixative solution for 2 hours, in course of the preparation for the light microscopy. Sections of tissues were stained with hematoxylin and eosin, according to the method of Van-Gizona. Light microscopy of the slides was conducted with 200x and 400x zoom using LOMO Biolam AU-12 (Russia) microscope. The microscope has an integrated digital camera ICC50 with the USB interface.

2.9. *Electron microscopy of tissues*
The study of the cytology structure was performed by transmission electron microscopy. The tissue was fixed in a 2.5% solution of glutaraldehyde for 2 hours at a temperature of 4°C. Then it was washed twice

IOP Conf. Series: Materials Science and Engineering **597** (2019) 012066 doi:10.1088/1757-899X/597/1/012066

with cacodylate buffer, was fixed in 1% solution of four osmium oxides (in 0.1 M cacodylate buffer) for 2 hours, and washed with cacodylate buffer again. Ultrathin tissue sections of 60–100 nm thick were prepared on an "Ultrotome III" ultratome (LKB, Sweden). The resulting sections were examined in a JEM-100 CXII electron microscope (JEOL, Japan) with an aperture diaphragm of 25–30 μm at an accelerating voltage of 80 kV.

2.10. Statistical processing of the research results

Statistical package IBM SPSS Statistics 20 was used for the statistical processing of the obtained results. A Student's t-test for unpaired values was used to analyze statistical significance between four groups. The Kruskal–Wallis test was used for the quantitative data. The Fisher's exact test was used for the nominal data. The dynamics analysis was conducted using the Wilcoxon signed rank test, which is used when comparing the mean value of two paired tests. Values of $p < 0.05$ were considered to be statistically significant.

3. Results

3.1. The surface topography of PLA films

The study of the structure and morphology of polylactic acid films showed that its roughness and topography depended on the side of the surface of the material: the outer side (contact with the atmosphere) had a more embossed surface as opposed to a smoother inner surface (contact with the Petri dish) (Figure 2, 3).

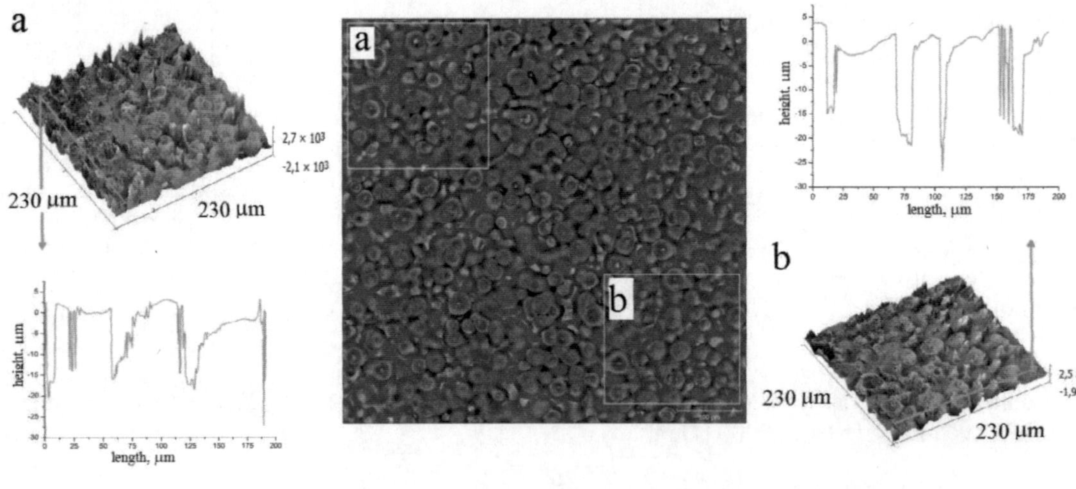

Figure 1. The surface topography of the PLA outer side.

Thus, the inner side of the material is formed smoother.

The surface roughness analysis showed that R_a of PLA films (inner) surface was 0.01±0.003 μm, R_q was 0.014 μm, R_{sk} was -0.001. PLA films (outer) surface had R_a = 0.17±0.06 μm, R_q=0.4 μm, R_{sk}=-1.0.

The sterilization increased the inner surface roughness of PLA films by 5 times. R_a of PLA films (inner) surface was 0.05±0.01 μm, R_q was 0.09 μm, R_{sk} was -0.66.

IOP Conf. Series: Materials Science and Engineering **597** (2019) 012066 doi:10.1088/1757-899X/597/1/012066

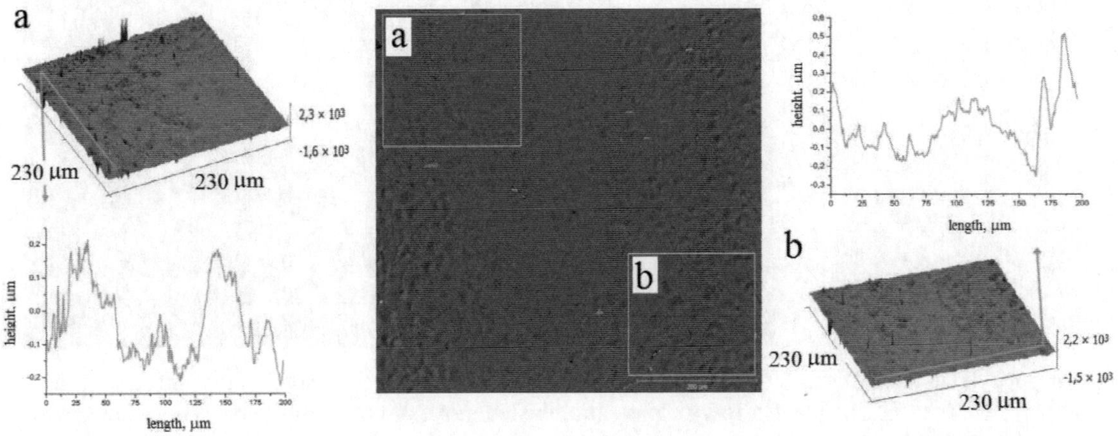

Figure 2. The surface topography of the PLA inner side.

3.2. The wetting angle of PLA films
The wettability of the surface analysis showed that the PLA inner side had a wetting angle of water of 80.5°±1.2. The PLA film properties are close to hydrophobic. The surface energy of the films varies within (26 ÷ 27) mJ/m². The polarity of PLA film (inner side) was 0.36. The PLA outer side had 78.5°±1.3. The polarity of PLA film (outer side) was 0.37.

The sterilization increased the PLA wettability and decreased the wetting angle of water. The wetting angle of PLA films after sterilization was 64.5°±2.3.

3.3. Visual check and Ocular tonometry
The implantation of the PLA film did not cause an inflammatory reaction and did not increase an intraocular pressure.

3.4. OCT of cornea
According to OCT the cornea had a normal thickness (430-450 µm) (Figure 3). There was observed a free position of the PLA film without contact with the iris and with the anterior chamber angle.

3.5. Histology and electron microscopy results
The following histological results were obtained. The anterior epithelium was represented by 4-5 layers of squamous epithelium with normochromic nuclei. Bowman's membrane was unchanged and visualized as a homogeneous eosinophilic strip. Collagen fibers were located compactly. In some places collagen fibers had increased twisted stroke (Figure 3a). Descemet's membrane was visualized throughout. The endothelial layer was represented by a single layer of cells. In some places proliferation of endothelial cells in the form of process cells was observed.

Electron microscopy showed that fibroblast had the normal structure (big oval shaped nucleus with dimensionally pronounced chromatins, normal structure mitochondrion, not extended tanks of rough endoplasmic reticulum) (Figure 4a). In most cases collagen fibers were located compactly. In some places collagen fibers had increased twisted stroke (Figure 4b).

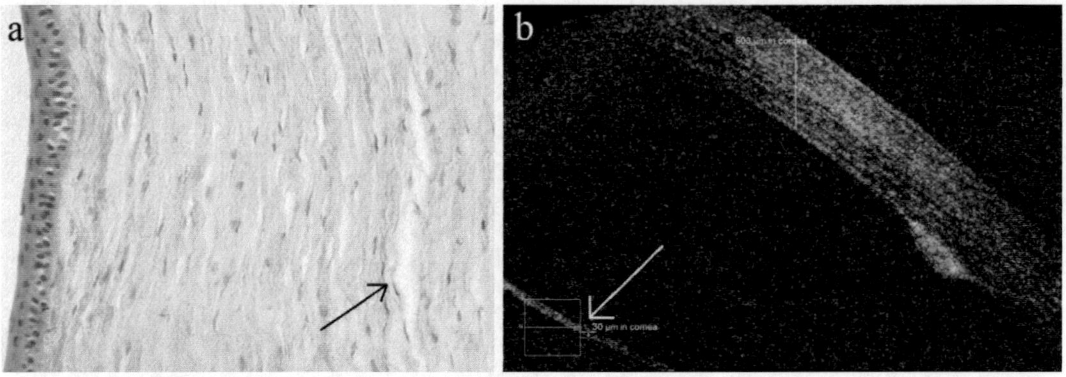

Figure 3. Histology tissue cross-section (a) and OCT of cornea (b) on day 21 after the start of the experiment *in vivo*. The black arrow shows extra twisted move of collagen fibers. The yellow arrow shows the PLA film into the eye anterior chamber. The magnification (a) is x200. Hematoxylin and eosin stain (a).

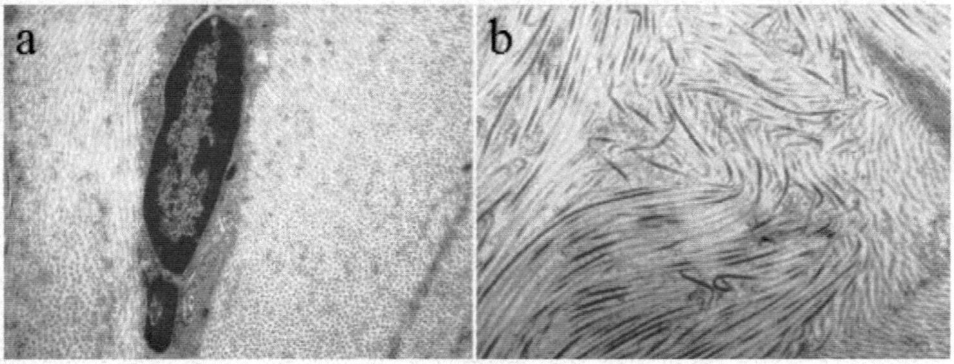

Figure 4. Electron microscopy pictures of cornea on day 21 after the start of the experiment *in vivo*. The magnifications are x7000 (a) and x12000 (b).

4. Conclusion

The implantation of the PLA films into the anterior chamber of eye does not induce an inflammatory reaction and does not increase an intraocular pressure. The study showed the possibility of PLA films using as a corneal implant.

Acknowledgments

The research was conducted with the financial support of the Russian Foundation for Basic Research (RFBR) as part of the project № 19-415-703005.

References

[1] Lopes MS, Jardini AL, Maciel Filho R 2012 *Procedia Eng.* **42** 1402
[2] Peesan M, Rujiravanit R, Supaphol P 2006 *J. Biomater. Sci .Polym. Ed.* **17** (5) 547
[3] Sangsanoh P, Waleetorncheepsawat S, Suwantong O, Wutticharoenmongkol P 2007 *Biomacromolecules* **8** (5) 1587
[4] Ramot Y, Haim-Zada M, Domb AJ, Nyska A 2016 *Adv. Drug. Deliv. Rev.* **107** 153
[5] Brunette I, Roberts CJ, Vidal F, Harissi-Dagher M, Lachaine J, Sheardown H 2017 *Prog. Retin. Eye Res.* **59** 97

IOP Conf. Series: Materials Science and Engineering **597** (2019) 012067 doi:10.1088/1757-899X/597/1/012067

Ultrashort pulse decomposition in reflection symmetric meander line of four cascaded half-turns

E B Chernikova, A O Belousov and T R Gazizov

Tomsk State University of Control Systems and Radioelectronics, 40, Lenina Ave., Tomsk, 634050, Russia

E-mail: chiernikova96@mail.ru

Abstract. The paper considers the protection of critical electronic equipment from an ultrashort pulse (USP) by means of modal filtering technology. Decomposition of the USP in a reflection symmetric meander line consisting of four cascaded half-turns is investigated for the first time. Quasi-static simulation of two half-turn connection diagrams has been performed. The output of the circuits shows the presence of negative polarity pulses, as well as additional pulses with time delay values non-multiple of per-unit-length time delays. It has been revealed that these values are linear combinations (arithmetic means) of per-unit-length time delays of the line modes.

1. Introduction

Currently, radio-electronic equipment (REE) is massively introduced in almost all domains of human activity. However, the vulnerability of electronic equipment to electromagnetic effects may entail undesirable consequences, for example, regular malfunctions of critical installations. This problem is regularly discussed at AMEREM EUROEM/ASIAEM conferences. For example, at the ASIAEM 2015 conference, the section "IEMI Threats, Effects and Protection" was allocated and even special sections were created (Design of Protective Devices and Test Methods. Evaluation of HEMP/IEMI Impacts on Critical Infrastructure). In addition, in some articles of the famous journal "IEEE Transactions on EMC", where the results of the newest EMC research in the world are published, there are many valuable results that can be used to create a noise immunity technology for critical REE. Thus, a new threat to civil society in the form of intentional electromagnetic interference was considered [1]. One of the reasons for REE mulfunctioning can be conducted interference, which can penetrate into REE directly through the conductors [2]. Recent investigations have shown the possibility of interrupting the normal functioning of IT networks using high-power electromagnetic interference through the low-voltage-power network [3] and the local area network cables [4]. The analysis of the impact of a commercially available powerful source of damped sinusoids on a computer network was presented [5]. A typical example of a dangerous conducted interference is an ultrashort pulse (USP) [6]. This is an ultrawideband pulse of high power and short duration, having both natural and man-made origin. The results obtained in the investigation of local area networks for Fast Ethernet and Gigabit Ethernet exposed to nanosecond electromagnetic interference were presented [7]. The devices for electronic system protection against natural or man-made electromagnetic interferences with high energies and amplitudes, in particular ultrawideband (UWB) pulses, based on nonlinear protection elements were presented [8].

Traditional devices that are used for pulse interference protection, for example, voltage suppressors, varistors, passive RC- and LC-filters are known. However, such protecting devices have a number of disadvantages (low radiation resistance, short service life, failure to operate at high voltages, insufficient operating speed, etc.), making it difficult to protect against powerful USP. Thus, there is currently no effective protection of REE against USP.

A technique of modal filtration has been proposed to protect REE against ultrashort pulses (USPs). This technique is based on modal decomposition of a pulse signal into pulses of lower amplitude and can be implemented through the use of modal filters (MF) and meander lines (ML) [9].

In addition, the approach to improve modal filtration using reflection symmetry is also well-known. This approach is implemented in a device called a reflection symmetric MF. In the cross section it consists of 5 identical and rectangular conductors on a dielectric layer, with the conductors 1 and 2 located on one side, two additional conductors placed reflection-symmetrically relative to the conductors 1 and 2 on the reverse side of the dielectric layer, and the grounded conductor 5 located in the dielectric layer at an equal distance from the external conductors [10] (Figure 1). The reflection symmetric MF is noteworthy inasmuch as, providing both the edge and broad-side couplings between the conductors at certain parameters, it allows to obtain USP decomposition pulses at the line output with nearly equal values of the amplitudes of the pulses and the time intervals between the decomposition pulses (Figure 1).

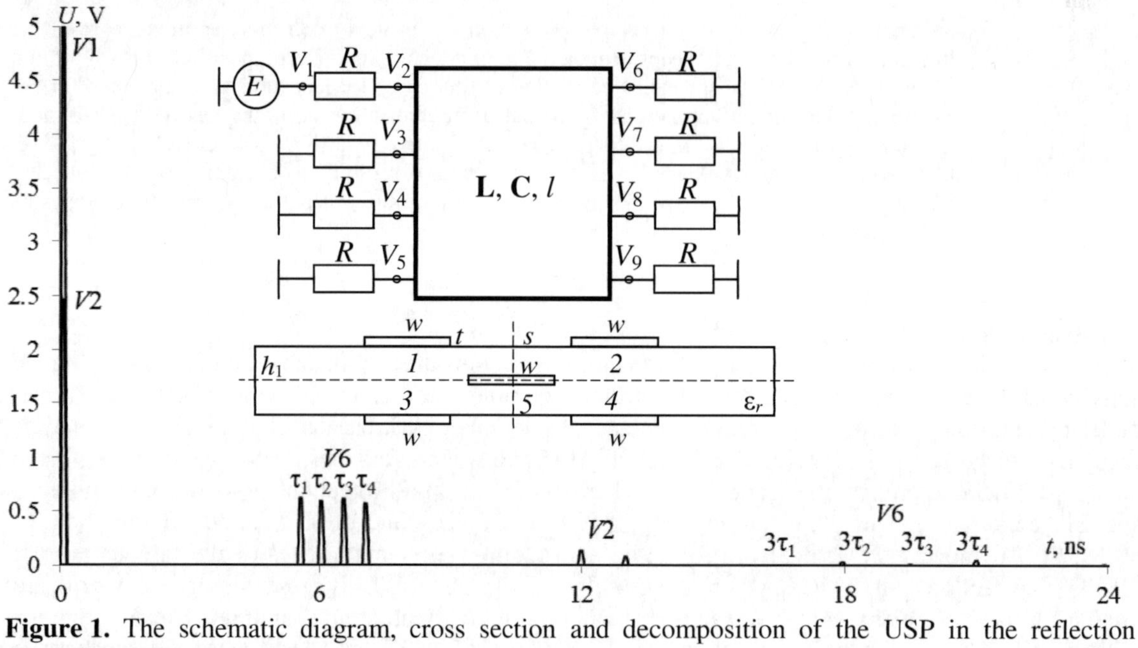

Figure 1. The schematic diagram, cross section and decomposition of the USP in the reflection symmetric MF.

Thus, it seems possible to investigate the ML on the basis of the reflection symmetric MF. So far, such configuration has not been considered as a protection device. Meanwhile, the analysis makes it possible to single out several variants of its circuit design, depending on the way how half-turns with the length l are connected. The aim of this work is to investigate the possibility to decompose the USP in reflection symmetric MLs, where the USP passes the maximum path (from the generator to the load) for a length of $4l$, i.e. along 4 cascaded half-turns.

2. Structures under investigation

Figure 2 shows the structures with the diagrams under investigation. Let us perform a quasistatic simulation with l=1 m and R=50 Ω in the TALGAT software [11] which gives acceptable accuracy and does not require high computational costs [12].

IOP Conf. Series: Materials Science and Engineering **597** (2019) 012067 doi:10.1088/1757-899X/597/1/012067

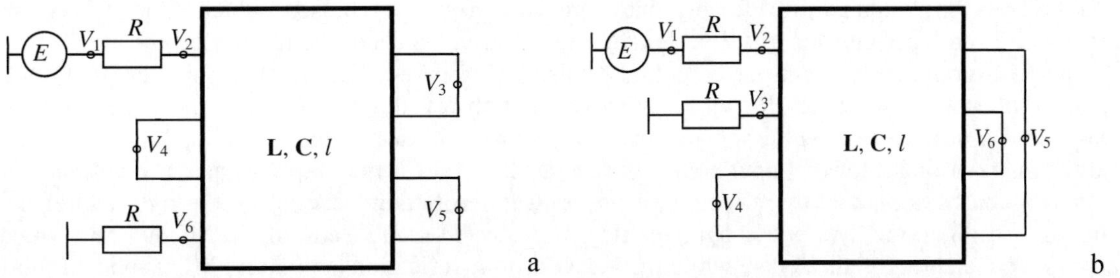

Figure 2. The schematic diagrams of half-turns connection: 1 (a) and 2 (b).

The losses in conductors and dielectrics at this investigation step were not taken into account. A trapezoid EMF source with an amplitude of 5 V with front, fall and flat top durations of 50 ps was used as an excitation, so the total duration was 150 ps (*V*1 in Figure 1). The cross section of these structures is shown in Figure 1, where *w* is the width of the conductors, *s* is the separation, *t* is the thickness of the conductors, *h* is the thickness of the dielectric and ε_r is the relative permittivity of the substrate. Simulation was performed with *s*=510 μm, *w*=1600 μm, *t*=18 μm, *h*=500 μm, ε_r=4.5.

3. Simulation results

Figures 3 and 4 show the obtained voltage waveforms at input (nodes V_2 in Figure 2) and output (nodes V_6 in Figure 2 and V_3 in Figure 2), respectively.

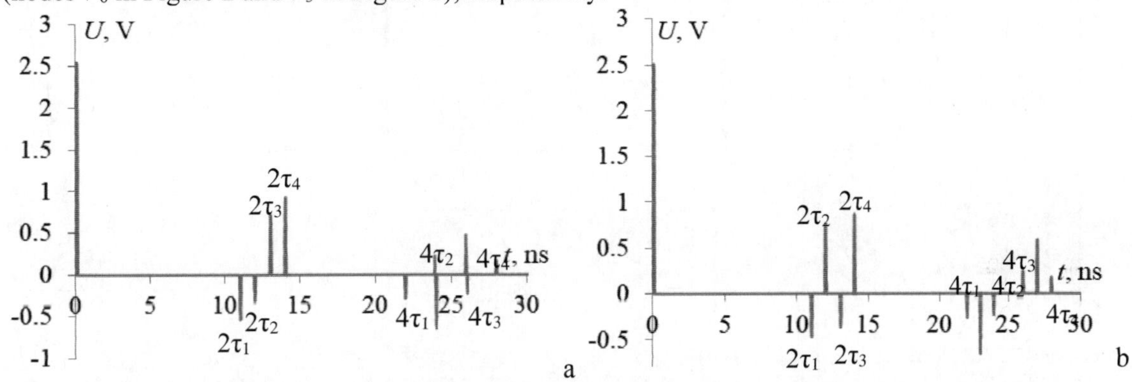

Figure 3. The voltage waveforms at the input of schematic diagrams 1 (a) and 2 (b).

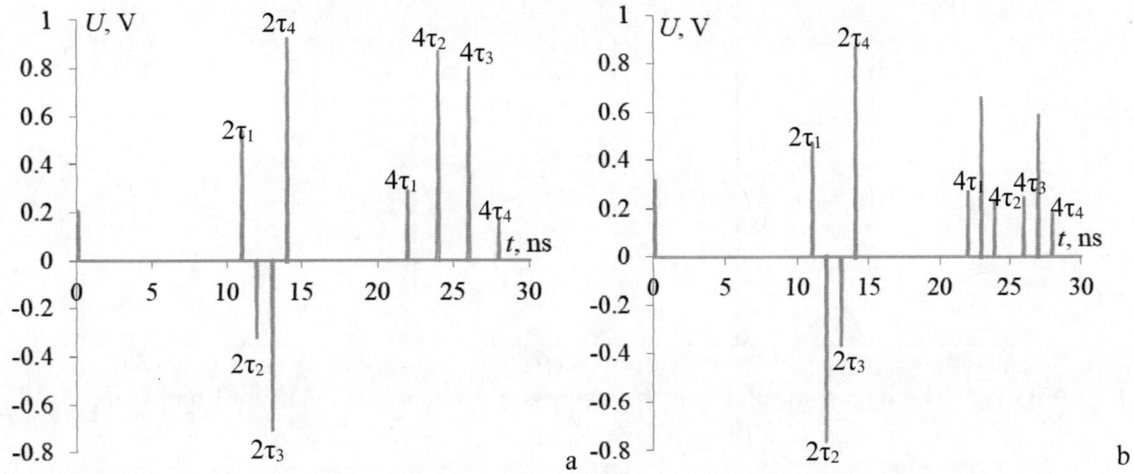

Figure 4. The voltage waveforms at the output of schematic diagrams 1 (a) and 2 (b).

On the resulting voltage waveforms, there are two groups of pulses, which time delays are multiples of 2 and 4 per-unit-lengths time delays, respectively. Consider them separately.

In the first group, a pair of negative polarity pulses is observed. Earlier, when simulating an ML [13] and a reflection symmetric MF, such pulses were not observed. For diagram 1, these are pulses of modes 1 and 2 of the input signal and modes 2 and 3 of the output signal. For diagram 2, these are pulses of modes 1 and 3 for the input signal and modes 2 and 3 for the output signal. The presence of such pulses can be explained by reflections from bridges (cross connections) on the right end of the circuit. In the reflection symmetric MF, resistors are located at all ends of the conductors, which improves the matching of all modes, while the bridges in the reflection symmetric ML provide a short circuit for some modes and an open circuit for others.

In the second group of pulses, among pulses with time delays that are multiples of 4 per-unit-lengths time delays for diagram 2, one can observe additional pulses with time delays non-multiple of per-unit-length time delays. It can be assumed that such pulses exist also for diagram 1. This can be judged by the increase in the total amplitude of modes 2 and 3, whose time delays are multiples of 4 per-unit-lengths time delays. To clarify this assumption, we will increase the l value from 1 to 8 m for diagram 1. The fragments of time responses with pulses whose delays are multiples of 32 per-unit-lengths time delays are shown in Figure 5.

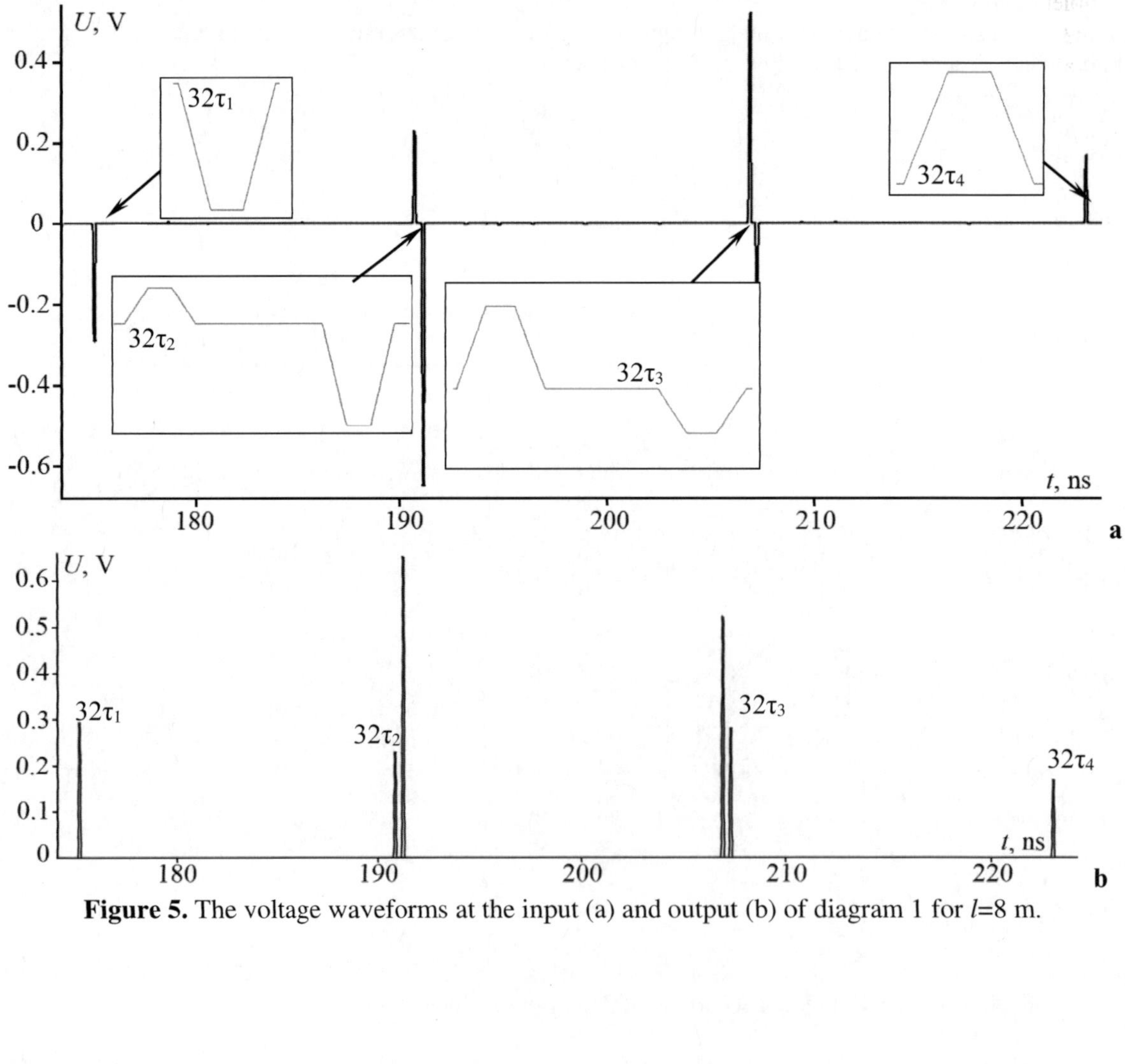

Figure 5. The voltage waveforms at the input (a) and output (b) of diagram 1 for l=8 m.

The simulation results confirm the assumption about the presence of additional pulses in diagram 1. Table 1 shows the values of per-unit-length time delays of modes 1–4 multiplied by 1, 2, 4 and 32, since the time delays of the pulses at the input and output of the reflection symmetric ML are multiples of 2 and 4 of per-unit-length time delays for *l*=1 m and 32 for *l*=8 m.

Table 1. Values of per-unit-length time delays (ns/m) of modes 1–4, multiplied by 1, 2, 4, 32

Multiplier	1	2	3	4
1	5.4698	5.9591	6.4746	6.9687
2	10.9398	11.9183	129.493	13.9376
4	21.8795	23.8366	258.987	27.8752
32	175.036	190.693	207.189	223.001

From the analysis of Table I and Figure 5 it can be seen that the time delays of the additional pulses are equal to the linear combinations (arithmetic means) of the per-unit-length time delays of the modes. Thus, for diagram 1, these delays are $(4\tau_1+4\tau_3)/2$=23.8891 ns and $(4\tau_2+4\tau_4)/2$=25.8559 ns, and for diagram 2 these delays are $(4\tau_1+4\tau_2)/2$=22.85805 ns and $(4\tau_3+4\tau_4)/2$=26.88695 ns. These completely coincide with the values obtained in the simulation of time responses.

Furthermore it can be seen in Figure 4 that the first impulse (with zero time delay) is a crosstalk, with an amplitude smaller than the amplitudes of the decomposition pulses. It follows from this that, by increasing the amplitude of the crosstalk pulse, it is possible to reduce the amplitudes of the subsequent decomposition pulses, which can be realized by means of optimization with the pulse amplitude equalization criterion. Obviously, the reflection symmetric ML must be optimized separately, despite the fact that it is based on the cross section of a reflection symmetric MF with optimal parameters. Moreover, since the time intervals between the decomposition pulses in the reflection symmetric MF were equalized, it was impossible to observe the presence of additional pulses in diagram 1. Thus, it is useful to optimize the reflection symmetric ML by the criteria of equalizing the time intervals between all decomposition pulses (including additional pulses) and minimizing the maximum amplitude at the ML output, in order to increase the USP attenuation. As for attenuation of the considered cases, the output voltage peak values are about 0.9 V, which is defined by the pulse amplitude with the time delay of $2\tau_4$.

4. Conclusion

For the first time, a quasistatic simulation of two half-turn connection diagrams for a reflection symmetric ML was performed in the time domain. In these structures, the possibility of additional pulses to appear in the output signal was revealed, whereas earlier they had not been detected in the reflection symmetric MF or the ML turn. It is shown that time delays of additional pulses, which are non-multiple of per-unit-length time delays, are linear combinations (arithmetic means) of per-unit-length time delays of the line modes.

In the future, it is planed to consider other diagrams of a reflection symmetric ML. In addition, multicriteria optimization of their parameters will be performed.

Acknowledgments

This study was funded by Russian Science Foundation (project №19-19-00424).

References

[1] Radasky W. A., Baum C. E., Wik M. W. 2004 *Special issue of IEEE Transactions on Electromagnetic Comconspatibility* **46(3)** 314–21.

[2] Gizatullin Z M, Gizatullin R M 2006 *Journal of Communications Technology and Electronics* **61(5)** 546–50.

[3] Brauer F., Fahlbusch S., Ter Haseborg J. L., Potthast S. 2012 *IEEE Transactions on*

IOP Conf. Series: Materials Science and Engineering **597** (2019) 012067 doi:10.1088/1757-899X/597/1/012067

Electromagnetic Comconspatibility **54(5)** 1055–65.
[4] Kreitlow M., Garbe H., Sabath F. *Proc. 2014 IEEE International Symposium on Electromagnetic Compatibility* 544–8.
[5] Kreitlow M., Sabath F., Garbe H. *Proc. 2015 IEEE International Symposium on Electromagnetic Compatibility* 1063–67.
[6] Mora N, Vega F, Lugrin G, Rachidi F, Rubinstein M 2014 *System and assessment notes* **41**.
[7] Gizatullin Z. M., & Gizatullin R. M. 2014 *Journal of Communications Technology and Electronics* **59(5)** 424–6.
[8] Weber T., Krzikalla R., Ter Haseborg J. L. 2004 *IEEE Transactions on Electromagnetic Compatibility* **46(3)** 423–30.
[9] Gazizov A T, Zabolotsky A M, Gazizov T R 2016 *IEEE Transactions on Electromagnetic Comconspatibility* **58(4)** 1136–42.
[10] Zabolotsky A M, Gazizov T R, Kuksenko S P Four-way mirror-symmetrically structure, protecting from ultrashort impulses. Patent RF, no 2624465, 2017.
[11] Kuksenko S P, Gazizov T R, Zabolotsky A M, Ahunov R R, Surovtsev R S, Salov V K, Lezhnin Eg.V *Proc. of the 2015 Int. Conf. on Modeling, Simulation and Applied Mathematics (MSAM2015)* 293–301.
[12] Orlov P E, Buichkin E N *18th International conference on micro/nanotechnologies and electron devices (EDM-2017)* 54–8.
[13] Surovtsev R S, Nosov A V, Zabolotsky A M, Gazizov T R 2017 *IEEE Transactions on Electromagnetic Compatibility* **59(6)** 1864–71.

IOP Conf. Series: Materials Science and Engineering **597** (2019) 012068 doi:10.1088/1757-899X/597/1/012068

Holographic formation of diffractive optical elements in photopolymer with two-beam interactions taken into account

V O Dolgirev, D I Dudnik, S N Sharangovich and A O Semkin

Tomsk State University of Control Systems and Radioelectronics, Tomsk, 634050 Russia

E-mail: vitial2@mail.ru, shr@tusur.ru

Abstract. This work presents the theoretical model of holographic formation of the diffraction optical element (DOE) in photopolymer material (PPM). DOE allows transforming Gaussian light fields to Bessel-like. Self-diffraction of writing light beams at low contrasts is taken into account. It is experimentally shown that influence of effect of self-diffraction in areas of low contrast leads to amplification of level of side's maxima of diffracted light beam.

1. Introduction

Bessel-like light beams are increasingly attracting the attention of scientists due to their ability to spread over a considerable distance without diffraction divergence [1-2]. These beams are used in medicine, biomedical optics, precision chemistry, pharmacology, etc. Thus, there is a need to search for available methods for the formation of such beams.

Different methods for the Bessel beams' formation are presented in [3 7]. However, presented methods are either difficult to implement or require expensive elements. In this paper, it is proposed to use a diffraction optical element (DOE) [8-10] as a transducer of transmitted light field into Bessel-like. One of the most effective methods for DOEs formation is holographic one using photosensitive media [2, 8–12]. Effectiveness of this method is determined by the fact that refractive index distribution of photosensitive medium completely depends on the holographic recording scheme and the physicochemical processes taking place in medium. Therefore, it is possible to create arbitrarily complex diffractive structures (DS), allowing to transform transmitted light beams. As a photosensitive medium, a photopolymer material (PPM) is used. It is characterized by an acceptable photosensitivity, and high diffraction efficiency (DE) [11-12].

2. Theory

Let two monochromatic light beams with amplitude profiles $E_0(\boldsymbol{r})$ (Gaussian distribution) and $E_1(\boldsymbol{r})$ (Bessel-like distribution), wave vectors $\boldsymbol{k}_0'$ and $\boldsymbol{k}_1'$ fall onto a flat sample containing an absorbing photopolymer layer at angles θ_0 and θ_1 (figure 1a). Vectors $\boldsymbol{k}_0'$ and $\boldsymbol{k}_1'$ lie in the XoY plane, and the width of the beams is much larger than the sample's thickness d . Then the optical field at the entrance to the sample can be represented as [13-15]:

$$E(t,\boldsymbol{r}) = \sum_{j=0,1} \boldsymbol{e}_j \cdot E_j \cdot \exp\left[i\left(\omega \cdot t - \boldsymbol{k}_j' \cdot \boldsymbol{r}\right)\right], \tag{1}$$

where e_j – polarization vectors; r – radius-vector; $k_j^{'} = k \cdot N_j$, $k = n \cdot \omega / c$ – wave number and N_j – normal to the wavefront, n – refraction index.

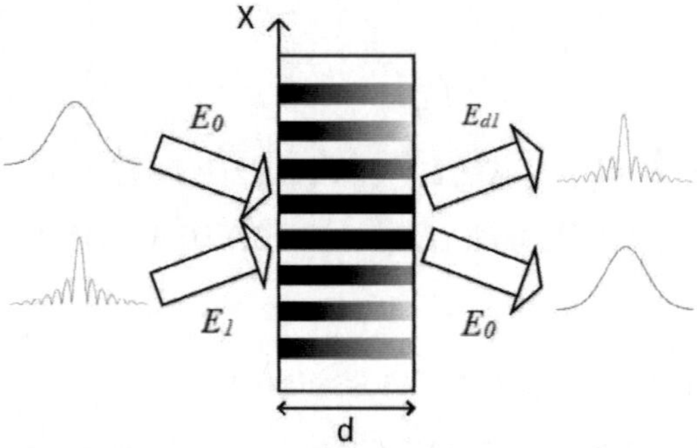

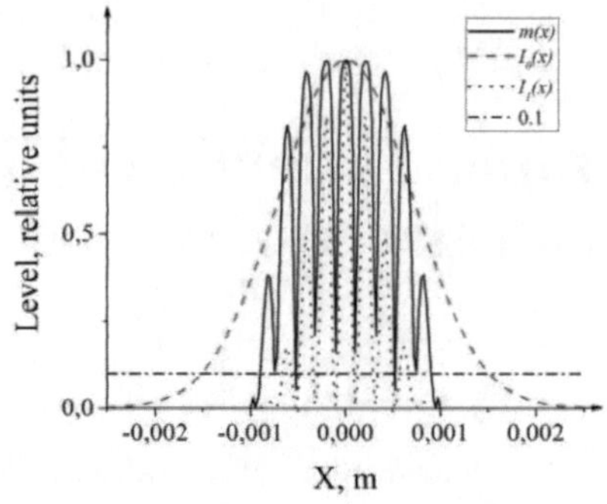

Figure 1. (a) Spatial recording geometry and (b) local contrast of interference pattern.

Under the influence of optical radiation the process of DOE holographic formation takes place in a PPM with a dye-sensitizer. The process is described in general form by kinetic equations of monomer concentration M and refractive index n [13-15]:

$$\frac{\partial M}{\partial t} = div\left[D(M)\,grad(M)\right] - h\left[I(t,\boldsymbol{r})\right]^k M, \tag{2}$$

$$\frac{\partial n}{\partial t} = \delta n_p h\left[I(t,\boldsymbol{r})\right]^k (M / M_n) + \delta n_i div\left[D(M)\,grad(M / M_n)\right], \tag{3}$$

where $M = M(t,x,y)$, $n = n(t,x,y)$, $I(t,\boldsymbol{r})$ – intensity distribution of the recording field inside the PPM, δn_p and δn_i – model parameters characterizing the n change due to polymerization and diffusion of material components respectively; M_n – initial concentration of the monomer; h – coefficient depending on the material parameters; k – degree of non-linearity of the photopolymerization process.

Under the action of diffraction of the recording light beams during recording process inside the sample with a PPM, the intensity distribution of the recording field $I(t,\boldsymbol{r})$ changes. The formation of the DOE at each time point continues in accordance with the changed distribution of the light field, and in areas of low contrast, under the action of self-diffraction, an additional structure is formed. It should be noted that, the effect of self-diffraction is most observed in areas of low contrast (<0.1, figure 1b).

The solution of the system of coupled waves equations, describing the interaction of light waves with the DOE being formed, is obtained in the approximation of a given field [16-17]. The change of the light wave E_1 in interaction area is written as:

$$\delta E_1(t,x,y) = E_1(t,x,y) - E_1(x) = -iGE_0(x)\int_0^y n_1(t,x,y^{'})dy^{'}, \tag{4}$$

where n_1 – first harmonic of the refractive index; $G = \pi / \left[\lambda \cos(\theta_0)\right]$, where λ, θ_0 – wavelength of light and the recording angle of the wave E_0 in the material.

The intensity distribution of the light field for expressions (2) and (3) can be written as [15]:

$$I(t,\mathrm{r}) = I^0(x)\left[1 + m(x)\cos(K_1\mathrm{r}) + \frac{E_0(x)\delta E_1^*(t,x,y)}{I^0(x)}e^{-iK_1\mathrm{r}} + \text{c.c.}\right], \tag{5}$$

where $I^0(x)=\left[I_0(x)+I_1(x)\right]$, $I_j(x)=\left|E_j(x)\right|^2$, $j=0{,}1$, $m(x)=2\sqrt{I_0(x)I_1(x)}\left(\boldsymbol{e}_1\cdot\boldsymbol{e}_2\right)/\left[I_0(x)+I_1(x)\right]$ – local contrast of interference pattern.

The solution of the system of kinetic equations for the amplitude of the first harmonic of the refractive index is the expression [15]:

$$n_1(\tau,x,y)=n_{1P}(\tau,x,y)+n_{1H}(\tau,x,y), \tag{6}$$

where $n_{1P}(\tau,x,y)$ is the contribution to amplitude of the first harmonic due to photopolymerization-diffusion processes, and $n_{1H}(\tau,x,y)$ is the contribution to amplitude of the first harmonic due to self-diffraction of the recording light beams. These expressions are described as [15]:

$$n_{1P}(\tau,x,y)=\delta n_p F_2(x)\sqrt{m_0(x)}\int_0^{\tau}R(\tau',x)d\tau', \tag{7}$$

$$n_{1H}(\tau,x,y)=\delta n_p F_2(x)\sqrt{m_0(x)}\int_0^{\tau}R(\tau',x)H(\tau',\tau,x,y)d\tau'. \tag{8}$$

where $\tau=t/T_m$ – relative time; $T_m=1/\left(K_1^2 D_m\right)$ – diffusion time; $K_1=\left|K_1\right|$ – wave number of the first harmonic; D_m – the initial value of the diffusion coefficient; $F_2(x)=\frac{2^k}{b_x}\cdot\frac{2k}{1+m_0(x)}$; $b_x=b(x)=T_p(x)/T_m$; $T_p(x)=h^{-1}/\left[I^0(x)\right]^2$ – local polymerization time; $m_0(x)=I_1(x)/I_0(x)$ – recording beam intensity ratio; $R(\tau',x)=\frac{M_0(\tau)}{M_n}-\left(\frac{2^k}{b_x}-C_n\right)\cdot\int_0^{\tau}\frac{M_0(\tau')}{M_n}\cdot e^{-\int_{\tau'}^{\tau}F_1(\tau'')d\tau''}d\tau'$; $M_0(\tau)$ – monomer concentration for zero harmonic; M_n – initial monomer concentration; $C_n=\delta n_i/\delta n_p$; $H(\tau',\tau,x,y)=\sqrt{\frac{i\cdot F_2(x)\cdot\Gamma}{y/d}\cdot\int_{\tau'}^{\tau}R\left(\tau''\right)d\tau''}\cdot J_1\left[2\sqrt{i\cdot F_2(x)\cdot\Gamma\cdot\frac{y}{d}\cdot\int_{\tau'}^{\tau}R(\tau'')d\tau''}\right]$; $J_1[x]$ – Bessel function; $F_1(\tau)=2^k/b_x+b_m(\tau,x)$; $\Gamma=\delta n_p\cdot G\cdot d=\omega\cdot d\cdot\delta n_p/2\cdot c\cdot\cos\left(\varphi_1\right)$ – normalized coupling coefficient characterizing the efficiency of the interaction of light waves with a structure.

The resulting expression (6) determines the temporal dynamics of the spatial distribution of DOE amplitude, taking into account self-diffraction effects. As a result of DOE recording by light beams with different amplitude and phase distribution, the amplitude and phase profile of the structure becomes non-uniform. The inhomogeneity of the amplitude-phase profile of the structure leads to the rotation of the effective vector of the grating, and as a result, the shift of the Bragg angle when reading, as well as to the exchange of energy between the beams.

3. Experiments

Figure 2 shows the experimental setup for the holographic formation and reading of photon structures in a sample with PPM [18-20]. A helium-neon (He-Ne) laser with a radiation wavelength of 633 nm forms a reference light beam with a Gaussian light distribution of 2 mm in diameter and a power of 2 mW. After reflection from the mirror (M), the beam is divided into two using a beam-splitting cube (BSC). Further, the signal beam with a Gaussian intensity distribution through an amplitude transparency (AT) is transformed into a one-dimensional Bessel-like [21-24]. AT has a slit width of 200 microns and a distance between slits of 900 microns. The distance from the AT to the lens (L) and from the lens to the sample corresponded to the focal length (18.4 cm). The angle of incidence of the reference and signal beam is 4 degrees. The reference beam after the mirror (M) was broadened using a collimator (C) to an aperture value of 4 mm. In the bulk of the PPM sample, the reference and signal beams interfere. Further, following to the holographic principle, a phase transmission hologram is formed in the sample.

Photopolymer films “GFPM633.5” produced by LLC Polymer Holograms-Novosibirsk with a layer thickness of 45 ± 5 μm on a glass substrate with a thickness of 1 ± 0.1 mm were used as PPM.

IOP Conf. Series: Materials Science and Engineering **597** (2019) 012068 doi:10.1088/1757-899X/597/1/012068

The laser beam analyzer (A) captures the intensity distribution of the transmitted signal and reference beam. To read the obtained hologram, the signal light beam was blocked by a shutter (B). At the output of a formed structure, the analyzer recorded the intensity distribution of the diffracted light field.

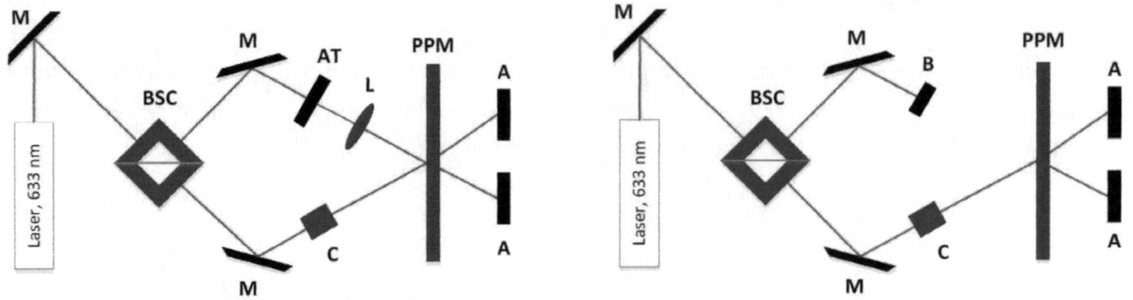

a b

Figure 2. (a) Experimental setup for PS recording and (b) Experimental setup for PS reading.

Self-diffraction effect has been experimentally confirmed by holographic formation of DOE by Bessel-like and Gaussian light beams in the PPM. Figure 3 shows dependences of the signal and diffracted light beams' intensity distributions and their normalized profile.

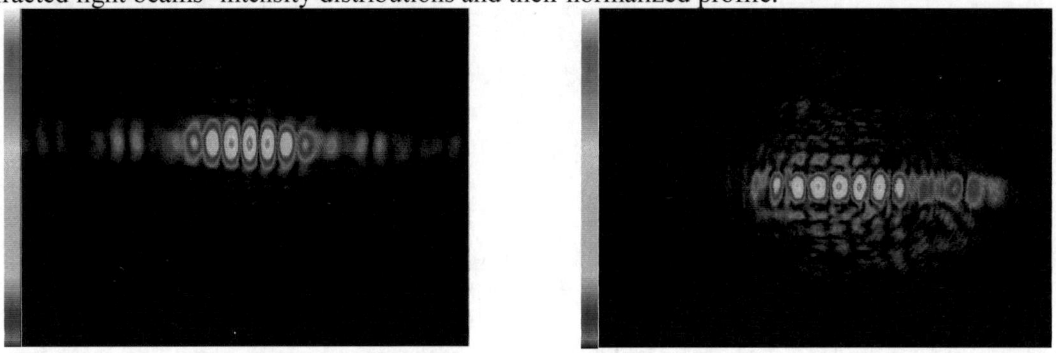

a b

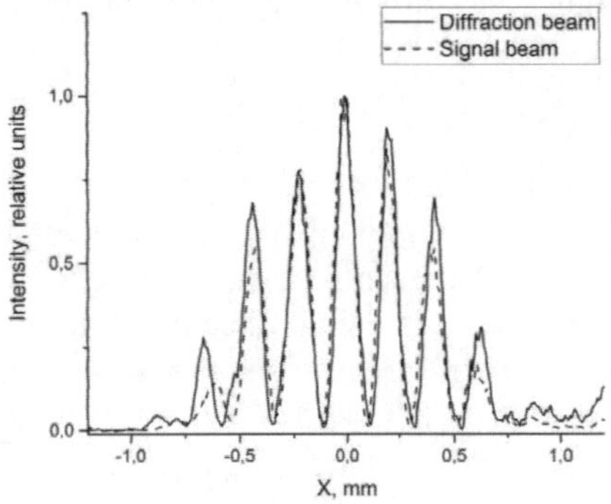

c

Figure 3. (a) Intensity distribution of signal light beam; (b) Intensity distribution of diffracted light beam; (c) normalized signal (E_l) and diffracted (E_{dl}) light beam profiles.

It can be seen from figure 3c that side maxima of diffracted light beam are enhanced in level. Moreover, these changes are most pronounced in areas of low contrast. This testifies to the influence of the self-diffraction effect on DOE holographic formation in PPM.

IOP Conf. Series: Materials Science and Engineering **597** (2019) 012068 doi:10.1088/1757-899X/597/1/012068

4. Conclusion

In this paper, we developed a theoretical model of the formation of a DOE for the conversion of light fields into one-dimensional Bessel-like, taking into account two-beam interaction. It has been shown by experiment that, due to the effect of self-diffraction in areas of low contrast (corresponding to the regions of the side maxima of the diffracted light beam), the amplitude of the diffracted field is higher relative to the recording one.

Thus, to determine the spatial distribution of PPM refractive index in the process of DOE holographic formation, it is necessary to take into account the two-beam interaction at low contrasts of recording light beams, which will make it possible to more accurately determine the diffraction characteristics of the formed optical elements.

Acknowledgement

The work was supported by the Ministry of Education and Science of the Russian Federation in the framework of the project part of the Government Task (Project No. 3.1110.2017 / 4.6).

References

[1] Siviloglou G A and Christodoulides D N 2007 *Optics letters* vol 32 p 979
[2] Khilo N A, Petrova E S and Ryzhevich A A 2001 *Quantum electronics* vol 31 p 85
[3] Dudutis J et al. 2018 *Optics Express* vol 26 No 3 pp 3627-2637
[4] Xhu X, Schulzgen A, Li L, Peyghambarian N 2009 Generation of controllable nondiffracting beams using multimode optical fibers *Applied Physics Letters* 201102
[5] Durnin J, Miceli J J, Eberly J H 2011 *Optics Express* vol 19 No 12 p 10
[6] Brunel M and Coetmellec S 2007 *Journal of the Optical Society of America A* vol 24 No 12 pp 3753-3761
[7] Milne G et al. 2008 Tunable generation of Bessel beams with a fluidic axicon *Applied Physics Letters* vol 92 261101
[8] Groshenko N A, Makalish O S, Volyar A V 1998 *Technical Physics* vol 43 No 12 p 1450
[9] Dallaire M, McCarthy N, Piche M 2009 *Optics Express* vol 17 No 20 pp 18148-18164
[10] Cheng H et al. 2019 *Applied optics* vol 58 No 13 pp 91-97
[11] Fernandez R F et al. 2018 Polymerizable materials for diffractive optical elements recording *Recent Research in Polymerization* ed Nevin Çankaya (IntechOpen) chapter 8 pp 161-185
[12] Barachevskiy V A 2017 *Optics and Spectroscopy* vol 123 No 3 pp 371-399
[13] Kovalenko E, Sharangovich S, Zelenskaya T 1996 *Synthetic metals* vol 83 p 293
[14] Sharangovich S N, Dovolnov E A 2004 *Organic Optoelectronics and Photonics* vol 5464 p 399
[15] Dovolnov E A and Sharangovich S N 2006 *Russian Physics Journal* vol 49 p 1189
[16] Ustyuzhanin S V, Sharangovich S N 2011 *Russian Physics Journal* vol 54 p 172
[17] Semkin A O, Sharangovich S N 2013 *International Conference on Advanced Optoelectronics and Lasers* p 48
[18] Kovalenko E S et al. 2001 *Optoelectronic Information Systems and Processing. – International Society for Optics and Photonics* vol 4513 p 75
[19] Dovolnov E A et al. 2007 *Russian Physics Journal* vol 50 p 342
[20] Dovolnov E A and Sharangovich S N 2008 *Optics and Spectroscopy* vol 105 p 310
[21] Inyushov A et al. 2016 *International Conference on Coherent and Nonlinear Optics* pp 159-160
[22] Safronova P K et al. 2017 *Photonics and information optics* pp 302-303
[23] Arlt J et al. 2001 *Opt. Commun* vol 197 pp 239-245
[24] McGloin D and Dholakai K 2004 *Contemporary Physics* vol 46 No 1 pp 15-28

Forevacuum plasma electron source of a ribbon electron beam with a multi-aperture extraction system

A S Klimov, A A Zenin, Chan Van Tu and I Yu Bakeev

Physics Department, Tomsk state university of control systems and radioelectronics 634050, Tomsk, Russia

E-mail: klimov@main.tusur.ru

Abstract. The article presents the results of using a multi-aperture extraction system in the source of a ribbon electron beam. The optimal size of the holes in the anode and extractor, as well as the distance between the anode and the extractor at which the beam losses are minimal, are investigated.

1. Introduction

Plasma technologies are used for etching and surface cleaning of semiconductors, dielectrics and metals [1-3], and are used in medicine for disinfection and sterilization of instruments [4]. Plasma can be used for the synthesis of biocompatible coatings [5]. As a rule, gas discharges of different frequency ranges are used to create plasma. In addition, it is possible to use an electron beam to create a plasma. When transported in a vacuum chamber, the electron beam ionizes the gas and thus creates a beam plasma. This method of plasma creation has a number of advantages over gas-discharge devices. The size of the plasma directly depends on the cross-section of the electron beam. Plasma parameters can be controlled by changing the parameters of the electron beam and the composition of the gas atmosphere. The formation of a plasma with a ribbon electron beam makes it possible to create a beam plasma with an area of several tens of square cm. [6]. In addition, the electron beam can be injected into almost any gas, as well as gas and vapor-gas mixtures. In this case, nonequilibrium plasma chemical reactions can occur in the plasma volume, including reactions not observed under other conditions. Choosing the type of gas, its pressure, and by changing the conditions of generation of beam-plasma it is possible to achieve a wide range of plasma chemical reactions. The most optimal pressure range for plasma chemical reactions is from 1 to 100 Pa. To create a beam plasma in this range, it is preferable to use the so-called forevacuum electron beam sources, and to form a large-volume plasma, use a ribbon electron beam source. The source of the ribbon electron beam forms an electron beam in the form of a ribbon with an aspect ratio in the cross section of more than 10:1. As a rule, in plasma sources, stabilization of the plasma boundary is carried out using a metal grid. In the case of cylindrical beams of small cross-section, this method works well, because due to the small cross-section of the beam, its homogeneity can be increased. In the case of an extended plasma boundary, as well as in the conditions of heating by a reverse ion flow, the grid can be bent, and the condition of uniformity cannot be fulfilled.

In the so-called multi-aperture extraction systems [7], which are used to generate beams of large cross-section, the electron beam is formed by adding separate beams. Such beams are formed in elementary acceleration cells, which are two or more electrodes with coaxial holes. In such systems, the beam losses on the electrodes are minimal and usually do not exceed a few percent. The coating of the emission (plasma) electrode of a multi-aperture system with a fine-structure grid stabilizes the position of the emission plasma boundary and provides a wider range of parameter changes both in the formation of ion [8] and electron [9] beams. In the electron source with arc discharge, the multi-aperture system was also used to stabilize the plasma boundary and increase the electrical strength of accelerating gap of the source [10]. In addition, the paper shows that the presence of a mask on the emitter electrode is a condition for stable operation of the electron source. With no mask, the average power of the beam in the acceleration gap was no greater than 3500 W, whereas with a mask placed in the emitter, it could be increased to 6500 W. The use of a multi-aperture system in sources of high-energy electron beams has no alternative and allows to obtain high-power beams of subrelativistic electrons with a cross-sectional area ~ 10^2 cm [11]. Such beams are used in a number of scientific and technical applications, particularly, in a material science and engineering, for surface modification of solids, modeling of high thermal fluxes on plasma facing components in fusion reactors, etc. [12]. In addition, the thermal load on the grid is reduced, since it is sandwiched between flat electrodes, the heating of which is much less.

The aim of this work was to study the influence of the parameters of a multi-aperture system on the formation of a homogeneous band electron beam in the forevacuum pressure region.

2. Experimental setup

The experiments were carried out using a forevacuum plasma source of a ribbon electron beam [13]. The scheme of the electron source, as well as its location and measuring instruments are shown in figure 1. The electron source consists of three main electrodes. The flat anode and the extended hollow cathode are the discharge plasma generator. The acceleration of electrons is carried out in the space between the anode and the extractor. The use of a hollow cathode makes it possible to provide conditions for the generation of a high-density discharge plasma, and the small distance between the anode and the extractor (2-10 mm) provides the electrical strength of the accelerating gap.

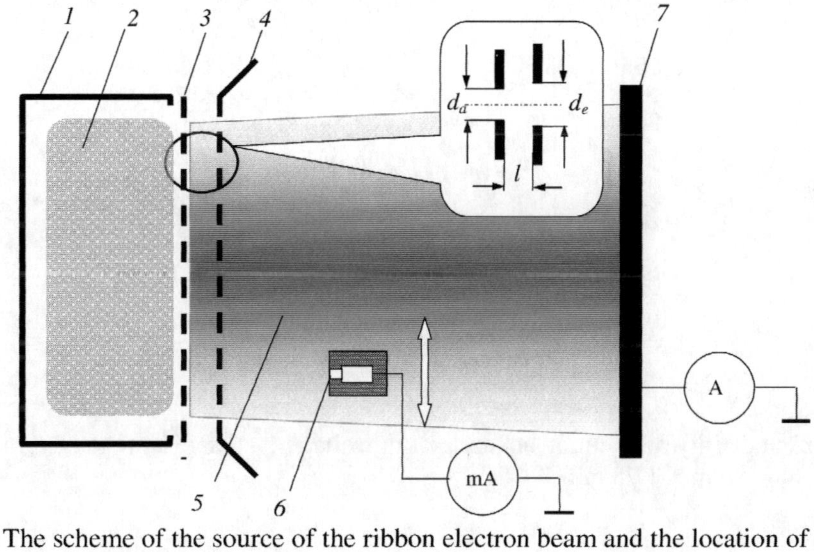

Figure 1. The scheme of the source of the ribbon electron beam and the location of the measuring probe in the chamber: 1 – Extended hollow cathode, 2 – glow discharge plasma, 3 – anode, 4 – extractor, 5 – electron beam, 6 – probe, 7 – beam collector.

The emission window in the anode was covered by a metal plate 1 mm thick containing 50 holes with a diameter d_a=3.5 mm located in the middle part of the plate along its long side. The distance between the centers of the holes was 6 mm. In the extractor it was possible to install a removable insert with holes arranged coaxially with each hole of the anode plate. The diameter of the holes in the d_e extractor plate was 3.5 mm, 4.5 mm, 5.5 mm. The distance between the plate in the anode and the extractor *l* varied from 2.5 mm to 10 mm. The work was carried out in an air atmosphere at a pressure of 7 Pa. The accelerating voltage in the experiments varied from 2 to 8 kV, the discharge current from 200 to 800 mA.

The current density distribution was investigated using a movable collector. The collector had a narrow slit width of 1 mm and allowed to measure the transverse distribution of the beam current density.

3. Experimental results and discussion

The current density distributions measured at a distance of 10 cm from the plane of the electron beam entry into the vacuum chamber are shown in figures 2-4. Since the emission is carried out from a set of elementary holes, the current distribution contains alternating areas with high current density. This form of distribution is obviously associated with incomplete overlap of current flows extracted from the elementary holes.

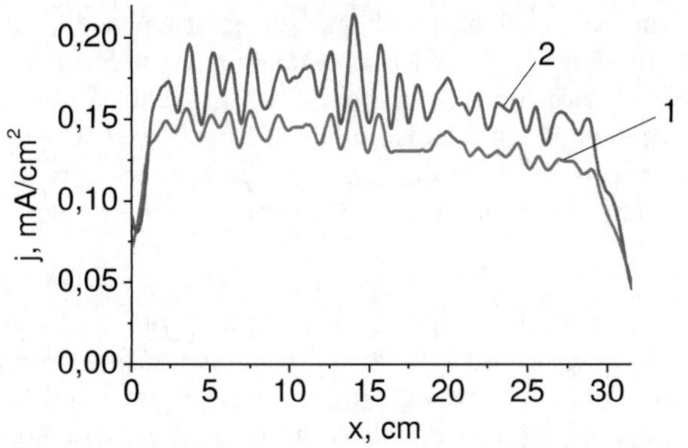

Figure 2 - Current density distribution at accelerating voltage 1 - 2 kV, 2 - 8 kV. Beam current 100 mA, d_a – 3.5 mm, d_e – 3.5 mm, l - 5 mm.

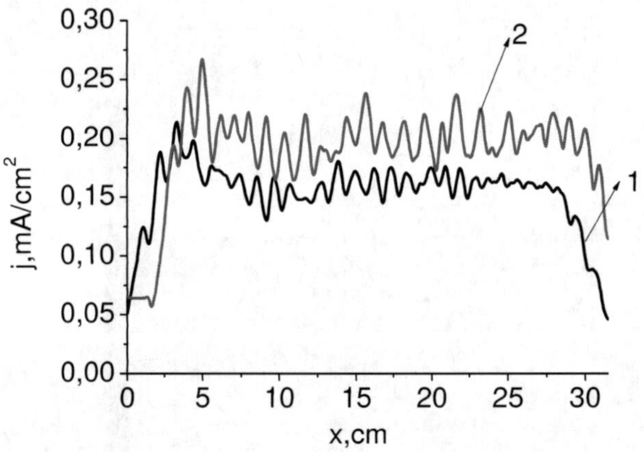

Figure 3 - Current density distribution at accelerating voltage 1 - 2 kV, 2 - 8 kV. Beam current 100 mA, d_a – 3.5 mm, d_e – 4.5 mm, l - 5 mm.

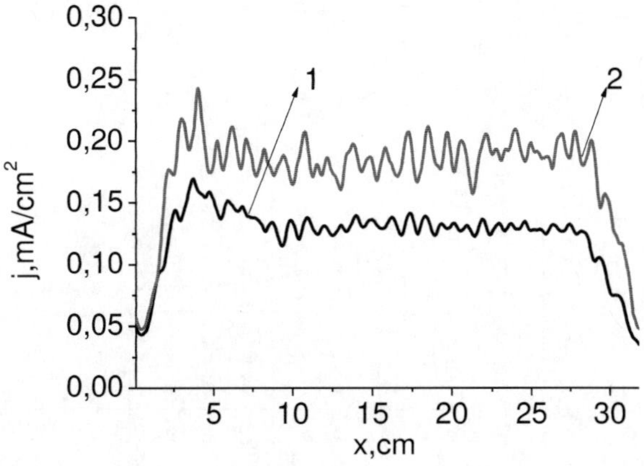

Figure 4 - Current density distribution at accelerating voltage 1 - 2 kV, 2 - 8 kV. Beam current 100 mA, d_a – 3.5 mm, d_e – 5.5 mm, l - 5 mm.

Indeed, at a distance of 30 cm from the plane of injection of the electron beam into the vacuum chamber, the distribution is more uniform, figure 5.

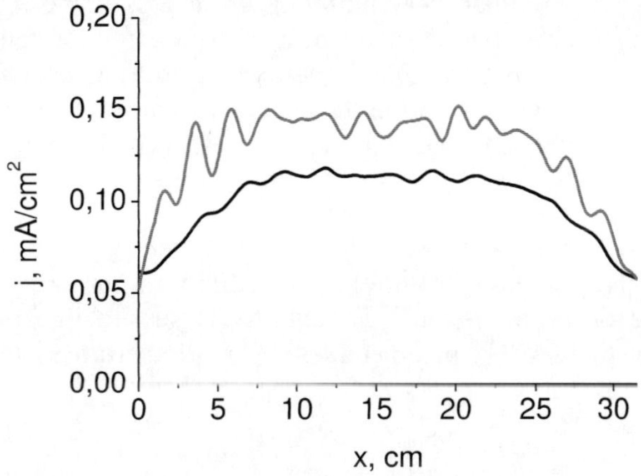

Figure 5 - Current density distribution at accelerating voltage 1 - 2 kV, 2 - 8 kV. Beam current 100 mA, da – 3.5 mm, de – 3.5 mm, l - 5 mm. The distance from the electronic source is 30 cm.

It should be noted that with the increase in the diameter of the hole, the change in the collector current depends on the accelerating voltage. When the accelerating voltage is low - 2 kV, the increase in the diameter of the hole in the extractor has little effect on the value of the extracted current. With an increase in the accelerating voltage of more than 8 kV, the collector current increases with the diameter of the hole in the extractor. Such an increase in the collector current with an increase in the accelerating voltage can be explained by a change in the configuration of the electric field in the accelerating gap. As the accelerating voltage increases, the divergence of elementary electron beams emitted from individual emission holes decreases. This is due to the reduction of the effect of the initial spread of the velocities of the electrons emitted from the plasma, as well as by reducing the scattering on the molecules of the residual atmosphere and, consequently, reducing the loss on the extractor.

Changing the distance between the plane of the anode and the extractor also affects the value of the beam current. So if you reduce the distance anode-extractor current losses in the extractor is reduced. This is due to the fact that at a smaller distance from the emission boundary, the beam divergence degree is less, and, consequently, the losses on the extractor will also be less. The current measurement of the extractor confirms it (figure 6).

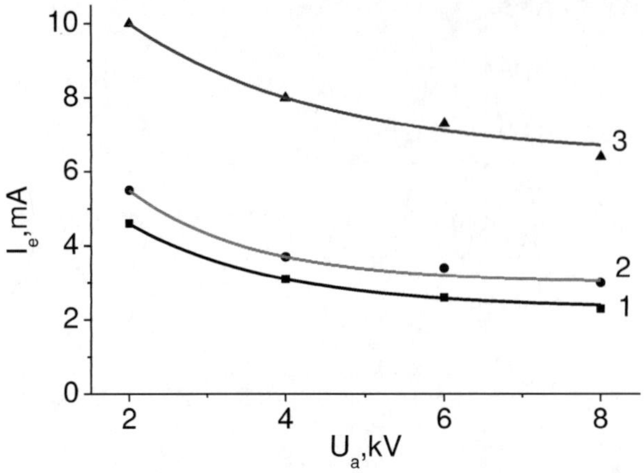

Figure 6 - Dependence of the extractor current on the accelerating voltage for different anode-extractor distances l: 1 – 2.5 mm, 2 - 5 mm, 3 - 10 mm.

Thus, the experiments show that when using a multi-aperture system with a diameter of elementary holes in the anode 3 mm and 5.5 in the extractor and at a distance between the anode and the extractor 2.5 mm losses on the extractor are minimal. The minimum current loss on the extractor indicates a high electron beam current flow and the efficiency of using a multi-aperture system to generate a ribbon beam. The result shows the difference between multi-aperture electron-optical acceleration systems and grid systems. In extraction systems using a grid in the anode, the current-carrying coefficient does not exceed 2/3, which leads to a loss of at least one third of all accelerated electrons on it.

4. Conclusion

Multi-aperture system used to extract electrons in the source of the electron beam, reduces the loss of current on the extractor, and thus increase the current flow. Minimum current losses are observed at an accelerating voltage of 8 kV, for low accelerating voltages the losses are higher due to the influence of the initial spread of electron velocities. The parameters of the multi-aperture system with minimal losses on the extractor are determined.

Acknowledgments

The work was supported by Russian Foundation for Basic Research, grant 18-48-700015

References

[1] Walton S, Leonhardt D, Murphy D, Meger R, and Fernsler R 1999 *Bull. Am. Phys. Soc.* **44** 58
[2] Leonhardt D, Muratore C and Walton S G 2005 *IEEE Trans. Plasma Sci.* **33** 783
[3] Alexeff I, Neidigh R V *et al.* 1963 *Phys. Rev. Lett.* **10(7)** 273
[4] Vasilyeva T M and Sokolov I V 2015 *Goren. Plazmokhim* **13(2)** 134
[5] Zolotukhin D B, Klimov A S and Zenin A A 2013 *Proc. TUSUR Univ* **30**(4) 79
[6] Burdovitsin V A, Oks E M and Fedorov M V 2005 *Russ. Phys. J.* **47**(3) 310
[7] Batalin V A *et al.* 2002 *Review of Scientific Instruments* **73**(2) 702
[8] Oks E M 2002 *IEEE Transactions on Plasma Science* **30**(1) 202
[9] Burdovitsin V A, Oks E M 2008 *Laser and Particle Beams* **26**(4) 619
[10] Vorobyov M, Koval N, Sulakshin S and Shugurov V 2015 *Journal of Physics: Conference Series* **652** 012048.
[11] Kandaurov I, Astrelin V, Avrorov A, Burdakov A, Bykov P, Derevyankin G, et al. 2011 *Fusion Science and Technology* **59** 1 67
[12] Kandaurov I., Kurkuchekov V and Trunev Yu 2017 *Journal of Physics: Conference Series* **830** 012032
[13] Klimov A S et. al. 2016 *Plasma Phys. Rep.* **42**(1) 96

IOP Conf. Series: Materials Science and Engineering **597** (2019) 012070 doi:10.1088/1757-899X/597/1/012070

Influence of electron-beam processing mode on the sintering of alumina ceramics

A S Klimov, I Yu Bakeev and A A Zenin

Physics Department, Tomsk state university of control systems and radioelectronics 634050, Tomsk, Russia

E-mail: klimov@main.tusur.ru

Abstract. The article presents the results of electron beam sintering of alumina ceramic in the so-called forevacuum pressure range (5-20 Pa). Ceramic was irradiated by a focused electron beam generated by the forevacuum plasma electron source. Various modes of irradiation are considered without and with scanning of the electron on the ceramic surface. It is shown that using electron beam scanning increases the homogeneity of the sintered material. The modes of irradiation allowing to obtain dense ceramics with a homogeneous density distribution over its volume are given.

1. Introduction

Ceramic based on aluminium oxide is widely used in power-stressed operation due to its high hardness, heat resistance, chemical inertness, as well as availability. The technology of creation of ceramic materials consists of the main operations: production of powders with the necessary composition, preliminary compaction (pressing) of blanks and sintering. Traditionally, micron powders of technical purity, mechanical compaction and thermal sintering are used. Obtaining ceramics with improved properties is possible by using nanoscale powders with chemical state control and using such technologies as sintering in HF and microwave fields, the Spark Plasma Sintering (SPS) method [1, 2], laser sintering [3], electron beam sintering [4-5] and using special methods of pre-compaction [6, 7]. As shown by the previous experiments [8], the usage of an electron beam allows produce quite successful sintering of aluminum oxide ceramics. However, the issue of influence of electron beam processing mode on sintered materials properties was not given enough attention. This article presents a study of the effect of electron beam scanning modes of the sintered surface on the properties of the sintered sample.

2. Experimental setup

For experiments on electron-beam sintering and optimization of electron-beam irradiation regimes a forevacuum plasma electron source was used [9]. Forevacuum source of electrons is capable process dielectric materials (in particular ceramics) without additional methods to reduce the charge on the irradiated surface [10]. The scheme of the experimental setup is shown in figure 1.

IOP Conf. Series: Materials Science and Engineering **597** (2019) 012070 doi:10.1088/1757-899X/597/1/012070

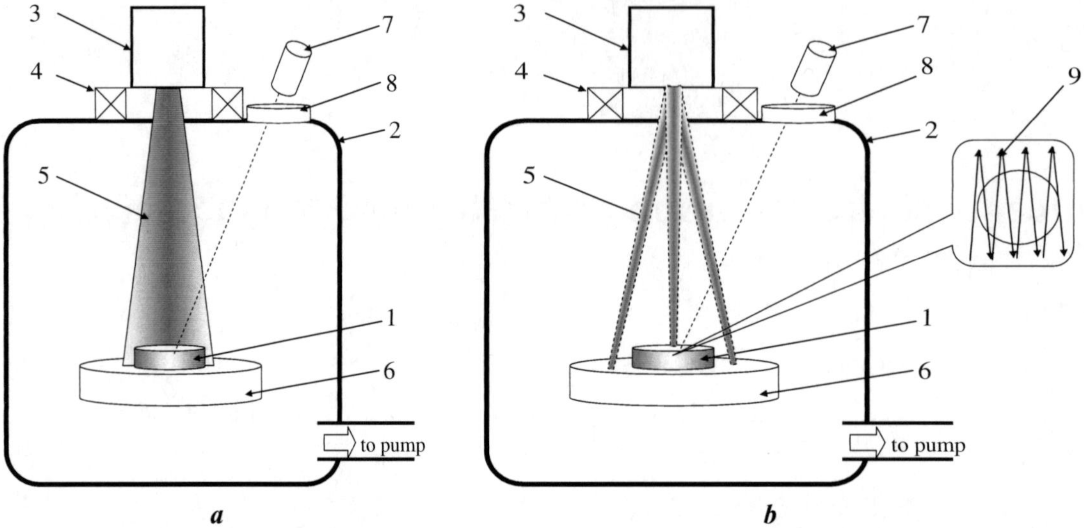

Figure 1. Scheme of electron-beam sintering of alumina ceramics in two regimes: a - without scanning, b - using scanning of electron beam. 1 – sintered sample, 2 – vacuum chamber, 3 – plasma electron source, 4 – focusing and deflection system, 5 – electron beam, 6 – collector, 7 – pyrometer, 8 - quartz glass, 9 – beam trajectory during scanning.

The sample of aluminum oxide ceramics 1 was a disk with the diameter of 14 mm and the thickness of 3 mm. The disk was pressed from an aluminum oxide nanopowder with the average grain size of 30 microns. The sample was installed in the holder placed in a vacuum chamber 2, which was pumped out to the pressure of 3 Pa. Then helium was injected into the chamber to the pressure of 30 Pa. The plasma electron source 3 was installed on the upper flange of the vacuum chamber and was equipped with a system of electromagnetic focusing and deflection 4 of the electron beam 5. The sample was heated for 25 minutes with increasing beam power. The accelerating voltage in the experiments increased from 5 to 12 kV. The deflection system control the position of electron beam on the sample surface. Also the deflection system allows to the electron beam scan the sample surface along the raster with the specified geometric dimensions (from 5*5 to 25*25 mm) at the frequency of 1 to 1000 Hz. The temperature control of the sintered sample was carried out by a non-contact method using an infrared pyrometer Marathon MM (Raytec, USA) 7 with a measurement range of 600-3000 °C. Thermal radiation from the sample was captured by the pyrometer located behind the quartz glass 8 with a bandwidth corresponding to the working band of the pyrometer wavelengths. Sintering of samples was carried out at the constant temperature of 1350 °C and the holding time of 15 minutes according to two different modes. Mode 1 – the sintering was carried out without the electron beam scanning by the electron beam with the diameter of 15 mm covered the all surface of the ceramic sample (Fig. 1 a). Mode 2 – the sintering was produced by the electron beam with the diameter of 0.5 mm with the included scanning system with a frame drawing frequency of 100 Hz. The scanning area 9 was 15*15 mm and was larger than the transverse dimensions of the sintered sample.

To study the microstructure of sintered samples, the JSM-7500FA scanning electron microscope (JEOL, Japan) was used. Mass measurement was carried out by electronic scales with the measurement accuracy of 0.01 gm. The sample density was determined from the measured value of the mass and the geometric dimensions of the sample measured by a micrometer with the accuracy of 0.1 μm.

3. Experimental results and discussion

During the electron beam sintering the volume of the samples decreases. The change in the sample dimensions is presented in table 1. The density increases by amount of 16-17% of the original value in the non-sintered state. The results presented in table 1 indicate the compaction of the ceramic material after electron beam sintering, that is an indirect confirmation of the sintering process.

Table 1. Geometrical parameters of samples.

	Mode 1 (no raster)		Mode 2 (100 Hz scan rate)	
	before	after	before	after
Weight, gm.	1.14	1.09	1.19	1.14
Thickness, mm	2,8	2.7	3	2,8
Diameter, mm	14.20	13.10	14.15	13.25
Density, g/cm3	2.57	2.99	2.54	2.97
Density, %	1	1.16	1	1.17

Despite the almost identical external parameters of ceramic samples, the internal structure of the samples after sintering is significantly different for various modes. The structure of the sintered samples in mode 1 is heterogeneous in the thickness of the sample. In the central part of the sample surface we observed the formation of large grains (figure 2), which may indicate overheating in this part.

Figure 2. Microphotography of the central part of the sintered sample in mode 1

Near the irradiated surface of the samples sintered in mode 1 large grains are not detected (figure 3), as they are not detected on the side not exposed by the electron beam (figure 4).

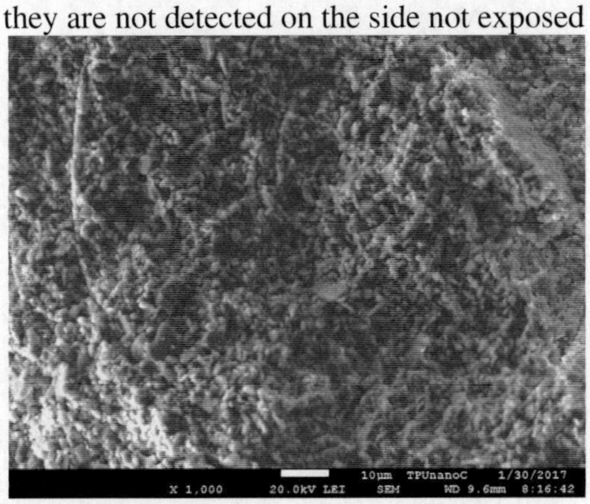

Figure 3. Microphotography of the irradiated part of the sintered sample in mode 1

Figure 4. Microphotography of the non-irradiated part of the sintered sample in mode 1

These facts indicate an inhomogeneous heating of the sample volume, which can lead to such significant differences in its internal structure. Microphotographs of the internal structure of the samples sintered in mode 2, i.e. which are made with scanning of the surface by the narrowly focused electron beam, are presented in figures 5-7.

Figure 5. Microphotography of the central part of the sintered samples in mode 2

Figure 6. Microphotography of the irradiated part of the sintered samples in mode 2

Figure 7. Microphotography of the non-irradiated part of the sintered samples in mode 2

In the samples sintered in mode 2 there is no formation of large grains and there is an initial stage of sintering, figure 5. The grain size distribution is more homogeneous, see figure 5 and 6, which seems to be related to the irradiation conditions and heating. In the case of irradiation by the electron beam with a diameter comparable to the diameter of the sintered sample, due to the Gaussian distribution of the current density across the beam section [11], the sample is heated unevenly with a maximum in its central part. Despite the existence of time intervals between beam hits one point of the sample surface, scanning on the surface by the narrow beam still contributes to a more homogeneous heating and, consequently, sintering.

4. Conclusion

Obtaining a homogeneous sintered sample of aluminum oxide ceramics is possible when a focused electron beam is used as a heating source and its scanning of the ceramic surface by this beam is ensured. It is shown that at a scanning frequency of 100 Hz it is possible to obtain a homogeneous sintered sample while maintaining a small grain size.

Acknowledgments

The work was funded by the grant of the President for doctors of Sciences MD-2649.2019.8

References

[1] Chaim P. R. [et al.] 2008 *Advances in Applied Ceramics* **107**(3) 159
[2] Yang D, Raj R and Conrad H 2010 *J. Am. Ceram. Soc.* **93**(10) 2935
[3] Tang Y et al. 2003 *Materials and Design* **24**(8) 623
[4] Ivanov Yu F et al. 2018 *Inorganic Materials: Applied Research* **9**(3) 437
[5] Gyngazov S A et al 2011 *Izvestiya vuzov. Fizika.* 54 1/3 355 [in Russian]
[6] Khasanov O et al. 2011 *IOP Conference Series: Materials Science and Engineering* **18** 082004
[7] Vogler T J et al. 2007 *International Journal of Solids and Structures* **44** 636
[8] Burdovitsin V, Zenin A, Klimov A et al. 2014 *Advanced Materials Research* **872** 150
[9] Bakeev Yu I, Klimov, A S, Oks E M and Zenin A A 2018 *Plasma Sources Science and Technology* **27** (7) 075002
[10] Burdovitsin V A, Klimov A S, Medovnik A V and Oks E M 2010 *Plasma Sources Science and Technology* **19**(5) 055003
[11] Kaur A et al. 2015 *Journal of Materials Processing Technology* **221** 225

Modeling threats to information security using IDEF0 methodology

A A Khristolyubova[1], A A Konev[2], A A Shelupanov[3] and M L Solovev[4]

[1]Joint stock company Central design bureau of automatics, Russia, Omsk, Kosmichesky pr., 24a, 644027, Russian Federation
[2] Department of Complex Information Security of Computer Systems, Tomsk State University of Control Systems and Radioelectronics,40 Lenin pr., Tomsk, 634050, Russian Federation
[3]University administration, Tomsk State University of Control Systems and Radioelectronics, 40 Lenin pr., Tomsk, 634050, Russian Federation
[4]University administration, Siberian state university of telecommunications and information sciences, 86 Kirova str., Novosibirsk, 630102, Russian Federation

E-mail: karjanna@mail.ru

Abstract. The paper describes the implementation of a method for modeling threats to information security using the methodology of IDEF0 functional modeling to solve problems of formalization of specific threat models. The formulation of the method for modeling threats to information security accounts for various media of information transmission and its carriers. In addition, a process approach to information handling is used.

1. Introduction

The analysis has shown [1–4] that the assessment of information system security, determination of a sufficient set of protection means and solution of subtasks, such as construction of the system model and the model of threats, predominantly involve methods that avoid the effects of real scenarios of an information system application. They rest upon probabilistic assessment results and reflect a subjective opinion of experts. The final result of the applied methods and approaches is a list of threats and their classification that are no way connected with the category of protected information, while accounting its state [5].

One of demanded directions of information protection system design is a justified comparison of real threats to information security with the means and mechanisms of information protection that neutralize the threats relying upon the description of real information flows described as generic elements [6]. Initial modeling of threats to information security, regardless of the chosen modeling method, is based on the determination of possible consequences caused by a destructive impact on the protected object [7].

By present, there are many methodologies for modeling information flows [8]. Among the most elaborated ones, that suit modeling of protected information flows is IDEF0. Utilization of a list of the information processes that handle information to be protected as the basis for the constructed formalized model of threats allows accounting for the information representation, transmission media and carriers of information to be protected [9]. There is an example of SADT methodology implementation to analyze the actions of intruders during an attack aimed at the collection of information system characteristics [10].

2. Materials and research methods

The method of decomposition of an information process and modeling of information flows is an element of the method for modeling the threats to the information security that is efficiently used at the stage of information system design to abate the threats caused by system architecture flaws [11]. The set of information transmission and processing media are given in Table 1.

Table 1. Description of the array of information medium and information processing environment.

	Set «Information transmission media» $S=\{s_1, s_2...s_5\}$		Set «Information carriers» $V=\{v_1, v_2...v_8\}$
s_1	Tangible	v_1	Tangible object (protected object).
		v_2	Human (personnel) and their biometric parameters.
s_2	Optical (visual)	v_3	Electromagnetic field (photons).
s_3	Acoustic	v_4	Acoustic waves.
s_4	Electromagnetic	v_5	Near electric field.
		v_6	Near magnetic field.
		v_7	Electromagnetic field.
s_5	Virtual	v_8	Various types of semiconductor-based memory.

Some studies on the methods for modeling threats to information security include—besides usual threats to confidentiality, integrity, availability—the threats of destruction, damage to, theft or loss, disclosure, waiver to use, elevation of privilege and illegal use of information [12]. In fact, the additional threat types characterize the way an intruder reaches their goals and do not reflect the goals of the protected information owner. Considering the approach to the construction of the model of threats to information security given in [9, 13], the threats to confidentiality, integrity and availability are divided into several classes given in Table 2.

Table 2. Classes of information security threats of confidentiality, integrity and availability.

Kind of information security threats	Classes of information security threats
Threats to confidentiality	Impersonation of recipient in medium state 1 Impersonation of recipient in medium state 2 Substitution of information transmission channel Control of information transmission channel
Threats to integrity	Impersonation of sender in medium state 1 Impersonation of sender in medium state 2 Substitution of information transmission channel Exposure on information transmission channel
Threats to availability	Unavailability of carrier in medium state 1 Unavailability of carrier in medium state 2 Disturbance of information transmission channel Inability of transformation

The analysis of information processes in various problem domains enables formulation of a finite number of desirable information protection results that are the main goal of protection system design. The multitude of carriers of protected information and medium of its processing can be correlated by identified goals of information protection (Table 3).

IOP Conf. Series: Materials Science and Engineering **597** (2019) 012071 doi:10.1088/1757-899X/597/1/012071

Table 3. Array of protected carriers and its transmission medium.

Protection goal	Transmission medium	Carriers
Hide semantic content of information	s_1, s_2,s_3, s_4, s_5	v_1,v_2,v_3,v_4,v_5,v_6,v_7,v_8
Hide linear dimensions of the protected object	s_1, s_2, s_3	v_1,v_2,v_3,v_4
Hide specific configuration of the protection object	s_1, s_2,s_4, s_5	v_1,v_2,v_3,v_7

According to the principles of model construction on the basis of IDEF0 methodology, the sources of threats to information integrity can be the resources providing the information flow (mechanism of information processing) [1] or the conditions of certain operation of the process (documents, regulating rules, strategies of processing [13, 14]) at any stages of the information system life. Eventually, for each typical information handling process, a graph of data transformation in various media can be constructed that will contain typical operations of data transformation. Table 4 presents possible typical operations of data transformation in terms of each protection goal and their examples.

Table 4. Typical data conversion operations for each protection target and their examples.

Protection goal	Typical transformation operations	Examples of typical operations
Hide semantic content of information	V1→S1→V2	Transmit paper document
	V2→S1→V1	Return paper document
	V2→S2→V3	Demonstrate content of paper document
	V3→S2→V2	Examine content of paper document
	V2→S3→V4	Voice information about content of paper document
	V4→S3→V2	Listen to information about content of paper document
	V2→S5→V8	Create file (text)
	V8→S3→V4→S3→V2	Listen to audio file
	V4→S3→V8	Create file (audio)
	V8→S2→V3→S2→V2	Watch video file
	V3→S2→V8	Create file (video)
	V8→S4→V5,6,7	Save files
	V5,6,7→S2→V3	Display information on monitor
	V5,6,7→S2→V3→S2→V2	Watch information on monitor
	V5,6,7→S1→V1	Print file (create paper document)
	V4→S3→V8→S4→ V5,6,7→S1→V1	Record audio information on carrier
	V3→S2→V8→S4→ V5,6,7→S1→V1	Record video information on carrier
	V1→S1→V2→S5→V8	Scan paper document
	V1→ S1→V1	Store paper document (carrier)
	V1→ S2→V3	Provide access to examine content of document to unlimited scope of persons
	V1→S5→V8→S5→V8→ V5,6,7	Provide information carrier to unlimited scope of persons
	V8→S5→V8	Process information

Protection goal	Typical transformation operations	Examples of typical operations
Hide linear dimensions of the protected object	V1→S1→V2	Give tangible object to person
	V2→S1→V1	Return tangible object
	V2→S2→V3	Demonstrate external view of tangible object
	V3→S2→V2	Examine external view of tangible object
	V2→S3→V4	Voice information about linear dimensions of object
	V4→S3→V2	Listen to information about linear dimensions of object
	V2→S5→V8	Create virtual model of tangible object (scan, photograph, draw, 3D-model, etc.)
Hide specific configuration of the protection object.	V1→S1→V2	Give tangible object to person
	V2→S1→V1	Return tangible object
	V2→S3→V4	Demonstrate external view of tangible object
	V4→S3→V2	Examine external view of tangible object
	V2→S5→V8	Create virtual model of tangible object (scan, photograph, draw, 3D-model, etc.)

As an example, let us consider the creation of a part using a machine with computer numerical control (CNC). The decomposition of functional block 4 “Integrate control program” yields a subprocess depicted in Figure 1.

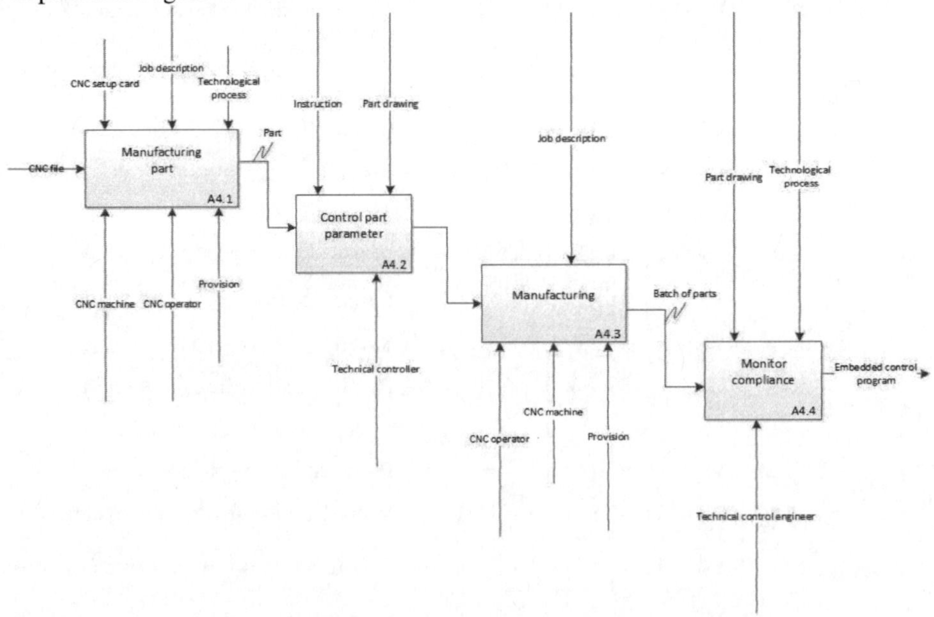

Figure 1. Subprocess "Deploy control program" in IDEF0 notation.

The process of initial data transformation on implementation of a control program is depicted in Figure 2.

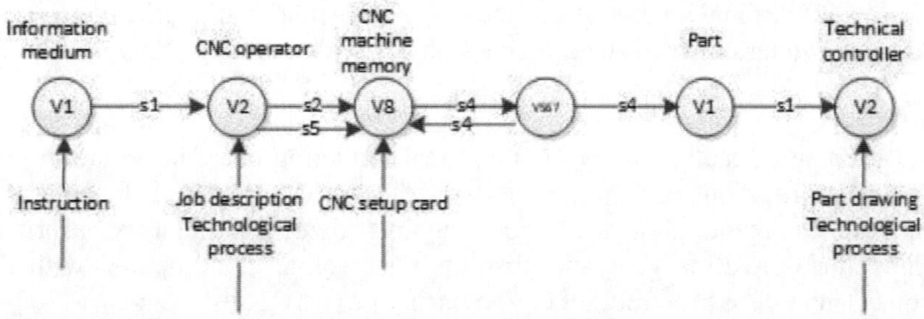

Figure 2. Converting the forms presenting information for subprocess "Deploy control program".

Table 5 contains possible threats to the integrity in terms of a chosen information protection goal that can occur during transmission of the developed program into the machine tool memory.

Table 5. Main information security threats of integrity to the selected protecting information.

Process	Conversion description	Class of integrity threat	Integrity Threat Description
V1→S1→V2	Handing over machine-readable medium with control program to CNC machine tool operator.	Impersonation of sender in medium state 1	
		Impersonation of sender in medium state 2	1 CNC machine tool operator is underqualified to manufacture such part.
		Substitution of information transmission channel	1 Machine-readable medium label is inaccurate and has no relation to stored information.
		Exposure on information transmission channel	1 Machine-readable medium contains control program for manufacturing another part.
V2→S1→V8	Operator connects machine-readable medium to CNC machine tool	Impersonation of sender in medium state 1	1 Operator is not authorized to use CNC machine tool.
		Impersonation of sender in medium state 2	1 Machine-readable medium misfits CNC machine tool. 2 Operator uses unauthorized machine-readable medium
		Substitution of information transmission channel	
		Exposure on information transmission channel	-
V2→S5→V8	Operator copies information from machine-readable medium into CNC machine tool memory	Impersonation of sender in medium state 1	1 There is no control program on machine-readable medium 2 The version of control program on machine-readable medium is outdated 3 Control program was compiled with errors 4 Control program contains errors in accuracy and configuration
		Impersonation of sender in medium state 2	3 Control program format on machine-readable medium is unreadable
		Substitution of information transmission channel	
		Exposure on information transmission channel	

3. Results

The level of details in the description of information handling process based on IDEF0 standard allows constructing a functional model, which will adequately describe real processes and account for significant factors affecting information. The factors of risk that pose threats to information security are estimated to formulate final values of the factors of risk that characterize the total level of threat, potential damage and vulnerability of the protected object (vulnerability scales) [15].

4. Conclusion

The results of implemented methodology of functional modeling of an information process, with due regard to typical transformations of data in various media, is the complete list of threats to information security. In this case, the elaboration of measures against identified threats to information security is aimed at reaching the protection goal and absolutely objective as compared with the methods for information protection system construction on the basis of expert assessments.

Acknowledgments

This research was funded by the Ministry of Education and Science of Russia, Government Order no.2.8172.2017/8.9 (TUSUR).

References

[1] Novokhrestov A, Konev A, Shelupanov A, Egoshin N 2017 Threat model of information security and its media *Bulletin of ISTU* **12** 131

[2] Shevchenko N, Chick T, O'Riordan P, Scanlon T, Woody C 2018 Threat Modeling: A Summary of Available Methods *Pittsburgh: Software Engineering Institute, Carnegie Mellon University* D M18-0894

[3] Mead N, Forrest S, Krishnamurthy V and Ole V 2018 A Hybrid Threat Modeling Method. CMU/SEI-2018-TN-002. *Software Engineering Institute, Carnegie Mellon University*.

[4] Alhabeeb M, Almuhaideb A, Le P and Srinivasan B 2010 Information Security Threats Classification Pyramid *IEEE 24th International Conference on Advanced Information Networking and Applications Workshops* p 208

[5] Konyavskaya S 2013 On the issue of classification of information security objects *Information Technology Security* **3** 14

[6] Shelupanov A, Evsyutin O, Konev A, Kostyuchenko E, Kruchinin D, Nikiforov D 2019 Information Security Methods – Modern Research Directions *Symmetry* p 150

[7] Bondar I, Kuznetsov M, Paklina A, Zolotarev V 2012 A method for modeling security threats to automated systems in a protected version *Actual problems of aviation and cosmonautics* 365

[8] Vernikov G. The main methodologies for survey organizations. IDEF0 standard. https://www.cfin.ru/vernikov/idef/idef0.shtml

[9] Konev A Approach to creation protected information model 2012 *Reports of Tomsk State University of Control Systems and Radioelectronics* pp 34–39

[10] Grant T, Burke I and van Heerde R 2012 Comparing Models of Offensive Cyber Operations *Proceedings of the 7th International Conference on Information Warfare and Security* p 108

[11] Shull F 2016 Evaluation of Threat Modeling Methodologies *Software Engineering Institute, Carne-gie Mellon University*

[12] Jouinia M, Ben Arfa Rabaia L, Ben Aissa 2014 Classification of security threats in information systems *Procedia Computer Science* **32** 489–496

[13] Novokhrestov A, Konev A 2016 Mathematical model of threats to information systems *Prospects of Fundamental Sciences Development, PFSD-2016. AIP Conf. Proc. 1772. 060015-1 – 060015-4.*

[14] Konev A, Mineeva T, Solovev M 2018 Organizational processes regulated in the protection of critical information infrastructure *Information and control systems in transport and industry* pp 149–156

[15] Mikov D 2016 Identification of information risk factors of the telemedicine system *Polytechnic youth magazine MSTU. N.E. Bauman* p 1–8

Applying the principle of distribution in the program complex for vocal recognition

A Yu Yakimuk, A A Konev, Yu V Andreeva and M M Nemirovich-Danchenko
Faculty of Security, Tomsk State University of Control Systems and Radioelectronics, Tomsk, Lenin str., 40, 634050, Russia

E-mail: yay@keva.tusur.ru

Abstract. This article is devoted to studying the possibility of improving the quality of the identification of notes in vocal performance. In the present work, the authors described the transition of the program complex to a distributed functioning model. At this stage, this allowed organizing the storage of results for students and the introduction of an assessment mode for teachers. A study was conducted to improve the quality of note recognition through the use of other algorithms. During the tests, it was determined that the presence of vibrato in singing makes the greatest contribution to the number of unidentified notes. In view of the findings scheduled implementation in a program complex the paalgorithm which is responsible for determining the quality of vibrato in singing. The purpose of this modification is to get rid of novice singers from problems in singing, such as tremor.

1. Introduction

Nowadays there are many applications designed to record various musical performances, but only a few of them provide the ability to identify notes of musical performances, in particular, vocal. Services that have the ability to identify notes involve the use of specific algorithms for determining notes without the possibility of replacing them with others algorithms. These services are used for various purposes. From getting information about audio recordings and organizing music information retrieval systems [1] to vocal training programs. The main task for all these systems is to obtain information from audio recordings and visualize [2] it for comfortable use by users.

In the developed algorithm, as well as in [3, 4], the assumption is taken that the speech signal is described by a harmonic model. The idea of an accurate estimate of the fundamental frequency, developed in [5], is based on the decomposition of a signal into narrowband components and using their instantaneous frequencies as input data. This approach has been used in the analysis of speech and singing voice [6], and in creating an error-resistant algorithm for determining fundamental frequency [7]. The idea of the algorithm for determining the fundamental frequency of the speech signal uses the effect of simultaneous masking, which consists in the following. Each point along the main membrane of the inner ear, which converts mechanical vibrations into nerve impulses, is assigned a sound frequency that causes the maximum response at a given point. The greater the distance from this point, the lower the amplitude of the response. And if the amplitude of the response to a component with a natural frequency is lower than to others, then this component will not be perceived by the auditory system. The main membrane is considered here as a set of frequency resonant filters.

The purpose of this work is to improve the quality of recognition notes sung by the singer. To achieve this, the study used a modification of the recognition score to give the ability to remotely perform studies vocal performances and apply various algorithms for recognizing notes. The modification will allow changes in the operation of the algorithms, without affecting the performance of the complex.

2. Problem statement

In a study [8], a program complex for recognizing notes was tested, which showed that in the range of 70-400 Hz, the algorithm used can correctly identify 85% of the notes sung by the singer. The used algorithm [9] for measuring the fundamental frequency, the results of which are used in determining notes, has an error in calculating fundamental frequency less than 1%, and the reliability of determining voiced sections during speech processing is 89-93%. The predicted accuracy of the note recognition algorithm with the above characteristics can exceed 95%. The resulting difference in the number of errors became the basis for additional research on finding the sources of errors in the work of the developed algorithm used in the program complex, which determines the notes in vocal performance.

It was decided to conduct a separate testing of the definition of fundamental frequency, segmentation and recognition of notes. In this way, it will be possible to understand at which of the problems arise that lead to 15% of errors in the recognition of notes.

The current limit of the upper threshold for determining the fundamental frequency of 400 Hz was used to study the signal parameters in continuous speech and in cancer diseases [10]. The increase in the upper threshold for determining the fundamental frequency is associated with the use of the considered approach for recognizing a melody sung, where the fundamental frequency can reach values of 1,400 Hz [8]. In order to increase the upper limit of determining the fundamental frequency, the algorithm was improved, but the study of the correctness of the work after the improvement was not carried out.

3. The advantages of a distributed model of the program complex

One of the changes in the program complex [8] was the transition to the use of the client-server model. This modification allowed to facilitate the introduction of changes in the applied algorithms. From the user using the program, you will not need to install new versions. As the user records the audio file in the client part of the application, the audio file conversion process is presented, as shown in Figure 1.

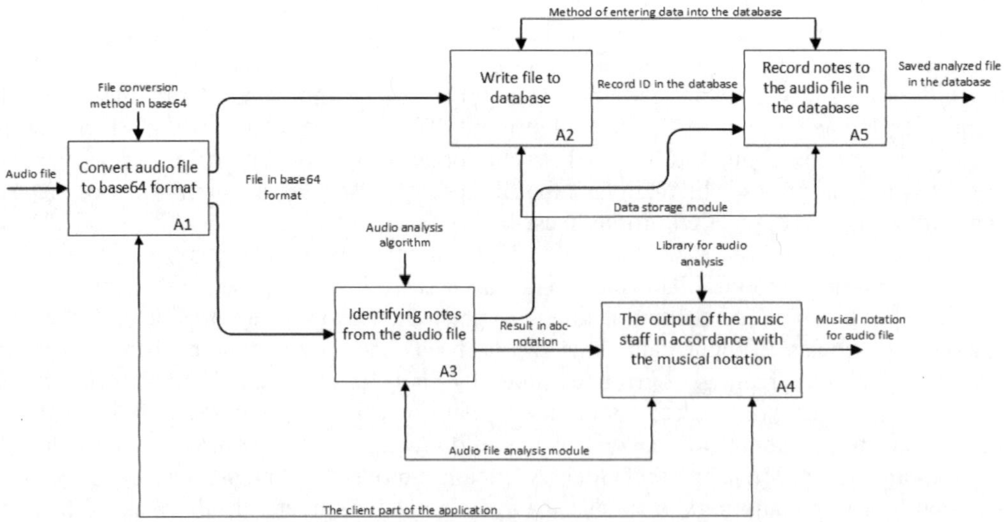

Figure 1. The process of converting the audio file for inclusion in the database

Next, the recorded file is sent to the server and processed there. After that, a message is sent to the client machine in abc notation with the results. The audio file is converted to base64 format for subsequent storage in the database using the built-in JavaScript method btoa (). In order to decode a base64 string on a server, you must likewise use the standard atob () method for the required string.

To determine the notes in the audio file, a library integrated into the project is used, which implements filtering, identifies voiced and unvoiced sections, determines the fundamental frequency and compares the received frequencies to the notes, and then outputs a string in the Abc format if the notes were found. After the user has recorded the audio, the file is converted to base64 and sent to the server. The mentor will be able to see this user in the list of his "students", and when disclosing the form, he will be able to download an audio file, rate it and leave a review. In turn, the user, after receiving feedback from the mentor, will be able to see him.

4. Conducting an experiment

The first task was to test the hypothesis of the presence of problems in the identification of notes. In the current version, the basis is the application of the Fourier transform. The hypothesis was that the transition to a wavelet transform or a hybrid algorithm can improve note recognition quality. The behavior of the analysis on the studied program complex using a hybrid algorithm showed recognition accuracy in the range of 65-70%. It was experimentally determined for test recordings that the algorithm worked better for some audio recordings. However, in most cases, the quality of identification has decreased.

The material used for experiment was audio recordings with a sequence of notes sung by the singer in the range covered by the program complex. The requirements for testing files were as follows: a duration of no more than 5 seconds, the *.wav format.

The crucial point in determining the quality of the identification of notes was the discovery in the vocal of the speakers of vibrato, perceived by experts as a correctly played note, and the program as noise. The specificity of one of the stages of the note identification algorithm, which is responsible for determining whether a voiced section belongs to a note, is a focus on pure performance.

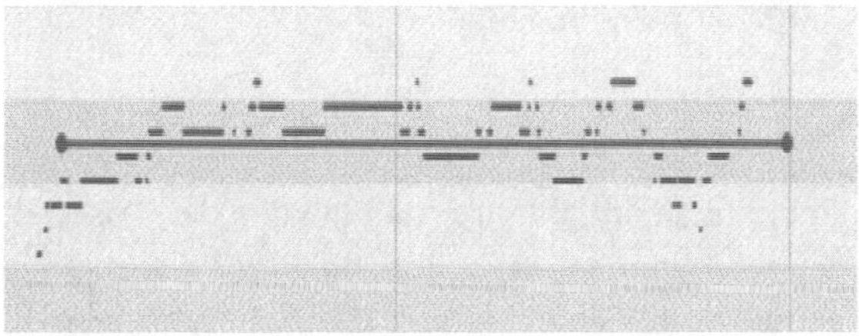

Figure 2. Note sung without vibrato in voice

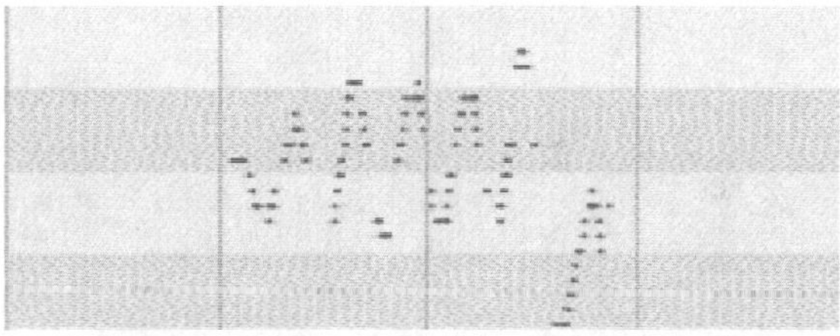

Figure 3. Note sung with vibrato in voice

To execute a note so that all frequencies at the moment of singing are within the band reserved for it is quite difficult. For this reason, the adjacent notes above and below the executable are taken into account. In Figures 2-3, the frequency spectrum is divided into sections corresponding to the boundaries of the notes. Each section is highlighted by a horizontal bar on the background of the frequency graph. On the X axis - the time, on the Y axis - the frequency.

As can be seen in Figure 2, a segment sung within a single note was recognized by the program. The presence of moments caught in the neighbouring notes is insignificantly large and did not affect the result of identification. If the fragments in each of the 3 sections turned out to be approximately equal, the algorithm perceived the entire segment as noise. This situation can be seen in Figure 3. The note sung by the speaker was played with a vibrato in the voice. By definition, vibrations within a semitone (neighbouring notes) with an oscillation frequency of 5 to 7 Hz are a sign of presence in vibrato singing [11]. As you can see on the plot, sung with the use of vibrato, there are vibrations in 4 adjacent notes. In addition, the assessment of the contribution to each of the notes in the sung segment is less than the required amount.

It should be noted that the presence of fluctuations in the voice when singing may indicate not only the professionalism of the singer. If fluctuations occur more often (from 8 to 9 Hz) - this effect is considered tremor. Also, the disadvantage is that the frequency variation (3-4 Hz) in singing is too slow, perceived as "sound swinging". One of the objectives of improving the quality of singing can be the refinement of the algorithm in order to determine the frequency of changes in the singing of a note.

To assess the accuracy of the algorithm, sinusoidal signals were generated with a known frequency in the range from 70 to 700 Hz. Graphs for signals with frequencies from 70 Hz up to 600 Hz inclusive looks like a straight line at generated frequency level.

The results of the algorithm for sinusoidal signals with different frequencies were analyzed for errors. Table 1 presents the values of the relative errors in determining the fundamental frequency. It is seen that the relative error increases with increasing frequency of the signal.

Table 1. The relative errors in determining the fundamental frequency

Fundamental frequency of a sinusoidal signal, Hz	Relative error, %
70	0
90	0
120	0
150	0
200	0
250	0.4
300	0.33
350	0.29
400	0.5
450	0.44
500	0.6
550	0.73
600	0.83
650	50
700	48.14

As for signals with frequencies of 650 and 700 Hz, the errors for them are calculated on the basis of a specific channel number, since the frequency at each time reference is defined as different (Fig. 4). The channel number for 650 Hz approximately corresponds to the channel number for a signal of 325 Hz, and the channel number for 700 Hz approximately corresponds to the channel number for a signal of 363 Hz. The abscissa is the time in milliseconds, and the ordinate is the fundamental frequency in Hz.

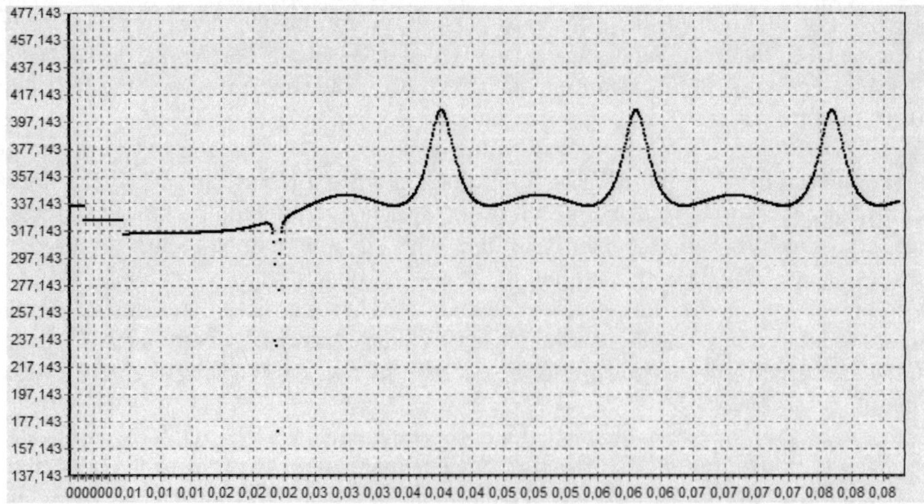

Figure 4. Graph of the fundamental frequency definition for a sinusoidal signal 650 Hz

It is worth adding that only voiced sections (having a periodic structure) are subject to analysis. This implies that the presence of a "dotted line" on the graphs is explained by the fact that the algorithm determines some parts of the signal as unvoiced.

5. Conclusion

The transition to a distributed model of the program complex made it possible to integrate new libraries that implement note recognition for the study of vocal performances without processing the source code of the program. Such an approach can become more effective in the vocal mastery model using the algorithms used in the work due to the possibility of entering new data without losing the ability to work with the previous ones. This is achieved by separating the modules responsible for data storage and audio file processing. This change will allow changing the algorithm for identifying notes without affecting the rest of the complex in order to determine the optimal parameters that will improve the quality of note recognition. One of the main directions of development of the note identification algorithm was the expansion of the existing range to 1,400 Hz. In the range of 70-400 Hz, the algorithm used can correctly identify 85% of the notes sung by the singer. Algorithms were tested at higher frequencies. As a result of the research, the algorithm showed incorrect operation for signals with frequencies from 650 Hz inclusive and requires adjustment. The magnitude of the errors tends to 50% for frequencies higher than 600 Hz, which is obviously unacceptable and requires further study.

Acknowledgments
This research was funded by the Ministry of Education and Science of Russia, Government Order no.2.8172.2017/8.9 (TUSUR).

References

[1] Dittmar C, Cano E, Abeßer J, Grollmisch S 2012 *Multimodal music processing* **3** 95–120
[2] McLeod P, Wyvill G 2003 *Computer Graphics International 2003*
[3] McAulay R, Quatieri T 1986 *IEEE Transactions on Acoustics, Speech and Signal Processing* **34(4)** 744–754
[4] Laroche J, Stylianou Y, Moulines E 1993 *IEEE International Conference on Acoustic, Speech, and Signal Processing* 550–553
[5] Abe T, Kobayashi T, Imai S 1995 *IEEE International Conference on Acoustic, Speech, and Signal Processing* 756–759
[6] Azarov E, Vashkevich M, Petrovsky A 2014 *IEEE International Conference on Acoustic, Speech, and Signal Processing* 7919–23
[7] Hotta K, Funaki K 2014 *Annual Summit and Conference AsiaPacific Signal and Information Processing Association* 1–4
[8] Konev A, Kostyuchenko E, Yakimuk A 2017 *Journal of Physics: Conference Series* **803(1)** 012077
[9] Bondarenko V, Konev A, Mescheriakov R 2007 *Proceedings of the XIIth International Conference "Speech and Computer" SPECOM'2007* 562-565
[10] Bondarenko V, Choinzonov E, Balatskaya L, Chizhevskaya S, Konev A, Meshcheryakov R 2007 *Biomedical Engineering* **41(4)** 154-156
[11] Leydon C, Bauer J, Larson C 2003 *Acoustical Society of America* **114(3)** 1575-1581

IOP Conf. Series: Materials Science and Engineering **597** (2019) 012073 doi:10.1088/1757-899X/597/1/012073

Influence of the long-range of compact Jets in the Aspiration Pipeline during Dustdeposing

N N Akimbekova[1], G T Danenova[2], G S Sattarova[1] and A D Nurgaliyeva[1]

[1]Department of Mining Aerology and Labour Protection, Karaganda State Technical University, 56 Nursultan Nazarbayev Avenue, Karaganda M00A1T8, Kazakhstan
[2]Departmentof Information Technologies and Security, Karaganda State Technical University, 56 Nursultan Nazarbayev Avenue, Karaganda M00A1T8, Kazakhstan

E-mail: g.danenova@kstu.kz

Abstract. There is the relevant issue of cleaning the ducts walls from dust deposits on industrial plants, workshops and production lines equipped with exhaust ventilation systems. Improving the efficiency of the process of cleaning air ducts from dust depends on the long-range of compact jets. The mathematical model is based on the classical solution of the "submerged jet" problem propagating in a stationary medium. The use of this model made it possible with sufficient accuracy to estimate the parameters of the range of compact jets when operating aspiration systems. Studies have shown that the presented model can be used to analyze the state of the pipelines of the suction networks and dust collection plants.

1. Introduction

Ventilation is the set of activities and devices that provide the calculated air in the premises of residential, public and industrial buildings. The production process is accompanied by release of gases and vapors harmful to human health into the air of workplaces. In addition, large amounts of heat, moisture, and dust may be released into the air in production areas, increasing its temperature and humidity, as well as increasing its dustiness [1, 2]. Ventilation is provided to maintain normal air parameters in the premises that meet sanitary, hygienic and technological requirements. Sanitary-hygienic purpose of ventilation is to remove harmful gases, vapors and dust [3, 4].

2. Research problem

Considering the pipeline of the aspiration system, in which the air moves in the turbulent mode, with a certain average velocity U_1 (Figure 1). To prevent the accumulation of dust on the pipeline walls, air is additionally supplied through the nozzles to the system in the form of a compact jet with an initial velocity U_2. In this case, the vectors $\overline{U}_1$ and $\overline{U}_2$ directed in one direction and parallel to the axis of the pipeline. The purpose is to determine the jet's long range, that is, a function of the maximum speed of the air flow $U_{\max}(z)$ in the pipeline during the nozzle work, as well as the distribution of the flow velocity in the section plane of the pipeline.

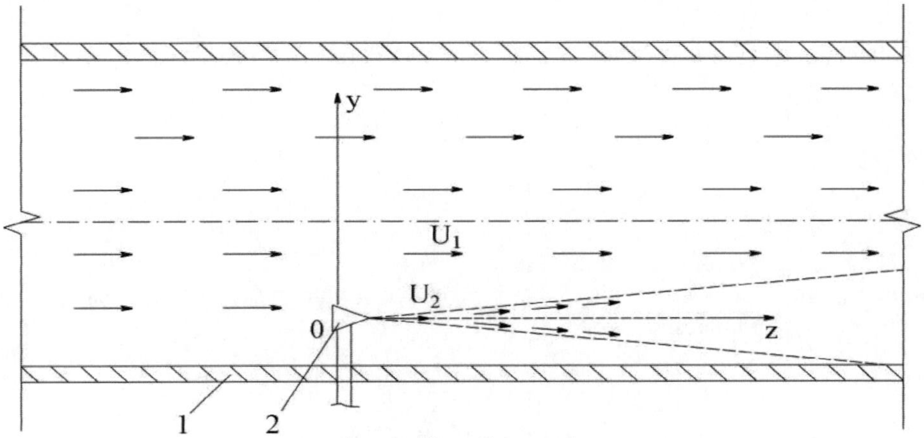

1 - pipeline; 2 - nozzle

Figure 1. Scheme of operation of the air nozzle in the aspiration pipeline.

3. Solution methods

The work is based on the classical version of the solution of the "submerged jet" problem propagating in a stationary medium with the same physical constants as that of the jet itself. An extremely thin (almost point) source of a jet with an initial impulse (second moment of movement) J_o is considered (Figure 2).

The initial impulse is related to the air flow Q through the expression $J_o = \rho Q$, (ρ is the air density). The axis Oz is directed along the axis of symmetry of a flat jet. When the external velocity is equal to zero, the equation of motion of a viscous incompressible fluid (air here is considered as an incompressible fluid) will have the following form:

$$\begin{cases} U\dfrac{\partial U}{\partial z} + \upsilon\dfrac{\partial U}{\partial y} = \nu\dfrac{\partial^2 U}{\partial y^2}, \\ \dfrac{\partial U}{\partial z} + \dfrac{\partial \upsilon}{\partial y} = 0, \end{cases} \tag{1}$$

where $U = \overline{V}_z$, $\upsilon = \overline{V}_y$, ν - is the kinematic viscosity of air.

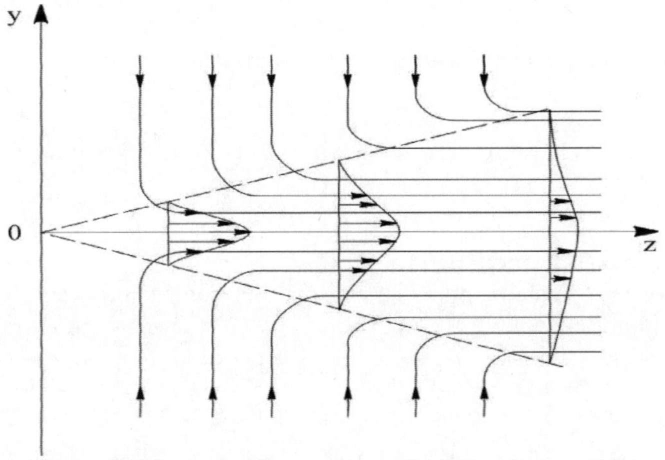

Figure 2.Scheme to the problem of a flat submerged jet.

The system of equations (1) for the case of a flat laminar submerged jet was solved by H. Schlichting [5].
For a nozzle having a final radius r_o, the initial impulse is equal to:

$$J_o = \rho \pi r_o^2 U_o^2, \tag{2}$$

Where U_o - is the relative initial velocity of the jet.
The system of equations (1) for an axisymmetric uncoiled jet having a circular cross section will be written in cylindrical coordinates:

$$\begin{cases} U\dfrac{\partial U}{\partial z} + \upsilon\dfrac{\partial U}{\partial r} = \left(\nu\dfrac{\partial^2 U}{\partial r^2} + \dfrac{1}{r}\dfrac{\partial U}{\partial r}\right), \\ \dfrac{\partial(rU)}{\partial z} + \dfrac{\partial(r\upsilon)}{\partial r} = 0. \end{cases} \tag{3}$$

The solution of system (3) will be self-similar, since equality of each section of the jet is maintained:

$$J = \rho\int 2\pi\pi r^{\,2}\partial r = J_o = \text{const}, \tag{4}$$

The solution of system (3), with (4) taking into account, is obtained as the distribution of the maximum velocity U_{max} along the z axis.

$$U_{max} = \frac{3}{8}\frac{J_o}{\pi\rho\nu z}, \tag{5}$$

For a turbulent flow, the problem of the "submerged jet" was solved using the semi-empirical theories of L. Prandt [6], in which the turbulent air flow was compared with the laminar one. In this case, the jet propagation equation has the form:

$$U\frac{\partial U}{\partial z} + \upsilon\frac{\partial U}{\partial r} = \varepsilon_\tau \frac{1}{r}\frac{\partial}{\partial r}\left(r\frac{\partial U}{\partial r}\right), \tag{6}$$

where ε_τ - is, by analogy with laminar, the coefficient of turbulent viscosity

$$\varepsilon_\tau = \sigma\sqrt{\frac{J_o}{\rho}}, \tag{7}$$

where σ - is the empirical coefficient $\sigma \approx 0.21$.
The maximum speed for an asymmetric non-swirl jet spouting from an infinitely small orifice is obtained from (6) with (7) taken into account:

$$U_{max} = \frac{3}{8\pi\pi}\cdot\sqrt{\frac{J_o}{\rho}}\cdot\frac{1}{z}, \tag{8}$$

In the case of a non-swirling turbulent jet flowing from a hole of finite diameter $d(d = 2r_o)$ with a finite initial flow rate, as well as a swirling turbulent jet, the correction is applied to formula (8):

$$U_{max} = \frac{3}{8\pi\pi} \cdot \sqrt{\frac{J_o}{\rho}} \cdot \frac{1}{z}\left(1 - \frac{1}{16\sqrt{\pi\sigma}} \cdot \frac{d}{z}\right), \tag{9}$$

4. Results and discussion

Comparison of values U_{max} calculated by formulas (8) and (9) showed that deviations of U_{max} are observed only at small values of z (Table 1). The numerical experiment conducted for the values of the constants $r_o = 0.01$ m and $U_o = 150$ m/s showed that when the $z \approx (1\ldots3)r_0$ relative deviations vary within $\Delta = 33.33 - 11.11\%$, and when $z \approx 5r_o$,is $\Delta \approx 7\%$.

Table 1. Results of comparison of calculated values

z, m	0.01	0.02	0.04	0.10	0.20	0.40	1.00	2.00	4.00	10.0
$\frac{z}{r_o}$	1	2	4	10	20	40	100	200	400	1000
$U^{(1)}_{max}$ m/s	150	75	37.5	15	7.5	3.75	1.5	0.75	0.375	0.15
$U^{(2)}_{max}$ m/s	100	62.5	34.38	14.5	7.38	3.72	1.50	0.75	0.37	0.15
Δ, %	33.33	16.67	8.33	3.33	1.67	0.83	0.33	0.17	0.08	0.03

The value $U^{(1)}_{max}$ was calculated by the formula (8), and $U^{(2)}_{max}$ - by the formula (9), and the deviation of these values is $\Delta = \frac{\left(U^{(1)}_{max} - U^{(2)}_{max}\right)\cdot 100\%}{U^{(1)}_{max}}$.

Thus, we have adopted formula (8) for calculating the parameters of a non-swirling turbulent jet flowing from an orifice of a finite diameter.

The velocity distribution U(r) over the jet section is calculated by the formula:

$$U(r) = \frac{U_{max}}{\left[1 + \frac{1}{3}\left(\frac{r}{z}\right)^2\right]^2}, \tag{10}$$

To go to a compact jet propagating in a stream of moving air (medium), consider two inertial reference systems described by cylindrical coordinates. The reference system rOz is stationary and has an origin at the exit of the nozzle. The reference system r'O'z 'is tied to the medium and moves with the speed U_1 along with the air flow along the z axis (Figure 3). At time t, the origin of O and O' are combined, and r = r'. At the same time relative to the system the speed of the main air flow is $U_1^{\prime} = 0$. Taking into account that the speed of the jet flowing from their nozzles in the reference system rOz is equal to U_2, we find the speed of the jet $U_2^{\prime}$ relative to the inertial reference system:

$$U_2^{\prime} = U_2 - U_1, \tag{11}$$

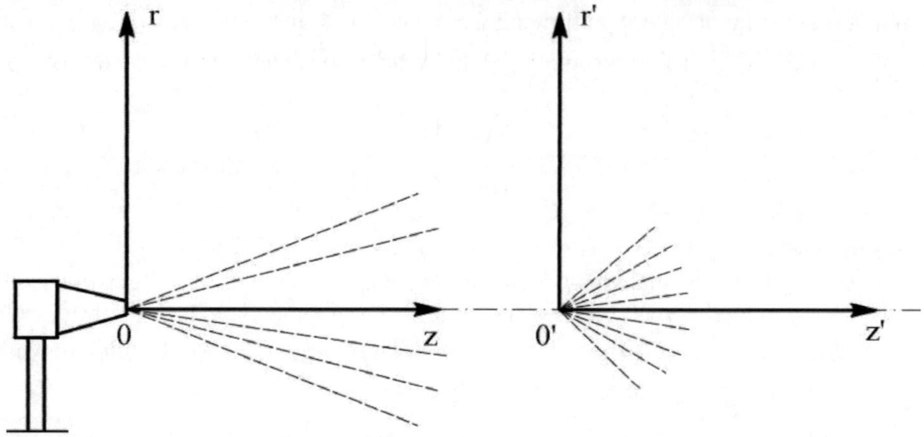

Figure 3. Scheme of the inertial model of a turbulent jet

Therefore, if we accept $U_o = U_2$, then the jet having the initial impulse (in the r'O'z' system) $J_o = \rho\pi r_o^2 (U_2^\backslash)^2$ can be considered as a submerged jet propagating in a fixed medium. By substitution (11) in formula (8) by replacing z with z' we get:

$$U_{max}(z^\backslash) = \frac{3}{8\pi\pi}\sqrt{\frac{\rho\pi r_o^2 (U_2^\backslash)^2}{\rho}} \cdot \frac{1}{z} = \frac{3 r_o U_2^\backslash}{8\sqrt{\pi\sigma z^\backslash}} \approx \frac{r_o(U_2 - U_1)}{z^\backslash}; \; z^\backslash \in [r_o, \infty], \tag{12}$$

Here $U_{max}(z')$ is the maximum speed of the jet relative to the reference system r'O'z'.
For the transition from $U_{max}(z')$ to the function $U_{max}(z)$ a transition from a moving reference system r'O'z' to a stationary rOz is necessary:

$$\begin{cases} U_{max}(z) = U_{max}(z') + U_1, \\ z = z' + U_1 T, \end{cases} \tag{13}$$

where T is the time during which in the r'O'z' system the jet velocity $U_2^|$ is equal to $U_{max}(z')$. To determine the value of T, we consider the differential:

$$U_{max}(z') = \frac{dz'}{dt}, \tag{14}$$

then

$$T = \int_0^T dt = \int_{r_o}^{z'} \frac{dz'}{U_{max}(z')} = \int_{r_o}^{z'} \frac{z'\,dz'}{r_o(U_2 - U_1)} = \left.\frac{(z')^2}{2r_o(U_2 - U_1)}\right|_{r_o}^{z'} = \frac{(z')^2 - r_o^2}{2r_o(U_2 - U_1)}. \tag{15}$$

By taking into account (15) from formula (13) we get:

$$z = z' + \frac{U_1\left[(z')^2 - r_o^2\right]}{2r_o(U_2 - U_1)}, \tag{16}$$

Equation (16) is a quadratic equation with respect to z'. We need to find the roots of this equation $z'_{1,2}$. Since from the condition of the problem $z' > 0$ we have:

$$z' = \frac{r_o}{U_1}\left\{\sqrt{(U_2 - U_1)^2 + 2zr_o^{-1}U_1(U_2 - U_1) + U_1^2} - (U_2 - U_1)\right\}, \tag{17}$$

By substituting (17) into formula (12), and moving from $U_{max}(z')$ to $U_{max}(z)$ we get:

$$U_{max}(z) = U_1 + \frac{U_1(U_2 - U_1)}{\sqrt{(U_2 - U_1)^2 + 2zr_o^{-1}U_1(U_2 - U_1) + U_1^2} - (U_2 - U_1)}, \tag{18}$$

The calculation function $U_{max}(z)$ is given in tabular form (table 2) for the following conditions: $r_o = 0.01$ m and $U_1 = 12$ m/s ($Q_1 = 204$ m³/min); $U_2 = 150$ m/s ($Q_2 = 2.83$ m³/min).

Table 2. $U_{max}(z)$ function

z, m	0.2	0.4	0.6	0.8	1.0	1.0	1.2	1.5	2.0
$U_{max}(z)$	22.7	18.6	17.0	16.2	15.6	15.6	15.3	14.9	14.4

The radial distribution of air velocity U(r) is calculated by the formula:

$$U(r,z) = U(r,z') + U_1, \tag{19}$$

$$U(r,z') = \frac{U_{max}(z')}{\left[1 + \frac{1}{3}\left(\frac{r}{z'}\right)^2\right]^2}, \tag{20}$$

5. Conclusion

The developed mathematical model using the equations systems of H. Schlichtingand semi-empirical theories about the "submerged jet" of L. Prandtl allows us to estimate the range of compact jets for cleaning the internal surfaces of the walls of the suction pipelines around the perimeter from dust deposits. The presented mathematical model was used in the development of the device, which eliminates the formation of stagnant and closed zones throughout the dust-gas flow through the duct and provides low aerodynamic resistance of pipelines with dust-cleaning equipment.

References

[1] Sun B, Cheng W M, Wang J Y and Wang H 2018 *J. Clean. Prod.* **201** 308-324
[2] Wang Y C, Luo G, Geng F, Li Y B and Li Y L 2015 *J. Loss. Prevent. Proc.* **36** 146-157
[3] Shao A L, Liu L P, Song Q L, Cui Z J 2015 *Advances in Energy Science and Equipment Engineering* **1** 1591-1595
[4] Wang K, Jiang S G, Ma X P, Wu Z Y, Zhang W Q and Shao H 2015 *Powder technol.* **286** 401-411
[5] Schlichting H 1979 *Boundary layer theory*, (New York:McGraw-Hill)
[6] Candra K J, Pulung S A and Sadashiv M A 2014 *International Journal of Mining Science and Technology* **24** 39-44

Numerical simulation of severe plastic deformation of aluminum specimens under dynamic groove pressing

L R Akhmetshin[1, 2], **E N Moskvichev**[1, 2] **and V V Skripnyak**[1]

[1] National Research Tomsk State University, 36 Lenina ave., Tomsk, Russia
[2] Institute of Strength Physics and Materials Science SB RAS, 2/4 Academic ave., Tomsk, Russia

Akhmetshin.lr@gmail.com

Abstract. The process of deformation of aluminum samples under dynamic groove pressing for a dynamic loading scheme is studied numerically in a two-dimensional formulation. Areas of intense plastic deformation of the sample are revealed. Localization of plastic deformations is similar for the static loading pattern. The results of damage accumulation in the material are given. The finite element results are in good agreement with the experimentally obtained results in the field of strain rates considered in this study.

1. Introduction

Currently the bulk nanostructured materials are considered as promising, structural and functional materials of a new generation. The main method for the preparation is the formation of nanostructures with severe plastic deformation (SPD). Modification of coarse-grained crystalline structure of light aluminum alloys to an ultrafine-grained state by methods of severe plastic deformation is necessary to improve their physical-mechanical properties [1 – 5].

Unlike the bulk samples, the use of widely developed processing techniques, such as the equal-channel angular pressing and the torsion under pressure [6, 7] is unacceptable for the sheet metal samples from due to their shape. The constrained groove pressing (CGP) can be a new alternative to these methods of processing [8, 9]. The method of CGP for modification of sheet metal samples from light structural alloys has proved to be a promising approach for changing the internal structure of materials and changing their physical and mechanical properties [9, 10].

The study of the ultrafine-grained (UFG) metals, obtained by the SPD, showed that they are characterized by a number of unique properties. There is the increased strength by several times, which is combined with the good ductility, the low and the high-temperature superplasticity, the cyclic and radiation resistance.

The principle of the method of the grooving method and analytical solution are described in [10]. The results of processing by the CGP of the aluminum, magnesium, copper, titanium and nickel alloys are described in detail in [10].

It is of interest to expand the alloys range is capable to undergo structural changes while the physical and mechanical properties are improved. In addition, there is the pressing issue of creating more accurate predictive physical-mathematical models are describing the process of this materials class. In this question, the mathematical modeling is a convenient theoretical tool that allows realizing the identification of the main mechanisms and patterns. Additionally, an adequate physical-mathematical model will allow estimating the magnitude of the forces during pressing, and it will help to avoid the mistakes in the design of molds and optimization of processing modes. The development of the groove pressing method leads to the use of dynamic pressing techniques. The method of dynamic groove pressing (DGP) will increase the plastic deformation rate of the sample.

The trends in the development of the SPD methods show that the large-scale numerical studies of the DGP processes are needed to identify the features of intense plastic deformation and to establish effective parameters of these processes.

Numerical simulation of the DGP process is carried out in a two-dimensional setting. The main advantage of this method in comparison with the CGP is that the plastic deformation rate increases, as well as shock-wave deformation is added, which increases the overall result of the action.

The purpose of the work was to study the process of the plastic deformation localization and the destruction localization in the material.

2. The mission statement

The Johnson-Cook material model was used to describe the Al-Mg alloy mechanical behavior. The model considers the temperature and strain rate effect on the flow stress and the ultimate degree of material strain:

$$\sigma = \left[A + B\left(\varepsilon^{pl}\right)^n\right]\left[1 + C\ln\dot{\varepsilon}^{pl}\right]\left[1 - \left(\frac{T-T_r}{T_m-T_r}\right)^m\right] \quad (1)$$

where ε^{pl} is the effective plastic deformation, $\dot{\varepsilon}^{pl} = \frac{\dot{\varepsilon}}{\dot{\varepsilon}_0}$ is the dimensionless plastic strain rate for $\dot{\varepsilon}_0 = 1\,s^{-1}$, T_m is the melting point, T_r is the room temperature, A is the corresponds to the yield stress ($\sigma_{0,2}$) of the material, B – is a strain hardening factor, C – is a strain rate constant of the material, n, m are the strain hardening index and temperature softening index, respectively.

After analyzing the obtained numerical data, it was found that in the area before the maximum load was achieved, the results obtained using the Johnson-Cook model are in close agreement with the experimental data. There is a need to select a failure numerical model, because of the sample destruction during the experiment. Description of the sample destruction process was carried out by the Johnson-Cook failure model:

$$D = \frac{1}{\varepsilon_f}\sum_i \Delta\varepsilon_p^i \quad (2)$$

where ε_f – is the material ultimate strain, $\Delta\varepsilon_p^i$ – is the element effective plastic strain increment at the i-th integration step over time.

The $\Delta\varepsilon_p^i$ value is calculated by the formula:

$$\Delta\varepsilon_p^i = \left(D_1 + D_2\,exp\left(D_3\frac{P}{\sigma_{ef}}\right)\right)\left(1 + D_4\,ln\frac{\varepsilon_p}{\varepsilon_0}\right)\left(1 + D_5\frac{T-T_r}{T_m-T_r}\right) \quad (3)$$

where $D_1,\ldots,D_5$ – are the material parameters, σ_{ef} – is the effective stress, P – is the considered element pressure. According to the selected criterion, the final element destruction occurs if the damage parameter $D=1$.

Constants values for the material and fracture models were determined experimentally. The experimental data were obtained earlier during tensile tests on flat specimens from the same material. Experimental data and numerical data obtained using the Johnson-Cook deformation model and the J-C fracture model show qualitative and quantitative agreement throughout the all deformation area. This suggests the correct choice of the determining equation, the failure model, the models constants and the mesh density.

An erosion model was used to visualize the destruction process. The erosion model operation principle was as follows: the element was automatically excluded from the solution when the element's damage parameter D reached a value of 1, however, it continued to be present in the grid model, creating the illusion of integrity. When the erosion model was turned on, the elements in which the fracture criterion was achieved were removed. Thus, the erosion model inclusion to the calculation in conjunction with the Johnson – Cook failure model made it possible to obtain a fracture image as close to reality as possible.

The mathematical model obtained in [11] was tested for the high-speed plate penetration.

3. Initial and boundary conditions

Figure 1 illustrates the geometrical models of dies with trapezoidal teeth for pressing of flat specimens and their further flattening. The dimensions of teeth height, interface distance between teeth, top horizontal width of teeth, and projection of sloping plane on x axis and the thickness of the specimen were equal to 1.5 mm.

Initial conditions for dynamic groove pressing at the moment of time $t_0 = 0$ looked as

$$\sigma_{ij}(x, y, t_0) = E(x, y, t_0) = D(x, y, t_0) = 0; \rho(0, x) = \rho_0;$$

where σ is a stress, E is energy, D is a damage parameter and ρ is density of a sample.

Boundary conditions for the models are given as

$$U_x|_{AD} = U_y\big|_{AD} = 0; U_x|_{AB} = U_x|_{CD} = 0;$$

$$V_x|_{BC} = 0, \quad V_y\big|_{AD} = -V_0 = -1000 \left(\frac{mm}{s}\right);$$

where *U, V* are components of displacement vector and speed of the mold elements. The x and y indices of the components of the boundary conditions are the projection on the corresponding axis.

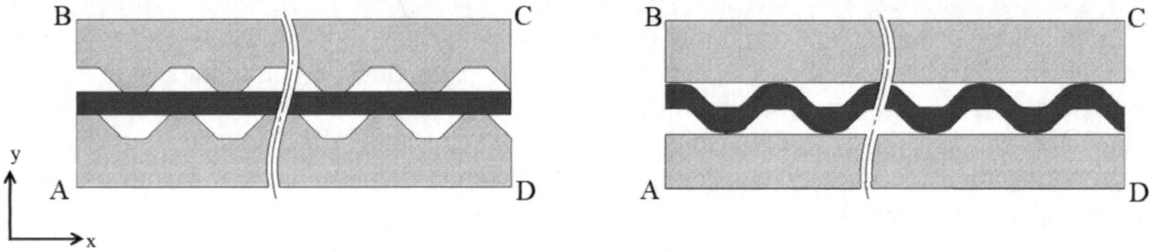

a b

Fig. 1. Schematic illustration of: a – grooved dies, b – flat dies.

4. Results and discussion

The sample is placed in between the dies and the pressing is performed such that the gap between the upper die and the lower die is same as that of the sample thickness. This results in pure shear deformation under plane strain condition in the inclined region of the sample. However, the flat region remains undeformed. In the second pressing, the corrugated sample is straightened by means of flat dies. This straightening under the constrained condition ensures that the already deformed region is subjected to reverse shear deformation while the undeformed region remains unaffected. Then the sample is rotated by 180° about the axis perpendicular to plane of the sample. This allows the deformation of the undeformed region due to asymmetry of the grooved die. These successive pressings with a pair of grooved dies and a pair of flat dies result in a homogeneous effective strain throughout the sample.

In the result of the numerical modeling of sheet metal, the distributions of effective plastic deformations in the sample were obtained. During the groove pressing in the sample, plastic deformations are local and uneven. The effective plastic deformation is accumulated in a shearing area. The maximum limit of the cumulative plastic deformation is equivalent to 0.54 (figure 2a). When the samples are straightening, the deformation is summed in the shearing area, that is, after a half cycle, its value increases approximately to 1.12 in local zones (figure 2b).

In the formulation of this problem when analyzing the destruction of the material, it was revealed that there is no destruction in the sample of sheet metal. This result may be associated with good ductility of 1560 alloy.

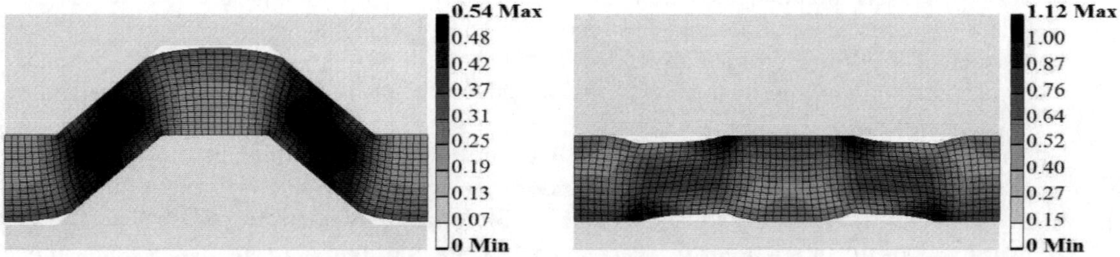

a b

Fig. 2. Equivalent plastic strain during: a – grooving, b – flattering.

The numerical experiments revealed that the highest values of the damage parameter is equivalent to 0.33 (figure 3a) during one DGP cycle. It is shown, that during the half-cycle of the CGP the damage growth is observed in the action zone of tensile stresses, which are realized in the zone of the geometry change of the mold protrusions.

Over time, the irreversible deformation is developing. The development of the inelastic deformation may lead to the formation of the main cracks, which cross the entire sample, like test sample (Figure 3б). In a numerical experiment, the configuration of the crack formation is the same as the test sample.

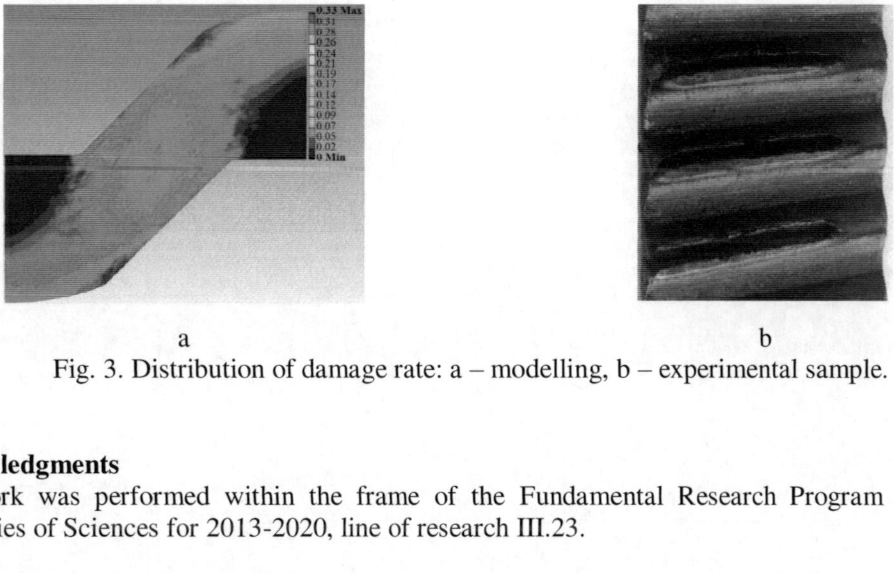

a b

Fig. 3. Distribution of damage rate: a – modelling, b – experimental sample.

Acknowledgments
This work was performed within the frame of the Fundamental Research Program of the State Academies of Sciences for 2013-2020, line of research III.23.

References

[1] Skripnyak N V 2015 The features of fracture behavior of an aluminum-magnesium alloy AMg6 under high-rate straining *Russian Physics Journal* **58** **5** 691-697

[2] Moskvichev E N, Skripnyak V A, Skripnyak V V, Kozulin A A and Lychagin D V 2016 Influence of structure to plastic deformation resistance of aluminium alloy 1560 after groove pressing treatment *Letters on materials* **6** **2** 141-145

[3] Filippov A V and Filippova E O 2018 Creation and Shaping of Three-Dimensional Ultrafine-Grain Materials *Russian Engineering Research* **38** **7** 540-543

[4] Filippov A V, Tarasov S Y, Shamarin N N, Podgornykh O A and Filippova E O 2018 Surface

IOP Conf. Series: Materials Science and Engineering **597** (2019) 012074 doi:10.1088/1757-899X/597/1/012074

Quality of AMg2 Aluminum Alloy with Ultrafine Grain Structure after Machining. 1. Turning *Russian Engineering Research* **38** **12** 1067-70

[5] Filippov A V, Tarasov S Y, Fortuna S V, Podgornykh O A, Shamarin N N and Vorontsov A V 2019 Wear, vibration and acoustic emission characterization of sliding friction processes of coarse-grained and ultrafine-grained copper *Wear* **424-425** 78-88

[6] Markushev M V 2011 K voprosu ob effektivnosti nekotorih metodov intensivnoy plasticheskoy deformatcii, prednaznachennih dlya polucheniya ob'emnih nanostrukturnih materialov *Pisma o materialax* **1** **1** 36-42

[7] Estrin Y, Murashkin M Yu and Valiev R Z 2010 Ultrafine-grained aluminium alloys: Processes, structural features and properties *Fundamentals of Aluminium Metallurgy* 468-503

[8] Kozulyn A A, Skripnyak V A, Krasnoveikin V A, Skripnyak V V and Karavatskii A K 2015 An investigation of physico-mechanical properties of ultrafine-grained magnesium alloys subjected to severe plastic deformation *Russian Physics Journal* **57** **9** 1261-67

[9] Moskvichev E N, Skripnyak V A, Skripnyak V V, Kozulin A A and Lychagin D V 2018 Structure and Mechanical Properties of Aluminum 1560 Alloy after Severe Plastic Deformation by Groove Pressing *Physical Mesomechanics* **21** **6** 515-522

[10] Gupta A K, Maddukuri T S and Singh S K 2016 Constrained groove pressing for sheet metal processing *Progress in Materials Science* **84** 403-462

[11] Akhmetshin L R et al 2019 *IOP Conf. Ser.: Mater. Sci. Eng.* **511** 012013

Direct integrators of modified multistep method for the solution of third order boundary value problem in ordinary differential equations

U Mohammed[1], M Semenov[2], A I Ma'ali[3], A Abubakar[3], and H Abdullah[1]

[1]Federal University of Technology, Minna, Nigeria
[2]National Research Tomsk Polytechnic University, Lenin Avenue, 30, Tomsk, Russia, 634050
[3]Ibrahihim Badamasi Babangida University Lapai, Niger State, Nigeria

E-mail: [1]umaru.mohd@futminna.edu.ng, [2]sme@tpu.ru

Abstract. In this paper, we propose an efficient modified multistep method for direct solution of boundary value problems (BVPs) using multistep collocation approach. The continuous form was evaluated at grid and off-grid points to obtain the multiple finite difference schemes. The basic properties, such as order and error constants, zero stability and convergence analysis of the proposed methods were investigated. Numerical experiment were performed to show the efficiency of the method and the results were compared with the existing methods in the literature.

1. Introduction

The application of mathematical formulation in science and engineering is given by a boundary value problem (BVP):

$$y''' = f(x, y, y', y''), \quad y(a) = y_0, \quad y'(a) = \delta_0, \quad y(b) = y_M, \tag{1a}$$

$$y''' = f(x, y, y', y''), \quad y(a) = y_0, \quad y'(a) = \delta_0, \quad y'(b) = y_M. \tag{1b}$$

There are many methods to solve third order ordinary differential equations (ODEs) (1). Most of these methods are solved BVP by reducing a higher ODEs to an equivalent system of first order ODEs which take a lot of time and human effort. Alternative approach is to solve higher ODEs directly. In the paper [1] authors investigated two- and three-stage Runge-Kutta type methods for special third order ODEs. Higher order linear multi-step methods were proposed by Jator [2] to the numerical integration of third order BVP. The case of the four-points block hybrid collocation method with two off-step points to solve general third order ODEs directly was studied by Yap and Ismail [3]. In the paper by Jikantoro et al. [4] presented the theory of B-series and the associated rooted trees through which order conditions of the hybrid methods for direct integration of special third order ODEs are derived.

In this research, we develop a continuous hybrid liner multistep method (HLMM) for direct solution of BVPs without reducing the problem to a lower order system or to an IVP equivalent. The proposed HLMM is zero stable, consistent and more accurate than the existing one. Experimental results confirm the superiority of the new schemes over the existing methods.

IOP Conf. Series: Materials Science and Engineering **597** (2019) 012075 doi:10.1088/1757-899X/597/1/012075

2. Development of the Method

Collocation solutions are desirable from practical and theoretical considerations and their advantages are now creating growing interest in continuous integration algorithms for numerical solution of ODEs. In particular, collocation solutions of the ODEs by their nature are continuous.

In the spirit of Onumanyi et al. [5] and Mohammed [6] we consider briefly the derivation of the continuous formula by the multistep collocation using constant mesh spacing h and give explicit representation for the coefficients.

The values of k and m are arbitrary except for collocation at the mesh points, where $0 < m \leq k+1$.

Let y_{n+j} be approximations to $y(x_{n+j})$ where $y_{n+j} = y(x_{n+j})$, $j = 0, 1, \ldots, k$, then a k-step multistep collocation formula with m collocation points is constructed as follows:

$$y(x_{n+k}) = \sum_{j=0}^{r-1} \bar{\alpha}_j y_{n+j} = h^3 \sum_{j=0}^{m-1} \bar{\beta}_j f(\bar{x}_j, y(\bar{x}_j)) + h^3 \beta_\eta f(x_\eta, y(x_\eta))$$

where $\bar{\alpha}_j$, $j = 0, 1, \ldots, r-1$, $\bar{\beta}_j$, $j = 0, 1, \ldots, m-1$, and β_η, $\eta \in \mathbb{R}$ are unknown constants of the discrete scheme. To obtain a continuous form of Eq. (2) we find the polynomial $y(x)$ of degree $p = r + m - 1$, $r > 0$, $m > 0$ of the form

$$y(x) = \sum_{j=0}^{r-1} \bar{\alpha}_j(x) y_{n+j} + h^3 \sum_{j=0}^{m-1} \bar{\beta}_j(x) f(\bar{x}_j, y(\bar{x}_j)) + h^3 \beta_\eta(x) f(x_\eta, y(x_\eta)), \tag{2}$$

such that it satisfies the conditions

$$\begin{aligned} \bar{\alpha}_j y(x_{n+j}) &= \bar{\alpha}_j y_{n+j}, & j = 0, 1, \ldots, r-1, \\ \bar{\beta}_j y'''(x_j) &= \bar{\beta}_j f(\bar{x}_j, y(\bar{x}_j)), & j = 0, 1, \ldots, m-1, \end{aligned}$$

where $\alpha_j(x)$ and $\beta_j(x)$ are assumed polynomials of the form

$$\alpha_j(x) = \sum_{i=0}^{r+m-1} \alpha_{j,i+1} x^i, \quad h^3 \beta_j(x) = \sum_{i=0}^{r+m-1} \beta_{j,i+1} x^i.$$

Points x_{n+j} in Eq. (2) are r arbitrarily chosen *interpolation* points taken from the range $\{x_n, \ldots, x_{n+k-1}\}$, $0 < r \leq k$, and the *collocation* points $\bar{x}_j$, $j = 0, 1, \ldots, m-1$ belongs to the extended set $Q = \{x_n, \ldots, x_{n+k}\} \bigcup \{x_{n+k-1}, x_{n+k}\}$.

From the interpolation conditions and the expression for $y(x)$ in Eq. (2) the following conditions are imposed on continuous coefficients $\alpha_j(x)$ and $\beta_j(x)$:

$$\begin{aligned} \alpha_j(x_{m+i}) &= \delta_j^i, \quad j = 0, 1, \ldots, r-1, \quad i = 0, 1, \ldots, r-1 \\ h^3 \beta_j(x_{n+i}) &= 0, \quad j = 0, 1, \ldots, m-1, \quad i = 0, 1, \ldots, r-1 \end{aligned}$$

and

$$\begin{aligned} \alpha_j'''(\bar{x}_{n+i}) &= 0, \quad j = 0, 1, \ldots, r-1, \quad i = 0, 1, \ldots, m-1 \\ h^3 \beta_j(x_{n+i}) &= \delta_j^i, \quad j = 0, 1, \ldots, m-1, \quad i = 0, 1, \ldots, m-1, \end{aligned}$$

here $\delta_j^i = [i = j]$ is the Kronecker delta.

IOP Conf. Series: Materials Science and Engineering **597** (2019) 012075 doi:10.1088/1757-899X/597/1/012075

We note that since the general third order ODEs (1) involves the first and second derivatives, the first and second derivative formulas are

$$y'(x) = \frac{1}{h}\left(\sum_{j=0}^{r-1} \alpha'_j(x)y_{n+j} + h^3\sum_{j=0}^{k}\beta'_j(x)y_{n+j} + h^3\beta'_\eta(x)f_{n+\eta}\right),$$

$$y''(x) = \frac{1}{h^2}\left(\sum_{j=0}^{r-1} \alpha''_j(x)y_{n+j} + h^3\sum_{j=0}^{k}\beta''_j(x)y_{n+j} + h^3\beta''_\eta(x)f_{n+\eta}\right),$$

and to obtain additional equations by imposing that

$$y'(x) = \delta(x), \quad y''(x) = \gamma(x), \quad y'(a) = \delta_0, \quad y''(a) = \gamma_0.$$

The method of continuous approximation can be expressed as

$$y(x) = \sum_{j=0}^{r-1} \alpha_j(x)y_{n+j} + h^3\sum_{j=0}^{k}\beta_j(x)y_{n+j} + h^3\beta_\eta(x)f_{n+\eta}. \tag{3}$$

3. Three-step hybrid method with one off-step collocation point

We use Equation (3) to obtain a three-step HLMM with the following specification: $r = 3$, $m = 5$, $\eta = \frac{8}{3}$, $k = 3$, $\alpha_j(x)$, $\beta_j(x)$, $\beta_\eta(x)$ can be expressed as functions of $t = \frac{x-x_n}{h}$. The HLMMs are usually represented in the form of a single block r-point multistep method [7]

$$\mathbf{A}Y_m = \mathbf{B}Y_{m-1} + h^3\mathbf{C}F_m + h^3\mathbf{D}F_{m-1} \tag{4}$$

where h a fixed mesh size within a block hybrid, $\mathbf{A}$, $\mathbf{B}$, $\mathbf{C}$, and $\mathbf{D} \in \mathbb{R}^{(k+1)\times(k+1)}$ are the coefficient matrices, Y_m, Y_{m-1}, F_m and $F_{m-1} \in \mathbb{R}^{k+1}$ are vectors of numerical approximation.

The hybrid method can be significantly shown in the form of Equation (4) to give

$$\begin{pmatrix} 3 & -3 & 0 & 1 \\ \frac{16}{9} & -\frac{20}{9} & 1 & 0 \\ -2 & \frac{1}{2} & 0 & 0 \\ 2 & -1 & 0 & 0 \end{pmatrix}\begin{pmatrix} y_{n+1} \\ y_{n+2} \\ y_{n+\frac{8}{3}} \\ y_{n+3} \end{pmatrix} = \begin{pmatrix} 0 & 0 & 0 & 1 \\ 0 & 0 & 0 & \frac{5}{9} \\ 0 & 0 & 0 & -\frac{3}{2} \\ 0 & 0 & 0 & 1 \end{pmatrix}\begin{pmatrix} y_{n-3} \\ y_{n-2} \\ y_{n-1} \\ y_n \end{pmatrix}$$
$$+ h^3\begin{pmatrix} \frac{47}{100} & \frac{23}{40} & -\frac{81}{800} & \frac{1}{20} \\ \frac{7}{27} & \frac{203}{729} & -\frac{25}{324} & \frac{65}{2187} \\ \frac{1399}{4200} & -\frac{23}{168} & \frac{783}{5600} & -\frac{17}{2800} \\ -\frac{109}{120} & \frac{61}{120} & -\frac{81}{160} & \frac{79}{360} \end{pmatrix}\begin{pmatrix} f_{n+1} \\ f_{n+2} \\ f_{n+\frac{8}{3}} \\ f_{n+3} \end{pmatrix} + h^3\begin{pmatrix} 0 & 0 & 0 & \frac{1}{160} \\ 0 & 0 & 0 & \frac{31}{8748} \\ 0 & 0 & 0 & -\frac{13}{224} \\ 0 & 0 & 0 & -\frac{451}{1440} \end{pmatrix}\begin{pmatrix} f_{n-3} \\ f_{n-2} \\ f_{n-1} \\ f_n \end{pmatrix}. \tag{5}$$

4. Order and Error Constant of Hybrid Linear Multi-step Method

With specific reference to the works of Fatunla [7] and Lambert [8], the local truncation error attributed to the conventional form of Equation (3) is defined as the linear difference operator

$$L[y(x);h] = \sum_{j=0}^{k}\left\{\alpha_j y(x+jh) - h^3\beta_j y'''(x+jh)\right\} - h^3\beta_\eta y'''(x+\eta h). \tag{6}$$

Suppose it is assumed that $y(x)$ can be adequately differentiated. It is possible to expand Equation (6) in the form of Taylor series about the point x to arrive at the expression

$$L[y(x);h] = C_0 y(x) + C_1 y'(x) + \ldots + C_q h^{(q)} y^{(q)}(x) + \ldots$$

where the constant coefficients C_q, $q = 0, 1, \ldots$ are given as shown below:

$$C_0 = \sum_{j=0}^{k} \alpha_j, \quad C_1 = \sum_{j=1}^{k} j\alpha_j, \ldots, C_q = \frac{1}{q!} \sum_{j=1}^{k} j^q \alpha_j - q(q-1)(q-2) \left(\sum_{j=1}^{k} j^{q-3}\beta_j + \eta^{q-3}\beta_\eta \right).$$

According to the paper by Henrici [9], the method (2) has the order p if

$$C_0 = C_1 = \ldots = C_p = C_{p+1} = 0, \quad C_{p+2} = 0, \quad \text{and} \quad C_{p+3} \neq 0.$$

Therefore, C_{p+3} is the error constant. The proposed method (5) has a uniform order $p = 5$ and the error constants as

$$C_{p+3} = \left(-\frac{7}{7200}, -\frac{85}{157464}, -\frac{811}{18}, -\frac{197}{14} \right)^\top.$$

In order to analyze the method (5) for zero stability, we normalize the scheme and write it as a block method from which we obtain the first characteristic polynomial $\rho(R)$ given by

$$\rho(R) = \det(R \cdot \mathbf{A}^{(0)} - \mathbf{A}^{(1)}) = R^k(R-1).$$

It is easily shown that method (5) is normalized to give the first characteristic polynomial $\rho(R)$ given by

$$\rho(R) = \det(R \cdot \mathbf{A}^{(0)} - \mathbf{A}^{(1)}) = R^3(R-1)$$

where $\mathbf{A}^{(0)} = \mathbf{1}_{4\times4}$ is the identity matrix of dimension 4, $\mathbf{A}^{(1)} = \mathbf{1}_{4\times1} \cdot \mathbf{i}_{4,4}^{\mathrm{T}}$ is the matrix of dimension 4, $\mathbf{i}_{4,4}$ is the 4-th column of $A^{(0)}$, and roots of the characteristic polynomial $\rho(R)$ are $R = (0, 0, 0, 1)$. Therefore, the proposed method is zero-stable.

5. Region of Absolute Stability

The boundary locus method is used to obtain the stability region of the main method of Equation (5). The boundary locus curve is obtained by setting

$$\bar{h} = \lambda h = \frac{\rho(z)}{\sigma(z)}, \quad \lambda > 0, \quad z = \exp\{i \cdot \theta\} \in \mathbf{C}, \quad \theta \in [0, 2\pi], \tag{7}$$

where $\rho(z)$, $\sigma(z)$ are first and second characteristic polynomial of liner multistep method. The main method from Equation (5) is written as

$$y_{n+3} - 3y_{n+2} + 3y_{n+1} - y_n = h^3 \left(\frac{1}{160} f_n + \frac{47}{100} f_{n+1} + \frac{23}{40} f_{n+2} - \frac{81}{800} f_{n+\frac{8}{3}} + \frac{1}{20} f_{n+3} \right). \tag{8}$$

The first and second characteristics polynomials are written as

$$\rho(z) = z^3 - 3z^2 + 3z - 1, \quad \sigma(z) = \frac{1}{160} + \frac{47}{100} z + \frac{23}{40} z^2 - \frac{81}{800} z^{\frac{8}{3}} + \frac{1}{20} z^3.$$

Substituting $\rho(z)$ and $\sigma(z)$ into the Equation (7), the values of $\bar{h}$ is obtain which is plotted to produce the region of absolute stability of the method (Figure 1). The stability region for the method (8) turns out to be the inside part of the complex plane shown in Figure 1, including the boundary colored by green [11]. From the Figure 1 we found that the interval of absolute stability is $(-1.5, 0)$.

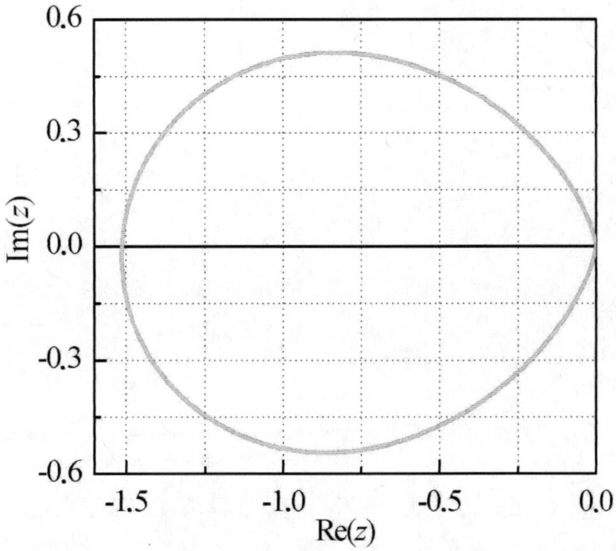

Figure 1. Region of absolute stability for the proposed method

Table 1. Numerical Results for Problem 2

x	Exact Solution	Numerical Solution	Error	Tirmizi et. al [12]
0.0	-0.01210709	-0.012107056410	3.36×10^{-8}	6.6530×10^{-5}
0.1	-0.01126851	-0.011268451800	5.82×10^{-8}	6.5000×10^{-5}
0.2	-0.00922221	-0.009222146339	6.37×10^{-8}	5.2254×10^{-5}
0.3	-0.00646687	-0.006466811798	5.82×10^{-8}	3.6300×10^{-5}
0.4	-0.00332019	-0.003320153971	3.60×10^{-8}	1.8750×10^{-5}
0.6	0.00332019	0.003320118521	7.15×10^{-8}	1.7340×10^{-5}
0.7	0.00646687	0.006466679335	1.91×10^{-7}	3.4050×10^{-5}
0.8	0.00922221	0.009221906116	3.04×10^{-7}	4.9801×10^{-5}
0.9	0.01126851	0.011268057530	4.52×10^{-7}	6.2020×10^{-5}
1.0	0.01210709	0.012106500950	5.89×10^{-7}	6.3480×10^{-5}

6. Numerical Experiments

The accuracy of the proposed method was implemented for direct solution of BVPs of third order ODEs of linear and non-linear equations. The implementation of the method was coded using the Maple Software.

Problems 1. Non-linear Blasius Equation [10]

$$2y''' + yy'' = 0, \quad y(0) = 0, \quad y'(0) = 0, \quad y'(\infty) = 1.$$

The exact solution does not exist. Comparison numerical solutions using the proposed method and the fourth order Runge-Kutta (RK) method for the Problem 1 is shown in Fig. 2.

Problems 2. Sandwich Beam Problem [12]

$$y''' - k^2y' + r = 0, \quad y(0.5) = 0, \quad y'(0) = 0, \quad y'(1) = 0.$$

The proposed schemes for the values $k = 1$ and $r = 5$ are relatively more accurate than the schemes of Tirmizi et al. [12] for Problem 2. Absolute errors are presented in Table 1.

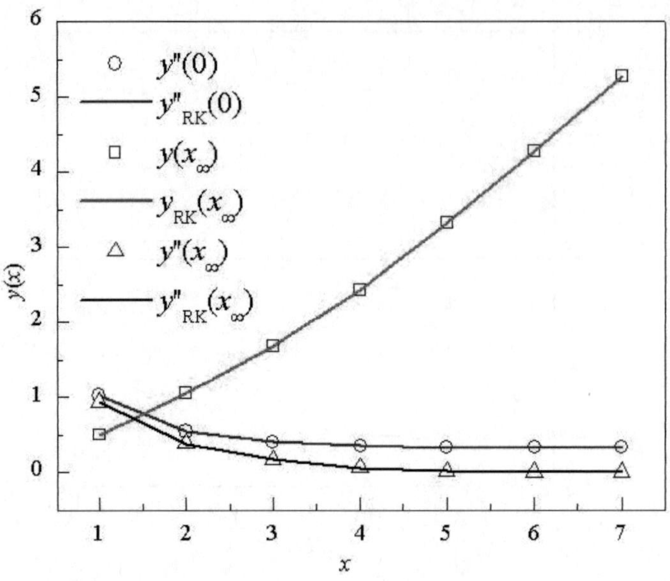

Figure 2. Comparison numerical solution of proposed method with Runge-Kutta method for the Problem 1

7. Conclusion

This research describes the development, analysis and implementation of block methods for solving third order ordinary differential equations directly. The development and/or construction of class of hybrid linear multi-step methods for direct solution of initial value problems and boundary value problems arising from third order ODEs have been presented. The derived schemes which are of block form were analyzed and applied to some selected and standard problems from literature.

References

[1] You X and Chen Z 2013 Direct integrators of Runge-Kutta type for special third-order ordinary differential equations, *Applied Numerical Mathematics* **74** 128–150

[2] Jator S N 2008 On the numerical integration of third order BVP by linear multi-step methods, *International Journal of Pure and Applied Mathematics* **46** (3) 375–388

[3] Yap L K and Ismail F 2018 Four point block hybrid collocation method for direct solution of third order ordinary differential equations, *AIP Conference Proceedings* **1974** (020055)

[4] Jikantoro Y D, Ismail F, Senu N, Ibrahim Z B 2018 Hybrid methods for direct integration of special third order ordinary differential equations, *Applied Mathematics and Computation* **320** 452–463

[5] Onumanyi P, Awoyemi D O, Jator S N and Siresina U W 1994 New Linear multistep methods with continuous coefficients for first order initial value problems, *Journal of Nigerian Mathematical society* **13** 37–51

[6] Mohammed U, Oyelami O, Semenov M 2019 An Orthogonal-based Self-starting Numerical Integrator for Second Order Initial and Boundary Value Problems ODEs, *Journal of Physics: Conference Series* **1145** (012040)

[7] Fatunla S O 1991 Block method for second order initial value problem (IVP), *International Journal of Computer Mathematics* **41** 55–63

[8] Lambert J D 1973 *Computational method in ordinary differential equation*, John Wiley and Sons, London, U.K.

[9] Henrici P 1962 *Discrete Variable Methods for ODEs*, New York, John Wiley and Sons

[10] Shampine L F and Thompson S 2007 Initial value problems, *Scholarpedia* **2** (3) 2861

[11] Butcher J C 2016 *Numerical Methods for Ordinary Differential Equations*, New York, John Wiley and Sons

[12] Tirmizi I A, Twizell E H and Islam S 2005 A numerical method for third-order non-linear boundary value problems in engineering, *International Journal Computer Mathematics*, **82** (1) 103–109

Mathematical modelling of a single supersonic jet impingement with a flat obstacle

A M Kagenov[1], K V Kostyushin, V A Kotonogov, A S Pomiluyko, K L Aligasanova, R Yu Burgomistrenko and I V Eremin

Research Institute of Applied Mathematics and Mechanics of *Tomsk State University*, 36 Lenin Avenue, Tomsk 634050, Russian Federation.

[1]Corresponding author: anuar@ftf.tsu.ru

Abstract. In this paper the mathematical modeling of the interaction of a single supersonic jet with a flat obstacle in a rarefied atmosphere and Mach numbers at the nozzle exit 4 are considered. Mathematical modeling using free OpenFOAM-Extended software and authors' program are done. Comparison of calculation methods: PISO algorithm, Rusanov and Godunov schemes was carried out. The comparison results show a qualitative coincidence in the shape of the supersonic jet, wall jet and the system of shock waves. But the numerical results are quantitatively different, pressure on flat obstacle varied within 7-14 kPa. The peak of pressure is on the periphery for results of sonicFoam and dbnsFoam. For all solvers average pressure at axis of jet is 8 kPa and the first barrel of wall jet is started about at 0.4 meter from axis was obtained.

1. Introduction

Now organizations and institutes involved in the design of rocket and space equipment components are carrying out experimental and theoretical studies of the physical processes at the launch of rockets, descent and landing of spacecraft. A large number of experimental and theoretical works to the study of jet flows and their interaction with obstacles are devoted [1-12]. In experimental studies, the main study was the leakage of jets (Mach numbers at the nozzle section M<3) onto flat (inclined) obstacles under atmospheric conditions. Complication of experiments (modeling the conditions of other planets) requires a significant increase in the scope of work, its cost and the solution of a number of additional technical problems (creation of specialized test benches). In theoretical studies, cycles of work on the simulation of single and multi-block turbulent jets, and their interaction with flat obstacles when varying various parameters (Mach numbers at the nozzle section, number of jets, degrees of non-calculation, distances to obstacles, oblique inclination angles, etc.) were carried out.

The aim of the work is a study of the interaction of a supersonic jet with obstacles in a wide range of applicability, namely with a rarefied atmosphere and Mach numbers at the nozzle exit 4-5; as well as to investigate the shock-wave structure of the impingement jet and its force action on obstacle using deference solver and schemes.

2. Physical and mathematical model

The supersonic jet emanates from the nozzle and flows onto a flat, impenetrable barrier. The jet leakage, the main gas flow is inhibited and spreads around the periphery. At high pressure ratio, a fan jet appears. Several cases of jet interaction with a flat obstacle are possible. First, when the maximum pressure falls on the jet axis. The second, when the maximum pressure is formed at the periphery, in this case, the formation of self-oscillatory modes.

IOP Conf. Series: Materials Science and Engineering **597** (2019) 012076 doi:10.1088/1757-899X/597/1/012076

For the mathematical description of the interaction of supersonic jet with a flat obstacle, the turbulent flow of a viscous compressible gas in the three-dimensional approximation is considered [13-16]. The system of equations in a rectangular Cartesian coordinate system x_1, x_2, x_3 for nonstationary gas flow has the form:

The continuity equation

$$\frac{\partial \rho}{\partial t}+\frac{\partial}{\partial x_j}\left[\rho u_j\right]=0 \tag{1}$$

where ρ – density, kg/m^3; t – time, s; u_j – velocity, m/s; x_j – coordinate, m.

The momentum equation

$$\frac{\partial}{\partial t}\left(\rho u_i\right)+\frac{\partial}{\partial x_j}\left[\rho u_i u_j+p\delta_{ij}-\tau_{ji}\right]=0 \tag{2}$$

where p – pressure, Pa; δ_{ij} – Kronecker delta.

The viscous stress tensor has the following form

$$\tau_{ij}=2\mu\left[\frac{1}{2}\left(\frac{\partial u_i}{\partial x_j}+\frac{\partial u_j}{\partial x_i}\right)-\frac{1}{3}\frac{\partial u_k}{\partial x_k}\delta_{ij}\right] \tag{3}$$

The energy equation

$$\frac{\partial}{\partial t}\left(\rho E\right)+\frac{\partial}{\partial x_j}\left[\rho u_j E+u_j p+q_j-u_i\tau_{ij}\right]=0 \tag{4}$$

where E – total internal energy, J; q_j – heat-flux, J/(m^2·s).

The ideal gas equation

$$p=\rho RT \tag{5}$$

where T – temperature, K; R – gas constant, J/(kg·K), indexes $i, j, k=1,2,3$.

To close the system of equations (1)-(5) for viscous gas $SST\ k-\omega$ turbulent model is used [17]. In figure 1 computational domain and boundaries are shown. At the inlet section of the nozzle (boundary BC1), the gas flows into the computational domain at a subsonic speed. The following boundary conditions are set on the boundary of BC1: $p=p_0$, $T=T_0$, $k=k_0$, $\omega=\omega_0$, $u_x=0$, $u_z=0$. For the boundaries BC2 and BC5, the following conditions are specified: for velocity, the sticking condition $u_x=u_y=u_z=0$, for temperature $grad\left(T\right)=0$. For the kinetic energy of turbulence and the specific dissipation of the kinetic energy of turbulence, the wall functions are used. On the boundaries of BC3 and BC4 exhibited "soft" boundary conditions.

To implement the physical and mathematical model (1)-(5) and conduct numerical research free software OpenFOAM-Extended [18-19] and authors' computational program [20] was used. For viscous gas sonicFoam solver of second order accuracy was used, where pressure and velocity as dependent variables through the PISO method.

IOP Conf. Series: Materials Science and Engineering **597** (2019) 012076 doi:10.1088/1757-899X/597/1/012076

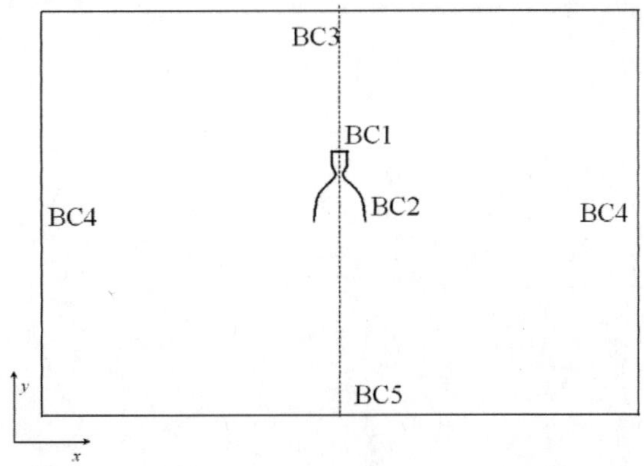

Figure 1. Computational domain and boundaries

For inviscid gas density based solver dbnsFoam and authors program of first order accuracy was used. In dbnsFoam solver was the approximate solution of the Riemann problem according to the Rusanov scheme, and in authors program Godunov method. System of linear of algebraic equations was solved using the Gauss-Seidel method. The time sampling was carried out by a four-step Runge-Kutta method. Numerical investigation were done using SKIF Cyberia supercomputer of Tomsk State University.

3. Numerical results

The geometric characteristics of the nozzle were as follows: nozzle throat diameter 0.03613 m, nozzle diameter 0.19395 m, and the angle of the semi-solution of the nozzle is 10 degrees. For numerical calculations at the initial time unmoved environment was assumed. The ambient pressure was 650 Pa, and temperature was 250 K. The parameters at the nozzle inlet was: pressure – 0.28 MPa, temperature – 1180 K, adiabatic index – 1.33719. The distance from nozzle exit to flat obstacle was 0.5 meter.

The results of calculations are shown in figure 2 of Mach number and in figure 3 the distribution of pressure on flat obstacle for calculation results of sonicFoam, dbnsFoam solvers and of authors' program.

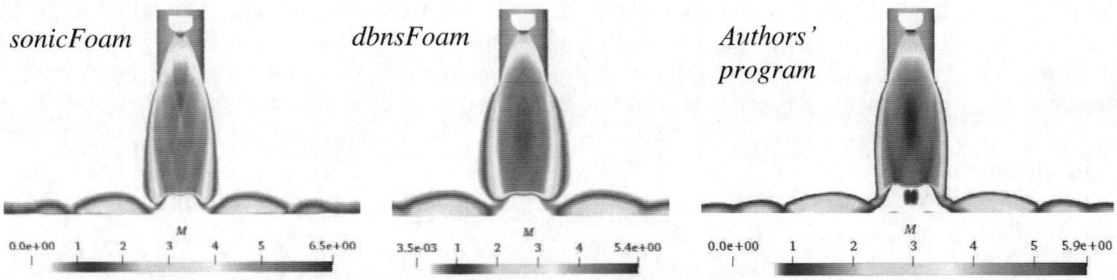

Figure 2. Distribution of Mach numbers

The qualitatively pattern of the distribution of Mach numbers are coincide. Local maximum of Mach numbers for sonicFoam – 6.5, dbnsFoam – 5.4, and authors' program – 5.9. Due to second order accuracy of sonicFoam solver, it gave system of shock waves.

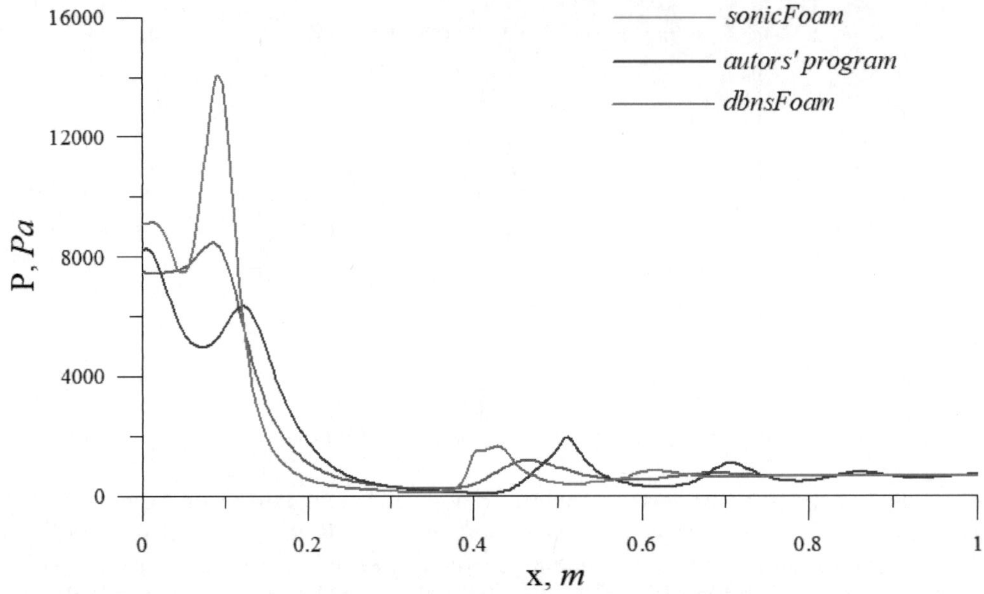

Figure 3. Distribution of pressure on flat obstacle

For dbnsFoam and authors' program blurred shock wave patterns. From distribution of pressure on flat obstacle in axis of jet maximum for sonicFoam – 9 kPa, dbnsFoam – 7.5 kPa, and authors' program – 8.2 kPa are obtained. For sonicFoam and dbnsFoam peak pressure is on the periphery: sonicFoam – 14 kPa, and dbnsFoam – 8.4 kPa. For all solvers wall jet started about at 0.4 meter. The second barrel of wall jet for sonicFoam and authors' program is observed. The difference in the results to the applicable scheme of calculations and orders of accuracy is related.

4. Conclusion

Mathematical modeling of the interaction of a single supersonic jet with a flat obstacle in rarefied atmosphere and Mach number at the nozzle exit 4 are done. Comparison of numerical results of sonicFoam, dnsFoam, and authors' program are done, which are used PISO algorithm, Rusanov scheme, and Godunov method. A qualitative coincidence was obtained in the shape of the supersonic jet, wall jet and the system of shock waves, but the results are quantitatively different. The average pressure at axis of jet is 8 kPa, for sonicFoam and dbnsFoam peak of pressure is on the periphery, and the first barrel of wall jet for all solvers is start about at 0.4 meter from axis was obtained.

Acknowledgments
The reported study was funded by RFBR and Tomsk region according to the research project № 19-41-703005

References

[1] Zapryagaev V I and Kiselev N P 2009 *J. Appl. Mech. Tech. Phys.* **50 (3)** 447–453
[2] Zapryagaev V I, Kavun I N and Kiselev N P 2010 *Int. J. of Aerospace Innovations* **2 (4)** 277–294
[3] Zapryagaev V I, Kavun I N and Kiselev N P 2010 *J. Appl. Mech. Tech. Phys.* **51 (2)** 202–210
[4] Zapryagaev V I, Kavun I N and Solotchin A V 2015 *J. Appl. Mech. Tech. Phys.* **56 (3)** 406–413
[5] Zapryagaev V I, Kiselev N P and Kundasev S G 2016 *PNRPU Aerospace Engineering Bulletin* **45** 32–49
[6] Zapryagaev V I, Gubanov D A, Kavun I N, Kiselev N P, Kundasev S G and Pivovarov A A 2017 *AIP Conference Proceedings* **1893** 030058
[7] Melnikova M F and Nesterov Yu N 1971 *TSAGI* **2 (5)** 44–58
[8] Krothapalli A, Rajkuperan E, Alvi F and Lourenco L M 1999 *J. Fluid Mech.* **392** 155–181
[9] Kudimov N F, Safronov A V and Tretyakova O N 2013 *MAI* **69** 1–11

[10] Kudimov N F, Safronov A V and Tretyakova O N 2013 *MAI* **70** 1–14
[11] Kudimov N F, Safronov A V and Tretyakova O N 2014 *Applied problems of gas dynamics and heat transfer in power plants of rocket technology* (Moscow: MAI) p 168
[12] Degtyar V G, Merkulov E S, Hlibov V I and Safronov A V 2013 *Cosmonautics and Rocket Engineering* **70 (1)** 37–45
[13] Glaznev V N, Zapryagaev V I, Uskov V N and et al. 2000 *Jet and nonstationary flows in gas dynamics* (Novosibirsk: Publishing Firm of Siberian Branch of Russian Academy of Sciences) p 200
[14] Dulov V G and Lukianov G A 1984 Gas Dynamics of Outflow Processes (Novosibirsk: Nauka) p 226
[15] Godunov S K, Zabrodin A V, Ivanov M Y, Kraiko A N, Prokopov G P 1976 *Numerical Solution of Multidimensional Problems of Gas Dynamics* (Moscow: Nauka) p 400
[16] Toro E F 2009 *Riemann Solvers and Numerical Methods for Fluid Dynamics* (Berlin: Springer-Verlag) p 724
[17] Menter F R, Kuntz M, Langtry R 2003 *Proc. of the 4th Int. Symposium on Turbulence, Heat and Mass Transfer* (Begell House Inc., West Redding) 625–632
[18] Kagenov A, Glazunov A, Kostyushin K and Eremin I 2017 *AIP Conference Proceedings* **1899** 060010
[19] Kagenov A M, Glazunov A A, Kostyushin K V, Eremin I V, Tyryshkin I M and Zhilcov K N 2018 *AIP Conference Proceedings* **2027** 030071
[20] Kostyushin K, Kagenov A, Eremin I, Zhilcov K and Shuvarikov V 2017 *AIP Conference Proceedings* **1899** 060012

IOP Conf. Series: Materials Science and Engineering **597** (2019) 012077 doi:10.1088/1757-899X/597/1/012077

Method of gas flows calculation in solid propellant rocket engines taking into account the combustion of solid fuel charge

K V Kostyushin[1], I V Eremin, A M Kagenov, K N Zhiltsov, I M Tyryshkin, M S Sereda, A V Chervakova

Research Institute of Applied Mathematics and Mechanics of *Tomsk State University*, 36 Lenin Avenue, Tomsk 634050, Russian Federation.

[1]Corresponding author: kostushink@niipmm.tsu.ru

Abstract. The paper presents a method for calculating the local and integral characteristics of the flow in the axisymmetric gas-dynamic paths of solid propellant rocket motors, taking into account the combustion of a charge of solid fuel. The numerical method of calculation is based on the use of the Godunov scheme, formulated for moving computational grids. The speed of movement of the combustion surface is defined locally on the edge of each calculation boundary cell. This approach allows us to take into account the uneven distribution of the pressure of the combustion products in the free volume of the combustion chamber. In test calculations, the power law of burning rate is used. Calculations of the gas flow in the solid propellant combustion chamber with cylindrical charge of solid fuel are carried out. Unsteady pressure curve in the combustion chamber is obtained. The method allows to determine all integral characteristics of the developed solid propellant rocket motors as a function of the engine running time.

1. Introduction

Currently in Russia and in the world there is the task of developing heavy launch vehicles. One of the concepts is to equip the launch vehicle with solid-fuel launching booster. Solid propellant rocket engines (SPRE) have a number of advantages: simple design, reliability and a quick preparation for launch. The control of engine thrust in SPRE is usually carried out in the "passive" mode and is determined by the form of solid fuel charge and fuel characteristics. The form of fuel charge in this case should provide the necessary burning law of the combustion surface depending on the pressure in the combustion chamber. The transient pressure curve in the general case determines the behavior of propulsive power. The law of change of the burning surface as a function of time is, as a rule, unregulated and is designed for a specific cycle of engine operation and task. As a rule, in modern SPRE use charges of a complex spatial or axisymmetric form such as the "cocoon", "umbrella", "star", "slot". Structural design of such engines is made using physical and mathematical modeling. Mathematical modeling is used to create a general layout with the aim of reducing costs at conducting field experiments [1-3].

In engineering practice, to solve such the problems, approximate approaches based on averaging the pressure over the chamber volume and changing the burning surface in parallel layers are often used. It is known that the operation process in a combustion chamber under actual usage conditions is characterized by the spatial distribution of the combustion products pressure in the free volume of combustion chamber. This leads to distortion changes of a charge burning surface of solid fuel during engine operation. Therefore, when calculating non-stationary performance characteristics, it is necessary

to take into account the change in the filler geometry according to the local pressure at the combustion surface, and not the average pressure in the combustion chamber volume. Minkov et al. ([4, 5]) were developed a calculation method taking into account non-stationary changes in the combustion surface based on the use of ghost point and the Lax-Wendroff method. This approach allows calculations to be carried out without rebuilding the grid even for complex combustion surfaces. However, the assumption that the combustion of propellant occurs in a quasi-stationary mode, and the pressure field has time to adjust to the change in the burning surface does not allow for the effects of chamber relaxation [6]. This feature makes it impossible to simulate unstable modes of solid propellant in SPRE.
The aim of this work is to develop an algorithm and methodology for calculating the local and integral characteristics of a solid propellant rocket motor, taking into account the change in the geometry of the flow path of the combustion chamber depending on the local pressure distribution at the combustion surface.

2. Calculation Methodology

The Euler equations, which govern the compressible, inviscid gas flow in three-dimensions, are [7]:

$$\frac{d}{dt}\left(\iiint_G \rho dG\right) + \oint_S \rho \mathbf{v} \cdot d\mathbf{S} = 0, \tag{1}$$

$$\frac{d}{dt}\left(\iiint_G \rho \mathbf{v} dG\right) + \oint_S \left(\rho \mathbf{v}\mathbf{v} + p\hat{\mathbf{I}}\right) \cdot d\mathbf{S} = 0, \tag{2}$$

$$\frac{d}{dt}\left(\iiint_G e dG\right) + \oint_S \left(e + p\right) \mathbf{v} \cdot d\mathbf{S} = 0, \tag{3}$$

where: ρ – density; t – time; p – pressure; $\mathbf{v} = \left[u, v, w\right]^T$ – gas velocity; $\hat{\mathbf{I}}$ – 3×3 identity tensor; ε – specific internal energy; $e = \rho\varepsilon + \rho\left(u^2 + v^2 + w^2\right)/2$ – total energy per unit volume; G – final volume in three-dimensional space; $dG = dxdydz$ – element of volume; S – surface limiting the volume G; $d\mathbf{S} = \mathbf{n}dS$ – orientable surface element S, where $\mathbf{n}$ – external normal unit vector to the surface S; dS – differential surface element.
System (1) – (3) is closed by equation of state:

$$\varepsilon = \frac{p}{\left(k-1\right)\rho}. \tag{4}$$

In order to build difference scheme the finite volume method is applied. Therefore, we will cover all computational domain with discrete cells consisting of arbitrary convex polyhedrons with volume G_i, where $i = 1,2,...$ – volume number, and $m(i)$ – face number of i finite volume. Each face of i finite volume has surface S_j, where $j = 1,2,...,m(i)$. Approximation of integral equations in each of polyhedrons obtains in the following way:

$$G_i \frac{\rho_i^{k+1} - \rho_i^k}{\Delta t} + \sum_{j=1}^{m(i)} R_j \left(\mathbf{V}_j \cdot \mathbf{S}_j\right) = 0, \tag{5}$$

$$G_i \frac{\left(\rho\mathbf{v}\right)_i^{k+1} - \left(\rho\mathbf{v}\right)_i^k}{\Delta t} + \sum_{j=1}^{m(i)} \left(R_j \mathbf{V}_j\right)\left(\mathbf{V}_j \cdot \mathbf{S}_j\right) + \sum_{j=1}^{m(i)} P_j \mathbf{S}_j = 0, \tag{6}$$

$$G_i \frac{e_i^{k+1} - e_i^k}{\Delta t} + \sum_{j=1}^{m(i)} \left(E_j + P_j\right)\left(\mathbf{V}_j \cdot \mathbf{S}_j\right) = 0. \tag{7}$$

As a result we will get Godunov's scheme for custom computational mesh. Here $\mathbf{S}_j = \mathbf{n}_j S_j$, and Δt – time step. Lower index i signifies value of functions, which are referred to mass center of i polyhedrons, and lower index j signifies values referred to center of j face of discrete cell. Upper index k signifies the number of time step. Flux values of R, $\mathbf{V}$, P, E signify correspondingly density, speed, pressure and total energy at faces of control volume. These values are defined from the solution of Riemann problem [14] in the direction of outward normal. Also for the flow calculations can be used other methods, based on approximate solutions of Riemann problem: Roe, Osher, HLL, HLLC, HLLE and others [8].

For moving grids, the Godunov scheme looks like:

$$\frac{(\rho G)_i^{k+1} - (\rho G)_i^k}{\Delta t} + \sum_{j=1}^{m(i)} R_j \left([\mathbf{V} - \mathbf{D}] \cdot \mathbf{S}^{k+1/2}\right)_j = 0, \tag{8}$$

$$\frac{(\rho \mathrm{v} G)_i^{k+1} - (\rho \mathrm{v} G)_i^k}{\Delta t} + \sum_{j=1}^{m(i)} (R\mathbf{V})_j \left([\mathbf{V} - \mathbf{D}] \cdot \mathbf{S}^{k+1/2}\right)_j + \sum_{j=1}^{m(i)} P_j \mathbf{S}_j^{k+1/2} = 0, \tag{9}$$

$$\frac{(eG)_i^{k+1} - (eG)_i^k}{\Delta t} + \sum_{j=1}^{m(i)} E_j \left([\mathbf{V} - \mathbf{D}] \cdot \mathbf{S}^{k+1/2}\right)_j + \sum_{j=1}^{m(i)} P_j \mathbf{S}_j^{k+1/2} = 0, \tag{10}$$

$$\frac{G_i^{k+1} - G_i^k}{\Delta t} - \sum_{j=1}^{m(t)} \left(\mathbf{D} \cdot \mathbf{S}^{k+1/2}\right)_j = 0. \tag{11}$$

Here $\mathbf{D}_j$ – velocity of the center of the j cell face. Top index $k+1/2$ signifies values at time $t+\Delta t/2$. Equation (11) describes changes in the volume of G_i discrete cell. To determine the flow through the moving faces of a discrete cell, a modified solution of the Riemann problem is used [7]. Scheme (8) – (11) is first-order accurate in space and time.

The geometrical parameters of the cells on the burning surface and the local mass flow from the edge of the computational cell at the current time step are calculated using the power law of the burning rate:

$$u_L = u_0 p_L^{\nu}, \tag{12}$$

$$\mathbf{x}^{k+1} = \mathbf{x}^k + \mathbf{n}_x u_L \Delta t, \tag{13}$$

$$m = \rho_T u_L, \tag{14}$$

where: u_L – local burning rate at the edge of the cell; u_0, ν – constants in the law of burning rate; p_L – local pressure in the neighborhood of the vertex of moving face; $\mathbf{x}$ – the coordinate of the vertex of the movable edge; $\mathbf{n}_x$ – external normal unit vector to the burning surface at $\mathbf{x}$ point; m – local mass flow; ρ_T – solid fuel density.

The solution algorithm at the current time step on a moving computational grid is constructed as follows:

1. Using relations (12) – (14), new coordinates of the vertices of the moving faces of the cells are determined.
2. For each moving cell, geometric parameters (cell volume and face areas) are recalculated.
3. Using (14), the local mass flow for each face of the moving computational grid is determined.

3. Numerical results

Based on the above calculation method, a software package for calculating flow characteristics in the combustion chambers and nozzles was developed. To verify the methodology and program for calculating, numerical studies in a combustion chamber with a cylindrical charge of solid fuel are carried out. The scheme of the engine is shown in Figure 1. The linear dimensions are related to the radius of the critical section. Here $R^{*}=1$, $L_c=28.93$, $R_c=2.24$, $R_{sf}=1.02$, $R_e=3$, $L_{sub}=1.6$, $L_{sup}=7.1$. The computational domain was divided into 2 subdomain: the subdomain of the combustion chamber, which includes the combustion surface; subdomain of the nozzle block. For each subregion, a structured computational grid was constructed, in accordance with the methodology [9]. At the initial time, in the first subregion, the size of the computational grid consist of 4×50 cells, and in the subregion of the nozzle unit - 10×20 cells. During the calculation the number of cells remains unchanged.

On all impermeable boundaries the solid wall boundary condition was specified, and in the exit section of the nozzle, a supersonic output. The boundary conditions on the burning surface were determined through a local gas input through the edge of the computational cell according to scheme (12) - (14). The coefficients in the law of burning rate of the model fuel is: $u_0 = 0.006$ m/s, $\nu = 0.3$. The heat capacity ratio of combustion products $k = 1.19$. The combustion temperature of the combustion products is 3400 K. In the combustion chamber at the initial time, the pump pressure was set to 5 atm.

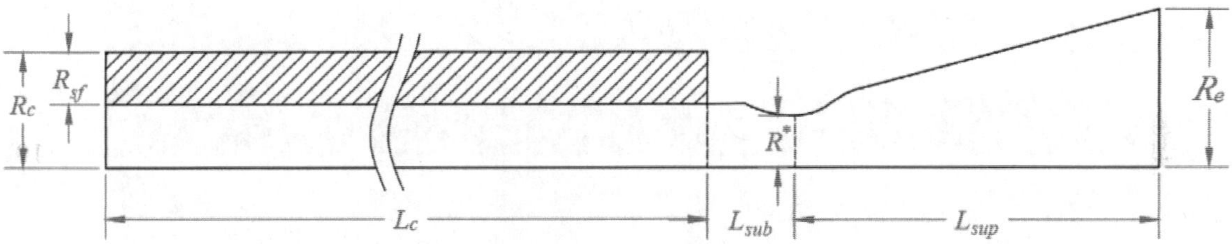

Figure 1. SPRE scheme

Figure 2 shows the complete pressure curve in the combustion chamber with a cylindrical charge from the start of engine operation to the moment of shutdown. We can see the sections of the starting operation, the main mode of engine operation and pressure drop. The pressure peak at 0.12 sec is caused by the effect of chamber relaxation. Figure 2 a) - e) shows the change in the distribution of the pressure field and the evolution of the computational grid at 0 s, 0.12 s, 0.54 s, 10.0 s, 15.0 s, 20.0 s, 22.5 s.

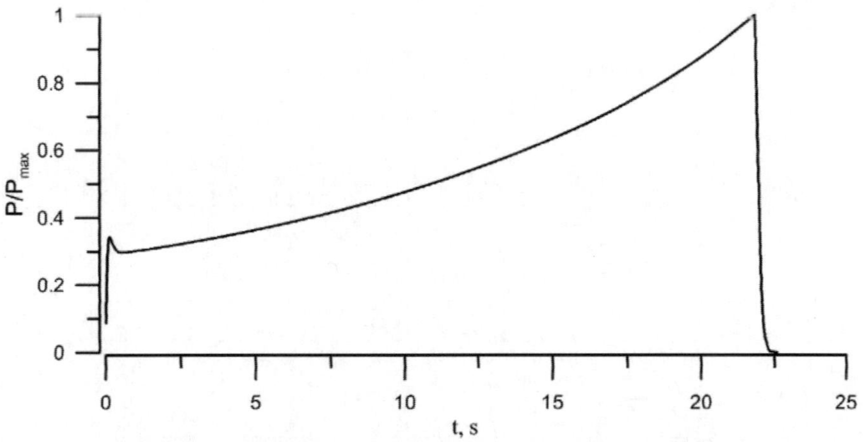

Figure 2. Pressure curve of engine operation process

Figure 4 shows the distribution of local values of the burning rate along the burning surface for time t=5 s (a), 10 s (b), 15 s (c), 20 s (d), 21.9 s (e). It can be seen that the rate of combustion along the surface of the charge of solid fuel is substantially uneven, higher at the front bottom, and lower at the entrance to the nozzle. This correlates with the pressure distribution along the axis of the combustion chamber. Figure 4 (d) shows an uncharacteristic increase of burning speed at the front of the chamber. This is due to a local increase of pressure during the flow around a ledge at the entrance to the nozzle, which is the element of the combustion chamber armor. The time t = 21.9 s. (Figure 4 (e)) is characterized by a partial burnout of the charge of solid fuel in the region starting from the front bottom to L = 18.5.

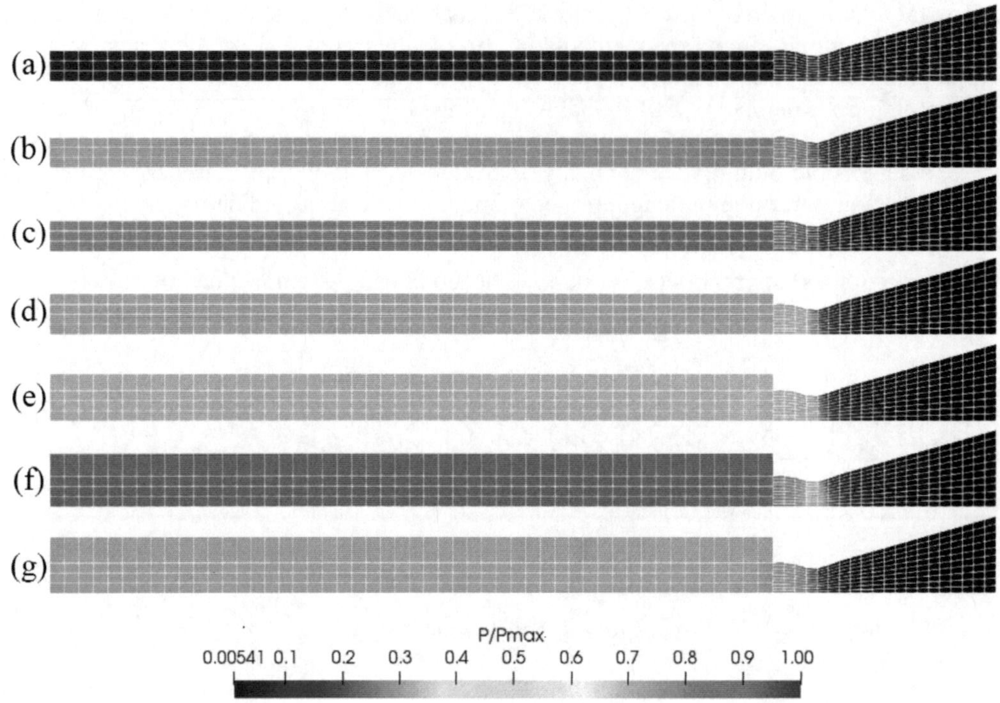

Figure 3. Pressure fields distribution and calculation grid change at various time a) 0 s, b) 0.12 s, c) 0.54 s, d) 10.0 s, e) 15.0 s, f) 20.0 s, g) 22.5 s

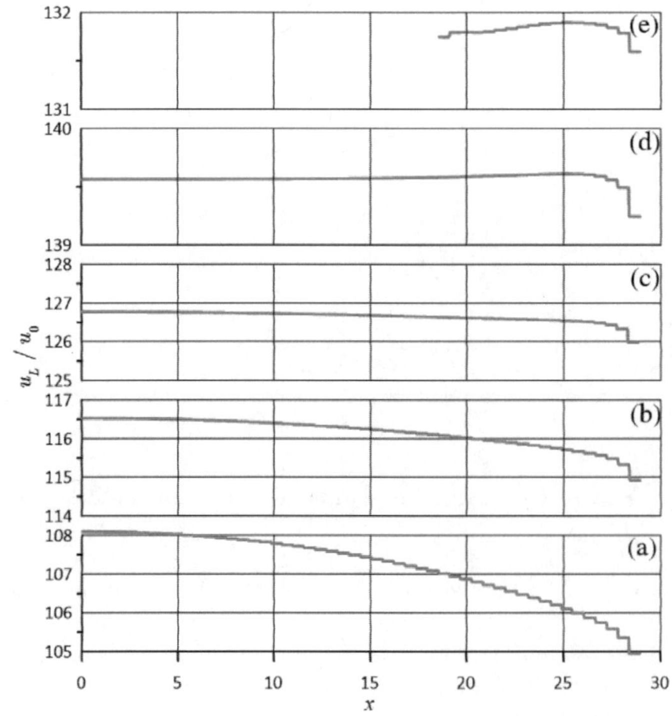

Figure 4. Local burning rate at time a) 5 s, b) 10 s, c) 15 s, d) 20 s, e) 21.9 s.

4. Conclusion

The method for calculating the local and integral characteristics of axisymmetric solid propellant rocket motors, taking into account the combustion of a charge of solid fuel, has been developed. The calculation method is based on the application of the Godunov scheme, formulated for moving computational grids. The speed of movement of the burning surface is determined by the local pressure values at the edges of each boundary computational cell according to the empirical law of the burning rate. Such an approach makes it possible to determine the combustion surface and obtain a pressure curve of the combustion chamber, taking into account the uneven distribution of pressure in the free volume. The technique allows to determine the intraballistic and integral characteristics of the developed solid propellant rocket motors as a function of the engine running time, for example, launching booster with cylindrical sectioned charges.

Acknowledgments

The work executed at financial support of the Ministry of education and science of the Russian Federation in the framework of the state assignment, project No 9.9063.2017/8.9.

References

[1] Arkhipov V A, Bondarchuk S S, Zhukov A S, Zolotorev N N and Perfil'eva K G 2017 *Combust Explos Shock Waves* **53** 634–640 DOI: 10.1134/S001050821706003X

[2] Arkhipov V A, Zarko V E, Zharova I K, Zhukov A S, Kozlov E A, Aksenenko D D and Kurbatov A V 2016 *Combust Explos Shock Waves.* **52 (5)** 497–513 DOI: 10.1134/S0010508216050014

[3] Vasenin I M, Shrager E R, Glazunov A A, Krainov A Yu, Krainov D A 2013 *Russ Phys J.* **56** pp 908-913. DOI: 10.1007/s11182-013-0117-2

[4] Minkov L L, Shrager E R, Kiryushkin A E 2016 *Key Engineering Materials.* **685** pp 114-118. DOI: 10.4028/www.scientific.net/KEM.685.114

[5] Kiryushkin A E and Minkov L L 2017 *J. Phys.: Conf. Ser.* **894** 012041 DOI: 10.1088/1742-6596/894/1/012041

[6] Glazunov A A, Eremin I V, Kostyushin K V, Zhiltsov K N and Tyryshkin I M 2019 *J. Phys.: Conf. Ser.* **1145** 012036 DOI: 10.1088/1742-6596/1145/1/012036

[7] Kulikovskiy A G, Pogorelov N V, Semenov A. J 2001 *Mathematical Aspects of Numerical Solution of Hyperbolic Systems* (London: Chapman & Hall/CRC)

[8] Toro E F 2009 *Riemann Solvers and Numerical Methods for Fluid Dynamics* (Berlin: Springer-Verlag)

[9] Kostyushin K V, Kotonogov V A, Kagenov A M, Tyryshkin I M, Glazunov A A, Eremin I V and Zhiltsov K N 2019 *J. Phys.: Conf. Ser.* **1145** 012048 DOI: 10.1088/1742-6596/1145/1/012048

IOP Conf. Series: Materials Science and Engineering **597** (2019) 012078 doi:10.1088/1757-899X/597/1/012078

Numerical investigation of a single supersonic jet interaction with a wedge barrier

A M Kagenov[1], K V Kostyushin, A A Glazunov, I V Eremin, I M Tyryshkin and K N Zhiltsov

Research Institute of Applied Mathematics and Mechanics of *Tomsk State University*, 36 Lenin Avenue, Tomsk 634050, Russian Federation.

[1]Corresponding author: anuar@ftf.tsu.ru

Abstract. This article presents the results of mathematical modeling of the interaction of a single supersonic jet with a wedge barrier. The angles of the wedge were 0, 5, and 10 degrees. The distance from the nozzle to the barrier was 0.5 meters. The parameters of the gas of hydrazine combustion products at the nozzle inlet varied. Mach number at the nozzle exit was 4 and 5. It was found that the wedge angle of the barrier slightly affects the shape of the jet, the shape of the wall jets and the pressure along the barrier at an ambient pressure of 650 Pa and a temperature of 250 K. With an increase in the barrier angle, the pressure on the jet axis decreases and the first barrel of the wall jet is displaced.

1. Introduction

In the design and development of rocket and space technology, experimental and theoretical studies are conducted during the launch and landing of rockets and spacecraft. Attention is paid here to the interaction of supersonic jets with various obstacles for studying the occurrence of various negative effects. Lots of experimental and theoretical studies have been devoted to explore of jet flows and their interaction with barriers. Mathematical modelling using modern numerical methods and high-performance computing systems plays a significant role in theoretical studies. The results of an experimental investigation of the leakage of supersonic underexpanded jets onto a flat obstacle established perpendicular to the axis of the jet in the immediate vicinity of the nozzle exit (0.92d, 2.14d, 3.5d) are presented in [1]. In this paper, empirical formulas for the approximate construction of pressure diagrams on a barrier and indicate the applicability limits are given. The interaction of a supersonic axisymmetric single jet with obstacle perpendicular obstruction was given in an experimental study [2]. Theoretical and experimental studies was carried out at the Institute of Theoretical and Applied Mechanics of the SB RAS, Russia [3-6]. Experimental work on a specially created vertical jet facility was performed. The facility is equipped with recording sensors and modern instruments, they collect information in real time and allow monitoring the ongoing experiment. The structures of supersonic jets and the leakage of underexpanded and overexpanded supersonic jets into impermeable perpendicular and inclined obstacles and obstacles of various permeability was investigated. The tasks of landing the returned spacecraft with 16 nozzles are considered. Experimental studies of the distribution of the averaged pressure and pressure pulsations on the surface of the returned spacecraft and the landing surface was carried out. Mathematical modeling was performed using ANSYS Fluent. In the experiments, two nozzles with a Mach numbers 1 and 1.5 at nozzle exit are considered. The results was presented in the form of shadow images and processed by the PIV method. A series of works on the calculation and experimental study of the interaction of single and multi-block supersonic turbulent jets

IOP Conf. Series: Materials Science and Engineering **597** (2019) 012078 doi:10.1088/1757-899X/597/1/012078

with obstacle was carried out in [7-11]. These studies are devoted to the study of the processes occurring at the launch of rocket and space technology. Various particular qualities of physical processes accompanying the outflow of jets of combustion products of rocket fuels was shown. The mathematical model consists of the three-dimensional Reynolds averaged Navier-Stokes equations with the Menter's SST turbulence model. Simulation was performed using LOGOS software. An analysis of the experimental and theoretical studies of the supersonic jets interaction with obstacles showed that despite the large volume of accumulated knowledge, the problem of supersonic turbulent jet interaction with obstacles at distances h/d=1.8-9.2 from the nozzle exit to obstacle and with Mach numbers on the nozzle exit M = 1.5÷3 was not considered in detail.

The aim of the work is a fundamental study of the interaction of a supersonic jet with wedge barrier in a wide range of applicability, namely with a rarefied atmosphere and Mach numbers at the nozzle exit 4-5.

2. Physical and mathematical model

In this study, the interaction of a single supersonic jet with a wedge barrier was considered. The supersonic jet expires from the nozzle with Mach numbers 5 and 6 at the nozzle exit. Hydrazine combustion products was used as a working fluid. Schematically, this interaction is shown in figure 1, where AA – nozzle exit; AB – jet boundary; AL – expansion fans; OT – barrel shock; TB – reflected shock; TD – contact discontinuity; TT – Mach disk (normal shock); T – triple point; C – critical point (braking point); 1 – continuation of the flow from the nozzle; 2 – expansion flow; 3 – near-axis region of free expansion flow; 4 – the annular area between the fronts of the branched shock waves; 5, 6 – flow in a stream behind a normal shock (impingement region); 7 – wall jet (peripheral flow) [10-12].

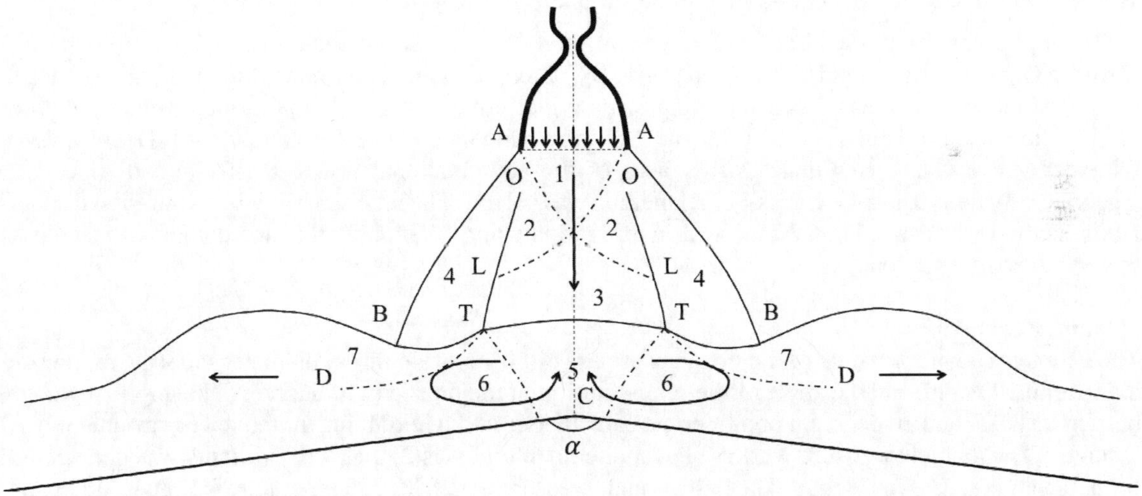

Figure 1. The structure of a single supersonic jet interaction with wedge barrier

The system of shock waves in front of the surface coincides with the flow in a free jet with the same parameters in the outlet section of the nozzle. If the obstacle is located in the sections between the central shock of the seal and the end of the first "cell" of the free jet, a large vacuum may form. Therefore, a relatively small part of the total gas flow will pass through the normal shock of the jet cross section. This part of the gas flow, spreading over the surface of the obstacle, forms a thin layer between the surface of the obstacle and the contact discontinuity coming off the contour of the normal shock (region 6). The bulk of the gas in the jet moves near the wall obstacle, passing through a system of shock waves. After the last shock wave, this mass of gas enters the area called the main peripheral flow or wall jet region (region 7). The nature of gas motion in region 7 has a significant effect on the distribution of pressure along the obstacle surface.

IOP Conf. Series: Materials Science and Engineering **597** (2019) 012078 doi:10.1088/1757-899X/597/1/012078

For the mathematical description of the interaction of supersonic jet with a flat obstacle the compressible Euler equations are used as a governing equations for ideal gas [12-15]. The system equations include continuity, momentum, and energy equations. In the integral conservative form given as:

$$\frac{\partial}{\partial t}\iiint_V \mathrm{U}\, d\mathrm{V} + \iint_{\partial S} \vec{\mathrm{F}}(\mathrm{U}) \cdot \vec{n} d\partial S = 0 \quad (1)$$

Where:

$$\mathrm{U} = [\rho,\ \rho u,\ \rho v,\ \rho w,\ \rho E]^T \quad (2)$$

$$\vec{u} = [u,\ v,\ w]^T \quad (3)$$

$$E = e + \frac{1}{2}\vec{u} \cdot \vec{u} \quad (4)$$

$$\vec{\mathrm{F}}(\mathrm{U}) \cdot \vec{n} = \begin{bmatrix} \rho\hat{u} \\ \rho\hat{u}u + pn_x \\ \rho\hat{u}v + pn_y \\ \rho\hat{u}w + pn_z \\ \rho\hat{u}(E + RT) \end{bmatrix} \quad (5)$$

Where $\hat{u}$ is velocity component normal to the surface ∂S, $\hat{u} = \vec{u} \cdot \vec{n} = un_x + vn_y + wn_z$.

To implement the physical and mathematical model (1)-(5) and conduct numerical research free software OpenFOAM-Extended was used [16-18]. Numerical method on finite volume method is based. Hexahedral numerical mesh was make using Salome Platform software. From the approximate solution of the Riemann problem according to the Rusanov scheme in dbnsFoam solver of OpenFOAM-Extended, there was fluxes through the faces of the computational grid cells. System of linear of algebraic equations using the Gauss-Seidel method was solved. The time sampling by a four-step Runge-Kutta method was carried out. Numerical investigation using SKIF Cyberia supercomputer of Tomsk State University was done.

3. Numerical results

The geometric characteristics of the nozzle were as follows: nozzle throat diameter 0.03613 m, nozzle diameter 0.19395 m, and the angle of the semi-solution of the nozzle is 10 degrees. The angle of wedge barrier was 0, 5, and 10 degree. For numerical calculations at the initial time unmoved environment was assumed. The ambient pressure was 650 Pa, and temperature was 250 K. Two variants was considered for parameters at the nozzle inlet. The first variant: pressure – 0.28 MPa, temperature – 1180 K, adiabatic index – 1.33719, the second variant: pressure – 1.962 MPa, temperature – 1336 K, adiabatic index – 1.29222. The distance from nozzle exit to flat obstacle was 0.5 meter.

The results of Mach number in figure 2 (first variant) and 3 (second variant) are shown. According to results from the figures the wedge angle of 5 and 10 degrees practically does not affect the shape of the main jet and the wall jets for the first and second variants. For the second variant, a change in the shape of the Mach disk in front of the barrier with increasing wedge angle is noticeable. The near-wall jet also hardly changes. In figure 4 and 5 distribution of pressure on wedge barrier for two variants with wedge angle 0, 5, and 10 degree are shown.

IOP Conf. Series: Materials Science and Engineering **597** (2019) 012078 doi:10.1088/1757-899X/597/1/012078

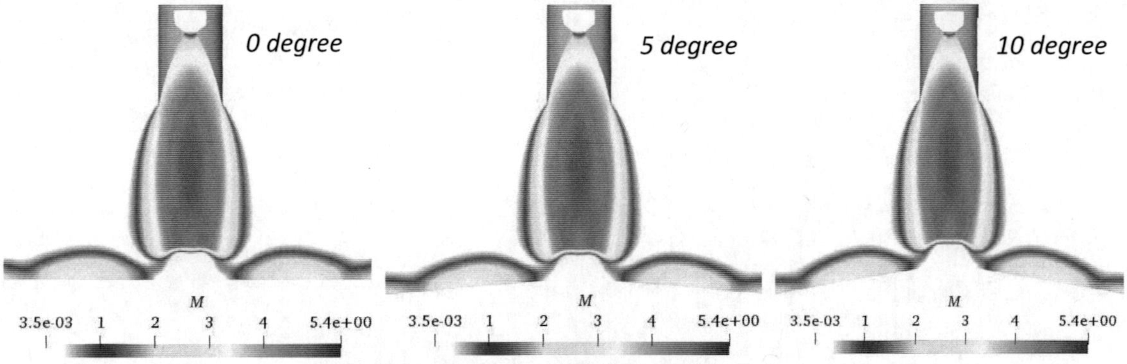

Figure 2. Mach number for the first variant

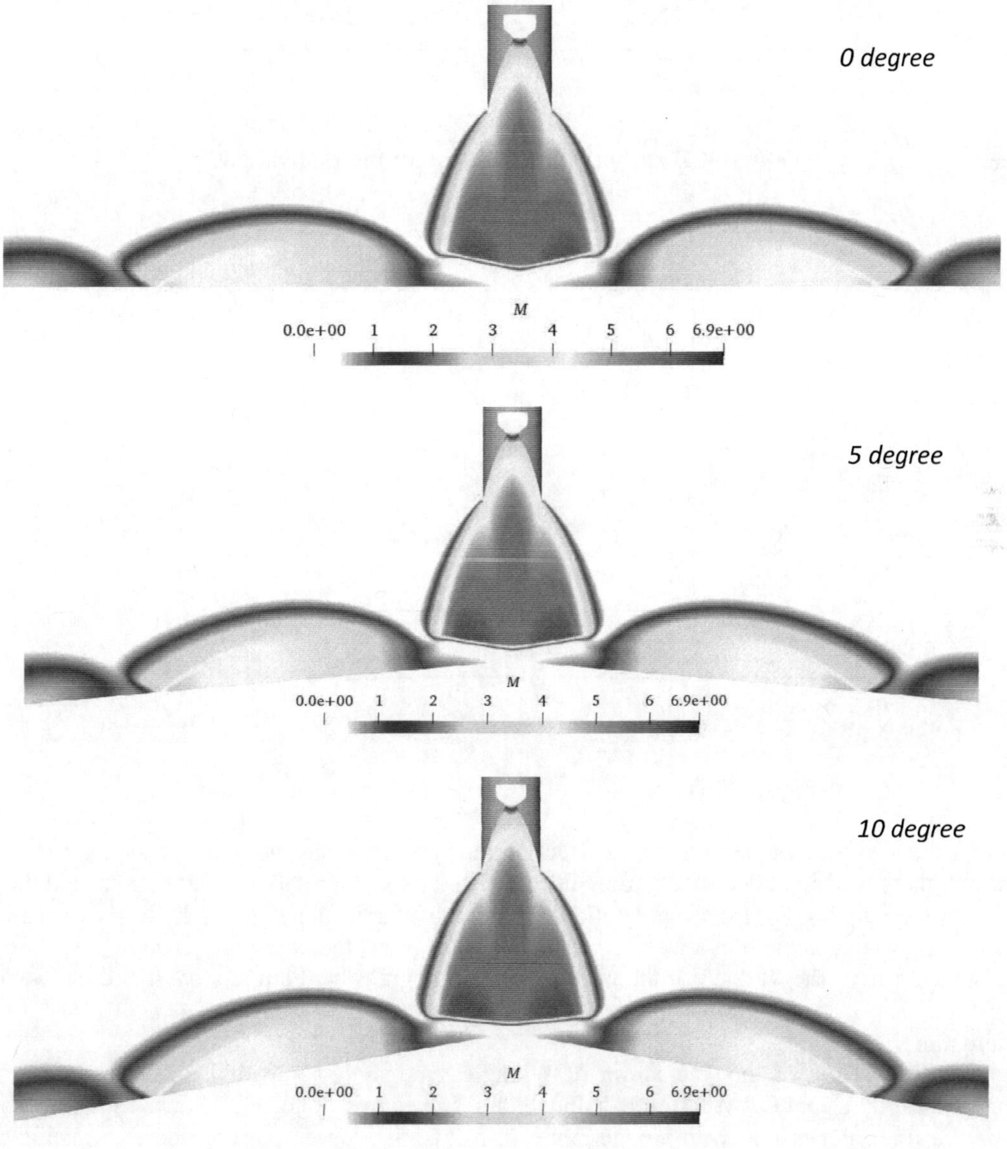

Figure 3. Mach number for the second variant

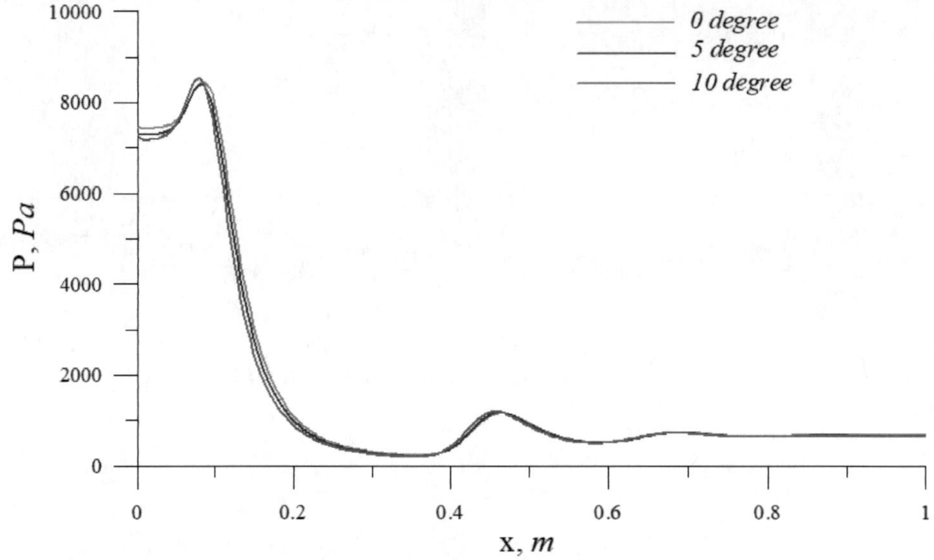

Figure 4. Distribution of pressure for the first variant

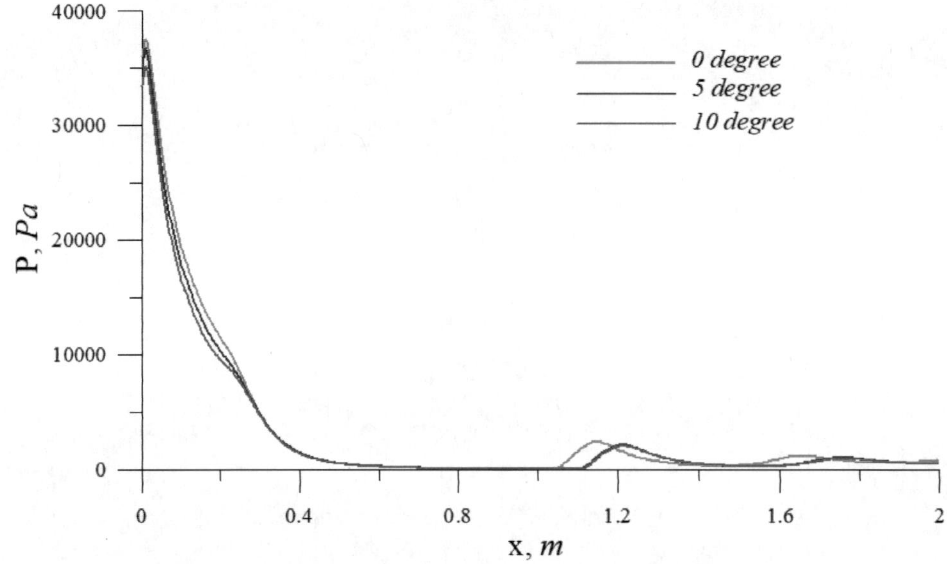

Figure 5. Distribution of pressure for the second variant

In Figure 4 the wedge angle has almost no effect on the pressure distribution along the barrier. For the first variant, there is a slight decrease in pressure on the jet axis of a relatively flat barrier, and the near-wall jet almost coincides. For the second option, also on the jet axis, there is a slight decrease in pressure with increasing wedge angle. For a wedge with 5 and 10 degrees, the near-wall jet coincides. There is a slight displacement of the wall jet for the wedge 5 and 10 degrees by 3 cm relative to the flat barrier.

4. Conclusion

Mathematical modeling of the interaction of a single supersonic jet with wedge barrier are done. Calculations using OpenFOAM-Extended dbnsFoam solver are done. Two variants with different parameters at the inlet nozzle with wedge angle 0, 5, and 10 degree are considered. Results of the calculations showed a weak influence of the wedge angle on the jet shape and pressure distribution along the barrier for two variants. A 3 cm deflection of the wall jet is noted for a flat barrier as opposed to wedge with 5 and 10 degree.

Acknowledgments
The reported study was funded by RFBR and Tomsk region according to the research project № 19-41-703005.

References
[1] Melnikova M F and Nesterov Yu N 1971 *TSAGI* **2 (5)** 44–58
[2] Krothapalli A, Rajkuperan E, Alvi F and Lourenco L M 1999 *J. Fluid Mech.* **392** 155–181
[3] Zapryagaev V I and Kiselev N P 2009 *J. Appl. Mech. Tech. Phys.* **50 (3)** 447–453
[4] Zapryagaev V I, Kavun I N and Solotchin A V 2015 *J. Appl. Mech. Tech. Phys.* **56 (3)** 406–413
[5] Zapryagaev V I, Kiselev N P and Kundasev S G 2016 *PNRPU Aerospace Engineering Bulletin* **45** 32–49
[6] Zapryagaev V I, Gubanov D A, Kavun I N, Kiselev N P, Kundasev S G and Pivovarov A A 2017 *AIP Conference Proceedings* **1893** 030058
[7] Kudimov N F, Safronov A V and Tretyakova O N 2013 *MAI* **69** 1–11
[8] Kudimov N F, Safronov A V and Tretyakova O N 2013 *MAI* **70** 1–14
[9] Kudimov N F, Safronov A V and Tretyakova O N 2014 *Applied problems of gas dynamics and heat transfer in power plants of rocket technology* (Moscow: MAI) p168
[10] Degtyar V G, Merkulov E S, Hlibov V I and Safronov A V 2013 *Cosmonautics and Rocket Engineering* **70 (1)** 37–45
[11] Glaznev V N, Zapryagaev V I, Uskov V N and et al. 2000 *Jet and nonstationary flows in gas dynamics* (Novosibirsk: Publishing Firm of Siberian Branch of Russian Academy of Sciences)
[12] Dulov V G and Lukianov G A 1984 *Gas Dynamics of Outflow Processes* (Novosibirsk: Nauka) p 226
[13] Godunov S K, Zabrodin A V, Ivanov M Y, Kraiko A N, Prokopov G P 1976 *Numerical Solution of Multidimensional Problems of Gas Dynamics* (Moscow: Nauka) p 400
[14] Toro E F 2009 *Riemann Solvers and Numerical Methods for Fluid Dynamics* (Berlin: Springer-Verlag) p 724
[15] Vasenin I M, Krainov A Y, Lipanov A M and Shrager E R 2015 *Comp. Math. Math. Phys.* **55** 883–890.
[16] Kagenov A, Glazunov A, Kostyushin K and Eremin I 2017 *AIP Conference Proceedings* **1899** 060010
[17] Kagenov A M, Glazunov A A, Kostyushin K V, Eremin I V, Tyryshkin I M and Zhilcov K N 2018 *AIP Conference Proceedings* **2027** 030071
[18] Kostyushin K, Kagenov A, Eremin I, Zhilcov K and Shuvarikov V 2017 *AIP Conference Proceedings* **1899** 060012

IOP Conf. Series: Materials Science and Engineering **597** (2019) 012079 doi:10.1088/1757-899X/597/1/012079

Exact solutions of initial boundary value problems for Sobolev type equations of elastic oscillations

S S Orlov, V V Shemetova, G K Sokolova

Institute of Mathematics, Economics and Informatics, Irkutsk State University, 1 Karl Marx Street, Irkutsk 664003, Russian Federation

E-mail: orlov_sergey@inbox.ru

Abstract. The article is devoted to the problem of the unique solvability of initial boundary value problems for partial differential equations. These initial boundary value problems are dissipative dynamic mathematical models from the field of linear elasticity. The considered equations are not resolved with respect to the highest time derivative. Therefore, these equations are referred to the so-called Sobolev type partial differential equations. The research of initial boundary value problems is carried out from general position in the form of the complete second order differential equation in abstract spaces. The theory of the distributions with values in Banach spaces and the concept of a fundamental solution of differential operator are used. Theorems of existence and uniqueness of the solutions of considered initial boundary value problems in the class of functions of finite smoothness with respect to time are proved. These solutions are obtained as an functional series of the Kummer confluent hypergeometric function.

1. Introduction

The paper is devoted to the study of the unique solvability of initial boundary value problems for Sobolev type differential equations. The name of these equations goes back to the pioneering work of S. L. Sobolev [1], where the problem of small oscillations of a rotating liquid was considered. This term means a partial differential equation that is not resolved with respect to the highest time derivative [2]. Partial differential equations of the Sobolev type, represented in an abstract differential operator form, are also called *degenerate* or *singular*. The results of research in the field of Sobolev type equations and a wide bibliography are given in monographs of Demidenko and Uspenskii [2], Carrol and Showalter [3], Favini and Yagi [4], Sidorov, Loginov, Sinitsyn and Falaleev [5], Sviridyuk and Fedorov [6], Al'shin, Korpusov, Sveshnikov [7] and others. Many partial differential equations of continuum mechanics can be represented as an abstract evolution equation of the second order with respect to time variable. *Complete* second order evolution equations are of particular interest in applied problems, because the terms with the first time derivative of an unknown function allow to formalize internal friction, energy dissipation, influence of gyroscopic forces, and other effects. Among the works devoted to the complete second order abstract differential equations and their applications in the form of the Sobolev type partial differential equations, let us cite the research papers, which are closest to our studies. They are [8] by Batty, Chill and Srivastava, [9] by Andronova and Kopachevskii, [10] by Bulatov and Lee, [11] by Keyantuo and Lizama, [12] by Falaleev and Grazhdantseva, [13] by Melnikova and Filinkov, [14] by Oka, [15] by Tijun and Jin.

2. Problem Statement

In this paper we consider an initial boundary value problem

$$\lambda' u_{tt} - u_{xxtt} - a(u_{xxt} - \lambda'' u_t) - b(u_{xx} - \lambda''' u) = f(x,t),\ x \in [0;\,h],\ t > 0; \tag{1}$$

$$u(x,t)|_{t=0} = u_0(x),\ u_t(x,t)|_{t=0} = u_1(x);\ u(x,t)|_{x=0} = u(x,t)|_{x=h} = 0,\ t \geq 0; \tag{2}$$

which describes small longitudinal oscillations of an elastic rod by taking inertia into account. The equation (1) is called the *Boussinesq–Love* equation [16]. The unknown function $u = u(x,t)$ is the displacement of points of the principal axis of the rod of length $h > 0$, and the given function $f = f(x,t)$ is the distribution of the external mass load per unit length of the rod. Real parameters a, b, λ', λ'', λ''' depend on properties of rod material such that density, Young modulus, Poisson ratio and Lamé coefficients. The initial boundary value problem (1), (2) has been studied thoroughly by Zamyshlyaeva, Bychkov and Tsyplenkova in [16]. Our research deals with construction of new exact solutions of this problem in terms of special functions.

Another object of the study is an initial boundary value problem

$$\lambda v_{tt} - \Delta v_{tt} - \Delta v_t + \Delta^2 v = f(\mathbf{r},t),\ \mathbf{r} \in \mathrm{K},\ t > 0; \tag{3}$$

$$v(\mathbf{r},t)|_{t=0} = v_0(\mathbf{r}),\ v_t(\mathbf{r},t)|_{t=0} = v_1(\mathbf{r});\ v(\mathbf{r},t)|_{\mathbf{r}\in\partial\mathrm{K}} = 0,\ \Delta v(\mathbf{r},t)|_{\mathbf{r}\in\partial\mathrm{K}} = 0,\ t \geq 0. \tag{4}$$

This is a linearized dissipative model of small transversal oscillations of a thin homogeneous isotropic thermoelastic plate [17]. Here Δ is the Laplace operator with respect to the spatial variables $\mathbf{r} = x\mathbf{i} + y\mathbf{j}$ defined on the square $\mathrm{K} = [0;\,h] \times [0;\,h]$, and Δ^2 is the biharmonic operator. The unknown function $v : \mathrm{K} \times [0;\,+\infty) \to \mathbb{R}$ describes the transversal displacements of a two-dimensional plate occupying K. The parameter $\lambda > 0$ is inversely proportional to the plate thickness, and the term $-\Delta u_{tt}$ describes the rotational inertia of the plate. The given function $f : \mathrm{K} \times [0;\,+\infty) \to \mathbb{R}$ is the distribution of the external mass load per unit area of the plate. The boundary conditions (4) defined on $\partial\mathrm{K}$ are called the *Navier boundary conditions* [18]. The boundary value problem with Dirichlet boundary conditions for equation (3) has been studied by Bisognin, Bisognin, Perla Menzala and Zuazua in [17]. The uniform exponential decay of the energy of the solutions of this problem has been proved.

3. Methods and Materials

Let E_1, E_2 be real Banach spaces. Consider the following second order linear differential operator

$$\mathcal{D}(u(t)) = Bu''(t) - A_1 u'(t) - A_0 u(t),$$

where B, A_1, A_0 are closed linear operators from E_1 into E_2 such that $D(B) \subseteq D(A_1) \cap D(A_0)$. In this section we consider an initial value problem

$$\mathcal{D}(u(t)) = f(t),\ t > 0;\ u(0) = u_0,\ u'(0) = u_1, \tag{5}$$

where $u : [0;+\infty) \to E_1$ is unknown function, and $f : [0;+\infty) \to E_2$ is given function.

Definition 1. A function $u = u(t)$ that is strongly continuously differentiable two times over semi-axis $[0;+\infty)$ and satisfies the differential equation (5) with initial conditions is called a *classical solution* of this initial value problem.

The initial boundary value problem (1), (2) is an example of the initial value problem (5) with the following function spaces $E_1 = \mathcal{L}_{2\,\circ\,[0;\,h]} = \{u(x) \in \mathcal{L}_{2\,[0;\,h]} : u(0) = u(h) = 0\}$, $E_2 = \mathcal{L}_{2\,[0;\,h]}$, operator coefficients $B = \lambda' - \partial_{xx}$, $A_1 = a(\partial_{xx} - \lambda'')$, $A_0 = b(\partial_{xx} - \lambda''')$, and their domains $D(B) = D(A_1) = D(A_0) = H^{\ell+2}_{\circ\,[0;\,h]} = \{u(x) \in W^{\ell+2}_{2\,[0;\,h]} : u(0) = u(h) = 0\}$. If we assume that

$$E_1 = \mathcal{L}_{2\,\circ\,\mathrm{K}} = \{v(\mathbf{r}) \in \mathcal{L}_{2\,\mathrm{K}} : v(\mathbf{r})|_{\mathbf{r}\in\partial\mathrm{K}} = 0\},\ E_2 = \mathcal{L}_{2\,\mathrm{K}};\ B = \lambda - \Delta,\ A_1 = \Delta,\ A_0 = -\Delta^2;$$

IOP Conf. Series: Materials Science and Engineering **597** (2019) 012079 doi:10.1088/1757-899X/597/1/012079

$$D(B) = D(A_1) = D(A_0) = H^{\ell+4}_{\circ \mathrm{K}} = \{v(\mathbf{r}) \in W^{\ell+4}_{2\mathrm{K}} : v(\mathbf{r})|_{\mathbf{r}\in\partial\mathrm{K}} = 0,\ \Delta v(\mathbf{r})|_{\mathbf{r}\in\partial\mathrm{K}} = 0\};$$

then the initial value problem (5) becomes the initial boundary value problem (3), (4). By $W^{\ell}_{2\mathrm{K}}$ or H^{ℓ}_{K} denote the Sobolev space, where $\ell \in \{0\} \cup \mathbb{N}$, and by $\mathcal{L}_{2\mathrm{K}}$ denote the Lebesgue space of square-integrable on K functions.

To study the unique solvability of the initial value problem (5) we used the Sobolev–Schwartz theory of distributions with values in Banach spaces [5]. Denote by $\theta(t)$ the Heaviside step function, and by $\delta(t)$ the Dirac delta function. In the space $K'_{+}(E_1)$ of distributions with left-bounded support and values in E_1 the initial value problem (5) has the convolutional form

$$\mathcal{D}(\delta(t)) * \tilde{u}(t) = \tilde{g}(t), \tag{6}$$

with given distribution $\tilde{g}(t) \in K'_{+}(E_2)$ such that

$$\tilde{g}(t) = f(t)\theta(t) + (Bu_1 - A_1 u_0)\delta(t) + Bu_0\delta'(t),$$

and $\tilde{u}(t) = u(t)\theta(t)$, where $u = u(t)$ is the classical solution of the initial value problem (5).

Definition 2. A solution $\tilde{u}(t) \in K'_{+}(E_1)$ of the equation (6) is called a *generalized solution* of the initial value problem (5).

Definition 3. A generalized operator valued function $\mathcal{E}(t)$ is called a *fundamental solution* or *fundamental operator-function* of the second order differential operator $\mathcal{D}(\delta(t))$ if the following convolutional equalities $\mathcal{D}(\delta(t)) * \mathcal{E}(t) * v(t) = v(t)$ and $\mathcal{E}(t) * \mathcal{D}(\delta(t)) * w(t) = w(t)$ are true for any distributions $v(t) \in K'_{+}(E_2)$ and $w(t) \in K'_{+}(E_1)$.

If fundamental operator-function $\mathcal{E}(t)$ is known, then generalized solution of the initial value problem (5) is given by formula

$$\tilde{u}(t) = \mathcal{E}(t) * \tilde{g}(t). \tag{7}$$

By the first equality of the Definition 3, it follows that the distribution (7) is the solution of the equation (6). The existence of this solution explains by the existence of convolution in the class of distributions with left-bounded support. The uniqueness of the solution (7) of the equation (6) is easy to prove. Let $\tilde{v}(t) \in K'_{+}(E_1)$ such that $\tilde{v}(t) \neq \tilde{u}(t)$, and $\mathcal{D}(\delta(t)) * \tilde{v}(t) = \tilde{g}(t)$, i. e. there exists another solution of the equation (6). By the second equality of Definition 3, it follows that $\tilde{v}(t) = \mathcal{E}(t) * \mathcal{D}(\delta(t)) * \tilde{v}(t) = \mathcal{E}(t) * \tilde{g}(t) = \tilde{u}(t)$, which contradicts our previous assumptions that $\tilde{v}(t) \neq \tilde{u}(t)$.

Suppose that linear operator B is continuously invertible and $D(B) \subseteq D(A_1) \cap D(A_0)$. Let us denote $\mathcal{A}_1 = A_1 B^{-1}$ and $\mathcal{A}_0 = A_0 B^{-1}$. It follows from the closed graph theorem that mappings $\mathcal{A}_1$ and $\mathcal{A}_0$ are bounded linear operators from E_2 to itself. By $\exp(\mathcal{A}_1 t)$ denote a uniformly continuous semigroup of bounded linear operators with an infinitesimal generator $\mathcal{A}_1$. Consider an operator valued function $V(t) = \exp(-\mathcal{A}_1 t)\mathcal{A}_0 \exp(\mathcal{A}_1 t)$. By the Baker–Campbell–Hausdorff formula [19], it follows that

$$V(t) = \mathcal{A}_0 + [\mathcal{A}_0, \mathcal{A}_1]\frac{t}{1!} + [[\mathcal{A}_0, \mathcal{A}_1], \mathcal{A}_1]\frac{t^2}{2!} + [[[\mathcal{A}_0, \mathcal{A}_1], \mathcal{A}_1], \mathcal{A}_1]\frac{t^3}{3!} + \ldots,$$

where $[\mathcal{A}_0, \mathcal{A}_1] = \mathcal{A}_0\mathcal{A}_1 - \mathcal{A}_1\mathcal{A}_0$ is the commutator of linear operators $\mathcal{A}_0$ and $\mathcal{A}_1$. Let us remark that operator valued function $V(t)$ and linear operator $\mathcal{A}_1$ satisfy the Lax equation $V'(t) = [V(t), \mathcal{A}_1]$. In the theory of integrable systems, such operators are called a *Lax pair* [20]. The initial condition $V(0) = \mathcal{A}_0$ associates $V(t)$ and $\mathcal{A}_0$.

Theorem 1. Suppose B is continuously invertible linear operator; then a fundamental operator-function of differential operator $\mathcal{D}(\delta(t))$ is given by

$$\mathcal{E}(t) = B^{-1}\sum_{k=1}^{+\infty}\int_0^t \frac{(t-s)^{k-1}}{(k-1)!}\exp(\mathcal{A}_1 s)U_{k-1}(s)ds\,\theta(t),$$

where operator valued functions $U_{k-1}(t)$ are such that

$$U_k(t) = \int_0^t V(s)U_{k-1}(s)ds, \; U_0(t) = \mathrm{I}_2;$$

I_1 and I_2 are indentity operators in E_1 and E_2 respectively.

To prove this theorem, it is necessary to check two convolutional equalities of the Definition 3 using recurrence formula for $\{U_{k-1}(t)\}_{k\in\mathbb{N}}$ and the following asymptotic and differential relations

$$U_{k-1}(t) \sim \mathcal{A}_0^{k-1}\frac{t^{k-1}}{(k-1)!}, \; t \to 0; \quad U'_k(t) - [U_k(t), \mathcal{A}_1] = U_{k-1}(t)\mathcal{A}_0.$$

Corollary 1. Let the conditions of the Theorem 1 be satisfied and the composition of linear operators $\mathcal{A}_0$ and $\mathcal{A}_1$ is commutative; then a fundamental operator-function of differential operator $\mathcal{D}(\delta(t))$ is given by

$$\mathcal{E}(t) = B^{-1}\sum_{k=1}^{+\infty}\frac{t^{2k-1}}{(2k-1)!}\mathcal{A}_0^{k-1}\,{}_1F_1(k;\,2k;\,\mathcal{A}_1 t)\,\theta(t),$$

where ${}_1F_1(a;\,b;\,t)$ is the *Kummer confluent hypergeometric function* [21] in an integral form

$${}_1F_1(a;\,b;\,t) = \frac{\Gamma(b)}{\Gamma(a)\Gamma(b-a)}\int_0^1 \exp(t\tau)\tau^{a-1}(1-\tau)^{b-a-1}d\tau, \; b > a > 0. \quad (8)$$

Corollary 2. Let the conditions of the Theorem 1 be satisfied and $f(t) \in C([0;+\infty);E_2)$; then the initial value problem (5) has a unique generalized solution

$$\tilde{u}(t) = u(t)\theta(t) = B^{-1}\Bigg[\exp(\mathcal{A}_1 t)Bu_0 + \sum_{k=1}^{+\infty}\int_0^t\int_0^{t-s}\frac{(t-s-\tau)^{k-1}}{(k-1)!}\exp(\mathcal{A}_1\tau)U_{k-1}(\tau)f(s)d\tau ds +$$

$$+\sum_{k=1}^{+\infty}\int_0^t\frac{(t-s)^{k-1}}{(k-1)!}\exp(\mathcal{A}_1 s)U_k(s)ds\,Bu_0 + \sum_{k=0}^{+\infty}\int_0^t\frac{(t-s)^k}{k!}\exp(\mathcal{A}_1 s)U_k(s)ds\,(Bu_1 - \mathcal{A}_1 u_0)\Bigg]\theta(t).$$

As has been mentioned above, a unique generalized solution of the initial value problem (5) is given by (7). The form of this solution is obtained in the Corollary 2 under the assumption that $f = f(t)$ is strongly continuous. Generalized function $\tilde{u}(t) \in K'_+(E_1)$ is regular and coincides with function $u = u(t)$ that is strongly continuously differentiable two times and satisfies the differential equation (5) with initial conditions.

Corollary 3. Let the conditions of the Theorem 1 be satisfied and $f(t) \in C([0;+\infty);E_2)$; then the initial value problem (5) has a unique classical solution $u = u(t)$ from the Corollary 2.

4. Results and Discussion

Let us consider a homogeneous boundary value problem

$$\varphi''(x) - \lambda\varphi(x) = 0, \; x \in [0;\,h]; \; \varphi(0) = \varphi(h) = 0.$$

The eigenvalues and eigenfunctions of this problem are given by

$$\lambda_m = -\frac{\pi^2 m^2}{h^2},\ \varphi_m(x) = \left(\frac{2}{h}\right)^{1/2} \sin\frac{\pi m}{h}x,\ m \in \mathbb{N}. \tag{9}$$

The system of eigenfunctions is orthonormal in the sense of the scalar product $\langle\cdot,\cdot\rangle$ of the Lebesgue space $\mathcal{L}_{2\,[0;\,h]}$. Assume that $\lambda' \notin \sigma(\partial_{xx})$, i. e. $\lambda' \neq \lambda_m$ for any $m \in \mathbb{N}$, then the operator $B = \lambda' - \partial_{xx}$ is continuously invertible, and the following theorem is due to the Corollary 3.

Theorem 2. Suppose $\lambda' \notin \sigma(\partial_{xx})$, $f(x,t) \in C(t \geq 0; \mathcal{L}_{2\,[0;\,h]})$, $u_0(x)$, $u_1(x) \in H^{\ell+2}_{\circ\,[0;\,h]}$; then the initial boundary value problem (1), (2) has a unique solution of class $C^2(t \geq 0; H^{\ell+2}_{\circ\,[0;\,h]})$, and this solution is given by

$$u(x,t) = \sum_{m=1}^{+\infty}\Bigg[\langle u_0(x),\varphi_m(x)\rangle \exp(\alpha_m t) + \sum_{k=1}^{+\infty}\Bigg[\beta_m^k \langle u_0(x),\varphi_m(x)\rangle \frac{t^{2k}}{(2k)!}\,{}_1F_1(k+1;\,2k+1;\,\alpha_m t) +$$

$$+\beta_m^{k-1}(\langle u_1(x),\varphi_m(x)\rangle - \alpha_m\langle u_0(x),\varphi_m(x)\rangle)\frac{t^{2k-1}}{(2k-1)!}\,{}_1F_1(k;\,2k;\,\alpha_m t)+$$

$$+\frac{\beta_m^{k-1}}{\lambda'-\lambda_m}\int\limits_0^t \frac{(t-s)^{2k-1}}{(2k-1)!}\,{}_1F_1(k;\,2k;\,\alpha_m(t-s))\langle f(x,s),\varphi_m(x)\rangle\,ds\Bigg]\varphi_m(x)\Bigg],$$

where $\alpha_m = a\dfrac{\lambda_m - \lambda''}{\lambda'-\lambda_m}$, $\beta_m = b\dfrac{\lambda_m - \lambda'''}{\lambda'-\lambda_m}$, and ${}_1F_1(a;\,b;\,t)$ is a special function defined by (8).

Let us consider a homogeneous boundary value problem

$$\Delta\varphi(\mathbf{r}) = \lambda\varphi(\mathbf{r}),\ \mathbf{r} \in \mathrm{K};\ \varphi(\mathbf{r})|_{\mathbf{r}\in\partial\mathrm{K}} = 0,\ \Delta\varphi(\mathbf{r})|_{\mathbf{r}\in\partial\mathrm{K}} = 0.$$

The spectrum $\sigma(\Delta)$ consists of the eigenvalues $\lambda_{m,\,n} = \lambda_m + \lambda_n$, where λ_m, $\lambda_n \in \sigma(\partial_{xx})$ are given by (9). It necessary that $\lambda_{m,\,n}h^2 = -\pi^2 s$, where $s \in \mathbb{N}$. The multiplicity $d(\lambda_{m,\,n})$ of eigenvalue $\lambda_{m,\,n}$ is equal to the number of different solutions $(m,\,n) \in \mathbb{N}^2$ of the equation $m^2 + n^2 = s$ with given $s \in \mathbb{N}$. By the theorem on sums of two squares from [22], it follows that

$$d(\lambda_{m,\,n}) = \sum_{d\in\mathbb{N}\,:\,s|d} \chi_4(d) - \sum_{i\in\mathbb{N}} \delta_{s\,i^2},$$

where $\chi_4(d)$ is Dirichlet character modulo 4 of $d \in \mathbb{N}$, and $\delta_{i\,j}$ is Kronecker delta. The systems of eigenfunctions $\varphi_{m,\,n}(\mathbf{r}) = \varphi_m(x)\varphi_n(y)$ is orthonormal in the sense of the scalar product $\langle\cdot,\cdot\rangle$ of the Lebesgue space $\mathcal{L}_{2\,\mathrm{K}}$, where $\varphi_m(x)$, $\varphi_n(y)$ from (9). Assume that $\lambda \notin \sigma(\Delta)$, then the operator $B = \lambda - \Delta$ is continuously invertible, and the following theorem is due to the Corollary 3.

Theorem 3. Suppose $\lambda \notin \sigma(\Delta)$, $f(\mathbf{r},t) \in C(t \geq 0; \mathcal{L}_{2\,\mathrm{K}})$, $v_0(\mathbf{r})$, $v_1(\mathbf{r}) \in H^{\ell+4}_{\circ\,\mathrm{K}}$; then the initial boundary value problem (3), (4) has a unique solution of class $C^2(t \geq 0; H^{\ell+4}_{\circ\,\mathrm{K}})$, and this solution is given by

$$v(\mathbf{r},t) = \sum_{m=1}^{+\infty}\sum_{n=1}^{+\infty}\Bigg[\exp(\mu_{m,\,n}\,t)\langle v_0(\mathbf{r}),\varphi_{m,\,n}(\mathbf{r})\rangle +$$

$$+\sum_{k=1}^{+\infty}\Bigg[(-\lambda_{m,\,n}\,\mu_{m,\,n})^k\langle v_0(\mathbf{r}),\varphi_{m,\,n}(\mathbf{r})\rangle\frac{t^{2k}}{(2k)!}\,{}_1F_1(k+1;\,2k+1;\,\mu_{m,\,n}\,t)+$$

$$+(-\lambda_{m,\,n}\,\mu_{m,\,n})^{k-1}(\langle v_1(\mathbf{r}),\varphi_{m,\,n}(\mathbf{r})\rangle - \mu_{m,\,n}\langle v_0(\mathbf{r}),\varphi_{m,\,n}(\mathbf{r})\rangle)\frac{t^{2k-1}}{(2k-1)!}\,{}_1F_1(k;\,2k;\,\mu_{m,\,n}\,t)+$$

$$+\frac{(-\lambda_{m,n}\,\mu_{m,n})^{k-1}}{\lambda-\lambda_{m,n}}\int\limits_0^t \frac{(t-s)^{2k-1}}{(2k-1)!}\,{}_1F_1(k;\,2k;\,\mu_{m,n}(t-s))\langle f(\mathbf{r},s),\varphi_{m,n}(\mathbf{r})\rangle\,ds\Bigg]\varphi_{m,n}(\mathbf{r})\Bigg],$$

where $\mu_{m,n}=\dfrac{\lambda_{m,n}}{\lambda-\lambda_{m,n}}$, and ${}_1F_1(a;\,b;\,t)$ is a special function defined by (8).

5. Conclusion

Thus, the present paper deals with nonclassical equations of mathematical physics. Initial boundary value problems of the theory of elasticity are considered. The unique solvability of initial boundary value problems in the class of the functions of finite smoothness with respect to time are proved. Explicit formulas of the solutions are obtained. The object of direct study in the article is complete second order differential equation in Banach spaces. Original initial boundary value problems are considered as concrete examples of this abstract equation. The applied approach develops the theory and methods for solving not only these initial boundary value problems, but also any others, which are a particular case of the abstract differential equation under consideration.

Acknowledgments

The reported study was funded by Russian Foundation for Basic Research according to the research projects No. 18-01-00643 A and No. 18-51-54001 Viet_a.

References

[1] Sobolev S L 2006 *On a new problem of mathematical physics Selected Works of S.L. Sobolev* ed G V Demidenko and V L Vaskevich (Boston: Springer) pp 279–332
[2] Demidenko G V and Uspenskii S V 2003 *Partial Differential Equations and Systems not Solvable with Respect to the Highest-Order Derivative* (New York; Basel: Marcel Dekker Inc.) p 490
[3] Carrol R W and Showalter R E 1976 *Singular and Degenerate Cauchy Problems* (New York; San Francisco; London: Academic Press) p 333
[4] Favini A and Yagi A 1999 *Degenerate Differential Equations in Banach Spaces* (New York; Basel; Hong Kong: Marcel Dekker Inc.) p 313
[5] Sidorov N, Loginov B, Sinitsyn A and Falaleev M 2002 *Lyapunov–Schmidt Methods in Nonlinear Analysis and Applications* (Dordrecht; Boston; London: Kluwer Academic Publishers) p 548
[6] Sviridyuk G A and Fedorov V E 2003 *Linear Sobolev Type Equations and Degenerate Semigroups of Operators* (Utrecht; Boston; Koln; Tokyo: VSP) p 216
[7] Al'shin A B, Korpusov M O and Sveshnikov A G 2011 *Blow-up in Nonlinear Sobolev Type Equations* (Berlin; New York: Walter de Gruyter) p 648
[8] Batty C J K, Chill R and Srivastava S 2015 *Operator Theory: Advances and Applications* **250** 49
[9] Andronova O A and Kopachevskii N D 2010 *Journal of Mathematical Sciences* **164** 478
[10] Bulatov M V and Lee M G 2008 *Differential Equations* **44** 1353
[11] Keyantuo V and Lizama C 2006 *Mathematische Zeitschrift* **253** 489
[12] Falaleev M V and Grazhdantseva E Y 2006 *Differential Equations* **42** 819
[13] Melnikova I V and Filinkov A I 1999 *Journal of Mathematical Sciences* **93** 22
[14] Oka H 1996 *Proceedings of the American Mathematical Society* **124** 3143
[15] Tijun X and Jin L 1990 *Pacific Journal of Mathematics* **142** 175
[16] Zamyshlyaeva A A, Bychkov E V and Tsyplenkova O N 2014 *Applied Mathematical Sciences* **8** 5477
[17] Bisognin E, Bisognin V, Perla Menzala G and Zuazua E 1998 *Mathematical Methods in the Applied Sciences* **21** 393
[18] Musina R and Nazarov A I 2014 *Communications in Partial Differential Equations* **39** 1780
[19] Hall B C 2015 *Lie Groups, Lie Algebras, and Representations: An Elementary Introduction* vol 222, ed S Axler and K Ribet (Cham; Heidelberg; New York; Dordrecht; London: Springer) p 453
[20] Newell A C 1985 *Solitons in Mathematics and Physics* (Philadelphia: SIAM) p 244
[21] Slater L J 1960 *Confluent Hypergeometric Functions* (New York: Cambridge University Press) p 247
[22] Ireland K and Rosen M 1982 *A Classical Introduction to Modern Number Theory* vol 84, ed P R Halmos and F W Gehring et al (New York; Berlin: Springer–Verlag) p 341

Institute of Physics
Dirac House, Temple Back
Bristol BS1 6BE UK

ISSN: 1742-6588
ISBN 978-1-5108-9467-9
9 781510 894679

CURRAN ASSOCIATES INC.
proceedings.com